Advanced Nanoscale ULSI Interconnects: Fundamentals and Applications

Yosi Shacham-Diamand · Tetsuya Osaka ·
Madhav Datta · Takayuki Ohba
Editors

Advanced Nanoscale ULSI Interconnects: Fundamentals and Applications

Springer

Editors
Yosi Shacham-Diamand
Tel Aviv University
69978 Ramat Aviv, Tel Aviv
Israel
yosish@eng.tau.ac.il

Tetsuya Osaka
Department of Applied Chemistry
Waseda University
3-4-1 Okubo
Tokyo
Shinjuku-ku 169-8555
Japan
osakatet@waseda.jp

Madhav Datta
Cooligy, Inc.
2370 Charleston Road
Mountain View CA 94043
USA
mdatta@cooligy.com

Takayuki Ohba
Division of Corporate Relations
The University of Tokyo
7-3-1 Hongo
Tokyo
Bunkyo-ku 113-8654
Japan
ohba@oucr.u-tokyo.ac.jp

ISBN 978-0-387-95867-5 e-ISBN 978-0-387-95868-2
DOI 10.1007/978-0-387-95868-2
Springer New York Dordrecht Heidelberg London

Library of Congress Control Number: 2009934298

Printed on acid-free paper

Springer is part of Springer Science+Business Media (www.springer.com)

This book is dedicated to my wife Anat for all her support and patience.

Preface

In **Advanced ULSI interconnects – fundamentals and applications** we bring a comprehensive description of copper-based interconnect technology for ultra-large-scale integration (ULSI) technology for integrated circuit (IC) application. Integrated circuit technology is the base for all modern electronics systems. You can find electronics systems today everywhere: from toys and home appliances to airplanes and space shuttles. Electronics systems form the hardware that together with software are the bases of the modern information society. The rapid growth and vast exploitation of modern electronics system create a strong demand for new and improved electronic circuits as demonstrated by the amazing progress in the field of ULSI technology. This progress is well described by the famous "Moore's law" which states, in its most general form, that all the metrics that describe integrated circuit performance (e.g., speed, number of devices, chip area) improve exponentially as a function of time. For example, the number of components per chip doubles every 18 months and the critical dimension on a chip has shrunk by 50% every 2 years on average in the last 30 years. This rapid growth in integrated circuits technology results in highly complex integrated circuits with an increasing number of interconnects on chips and between the chip and its package. The complexity of the interconnect network on chips involves an increasing number of metal lines per interconnect level, more interconnect levels, and at the same time a reduction in the interconnect line critical dimensions.

The continuous shrinkage in metal line critical dimension forced the transition from aluminum-based interconnect technology, that was dominant from the early days of modern microelectronics, to copper-based metallization that became the dominant technology in recent years. As interconnect critical dimensions shrank to the nano-scale range (below 100 nm) more aggressive interconnect designs on smaller scale became possible, thus keeping "Moore's law" on pace. In addition to the introduction of copper as the main conducting material, it was clear that new dielectric materials, with low dielectric constant ("low-*k*" materials), should replace the conventional silicon dioxide interlevel dielectric (ILD). Thus the overall technology shift is from "aluminum–silicon dioxide" ULSI interconnect technology to "copper-low-*k*" technology. The Cu-low-*k* technology allows patterning of 45 nm wide interconnects in mass production and will probably allow further shrinkage in patterning of 15–22 nm lines in the next 10 years.

Copper metallization is achieved by electrochemical processing or processes that involve electrochemistry. The metal deposition is done by electrochemical deposition and its top surface is planarized (i.e., made flat or planar in the industry jargon) by chemical mechanical polishing (CMP). Electroplating is an ancient technique for metal deposition. Its application to ULSI technology with nano-scale patterning was a major challenge to scientists and engineers in the last 20 years. The success in the introduction of copper metallization so that it became the leading technology demonstrated the capability and compatibility of electrochemical processing in the nano-scale regime. In this book we will review the basic technologies that are used today for copper metallization for ULSI applications: deposition and planarization. We will describe the materials that are used, their properties, and the way they are all integrated. We will describe the copper integration processes and a mathematical model for the electrochemical processes in the nano-scale regime. We will present the way we characterize and measure the various conducting and insulating thin films that are used to build the copper interconnect multilayer structures using the "damascene" (embedded metallization) process. We will also present various novel nano-scale technologies that will link modern nano-scale electronics to future nano-scale-based systems.

Following this preface we bring an introduction where we bring the fundamentals of Cu electroplating for ULSI – when electrochemistry meets electrical engineering.

In Part II we give a historical review describing interconnect technology from the early days of modern microelectronics until today. It describes materials, technology, and process integration overview that brings into perspective the ways metallization is accomplished today. Further understanding of the scaling laws is presented next. Both semiconductor and interconnect progress are described, since they are interwoven into each other. Progress in interconnects always follows progress in transistor science and technologies. Although this book focuses on interconnect technology it should be clear that interconnects link transistors and the overall circuit operation is achieved by combined interaction of a highly complex network. The basic role of interconnects in such networks and how interconnects performance is linked to overall circuit performance are discussed next. One of the key issues in the increasing complex system is whether there are also other paradigms. One such paradigm is the 3D integration of ULSI components, also known as "3D integration."

In Part III we present a detailed review of interconnect materials. There is no doubt that the advancement in materials science and technology in recent years was the key to the advances in the ULSI technology. There are few groups of materials in ULSI interconnects: conductors (e.g., copper, silicides), barrier layers (e.g., Ta/TaN, TiN, WC), capping layers (dielectrics such as nitride-doped amorphous silicon or silicon nitride or electroless CoWP), and dielectrics with a dielectric constant less than that of silicon dioxide (i.e., low-k materials). We dedicate a special part to the material properties of silicides (metal–silicon compounds) that are used as the conducting interfacing material between the metallic interconnect network and the semiconductor transistors. The following parts bring an intensive review of low-k materials. They pose a major challenge since they should compete with the conventional silicon dioxide that, although its dielectric constant is higher, has excellent electrical and mechanical properties and whose process technology is well established and entrenched in the industry and research communities.

In Part IV we focus on the actual electrochemical processes that are used for ULSI interconnect applications. We will first present the copper plating principles and their application to sub-micron patterning. Additives will be described in light of their role in the fully planar embedded metallization technology (i.e , the damascene process). In addition to conventional process we also mention some novel processes. Among them, the atomic layer deposition is the most promising and is under intensive investigation due to its ability to form ultra-thin seed layers with excellent uniformity and step coverage. Other interesting nano-scale processes are the deposition of nano particles, either inorganic or organic, that yield nano-scale metal lines that may, one day, be used for nano-electronics applications.

A common approach that links basic modeling to actual structure is the use of computer-aided design (CAD) simulating the desired structure based on the fundamental physical and chemical models of the process. For example, the use of electrochemical deposition onto narrow features with critical dimensions below 100 nm and with aspect ratio (i.e., the ratio of height to width) more than 2 to 1 requires a special process that is called "superfilling." In such a process, the filling of the bottom of the feature is much faster than the deposition on its upper "shoulders." Rapid deposition and full deposition onto the feature is achieved without defects (e.g., voids, seams) and with relatively thin metal on the shoulders that can be reliably removed in the ensuing chemical mechanical polishing planarization step. The discovery of the "superfilling" process was a major breakthrough in the initial stages of the introduction of copper metallization. In Part V we give a detailed description of such modeling of copper metallization using electrochemical processes for nano-scale metallization.

Part VI links all the previous parts together and describes the actual fully planar embedded metal process that is known as the damascene process. Following a detailed description of the various damascene concepts and its associated process steps we discuss the process integration issues. The integration involves linking all the various components: starting at the lithography level, patterning the wafer, deposition of the barrier and seed layers followed by the copper plating and its chemical mechanical polish (CMP) planarization, and ending with capping layer deposition. In this part we focus on the basic roles of each one of the components in the overall integration issue and on the way we put them all together.

Part VII describes the basic principles of the tools that are used for the copper metallization. There are two families of tools that we describe here – tools for deposition and tools for chemical mechanical polishing (CMP). Plating tools, both for electroplating and for electroless plating, are described in detail emphasizing their relation to the damascene process as applied for ULSI applications, i.e., material properties and integration in the manufacturing line.

Another family of tools is the one used for metrology and inspection. We present in Part VIII the innovative and advanced tools that are being used for Cu nanotechnology. One of the most promising tools is the use of X-ray technology, especially

X-ray reflection (XRR), which has proven to be the only method suitable for ultra-thin barrier layers and for porous materials that are used for low dielectric constant insulators. Another interesting development in modern planarization technology is the capability for in-line metrology. We present recent innovations in this field using optical metrology that is integrated with chemical mechanical polishing processes.

Finally, in Part IX we present a full and comprehensive review of the most promising interconnect technologies for future nanotechnology. This part includes a complete review of novel nanotechnologies such as bio-templating and nano-bio interfacing. Another key issue is the role of interconnect with future computation and storage technology. In this part we review the role of interconnect and 3D hyper integration, spintronics, and moletronics. In summary this part and the following prolog lay forth the reasons why electroplating is considered as the key technology for nano-circuits interconnects.

Tel Aviv, Israel — Yosi Shacham-Diamand
Tokyo, Japan — Tetsuya Osaka
Mountain View, CA — Madhav Datta
Tokyo, Japan — Takayuki Ohba

Acknowledgments

We would like to thank all the contributors to this book. Each one of them is a leader in his field and the contributions are highly appreciated.

We also would like to thank Dr. Ragini Raj Singh and Ms. Rakefet Ofek-Almog from Tel Aviv University for the tedious work of editing, formatting, and communicating with the various authors. The devoted work of Dr. Singh allowed the successful completion of this book.

Tel Aviv, Israel — Yosi Shacham-Diamand
Tokyo, Japan — Tetsuya Osaka
Mountain View, CA — Madhav Datta
Tokyo, Japan — Takayuki Ohba

Contents

Contributors

Bulent M. Basol SoloPower Inc., 5981 Optical Court, San Jose, CA 95138, USA, bbasol@solopower.com

A.M. Bittner Group leader "Self-Assembly" Asociacion CIC nanoGUNE Tolosa Hiribidea, 76, 20018 Donostia – San Sebastian, Spain, a.bittner@nanogune.eu

David Brandon Faculty of MSE, Technion IIT, Haifa 32000, Israel, brandon@technion.ac.il

James Castracane College of Nanoscale Science and Engineering, The University at Albany-SUNY, Albany, NY 12203, USA, jcastracane@uamail.albany.edu

Schubert S. Chu Global Product Manager at Applied Materials, Applied Materials, Inc., Santa Clara, CA 95054-3299, USA, schubert_chu@amat.com

Madhav Datta Cooligy Inc., 800 Maude Avenue, Mountain View, CA 94043, USA, mdatta@cooligy.com

Kathleen Dunn College of Nanoscale Science and Engineering, The University at Albany-SUNY, Albany, NY 12203, USA, kdunn1@uamail.albany.edu

Y. Ein-Eli Department of Material Science and Engineering, Technion-Israel Institute of Technology, Haifa 32000, Israel, eineli@technion.ac.il

Eric Eisenbraun College of Nanoscale Science and Engineering, The University at Albany-SUNY, Albany, NY 12203, USA, eeisenbraun@uamail.albany.edu

Moseh Eizenber Faculty of MSE, Technion IIT, Haifa 32000, Israel, eizen@tx.technion.ac.il

Moshe Finarov Nova Measuring Instruments Ltd., Weizmann Science Park, Rehovot 76100, Israel, moshe-f@nova.co.il

Anand Gadre College of Nanoscale Science and Engineering, The University at Albany-SUNY, Albany, NY 12203, USA, agadre@uamail.albany.edu

Ehud Gazit Department of Molecular Microbiology and Biotechnology, Life Sciences faculty, Tel Aviv University, Tel Aviv 69978, Israel, ehudg@post.tau.ac.il

Ronald J. Gutmann RPI, Low Center for Industrial Innovation, Troy, New York 12180, USA, gutmar@rpi.edu

J.E. Guyer Materials Science and Engineering Laboratory, National Institute of Standards and Technology, Gaithersburg, MD 20899, USA, guyer@nist.gov

Madoka Hasegawa Faculty of Science and Engineering, Waseda University, 3-4-1 Okubo, Shinjuku-ku, Tokyo 169-8555, Japan, madoka-hasegawa@aoni.waseda.jp

Akira Hashimoto Institute of Science and Technology, Kanto Gakuin University, 1-50-1 Mutsuurahigashi, Kanazawa-ku, Yokohama, Japan, t0509001@kanto-gakuin.ac.jp

Yoshihiro Hayashi ULSI Fundamental Research Laboratory, Microelectronics Research Laboratories, NEC Electronics Corporation, 1120, Shimokuzawa, Sagamihara, Kanagawa 229, Japan, hayashi@mel.cl.nec.co.jp, y-hayashi@az.jp.nec.com

Paul S. Ho Microelectronics Research Center, The University of Texas at Austin, TX 78712-1100, USA, paulho@mail.utexas.edu

Z. Hu Semitool Inc., 655 W. Reserve Dr., Kalispell, MT 59901, USA, zhu@semitool.com

Jay Im Microelectronics Research Center, The University of Texas at Austin, TX 78712-1100, USA, jayim@mail.utexas.edu

K. Ito Department of Materials Science and Engineering, Kyoto University, Sakyo-ku, Kyoto, Japan

Igor Ivanov Blue29, 615 Palomar Avenue, Sunnyvale, CA 9408, USA, igor@intermolecular.com

Alain E. Kaloyeros College of Nanoscale Science and Engineering, The University at Albany-SUNY, Albany, NY 12203, USA, akaloyeros@uamail.albany.edu

Bernd Kastenmeier Freescale Semiconductor Inc., Austin, TX 78729-8084, USA, bernd.kastenmeier@freescale.com

Nobuyoshi Kobayashi Process Integration Technology, R&D, ASM Japan, n.kobayashi@asm.com

Amit Kohn Department of Materials, University of Oxford, Parks Road, Oxford OX1 3PH United Kingdom, amit.kohn@materials.ox.ac.uk

Junichi Koike Department of Materials Science, Tohoku University, Sendai 980-8579, Japan, koikej@material.tohoku.ac.jp

Ichiro Koiwa Institute of Science and Technology, Kanto Gakuin University, 1-50-1 Mutsuurahigashi, Kanazawa-ku, Yokohama, Japan, koiwa@kanto-gakuin.ac.jp

Avinoam Kolodny Faculty of EE, Technion IIT, Haifa 32000, Israel, kolodny@ee.technion.ac.il

Matthias Kraatz Microelectronics Research Center, The University of Texas at Austin, TX 78712-1100, USA, matthias.kraatz@mail.utexas.edu

Vincent LaBella College of Nanoscale Science and Engineering, The University at Albany-SUNY, Albany, NY 12203, USA, vlabella@uamail.albany.edu

Bill Lee Blue29, 615 Palomar Avenue, Sunnyvale, CA 9408, USA, bill.t.lee@lamrc.com

Jian-Qiang Lu RPI, Low Center for Industrial Innovation, Troy, NY 12180, USA, luj@rpi.edu

H. Miyajima Advanced ULSI Process Engineering Department, Process & Manufacturing Engineering Center, Toshiba Corporation Semiconductor Company, 8 Shinsugita-cho, Isogo-ku, Yokohama 235-8522, Japan, hideshi.miyajima@toshiha.co.jp

Masanori Murakami Department of Materials Science and Engineering, Kyoto University, Sakyo-ku, Kyoto, Japan, masanori.murakami@materials.mbox.media.kyoto-u.ac.jp

Osamu Nakatsuka Department of Crystalline Materials Science, Graduate School of Engineering, Nagoya University, Furo-cho, Chikusa-ku, Nagoya 464-8603, Japan, nakatuka@alice.xtal.nagoya-u.ac.jp

Takayuki Ohba The University of Tokyo, 7-3-1, Hongo, Bunkyo-ku, Tokyo 113-0033, Japan, ohba@ducr.u-tokyo.ac.jp

Tetsuya Osaka Faculty of Science and Engineering, Waseda University, 3-4-1 Okubo, Shinjuku-ku, Tokyo 169-8555, Japan, osakatets@waseda.jp

Hyun Park Memory Division, Samsung Electronics Co., LTD., 445-701, Korea, parkhyun@che.utexas.edu

T. Ritzdorf Semitool Inc., 655 W. Reserve Dr., Kalispell, MT 59901, USA, tritzdorf@semitool.com

Marc Van Rossum IMEC, Kapeldreef 75, B-3001, Leuven, Belgium, marc.vanrossum@imec.be

James G. Ryan Dean, JSNN, 2901 East Lee Street, Suite 2200, Greensboro, NC 27401, jgryan@uncg.edu

David Scheiner Nova Measuring Instruments Ltd., Weizmann Science Park, Rehovot 76100, Israel, david-s@nova.co.il

Jay J. Senkevich Brewer Science Inc., 2401 Brewers drive, Rolla, MO 65401 USA, jsenkevich@brewerscience.com

Y. Shacham-Diamand School of EE, Tel Aviv University, Tel Aviv 69978, Israel, yosish@eng.tau.ac.il

Gabi Sharon Nova Measuring Instruments Ltd., Weizmann Science Park, Rehovot 76100, Israel, gab-s@nova.co.il

D. Starosvetsky Department of Material Science and Engineering, Technion-Israel Institute of Technology, Haifa 32000, Israel, davstar@tx.technion.ac.il

Timothy Stoner College of Nanoscale Science and Engineering, The University at Albany-SUNY, Albany, NY 12203, USA, tstoner@uamail.albany.edu

Anna Topol IBM T. J. Watson Research Center, Yorktown Heights, NY 10598, USA, atopol@us.ibm.com

Manabu Tsujimura Ebara Corporation, Nissay Aroma Square, 5-3-7 Kamata, Ohta-ku, Tokyo 144-8721, Japan; 4-2-1 Honfujisawa, Fujusawa-shi 251-8502, Japan, tsujimura.manabu@ebara.com, tsujimura@fuj.ebara.co.jp

S. Tsukimoto Department of Materials Science and Engineering, Kyoto University, Sakyo-ku, Kyoto, Japan, tsukimoto@micro.mtl.kyoto-u.ac.jp

D. Wheeler Materials Science and Engineering Laboratory, National Institute of Standards and Technology, Gaithersburg, MD 20899, USA, daniel.wheeler@nist.gov

Bai Xu College of Nanoscale Science and Engineering, The University at Albany-SUNY, Albany, NY 12203, USA

Noriyuki Yamachika Faculty of Science and Engineering, Waseda University, 3-4-1 Okubo, Shinjuku-ku, Tokyo 169-8555, Japan, n.yamachika@asagi.waseda.jp

T. Yoda Advanced ULSI Process Engineering Department, Process & Manufacturing Engineering Center, Toshiba Corporation Semiconductor Company, 8 Shinsugita-cho, Isogo-ku, Yokohama 235-8522, Japan, takahashi.yoda@toshiba.co.jp

Boris Yokhin Jordan Valley Semiconductor, Ramat Gavriel, Migdal Haeemek, Israel, boris@jordanvalley.com

Masahiro Yoshino Faculty of Science and Engineering, Waseda University, 3-4-1 Okubo, Shinjuku-ku, Tokyo 169-8555 Japan, masahiro.yoshino@aoni.waseda.jp

Shigeaki Zaima Department of Crystalline Materials Science, Graduate School of Engineering, Nagoya University, Furo-cho, Chikusa-ku, Nagoya 464-8603, Japan, zaima@alice.xtal.nagoya-u.ac.jp

Part I
Introduction

Chapter 1
Challenges in ULSI Interconnects – Introduction to the Book

Y. Shacham-Diamand

Ultra large-scale integration (ULSI) technology is one of the most dominant and important technologies of the 21st century. It is the base for the global electronics system industry.

Our current information society is based on information technologies that have been developed in the last decade. Information technologies emerged as a result of developments in silicon-based integrated circuits (ICs) that are the "engines" that collect, process, and distribute information (Fig. 1.1). The rapid development in highly complex hardware, allowing both digital and analog signal processing, conditioning, and dissemination, was accompanied by developments in software allowing better command and control of the hardware. All of this became possible due to the invention of the integrated circuit, first on germanium (Kilby 1958) [1] and next on silicon (Noyce 1959) [1], and the rapid development toward ultra large-scale integration (ULSI) using complementary metal oxide semiconductor (silicon) (CMOS) technology.

The integration of few devices on chip was started as small-scale integration (SSI), followed by medium-scale integration (MSI), large-scale integration (LSI), very large-scale integration (VLSI), and finally ultra large-scale integration (ULSI). The integrated circuit includes transistors as the switching devices and interconnects that interface the switches between themselves and the external world. Typical integrated circuit (IC) dimensions are between few millimeters (length or width) and few centimeters. It is laid on a thin (slightly less than 1 mm) single-crystal silicon wafer forming arrays of chips. The chips are separated by narrow scribing lanes that define boundaries along them; the chips are later separated and packaged. For a detailed description of the ULSI processing the reader is referred, for example, to references [2] and [3].

The term ULSI was coined when the critical dimension of the patterns within the micro-chips reached the range of 0.25 μm. Current ULSI circuits have features with dimensions in the nanoscale region (smaller than ~100 nm). This reflects the transition from the era of micron-scale electronics into the nanoscale era. The

---

Y. Shacham-Diamand (✉)
School of EE, Tel Aviv University, Tel Aviv 69978, Israel
e-mail: yosish@eng.tau.ac.il

Y. Shacham-Diamand et al. (eds.), *Advanced Nanoscale ULSI Interconnects: Fundamentals and Applications*, DOI 10.1007/978-0-387-95868-2_1,

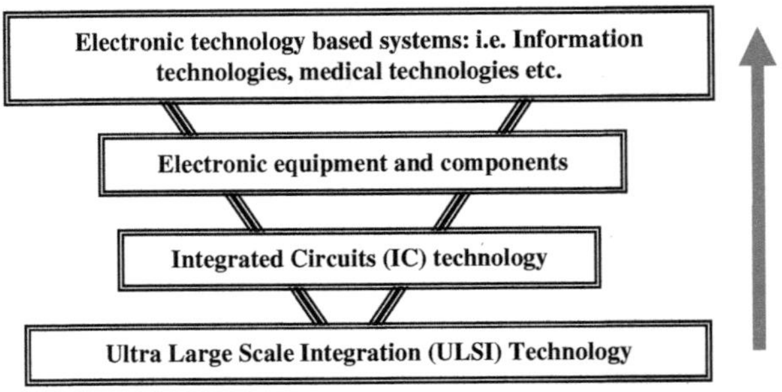

Fig. 1.1 The electronic technology-based hierarchy – from ULSI technology up to electronic technology-based systems

technology that is currently used for ULSI manufacturing is a "top-down" manufacturing, i.e., alternating patterning using optical lithography and ULSI process modules for removing materials (i.e., etching), adding materials (i.e., deposition, growing), or modifying materials (i.e., doping, silicidation). For a detailed description of ULSI technology status and forecast the reader is referred to the International Technology Roadmap for Semiconductors (ITRS) [4].

According to the ITRS 2006 report on interconnects [4], the most difficult and challenging issues for near future interconnects manufactured with dimensions of 32 nm and above are the following:

- Introduction of new materials to meet conductivity requirements and reduce the dielectric permittivity.
- Engineering manufacturable interconnect structures compatible with new materials and processes.
- Achieving necessary reliability.
- Three-dimensional control of interconnect features (with its associated metrology) is required to achieve necessary circuit performance and reliability.
- Manufacturability and defect management that meet overall cost/performance requirements.

Along with those specific issues there are always the issues of patterning, cleaning, filling at the nanoscale dimensions, integration of new processes and structures, and mitigating size effects. The challenges refer to both local interconnects and global wiring scaling issues which may affect circuit layouts and architecture to accommodate higher clock rate and better thermal engineering.

The ITRS 2006 report on interconnects [4] forecast for the year 2013 is that interconnect ½ pitch will be 32 nm for both memories and processing circuits. The total interconnect length will be up to 3125 m/cm^2 with 13 metallization levels and 4 optional levels for ground planes and capacitors. The metallization pitch of 64 nm will be with Cu lines with aspect ratio of 1.9 (height/width) for micro processor units (MPUs) that can carry a maximum current density of 8.08×10^6 A/cm^2. The contact aspect ratio is expected to be up to 20 for dynamic random access memories (DRAM) which have more regular and, hence, denser layouts than MPUs. The delay of such Cu lines depends on their length, their driving circuit, and the capacitive load for both their sides and at their end. For a 1 mm long line the delay of a minimum size wire will be ~3.45 ns assuming that the effective specific resistance of such lines will be in the range of 4.83 μΩ·cm.

Longer term forecast for the year 2020 assumes 28 nm pitch for the metallization, 14 metal layers, total interconnect length of 7243 m/cm^2, aspect ratio of 2 (height/width), maximum current density of 2.74×10^7 A/cm^2, and an RC delay for a 1 mm long line of ~23.4 ns, assuming that the effective specific resistance of such lines will be in the range of 8.2 μΩ·cm. Note that one of the key problems that will limit Cu interconnects in the far future is the rapid increase in the line-specific resistivity due to size-dependent scattering and the effect of barriers that reduces the effective cross section of the Cu lines.

1.1 Material Issues in Cu Interconnects

One of the key challenges for ULSI interconnect technology is related to the materials that are used for manufacturing the ULSI ICs. In Table 1.1 we summarize the most important conducting materials and some of the most important properties.

We classified the materials according to their main applications:

a. Conductors – used as conductors. The most common conductor for ULSI ICs today is copper, the main topic of this book.
b. Silicides – mainly used for contacts; however, they can also be used for local interconnect applications [5].
c. Barrier layers – used to protect the silicon and the interlevel dielectrics (ILD) from the hazardous effects of Cu [6]. They are also used to protect Cu against corrosion due to the interaction with air or with the following deposition processing steps.

One key problem in copper metallization is the effect of scaling on the resistivity. The effect of scaling becomes noticeable when the vertical and lateral dimensions of the lines become comparable to the electrons' mean free path (Table 1.2) in the metal [7].

The size effect also depends on the quality of the metal interface which affects the nature of the scattering, i.e., specular vs. diffusive. Improving the interface smoothness at the atomic level and reduction of surface defects may assist in keeping the effective resistivity and its distribution in a useable range. Otherwise, the line

Table 1.1 ULSI material properties (* – thin film)

	Resistivity (μΩ·cm)	Young's modulus (GPa)	TCR (%/°C)	Thermal conductivity (W/cm/°C)	CTE (ppm/°C)	Corrosion resistance in air	Melting point (°C)
Conductor							
Cu	*1.7–2.2	130	0.43	3.98	16.5	Poor	1084
Ag	1.59	83	0.41	4.25	19.1	Poor	
Au	2.35	78	0.4	3.15	14.2	Excellent	
Al	*2.7–3.0	70	0.45	2.38	23.5	Good	660
W	*8–15	411	0.48	1.74	4.5	Good	3410
Silicide							
PtSi	28–33					Excellent	1229
$TiSi_2$	13–16					Excellent	1540
WSi_2	30–70					Excellent	2165
$CoSi_2$	15–20					Excellent	1326
NiSi	14–20					Excellent	992
Poly Si (heavily doped)	500–1000					Excellent	1410
Barriers							
TiN	50–150					Excellent	2950
$Ti_{0.3}W_{0.7}$	75–200					Excellent	2200
TaN	>150					Excellent	
CoWP	25–80					Excellent	

Table 1.2 Electron mean free path in the conducting metals [7]

Metal	Mean free path, λ (nm)
Tungsten	14.2
Aluminum	14.9
Gold	35.5
Copper	39.3
Silver	52.7

resistance will be too high and circuits will not be able to operate at high enough frequency.

The second family of materials are the insulators that isolate between the interconnect materials. The most common material that is being used is the silicon dioxide that is deposited by chemical vapor deposition (CVD), which is typically enhanced by plasma (plasma-enhanced CVD – PECVD) allowing deposition at low temperatures (below 400°C). Silicon dioxide has excellent mechanical and electrical properties; it is compatible with CMOS processing and can be deposited with a very low defect density. However, as the critical dimension of interconnects had been scaled to the range of 90 nm and below the issue of interconnect delay became the limiting factor of the whole integrated circuit.

1.2 Interconnect Performance Issues

Interconnect affects the performance of the circuit. Here is a short list of the main performance variables that characterize ULSI ICs:

- Speed – it depends on the delay due to the loading of the parasitic capacitances of the lines and the series resistance of the conducting lines (interconnects) and the switching devices. We refer to this as "RC delays."
- Power dissipation – the dynamic power dissipation is proportional to the parasitic capacitance, to the power supply squared, and to the clock frequency. There is also some utilization factor that may take into consideration the average actual transition rate between logical states per unit time, which is only a fraction of the clock frequency that describes the maximum possible transition rates between logic states.
- Cost – interconnects define a significant part of the chip and they affect its area, thus the number of chips per wafer, and the wafer yield; both factors that affect cost. Interconnects also require many lithography and patterning steps that are a significant factor of the overall cost of manufacturing.
- Defect density – there are many interconnect layers and each one has a large number of components. Therefore, significant number of defects, such as shorts and opens, are related to the interconnect process. It also definitely affects the cost of manufacturing.

Lowering the RC delay is possible by one of the following ways:

a. reducing the interconnect resistance – reducing the metal-specific resistance;
b. reducing the parasitic capacitance – lowering the dielectric constant of the inter-level dielectrics;
c. optimizing the interconnect geometry – optimizing the aspect ratio improved layout on critical points;
d. improved device properties – higher current-driving capabilities;
e. improved circuit design – adding repeaters, use of transmission lines [8, 7].

Note that all of the above are a subset of the overall requirements which also include requirements for manufacturability, cost performance, and thermal management.

Reducing the interconnect resistance was achieved by the transition from aluminum to copper metallization. Improving the copper-effective resistance is achieved by lowering its barrier layer thickness and improving its interface properties. Lowering the parasitic capacitance is possible by using dielectric materials with a lower dielectric constant – low-k dielectrics. Intensive study of both organic and organic dielectrics with low dielectric constant has been underway and it yields few interesting materials that are considered as candidates to replace silicon dioxide. Low-k materials are classified as solid and porous materials, where the porous materials offer lower dielectric constant; however, their porosity generates some major processing challenges.

The absolute value of resistance and capacitance is also determined by the lines' actual dimensions. Therefore, the overall interconnects' layout plays an important role on the chip performance. Due to the high complexity of current ULSI chips, which may contain >10^8–10^9 devices and interconnects, it is difficult to simulate the overall chip performance. Therefore, although there are current "known best methods" for layout the problem is far from being solved and there is a dire need for better algorithms improving the layout and achieving better performance.

Improving the device properties has been achieved recently by the following ways:

a. Using bipolar devices with the CMOS circuits – this technology, known as BiCMOS, includes bipolar devices which have better current-driving capabilities than MOS transistors. It is used for driving high capacitive loads such as long buses and interfacing pads.
b. The introduction of silicon on insulator technology – lowering the source/drain parasitic capacitances eliminates significance of the devices' parasitic capacitance, therefore allowing better performance, i.e., higher speed at lower dissipated power.
c. The successful implementation of SiGe (Silicon–germanium) technologies. There are few possibilities that are being explored:

    c.1. p-type MOS transistors with improved current-driving capabilities and
    c.2. strained silicon transistors where the conduction is in single crystal silicon and the SiGe technology is used to apply the stress.

The last option improving interconnect performance is by improving the circuit design, for example, adding repeaters and choosing logic design with the least number of transitions between stages for the given application if possible (for reviews of on-chip interconnect the reader is referred to references [8, 7, 9]).

1.3 Interconnect Process Issues

A key development in modern ULSI manufacturing was the introduction of the Damascene process by P. Andricacos and his colleagues from IBM in 1998 [10]. The transition from the old aluminum process, which uses negative lithography, to the new Cu process (see Fig. 1.2), using positive lithography, revolutionized the industry.

Altogether Al-negative patterning was well established and yielded excellent results; it was impossible to apply it to copper patterning because Cu has no volatile species in conventional plasma etch processes; therefore, it was impossible to use conventional photoresist-based lithography for Cu patterning. In addition to this the transition to deep lithography preferred planar surface since the depth of focus

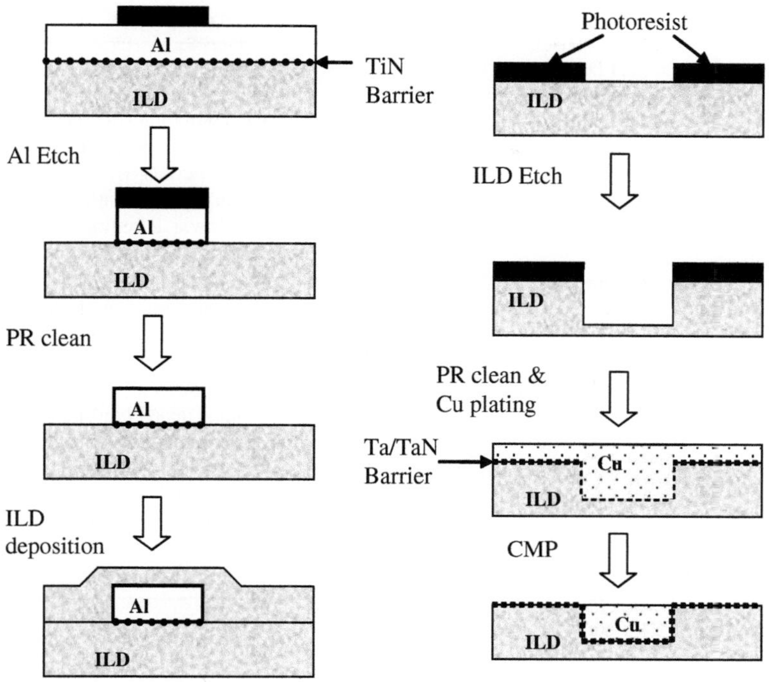

Fig. 1.2 The "old" Al-negative patterning (*left*) vs. the "new" single Damascene-positive patterning (*right*)

became comparable to the thin film thickness. This requirement leads to an additional planarization step of the interlevel dielectric.

Another issue that gave advantage to the fully planar Cu process was the ability to apply the Damascene process for both the lines and the contacts. This process is called "dual Damascene" where the lines and the contacts are etched first and the metal is deposited only once onto the whole metal level before it is planarized by chemical mechanical polishing (CMP). Compare this to standard Al process where the contact requires an additional tungsten deposition step followed by a planarization step. We summarize the Cu Damascene process on the conventional Al process in Table 1.3.

The need to use positive patterning for copper yielded a better process than Al patterning with superior resolution and less processing steps. This allowed lowering the cost of the metallization process which is a significant part of the overall cost of manufacturing since there many metallization levels require many tools and manufacturing facilities. Since the equipment cost and building are a major part of IC manufacturing, no wonder that the industry adopted the copper dual Damascene process although it was a new and immature process compared to the well-established and mature Al processing. This was at the end of the 1990s, while today most of the problems in the dual Damascene have been solved and it is the dominant Cu patterning technology.

Table 1.3 The advantage of the Cu dual Damascene process

	Cu dual Damascene	Al lithography
Lithography	Normally fully planar→ • better resolution	Requires an additional ILD planarization step to achieve full resolution
Etching	ILD etch→ • Only one material • Better process control	Etching two materials→ • Metal etch: Al and barrier • ILD etch: via definition process
Contacts	Cu contacts→ • Same material • Same deposition step	W contacts→ • Additional deposition step • Additional planarization step • Additional interface between the Al wire and the W plug
Planarization	One planarization step	Two planarization steps
Deposition	• One metal deposition step • One barrier deposition step • One ILD deposition step	• Two metal deposition steps: Al and W • One barrier deposition step • Two ILD deposition steps
Metal deposition	• Cu electroplating • Barrier sputtering • Capping layer by sputtering or electroless plating	• Al sputtering • W chemical vapor deposition

Among the few issues debated in the 1990s regarding the application of the dual Damascene process was the method of deposition of Cu. There were few possible options:

a. physical deposition – either by evaporation or sputtering;
b. chemical vapor deposition – CVD or PECVD;
c. electrochemical deposition methods – electroplating (EPD or ECD) and electroless plating (ELP).

Electrochemical deposition (ECD) emerged as the winning technology due to its superior filling capabilities of narrow trenches with high aspect ratio [10]. This was achieved due to the "superfilling" phenomena that were intensively investigated by many researchers. Commercial tools for Cu electroplating for ULSI metallization were conceived in the 1990s and became available recently along with high-quality deposition solution that contained Cu ions and additives allowing good superfilling [12]. The availability of sub-100 nm Cu nano-wires using electroplating for ULSI application was the result of efforts of many people, research institutes, and companies. It depicted the highlight of using electrochemical processes for nano-technologies applications. There are still many problems in process [11] and equipment [12]; however, it seems that scaling to 32 nm interconnect is possible. Recently few other methods have been introduced such as atomic layer deposition (ALD) of barriers [13, 14] and Cu seed layers [15] that may significantly improve Cu nano-wire processing for ULSI.

In this book we cover the latest issues in Cu nano interconnect technology for ULSI applications. We focus on the electrochemical issues related to Cu nano-wires applications. In this book the issues that are briefly described in this introduction are outlined one by one. While the majority of the Cu nano-technology issues are related to ULSI application, there are possible other applications. Therefore, we present in the last chapter a brief review and forecast of some promising electrochemical-based nano-scale metallization schemes for various applications mostly related to electronics.

References

1. Kilby, J. S.: Invention of the integrated circuit. IEEE Trans. Electron. Devices **23**, 648 (1976)
2. van Zant, P.: Microchip Fabrication: A Practical Guide to Semiconductor Processing. McGraw Hill, New York (2000)
3. Campbell, S. A.: The Science and Engineering of Microelectronic Fabrication. Oxford University Press, New York (2001)
4. www.itrs.net
5. Murarka, S. P.: Transition metal silicides. Ann. Rev. Mater. Sci. **13**, 117 (1983)
6. Shacham-Diamand, Y.: Barrier layers for Cu ULSI metallization. J. Electronic Mater. **30(4)**, 336–344 (2001)
7. Davis, J. A.; Venkatesan, R.; Kaloyeros, A.; Beylansky, M.; Souri, S. J.; Banerjee, K.; Member IEEE, Saraswat, K. C.; and Rahman A.: Interconnect limits on Gigascale Integration (GSI) In the 21st century. Proc. IEEE **89(3)**, (2001)
8. Havemann, R. H.; and Hutchby, J. A.: High-performance interconnects: An integration overview. Proc. IEEE **89(5)**, 586–601, (2001)
9. Edelstein, M. D.; Sai-Halasz, G. A.; and Mii, Y.-J.: LSl on-chip interconnection performance simulations and measurements. IBM J. Res. & Dev. **39(4)**, 383–401 (1995)
10. Andricacos, P. C.; Uzoh, C.; Dukovic, J. O.; Horkans, J.; and Deligianni, H.: Damascene copper electroplating for chip interconnections. IBM J. Res. & Dev. **12(5)**, 567–574 (1998)
11. Moffat, T. P.; Wheeler, D.; Edelstein, M. D.; and Josell, D.: Superconformal film growth: Mechanism and quantification. IBM J. Res. & Dev. **49(1)**, 19–36, (2005)
12. Ritzdorf, T. L.; Wilson, G. J.; McHugh, P. R.; Woodruff, D. J.; Hanson, K. M.; and Fulton, D.: Design and modeling of equipment used in electrochemical processes for microelectronics. IBM J. Res. & Dev. **49(1)**, 65–87, (2005)
13. Ritala, M.; Kalsi, P.; Riihela, D.; Kukli, K.; Leskela, M.; and Jokinen, J.: Controlled growth of TaN, Ta_3N_5, and TaO_xN_y thin films by atomic layer deposition. Chem. Mater. **11**, 1712 (1999)
14. Rossnagel, S. M.; Sherman, A.; and Turner, F.: Plasma-enhanced atomic layer deposition of Ta and Ti for interconnect diffusion barriers. J. Vac. Sci. and Technol. **B18**, 2016 (2000)
15. Pathangey, B. and Solanki, R.: Atomic layer deposition for nanoscale thin films. Vac. Technol. Coating **1**, 32 (2000)

Part II
Technology Background

Part II presents the technological background for interconnect in ULSI systems. It includes detailed description of MOS device and interconnects scaling physics (Chapter 2). Chapter 3 is about electrical performance of Cu interconnects for ULSI applications. It is well known that the steady growth in complexity of integrated electronic systems requires a growing number of metal interconnect layers in ULSI chips. The role of interconnects in integrated electronic systems is described in this chapter: Metrics for evaluating the quality of signal interconnections are presented, and simple electrical circuit models for interconnects are introduced. In light of these models, the fundamental problem of interconnect scaling is presented and design approaches for addressing the problem are briefly surveyed. As there are lots of deposition techniques, either electroless or electrodeposition, available elsewhere Chapter 4 presented a brief description of electrodeposition process and a focused discussion of copper electrodeposition for chip interconnects. It must be emphasized that different chip manufacturers have their unique combination of electrodeposition tools, proprietary tailored bath, and integration scheme for copper chip interconnects. Successful implementation of the copper electrodeposition process in high-volume chip manufacturing involves equal attention to metrology, process integration, and reliability issues. Besides void-free deposition of interconnect structures, selection of a reliable CMP process, excellent adherence of copper lines/vias to the dielectric, low resistivity, and resistance to electromigration are some of the key requirements for a high-yielding, reliable interconnect electrodeposition process. All these aspects are addressed in greater detail in different chapters of this book. Along with this Chapter 5, "Electrophorectic Deposition," defined the process of electrophoretic depositionand the process is summarized. The limitations of the process are explained and potential applications in the microelectronics sector are identified. Three areas in which industrial feasibility has been explored are then described in more detail. In addition to deposition technologies, proper device integration technology is important. As wafer-level 3D integration (i.e., 3D stacking prior to singulation of wafers into individual chips) has become an increasingly active research topic, without any large-scale IC manufacturing, Chapter 6 "Wafer Level 3D Integration for ULSI Interconnects" presents types of 3D integration, BEOL-based wafer-level 3D processing and wafer-level 3D design opportunities. Wafer-level interconnectivity is extended to multiple device levels.

Chapter 2
MOS Device and Interconnects Scaling Physics

Marc Van Rossum

2.1 Device Fundamentals

2.1.1 The MOSFET Transistor

2.1.1.1 Basic Device Physics

The metal-oxide-semiconductor field-effect transistor (MOSFET) is the most common active device in today's integrated circuits. Its basic structure consists of a doped silicon well, with at the opposite ends two highly doped contact regions (the source and drain junctions) allowing the current to pass close to the well surface (Fig. 2.1). In an n-type MOSFET, the well region is p-type doped and the source and drain are n+ doped, whereas the reverse polarity scheme applies for p-type devices. CMOS circuits contain both n-MOS and p-MOS transistors combined to form various logic gates. The transistor body is electrically isolated from the surrounding circuitry by a thick "field" oxide. A third electrode (the gate), to which the input signal is applied, is sitting on top of the well. It consists of an electrical contact layer (usually heavily doped polysilicon with a metallic top layer) separated from the silicon substrate by a thin insulator film made of thermally grown silicon dioxide. The substrate is thus capacitively coupled to the gate electrode, making the MOSFET a nearly ideal switch element due to the high isolation between input and output.

The output signal modulation takes place by varying the potential of the gate with respect to the substrate, which affects the charging of the MOS capacitor. In an n-MOSFET for instance, a negative gate voltage induces a positive (hole) charge accumulation region under the gate insulator. At positive gate voltages, holes are repelled into the substrate, creating a depletion region with fixed negative charges due to the ionized acceptor ions. At even more positive voltages, a negative charge

M.V. Rossum (✉)
IMEC, Kapeldreef 75, B-3001, Leuven, Belgium
e-mail: marc.vanrossum@imec.be

Y. Shacham-Diamand et al. (eds.), *Advanced Nanoscale ULSI Interconnects: Fundamentals and Applications*, DOI 10.1007/978-0-387-95868-2_2,

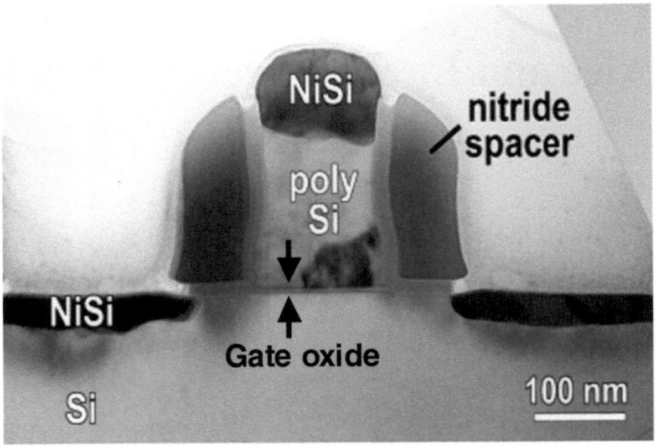

Fig. 2.1 The MOSFET transistor (IMEC)

Fig. 2.2 Charge regimes of a MOS capacitor in n-MOSFET

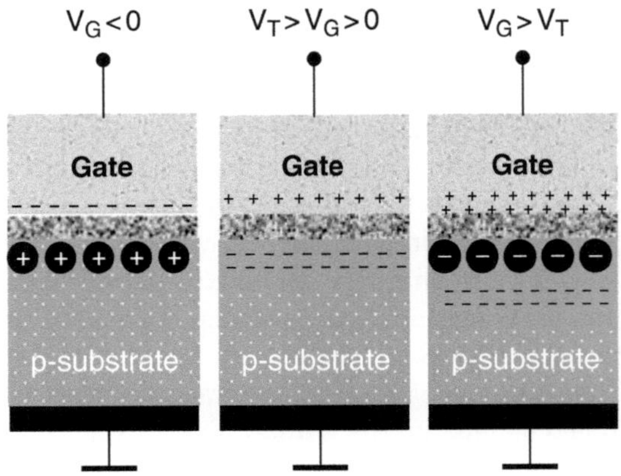

inversion layer (i.e., with a negative charge imbalance) starts forming at the top of the depletion region by the minority carriers (electrons) that are attracted to the surface. The gate voltage that corresponds with the transition between the depletion and the inversion regime is called the threshold voltage V_T (Fig. 2.2).

According to the MOS capacitor model, the charge density Q_S induced into the substrate per unit area is linearly proportional to the applied gate voltage V_G:

$$Q_S = -(V_G - \psi_S)C_{ox} \tag{2.1}$$

where ψ_S is the band bending potential at the silicon surface and C_{ox} is the gate oxide capacitance expressed as

$$C_{ox} = \frac{\varepsilon}{t_{ox}} \tag{2.2}$$

with ε as the dielectric constant of the gate insulator and t_{ox} its physical thickness.

When a bias voltage is applied between source and drain (with the source usually kept at ground potential), a current is allowed to flow through the inversion layer. The threshold voltage therefore separates the "off-state" and "on-state" of the transistor. In a p-MOS structure, similar mechanisms take place with the proper reversion of polarities.

2.1.1.2 Technology

Today, the most advanced commercial transistors have a physical gate length of about 30 nm. Since the gate is the narrowest feature on any IC, its formation involves the most demanding steps of the front-end process flow. First, the SiO_2 gate insulator is grown by thermal oxidation of a clean silicon wafer in a controlled atmosphere. Subsequently, a blanket layer of polysilicon is deposited by chemical vapor deposition (CVD), after which the gate fingers are defined by lithography and patterned by dry etching. In order to achieve the right threshold voltage for the transistor, the polygate on the NMOS is n-doped whereas on PMOS it is p-doped. Doping is performed by ion implantation followed by annealing of the implantation damage. The source and drain electrodes are defined by local implantation of suitable doping species (n for p-well and p for n-well) very close to the surface, thereby forming shallow p–n or n–p junctions in the well, depending on the transistor type. The depth of the junctions scales with the other dimensions of the transistor (see Section 2.2), and in today's advanced devices it is often less than 100 nm, in which case they are referred as "ultra-shallow junctions." Fine-tuning of the junction profiles may require several implantation steps followed by annealing.

Fabrication of the electrical contacts to the source, gate, and drain involves specialized metallurgy. The contact material must exhibit low electrical resistance and be chemically compatible with silicon in order to avoid interface degradation over time. For many years, metal silicides have been used extensively on source and drain, first titanium disilicide ($TiSi_2$), later replaced by cobalt disilicide ($CoSi_2$), and more recently by nickel monosilicide (NiSi). The silicide layers are formed by solid-state reaction of a deposited metal film with the underlying silicon; therefore, it is important that the reaction should not consume too much silicon. In the same way, a polycrystalline silicide layer or polycide is formed on top of the polygate in order to reduce the gate series resistance (Fig. 2.3).

2.1.2 Current Regimes

MOSFET can operate in three distinct current regimes, depending on the gate bias and the source–drain voltage V_{DS} [1]. Following are simple expressions for the source–drain current as a function of V_{DS} (drain–source bias) and V_{GS} (gate–source bias) in a long-channel n-MOSFET:

The linear region: in this region the MOSFET behaves as a linear resistor with a resistance modulated by the gate voltage. According to Ohm's law, current

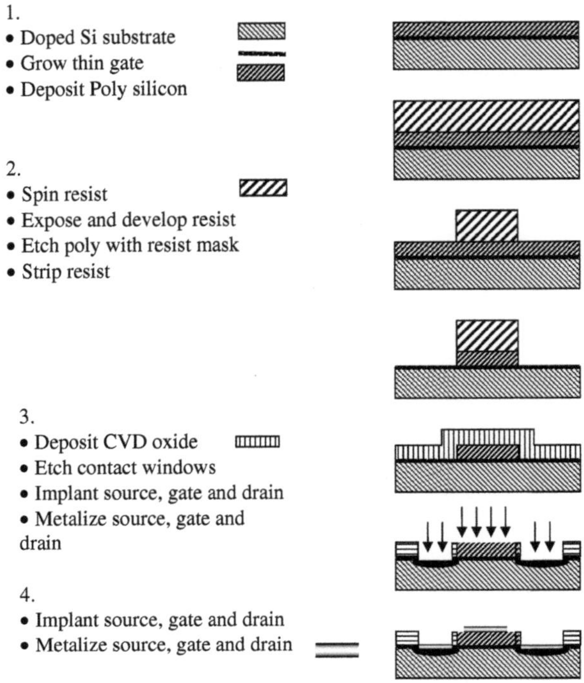

Fig. 2.3 Simplified process flow for MOSFET fabrication

modulation follows the variation of the channel resistance. The source–drain current is given by

$$I_{\mathrm{DS}} = \frac{\mu_n C_{\mathrm{ox}}}{2} \frac{W}{L} \left(2 \left(V_{\mathrm{GS}} - V_{\mathrm{T}} \right) V_{\mathrm{DS}} - V_{\mathrm{DS}}^2 \right) \tag{2.3}$$

where μ_n is the charge–carrier mobility, W is the gate width, L is the channel length, and V_{T} the threshold voltage. In analog circuits, devices often operate in this regime to take advantage of the linear amplification mode (Fig. 2.4a).

The saturated region: at fixed gate voltage, the channel width gradually narrows toward the drain with increasing source–drain voltage. Current saturation occurs when the channel nearly vanishes at the drain end ("channel pinch-off"). The saturation current depends on the gate bias but not on the source–drain bias; this behavior is usually referred to as the "long-channel characteristics." In this regime, I_{DS} is given by

$$I_{\mathrm{DS}} = \frac{\mu_n C_{\mathrm{ox}}}{2} \frac{W}{L} \left(V_{\mathrm{GS}} - V_{\mathrm{T}} \right)^2 \tag{2.4}$$

The MOSFET saturation current is usually written as I_{DSAT}. In digital circuits, the on-state of the device is normally set in the saturation region.

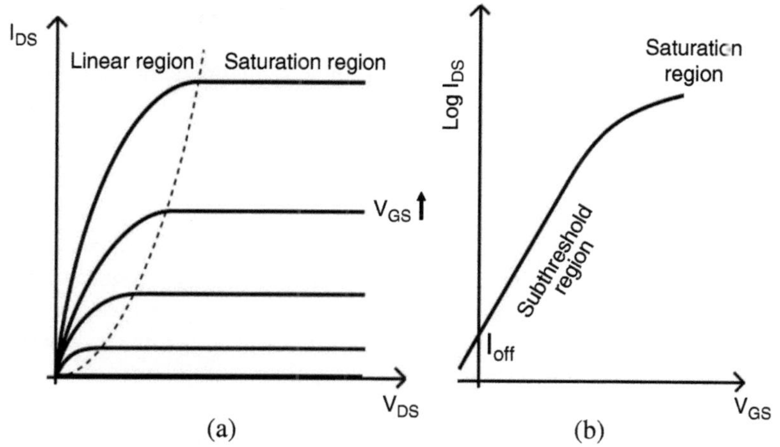

Fig. 2.4 MOSFET characteristics (**a**) linear and saturation regions; (**b**) subthreshold region

The subthreshold region: at fixed source–drain voltage, the inversion charge density decreases when the gate voltage approaches the threshold voltage. Below V_T, the inversion charge approaches zero asymptotically on a logarithmic scale. Therefore, the actual off-current reduces exponentially below the threshold voltage as

$$I_{DS} \propto \exp\left(\frac{V_{GS} - V_T}{\eta \cdot V_{Th}}\right) \tag{2.5}$$

where $V_{Th} = kT/q$ is the thermal voltage with k the Boltzmann constant, T the temperature, and q the electron charge. η is a nonideality factor which is approximately equal to $(1 + C_d/C_{ox})$ where C_d is the depletion layer capacitance at the onset of inversion:

$$C_d = \frac{\varepsilon_{Si}}{W_d} \tag{2.6}$$

with ε_{Si} the dielectric constant of the silicon substrate and W_d the depletion layer width. This is the so-called subthreshold regime (Fig. 2.4b), which can be characterized by the subthreshold slope S of I_{DS}, according to

$$S = \left(\frac{d(\log_{10} I_{DS})}{dV_{GS}}\right)^{-1} \cong 2.3 \cdot \frac{kT}{q} \cdot \eta = 2.3 \cdot \frac{kT}{q} \cdot \left(1 + \frac{C_d}{C_{ox}}\right) \tag{2.7}$$

Control of the subthreshold slope is an important aspect of the MOSFET scaling theory. It is important to notice that, to a first approximation, S does not depend on device dimensions. This creates extra complication for the scaling rules as will be shown below.

2.1.3 Mobility and Carrier Velocity

A simple estimate for carrier mobility μ in a semiconductor is given by the well-known Drude-type expression:

$$\mu = \frac{q\tau}{m^*} \tag{2.8}$$

where q is the electron charge, m^* the effective mass of the carrier (electron or hole), and τ the average relaxation time of the carriers, i.e., the average time lapse between successive collision events on the carrier's path. τ is a complex quantity including contributions from various scattering mechanisms, also taking into account the delocalization of the electron wavefunction.

Most scattering probabilities depend on the driving force experienced by the carriers and therefore on the electric field in the channel. The carrier mobility being thus electric field dependent will vary along the channel together with the effective field E_{eff} which follows from the applied source–drain bias. This complicated dependence can be approximated by a simple expression [2] for the effective mobility μ_{eff}, which holds for $E_{\text{eff}} < 5 \times 10^5$ V/cm:

$$\mu_{eff}\,(E) \approx 32500 \times \mathrm{E}_{eff}{}^{-1/3} \tag{2.9}$$

The effective mobility directly determines the field-dependent carrier velocity, and therefore the source–drain current, through the relationship

$$v_{\text{eff}} = \mu(E) \cdot E_{\text{eff}} \tag{2.10}$$

The decrease of the effective mobility with increasing field leads to the phenomenon of velocity saturation at high fields. This velocity saturation is caused by the increased scattering rate of highly energetic electrons, primarily caused by optical phonon emission. The overall proportionality between carrier velocity and electric field also changes with the device scale, since the effective field at constant bias increases at smaller channel dimensions. Moreover, the carrier mobility in the inversion layer is lower than in bulk material, because in this region the electron wavefunction penetrates into the gate oxide where higher scattering rates are experienced; high transverse electric fields at the channel surface – which typically result from device downscaling – shift the electron wavefunction even more into the oxide. The saturation velocity will ultimately depend on the balance between these effects.

The field-dependent mobility and the velocity saturation effect are some of the basic ingredients of the well-known drift-diffusion model, which computes the source–drain current under the assumption of a thermal equilibrium between the conducting electrons and the silicon lattice. However, this assumption no longer holds in very short gate devices, where high fields are present in the channel. In these devices, electrons will be driven to very high kinetic energies near the drain end of the channel, thereby effectively decoupling their energy from the lattice thermal bath. These "hot" carriers may acquire effective velocities that significantly exceed the saturation velocity, which is about 1×10^7cm/s in planar MOSFETs. This effect is called velocity overshoot [3] and is at the origin of the increase in current drive and transconductance experimentally observed in nanoscale MOSFETs. At these scales, one may thus expect a stronger impact of the channel dimensions on the transistor switching speed. In fact, this effect is not as strong as one could expect from estimates of the maximal electron velocity, as obtained from Monte-Carlo simulations [4]. This is mainly because the velocity overshoot regime only affects a small fraction of the total path of the electrons, which remain at the velocity saturation threshold for most of their trajectory.

2.2 Digital Signal Propagation

2.2.1 Gate Delay

In digital data processing, bits represented by fixed voltage levels are shifted from one logic gate to the next following the rules of binary Boolean logic. An "input switching threshold" is the point at which an input signal to a logic gate first records the occurrence of a voltage transition. Input switching thresholds are usually specified as a percentage of the voltage differential between logic 0 and logic 1 (Fig. 2.5). The speed at which this voltage signal is processed by the CMOS device is associated with the latter's gate propagation delay. The gate delay (or propagation delay) is divided into two terms: the intrinsic gate delay and the (external) gate load delay. The intrinsic gate delay depends on the physical characteristics of the MOSFET transistors. The load delay includes the slowing effect of the load on the gate propagation delay. Therefore, the intrinsic gate delay equals the propagation delay under zero load condition. It can be defined as the time needed for the saturated transistor current I_{DSAT} at drain voltage V_{DS} to charge the gate capacitance C_G:

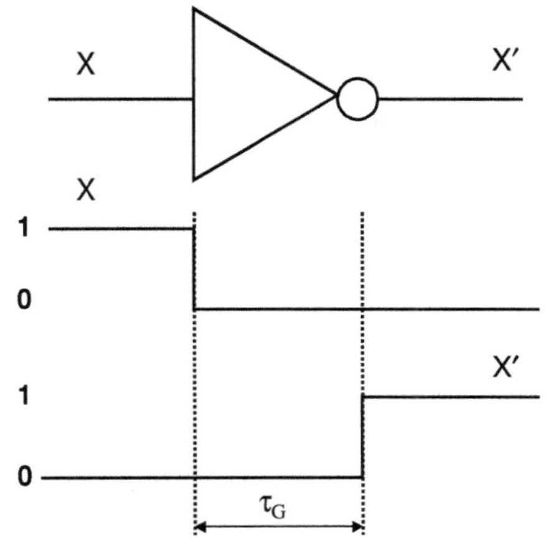

Fig. 2.5 Gate delay of logic inverter

$$\tau = C_G V_{DS}/I_{DSAT} \tag{2.11}$$

The *CV/I* characteristic time is an estimate of the intrinsic device switching speed, as it includes the intrinsic device capacitance, the voltage swing of the transistor, and its drive current. The dependence of τ on the device dimensions depends on the particular scaling rules being applied but until the latest generations the trend has been steadily downward (see Section 2.3). Moreover, there is also an intrinsic link between gate delay and gate length, through the scale-dependent carrier velocity (including possibly velocity overshoot) and the latter's influence on I_{DSAT}. This reinforces the tendency for τ to decrease with shrinking device sizes, as long as the downscaling has a positive influence on the drive current.

Since the *CV/I* figure does not take into account the external loads, it cannot provide a realistic estimate of the total propagation delay. From a practical point of view, the actual switching speed of a CMOS gate can be better derived from the inverter delay, which is defined as the time to propagate a digital signal through an inverter stage with a fan-out of one as, e.g., in a ring oscillator. This figure is correlated with the intrinsic gate delay, but will usually be an order of magnitude larger, because of the need to drive the next inverter stage.

2.2.2 Gate Delay Versus Interconnect Delay

In the past, the circuit delay was mainly determined by the gate delay figure. However, the latter has been continuously decreasing with device downscaling, due to the shrinking of the load capacitances. On the other hand, with shrinking circuit dimensions, the interconnect wire spacing (or pitch) becomes smaller, which must be compensated by the interconnect wire thickness in order to carry the required current for high-speed signal transmission. Overall, the interconnect resistance increases because of the smaller wire cross section, whereas the interconnect capacitance increases due to the combination of closer spacing and thicker wires. Since several layers of wiring are now required for interconnect and power distribution, this interconnect parasitic loading becomes the real limiting factor to speed.

Approximate expressions can be given for the latency of a single isolated interconnect that is *RC* limited within an ideal return path [5]

$$\tau\,(90\%) \cong r_{\text{int}}\, c_{\text{int}}\, L^2 + 2.3\, R_{\text{tr}}\, c_{\text{int}}\, L + 2.3\, C_{\text{L}}\, (r_{\text{int}}\, L + R_{\text{tr}}) \tag{2.12}$$

where r_{int} and c_{int} are the interconnect resistance and capacitance per unit length, R_{tr} is the source resistance, C_{L} is the load capacitance, and L is the interconnect length. Already at the 0.25 μm generation, the interconnect delay began to surpass the intrinsic gate delay. Figure 2.6 shows the gate delay with the corresponding interconnect delay for various CMOS nodes using aluminum interconnect technology. This rapid degradation has triggered the shift from Al wires to Cu technology around the 0.18 μm node.

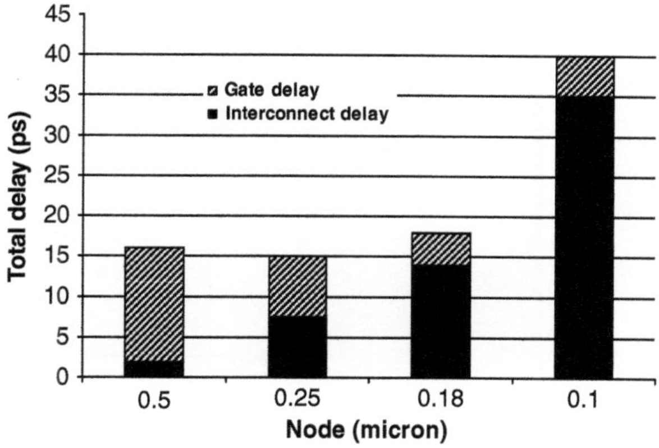

Fig. 2.6 Gate delay and interconnect delay (for Al wires) dependence on CMOS scaling

It is clear that, for state-of-the-art technologies, the gate delay is no longer the limiting factor for the circuit speed, and therefore the transistor switching speed can be traded for optimal overall performance against other device parameters such as the power dissipation. This is a very important consideration, since power, rather than speed, is becoming the main limiting factor for further miniaturization. As the technology proceeds into the nanometer era, the shift from device limited to interconnect limited design rules becomes a major trend, which is discussed at length in other chapters of this book.

2.2.3 Trends in CMOS Miniaturization

2.2.3.1 Moore's Law

Fast expansion of the semiconductor industry started very early after the invention of the first integrated circuits (1959–1960) and has since long been associated with Moore's law. Moore's original statement, issued in 1965, was modestly presented as an "educated guess" at the expected development of integrated circuits over the next 10 years. Or, to put it in his own words [6]: "With unit cost falling as the number of components per circuit rises, by 1975 economics may dictate squeezing as many as 65,000 components on a single silicon chip." Forty years later, unit cost is still falling with the number of components, and as long as this favorable trend persists, the "law" will remain firmly in place.

The mechanism underlying Moore's law can be understood using a simple model which we call "Moore's clock." Its two main features are found in any well-behaved watch, i.e., a spring and a pendulum. The spring provides the driving force that keeps the wheelwork running. In Moore's clock, this drive is essentially provided by the set of MOSFET scaling rules (see Section 2.2.8) which were first put forward some 6 years after Moore's initial paper, and which have shown almost the same remarkable endurance over time as the "law" itself (Fig. 2.7). With dimensional shrinking

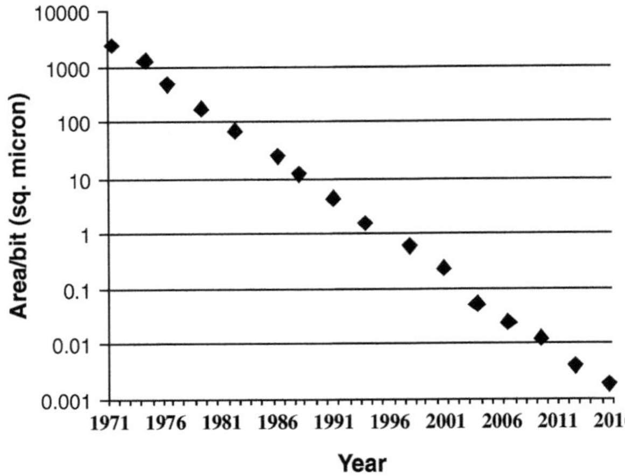

Fig. 2.7 Moore's law at work: scaling trend of DRAM cell area

now spanning more than two orders of magnitude, the persistence of scaling algorithms for CMOS ICs is a truly unique occurrence in the history of technology.

At this point, it should be realized that scaling rules apply to the definition of spatial dimensions, but do not define the pace of the miniaturization process. Therefore, Moore's clock also needs a pendulum to set the time with its periodic motion. In contrast to the spring, the pendulum is not only based on technical algorithms but also on business development factors. As such, it is closely linked with the microeconomic base cycle of the IC industry. The latter consists of four phases:

1. Downscaling of circuit components allows more functions to be integrated on the same area; therefore the average cost per function decreases.
2. Lower cost leads to market expansion and higher profit margins.
3. Profits are reinvested in R&D to prepare the next scale reduction.
4. In this way the cycle has repeated itself, on the average every 2–3 years, for almost four decades!

In spite of the apparent regularity of the pendulum, "setting the timescale" has always been the weak side of Moore's law. In reality, the speed of Moore's clock (Fig. 2.8) is not constant over time, but has gone through multiple stages [7]. In his 1965 article, Moore noted that the complexity of minimum cost semiconductor components had doubled every year since the first prototype IC (which did not contain MOSFETs but bipolar transistors) was produced in 1959. He then extrapolated the same trend until 1975, but at that time the cycle was already slowing down, as Moore himself later acknowledged. In the 1980s, Moore's law became stated as the doubling of number of transistors on a chip every 18 or 24 months. Later in the 1990s, it was widely associated with the claim that computing power at fixed cost is doubling every 18 months. In fact, none of these recent statements can be

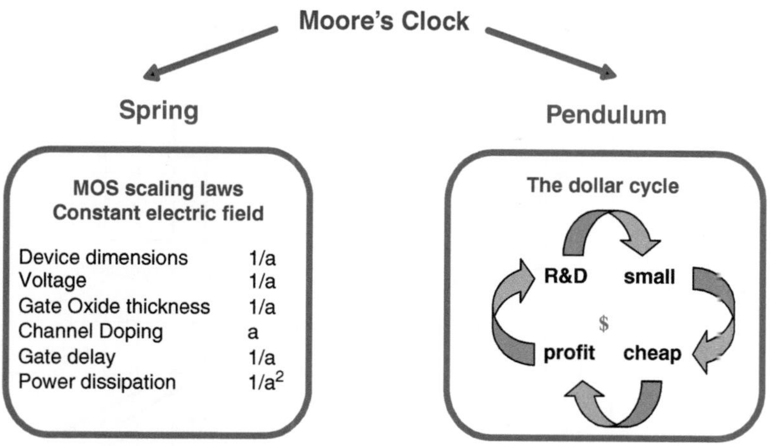

Fig. 2.8 Moore's clock

corroborated with accurate data. In its most general version, the law just points to regular doubling of "some measure of IC complexity."

2.2.3.2 Roadmaps

The National Technology Roadmap for Semiconductors (now the International Technology Roadmap for Semiconductors, or ITRS) was established in 1992 to codify the technological progress expected from Moore's law into a set of process targets and specifications, structured by the definition of future CMOS technology generations. The ITRS document is the product of a worldwide consensus building process in predicting the main trends of CMOS technology out to a 15 years horizon. The participation of experts representing the main actors in IC manufacturing and R&D ensures that the ITRS is a valid source of guidance for the global semiconductor industry.

The expected technological developments are timed with respect to the ITRS technology nodes, which are identified by the critical dimensions (e.g., the smallest half-pitch of contacted metal lines) of the circuits (90 nm, 65 nm...).[1] These numbers are "rounded off" figures derived from complex scaling formulas. Guidance for progress in the technological areas is provided by the definition of "grand challenges" to be met in moving to successive nodes.

Pressed by the champions of the semiconductor industry, the ITRS has regularly updated its forecasts of the CMOS scaling trends. At the end of the previous century, a phenomenon called "roadmap acceleration" was witnessed, by which the time window of each generation had gradually shortened toward a 2-year cycle (Fig. 2.9). For instance, the 1997 edition specified that the minimum device features of 100 nm

[1] The 2005 iteration of the ITRS roadmap has abandoned the simple concept of a unique node for all IC types, yet the technology generations are still labeled according to their critical feature sizes.

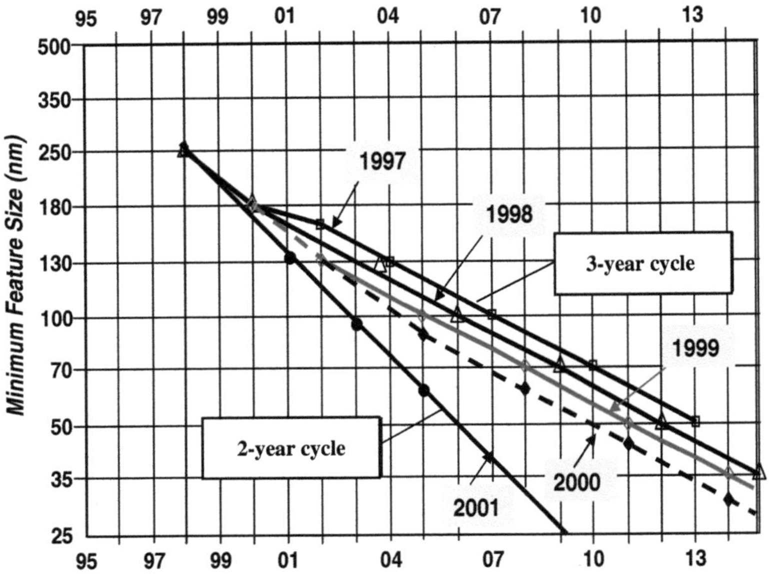

Fig. 2.9 Scaling acceleration of ITRS between 1997 and 2001 (IMEC)

would be reached in 2010. In fact, the 90 nm node was first delivered in 2003 and has a printed gate length (for high-performance devices) of 54 nm.

However, as a result of accumulating difficulties in the technological development of the latest generations, the ITRS has recently reversed this trend. Its most recent timescale (ITRS update 2005) assumes 3-year lapses between so-called "major" generations (see Table 2.1). In spite of this, some of the largest IC companies still attempt at "beating the roadmap" in an effort to secure a competitive advantage.

In spite of the recent slowdown, actual trends still clearly display the characteristic shape of an exponential growth rate. Exponential growth, however, also means that the fundamental limits of miniaturization are approaching rapidly. Many observers have therefore speculated about the "the end of Moore's law." The ITRS itself is putting a definite time horizon on each of its prognostics, but does not speculate on the endpoint position. In fact, up to this point the industry has been remarkably successful in keeping itself on schedule with the Roadmap timetables. In the last few years, however, the sustainability of the "Moore era" is being increasingly

Table 2.1 Near-term technology trend targets of ITRS 2005 (Source: ITRS).

Year of production	2006	2007	2008	2009	2010	2011	2012	2013
Microprocessor printed gate length (nm)	48	42	38	34	30	27	24	21
Microprocessor physical gate length (nm)	28	25	23	20	18	16	14	13

questioned among experts. In particular, there is a growing awareness within the ITRS groups that, within the next 10–15 years, most of the known technological capabilities will approach or have reached their limits. This threat was identified at the turn of the century and has since then been known as the "red brick wall."

2.2.3.3 Scaling Theory

For almost 40 years, shrinking the MOSFET transistor has been the dominant drive behind Moore's law. This process has been guided by the scaling laws, which in their original version were proposed as early as 1974 by Dennard et al. [8]. The basic idea of scaling is to reduce the dimensions of the MOSFET and associated interconnect wires, to produce a smaller transistor without degrading its performance. The original algorithm is based on the so-called "constant field" rule, whereby the applied voltages are scaled together with the geometrical dimensions of the device, such as to keep the internal electric fields constant. This can be achieved with a single scaling factor α, as can be seen in the second column of Table 2.2. This scaling algorithm has long been considered the most attractive, as it results in several simultaneous advantages:

1. The density of devices on the circuit increases by α^2, which is the basic claim of Moore's law.
2. The power dissipation per device scales like α^{-2}, which together with (1) results in a constant power dissipation density on the circuit.
3. The gate delay decreases by $1/\alpha$, due to the reduction of the device capacitance for a constant intrinsic transconductance.

The success of this model is largely due to the fact that, at least in the earlier stages, transistor performance as measured by its intrinsic gate delay would actually improve with scaling, whereas the potentially harmful high-field effects were avoided by decreasing supply voltages. However, the latter cannot be scaled down

Table 2.2 Scaling parameters for MOSFET transistors (see Refs. [8] and [10])

Physical parameter	Constant field scaling	Generalized scaling
Gate length	$1/\alpha$	$1/\alpha$
Gate width	$1/\alpha$	$1/\alpha$
Electric field	1	ε
Voltage	$1/\alpha$	ε/α
Gate oxide thickness	$1/\alpha$	$1/\alpha$
Channel doping	α	$\varepsilon\alpha$
Device area	$1/\alpha^2$	$1/\alpha^2$
Gate capacitance	$1/\alpha$	$1/\alpha$
Gate delay	$1/\alpha$	$1/\alpha$
Power dissipation	$1/\alpha^2$	ε^2/α^2

to arbitrary low levels. Indeed, at some point voltage reduction has to slow down due to following main reasons [9]:

- Reducing the threshold voltage increases the subthreshold slope of the MOSFET and the off current; this is mainly a consequence of the non-scaling of the thermal potential kT/q.
- As the power-supply voltage is reduced, the transistor performance degrades significantly at higher threshold voltages and also becomes more sensitive to tolerances in V_T.

For these reasons a switch was made in the 1980s from constant field scaling to a generalized scaling scheme [10], also shown in Table 2.2. In this new scheme, a second scaling constant ε is introduced to allow the electric field to scale independently of α. The supply voltage now scales with ε/α, the power dissipation with $(\varepsilon/\alpha)^2$, thus partially decoupling the electrical parameters from the dimensional scaling factor.

However, even this adjusted model has only a restricted validity range, as some limiting factors, generally known as short-channel effects, become stronger at smaller dimensions [11]:

1. Drain-induced barrier lowering (DIBL): the depletion barrier formed in the channel under the gate is lowered at higher source–drain voltages, which causes a degradation of the transconductance. This effect can be accompanied by the so-called punch-through that occurs when the depletion region surrounding the drain extends to the source.
2. Surface scattering occurs when electrons are accelerated toward the surface by the vertical component of the electric field. The scattering of the electrons by the surface potential causes a reduction in the mobility.
3. Hot electrons degradation, caused by electrons injected into the oxide at the Si–SiO_2 interface with high kinetic energy, can cause permanent damage to the gate insulator.
4. Velocity saturation has a stronger impact due to the upscaling of the electric fields with ε.

Other negative effects must also be taken into account. As a general consequence of physical scaling, bulk depletion charges are smaller than expected and the threshold voltage expression must be modified to account for this reduction. The scaling of physical dimensions is also limited in a practical sense by the discreteness of dopants, since present manufacturing techniques do not control the exact placement of dopant atoms. Consequently, since very small device volumes contain only a small number of dopants, large statistical variations become likely. In fact, the statistical distribution of dopants is only one of the sources of electrical variability that are likely to affect future circuits. Shift of device parameters also results from the increasing difficulty to control lithographic dimensions on a nanometer scale over the full circuit area. The first impact of the variability bottleneck can already be felt in today's circuit design, and the problem will likely get much worse for future generations. Below some critical dimensions single devices can still be built, but large functional circuits may be difficult to design and manufacture with available techniques.

2.2.3.4 Scaling and Power Dissipation

The scaling algorithms also impact on the power consumption of the IC. There are two main sources of power dissipation in a CMOS device: dynamic (or active) switching power in the on state due to the charging and discharging of circuit capacitances, and static dissipation from leakage currents in the off state.

When CMOS devices switch, the output is either charged up to the transistor bias voltage or discharged down to ground. The power dissipated during switching is therefore proportional to the switching speed and to the capacitive load. The dynamic power dissipation arising from normal circuit operation is given by

$$P_{\mathrm{on}} = C_{\mathrm{EFF}} V_{\mathrm{DS}}^2 f \tag{2.13}$$

where C_{EFF} is the effective output capacitance that is driven by the transistor and f the clock frequency of the circuit. Decreasing the clock frequency and/or the drain bias is therefore an efficient (although not always desirable) way to lower the dynamic power consumption.

The static power dissipation is taking place between switching events and is associated with source-to gate and source-to-drain leakage mechanisms. The source-to-gate leakage will be discussed in the next section. The source-to-drain leakage has two components: reverse-bias diode leakage on the transistor drains and subthreshold leakage through the channel when the transistor is turned off. Reverse-bias diode leakage must be tackled through process optimization, mainly by improving the quality of the junctions. Subthreshold current is a more complex issue. The subthreshold power dissipation formula is [12]

$$P_{\mathrm{off}} = W_{\mathrm{tot}} V_{\mathrm{DS}} I_{\mathrm{off}} = W_{\mathrm{tot}} V_{\mathrm{DS}} I_0 \exp\left(-\frac{qV_{\mathrm{T}}}{mkT}\right) \tag{2.14}$$

where W_{tot} is the total device width, I_0 the extrapolated drain current per unit device width at threshold voltage, and m the so-called body effect coefficient. A simple expression for m is

$$m = 1 + \frac{3t_{\mathrm{ox}}}{W_{\mathrm{dm}}} \tag{2.15}$$

where t_{ox} is the gate insulator thickness and W_{dm} the bulk depletion layer width under the gate, which itself depends on the doping level of the channel.

The parameter that predominantly affects the P_{off} value is the threshold voltage, which must therefore remain above a critical value corresponding to the power tolerances set by the circuit design. In this respect, a distinction is usually made between low power circuits, where power constraints are the main priority, and high-performance circuits allowing for more dissipation.

It can be deduced from previous formulas for P_{on} and P_{off} that both V_{DS} and V_T are important parameters in the setting of the overall power dissipation levels. A lowering of V_{DS} decreases the active power level, but at the same time it will have a deleterious effect on P_{off}, since V_T is limited by V_{DS} for efficient transistor operation [13]. Since lowering the threshold voltage leads to an exponential increase of the off current, it can only occur between narrow margins.

For very short MOSFETs, the gate-induced potential barrier between source and drain is so thin that direct source-to-drain tunneling becomes possible. Calculations of the source-to-drain tunneling current based on one-dimensional transport models have demonstrated the exponential dependence of the off state current on the depletion barrier width, which is a clear signature of direct tunneling phenomena. The effect of such tunneling first shows up in the degradation of the subthreshold slope of the device. Even if two-dimensional effects might worsen the picture somewhat, the general conclusion of these simulations is that source–drain tunneling should gradually become a major limiting effect for transistor operation below 10 nm gate lengths [14].

Below roughly 100 nm gate length, a major problem arises from the gate insulator, which in standard CMOS technology consists of a thin layer of thermally grown SiO_2. According to the scaling rules of Table 2.2, the thickness of this layer is reduced in the same proportion as the gate length. This is necessary to insure sufficient capacitive coupling ε/t_{ox} between the gate and the channel, and hence a good transconductance of the device. However, the direct quantum tunneling current from the channel to the gate electrode increases exponentially with decreasing oxide thickness (see Fig. 2.10) for a graphical estimate; accurate calculations are

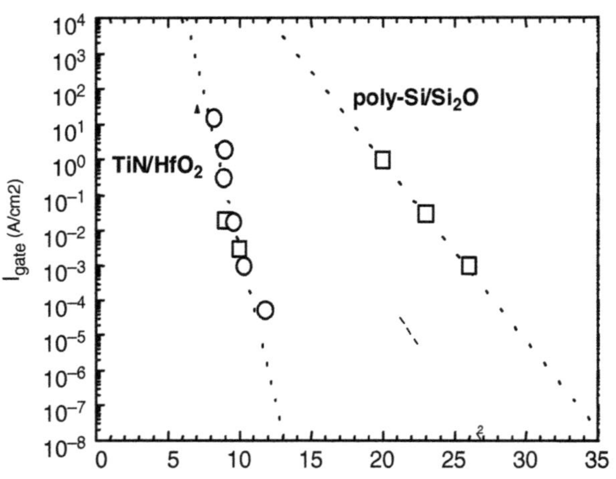

Fig. 2.10 Direct gate tunneling current density vs. effective oxide thickness for SiO_2 and for HfO_2-based gate stacks (M. Heyns, IMEC)

rather complex because they must take into account the band structure of the oxide). Although the tunneling current component is typically small as compared to the on current, it can become a sizable part of the off current in n-MOSFETs for gate oxide thicknesses below 2 nm, and will therefore contribute significantly to the static power dissipation of circuits below the 90 nm node. Assuming an acceptable gate leakage current of 10 A/cm^2, which at this time is a representative number for low operating power circuits, sets a practical lower limit of about 1.5 nm for the SiO_2 insulator thickness. Moreover, the direct tunneling current is only weakly dependent on the applied voltage and can therefore not be compensated by the scaling of V_{DS}.

The way out of this dilemma would be to replace SiO_2 by another insulator with a higher dielectric constant and therefore a lower equivalent SiO_2 thickness. This would suppress the direct tunneling current by increasing the gate insulator thickness, while keeping enough coupling between gate and channel to avoid short-channel effects. The challenge with high-*k* dielectrics is to find an insulator material resulting in an equivalent (SiO_2) oxide thickness (EOT) of at most 1 nm thick, and which would not degrade the operational properties of the transistor.

The key guidelines for selecting an alternative gate dielectric are

- Permittivity, band gap, and band alignment to silicon
- Thermodynamic stability; film morphology; interface quality
- Compatibility with other materials used in CMOS devices
- Process compatibility and reliability

Many dielectrics appear promising in some areas (see Table 2.3), but very few materials are capable of fulfilling all of these criteria. It must be kept in mind that replacing SiO_2 as the gate insulator will be a major milestone in the evolution of CMOS technology. Because it implies difficult changes in the fabrication process, the industry has opted for a gradual approach. The first alternative dielectrics to be introduced are based on silicon nitride or oxynitride, which are already well known and do not largely deviate from the standard technology. However, with shrinking

Table 2.3 Overview of high-*k* dielectrics (U. Berkeley)

High-*k* dielectric	*k* value
SiO_2	3.5
Si_3N_4	7
$Si_xN_yO_z$	4–7
Al_2O_3	9
Ta_2O_5	25
ZrO_2	25
HfO_2	40
TiO_2	50
$BaSrTiO_3$	300

gate dimensions a transition to more radical alternatives with higher *k* values is becoming more pressing.

As of today, the most promising candidates have been identified in the family of refractory metal oxides (mainly the Hf- or Zr-based ones) and their silicate compounds, such as $Hf_xSi_yO_z$, as well as their nitrided counterparts. The latter films are easier to etch than pure HfO_2. However, they must be deposited with sophisticated chemical vapor deposition (CVD) or atomic layer deposition (ALD) techniques. At this point, the main obstacle remains the poor quality of the high-*k*/silicon interface, resulting in gap states caused by Hf–Si bonds or oxygen vacancies. These local states lead to Fermi level pinning, V_T shifts, mobility degradation, and reliability problems. A common procedure nowadays in use is the intercalation of a thin SiO_2 interlayer between the high-*k* and the channel to improve the interface quality. Because of the difficult materials issues involved, introduction of high-*k* materials is likely to be pushed back to the 45 nm node, especially for high-performance circuits.

There are also problems to be solved with respect to the compatibility of high-*k* insulators with the gate electrode, which traditionally has been polysilicon ("poly"). Indeed, high-*k* materials and polysilicon gates are incompatible due to the above-mentioned Fermi-level pinning at the dielectric/poly interface. Therefore, many researchers believe that high-*k* layers will have to be used in conjunction with a metal gate or even two different metals for PMOS and NMOS devices for a better positioning of the respective threshold voltages. Moreover, there is evidence that metal gates by themselves offer some performance advantages, even with conventional dielectrics. One of the primary candidates is a metal gate made of NiSi, also known as "FUSI" (fully silicided gate). This approach can draw on the extensive knowledge of silicides processes, and especially of NiSi which is already in use for source and drain contacts.

2.3 New Device Structures and Materials

2.3.1 Strained-Silicon MOSFETs

As has been discussed above, many of the problems showing up in device miniaturization are related with the degradation of their transport properties. Strained silicon has recently been introduced in CMOS devices as a means to improve the carrier mobility in the channel, which should lead to shorter switching times. Evidence that transistors fabricated with strained-silicon channels were indeed faster accumulated during the 1990s, and was decisively demonstrated when the 90 nm node was reached. Therefore, strained-silicon channels have now become an integral part of the ITRS roadmap.

The first approach (so-called "global strain" and pioneered by IBM) for applying stress to the devices used a silicon germanium buffer layer between the substrate and the transistor channel (Fig. 2.11). $Si_xGe_{1-x}$is a near-ideal solid solution whose

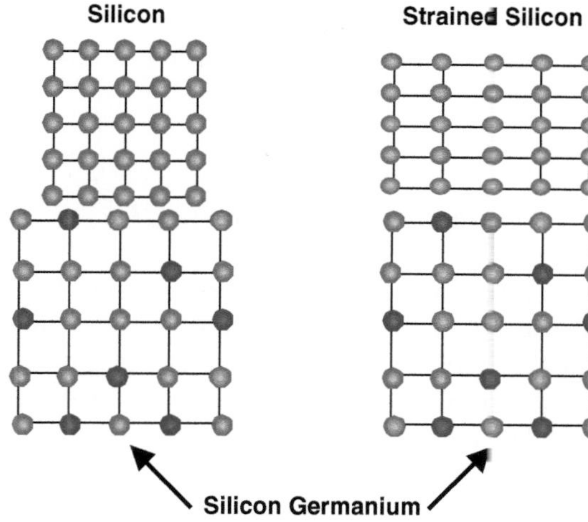

Fig. 2.11 Straining the silicon channel by growing it on a SiGe buffer layer

lattice constant follows Vegard's law to a good approximation. The lattice mismatch between Si and Si_xGe_{1-x} depends on the Ge content, but can easily reach 1% or more due to the 4% larger Ge lattice. When a thin silicon layer is grown epitaxially on top of a silicon germanium buffer, a pseudomorphic Si lattice results with a stretched in-plane lattice constant. The increase in lattice spacing produces biaxial strain in the silicon channel, which changes the shape of the energy bands both for electrons and holes. This deformation results in an increased mobility and channel drive current, which can be observed on n-type as well as on p-type devices. An alternative approach is the local strain method, developed by INTEL, which uses different processes for n- and p-MOSFETs. The n-channels are put under uniaxial tensile stress by depositing a thin silicon nitride film on the gate area, whereas the p-channels are compressed sideways (but also uniaxially) by growing local silicon germanium pockets under the source and drain areas (Fig. 2.12). Although the underlying solid-state mechanisms are basically the same as in the

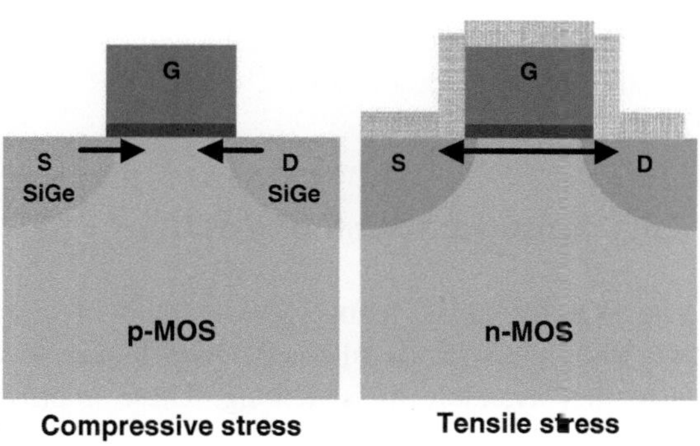

Fig. 2.12 Uniaxial strain applied to n-MOS and p-MOS devices

global scheme, the performance improvements at the circuit level tend to be better than with the global method, and therefore this local approach has now gained wider acceptance.

2.3.2 *Silicon-On-Insulator (SOI)*

At present, there is much interest in using Silicon-On-Insulator (SOI) wafers in advanced CMOS. SOI substrates consist of three layers: a thin surface layer of monocrystalline silicon on which the transistors are made, an underlying layer of amorphous SiO_2, and the bulk silicon support wafer underneath. The insulating silicon dioxide is referred to as the "buried oxide" or "BOX" and is typically a few thousand Ångström thick. There are several techniques for BOX fabrication, the most popular ones being buried oxygen implants and wafer bonding. The SOI wafer structure has several important advantages over bulk or epitaxial starting wafers. SOI wafers offer near-perfect transistor isolation (resulting in lower leakage currents and tighter transistor packing density), reduced parasitic drain capacitance (hence higher switching speeds and lower power consumption), and some process simplification relative to bulk or epitaxial silicon wafers. Due to these advantages, SOI wafers appear to be well suited for high-performance ICs requiring high-speed switches, high integration density, and low voltage/low power operation. Due to the reduced leakage levels, SOI should also be beneficial for battery operated systems. Moreover, SOI wafers offer an excellent platform for integrating RF and digital circuits on the same chip.

SOI substrates are used in two main application schemes; partially depleted (PD) and fully depleted (FD) SOI transistors, depending on the depth of the depletion layer with respect to the upper crystalline Si thickness. Both have specific advantages and drawbacks, but the general trend is toward the FD technology, because it allows higher circuit performance for a given power dissipation.

In the past, the main barriers to the widespread adoption of SOI wafers for mainstream CMOS fabrication have been the uncertain material quality and the higher cost of SOI wafers. The key materials quality issues are continuity and thickness uniformity of the BOX and defect density and thickness uniformity of the device-quality, single-crystal silicon layer. However, the wafers that are now commercially available are considered to be technically and economically ready for use in mainstream CMOS IC production.

2.3.3 *Strained Silicon and SOI*

The SOI MOSFET reduces the amount of current needed to switch a transistor, while the strained-silicon MOSFET increases the amount of current the transistor has available for switching. These improvements being complementary, it would seem natural to combine both in an effort to achieve maximum performance. Moreover, there are other technical arguments for integrating strained silicon and SOI. Strained-silicon channels are usually adjacent to thick layers of SiGe, so the source and drain junctions of a bulk strained-silicon MOSFET will penetrate into the SiGe. Since the latter has a lower energy gap and higher dielectric constant than bulk Si, this leads to higher junction capacitances as well as higher junction leakage. However, when a strained-silicon channel is formed on an SOI structure, the increased junction capacitance and leakage associated with SiGe are restrained by the silicon-on-insulator structure, thereby improving the transistor performance. Strained silicon can even be deposited directly on SOI, without SiGe interlayer (SSOI process). After the strained-silicon SOI substrate has been formed, the rest of the fabrication process can continue as for a normal SOI circuit flow.

2.3.4 Germanium and III–V Channel Devices

With continuous downscaling, the degradation of transistor transport properties is likely to become more acute. Over the last decade, device dimensions have been shrunk by an order of magnitude, but drive current has only doubled. Since planar silicon may be unable to accommodate the rigorous current scaling requirements of sub-22 nm geometries, recent research has identified Ge as a potential alternative. The higher carrier mobility in Ge makes it a candidate for high-performance CMOS devices, which could easily be integrated into the existing silicon manufacturing infrastructure. Since high-k materials are under development to replace thermal oxides, the problem of the gate insulator on Ge is potentially solvable. Using similar arguments as for silicon, there is also the potential of using germanium-over-silicon–germanium and its combination with germanium-on-insulator at some point in the future.

However, much research is still needed to remove the possible showstoppers before applying Ge semiconductor material to advanced CMOS scaling. The main bottlenecks for a future Ge device technology are the passivation of interface states, reduction of diode leakage, and availability of high-quality germanium-on-insulator substrates. Although progress has been made on these three issues, specific problems with the n-type activation of Ge channels, as well as disappointing mobility date in n-type Ge MOSFETs, have cast some doubts on the future use of Ge for nMOS. Therefore, alternatives for Ge nMOS in advanced CMOS are presently investigated. It is well known that several III–V materials show large electron mobilities. Moreover, GaAs has a lattice parameter very close to that of Ge, which allows nearly defect-free epitaxial growth of GaAs on Ge. This opens the possibility of making high-performance CMOS with a Ge pMOS device and a GaAs nMOS device on the same substrate. In the future other III–V materials with even higher mobility than the one of GaAs could be investigated. One of the main challenges of this approach is to optimize the gate stack for MOS devices on Ge as well as on

III–V compounds. At this time, it is too early to tell if the Ge/III–V scheme will be able to reach the CMOS mainstream technology integration level.

2.3.5 Novel MOSFET Devices

In the coming years, the major scaling challenges at the device level will be

- Controlling leakage currents and short-channel effects
- Increasing the drive current while reducing the overall power supply
- Reducing the variability of the device operational parameters across the chip and from chip to chip

Although new materials are expected to play an important role with CMOS scaling now entering into the nanometer regime, it is also expected that radical changes in device geometries will be necessary to solve the bottlenecks just mentioned. The majority of the new materials has been reviewed in the preceding sections and includes gate stack (high-*k* dielectric and metal gate) materials, channel materials with improved carrier transport properties, as well as some new materials for the source/drain regions with reduced resistance and carrier injection properties. New transistor structures seek to improve the electrical behavior of the MOSFET and accommodate the integration needs of new materials. The combination of new structures and new materials enables novel device operating conditions that may provide better performance by overcoming the physical constraints of bulk planar CMOS. A starting point for this evolution could be provided by the double-gate MOSFET architecture. In this structure, a second gate and gate insulator are inserted at the device bottom between the channel and the substrate, thereby substantially improving the gate to channel coupling. The better gate control over the channel region steepens the subthreshold slope in the off state; moreover, it also increases the on-state current by providing a second current path along the channel bottom, thereby improving the I_{on}/I_{off} ratio. However, the price to be paid is a considerable complication in manufacturing processes. Indeed, the back gate must be self-aligned with the source and drain junctions as well as with the front gate, in order to avoid excessive parasitic capacitances. Furthermore, both gates must be connected via a low-resistance path to minimize the parasitic resistance. Because these steps are very difficult to optimize by standard lithographical means, attention has recently shifted to a more manufacturable version of the original DG-FET, called the FinFET. This device eases the process requirements by placing the silicon channel (the "fin") perpendicularly on the substrate, thereby effectively creating three-dimensional device geometry. In this geometry, "top" and "bottom" become "front" and "back" gates, both of which can be easily accessed from the top of the wafer during processing (see Fig. 2.13).

The fin dimensions must be optimized to alleviate short-channel effects, which require the fin thickness to be no more than about a quarter of the gate length. In the

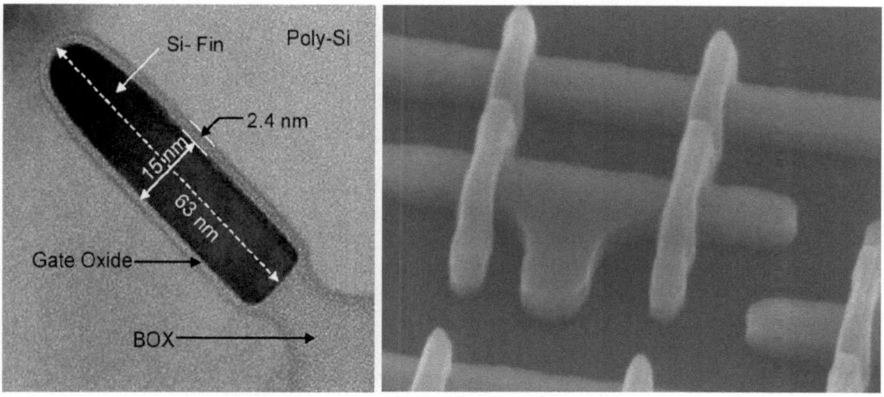

Fig. 2.13 FinFET structure and FinFET transistor top view (IMEC)

smallest devices, this ratio exceeds the present lithographic printing capabilities and thus requires special patterning techniques [15]. Another difficulty arises from the drive current requirements. In a planar MOSFET, the drive current can be increased just by making the device wider. In a FinFET, however, the effective width is limited by the height of the fin.

Increasing the drive current must be achieved by placing several fins in parallel and connecting them with bridge structures. This structure is known as the multi-gate FET or MuGFET. Among all available options, the multi-gate FET is considered as the most serious candidate due to the possibility of implementing the double-gated device concept with standard CMOS processing. Present development efforts focus on the modules specific for MuGFET topography such as fin and gate patterning, implementation of advanced gate stacks, ultra-shallow source and drain junctions, and mobility enhancement techniques.

As stated above, the device widths in the MuGFET architecture can be increased at a fixed lithographic scale by increasing the height of the silicon fins, thus providing more device area in a physical area than is possible to obtain with planar devices. While the MOSFET performance as measured by CV/I delay is not improved, since both C_G and I_{DSAT} increase in direct proportion to the fin height, interconnect contributions to delay may be decreased by allowing for closer placement of MOSFETs of the same drive capability and hence lower interconnect capacitance and resistance, [15,17] This is important since, as stated above, such interconnect delays already present major obstacles to scaling CMOS designs. Thus, one new direction (literally) for device scaling could become the vertical direction with respect to the wafer plane.

High parasitic resistance of source and drain regions are still obstacles on the way to reach high MuGFET performance. Selective epitaxial growth has been implemented into the MuGFET process flow to increase the fin width outside the spacers and lower the contact resistance. Successful implementation of this concept has already resulted in substantial drive current improvement for nMOS as well as pMOS devices [16].

References

1. Taur, Y. and Ning, T. H.: Fundamentals of Modern VLSI Devices. Cambridge: Cambridge University Press (1998)
2. Baccarani, G. and Wordeman, M. R.: Transconductance degradation in thin-oxide MOSFET's. IEEE Trans. El. Dev. **ED-30**, 1295 (1983)
3. Sai-Halasz, G. A.; Wordeman, M. R.; Kern, D. P.; Rishton, S.; and Ganin, E.: High transconductance and velocity overshoot in NMOS devices at the 0.1 mm-gate-length level. IEEE Electron Device Lett. **EDL-9**, 464 (1988)
4. Laux, S. E. and Fischetti, M. V.: Monte Carlo simulation of submicron Si n-MOSFETs at 77 and 300 K. IEEE Electron Device Lett. **9**, 467 (1988)
5. Meindl, J. D. et al.: Interconnect opportunities for gigascale integration. IBM J. Res. & Dev. **46**, 245 (2002)
6. Moore, G. E.: Cramming more components onto integrated circuits. Electronics **38(4)** (1965)
7. Tuomi, I.: The life and death of Moore's law, published on-line in First Monday, **7** (2002)
8. Dennard, R. H.; Gaensslen, F. H.; Yu, H. N.; Rideout, V. L.; Bassous, E.; and LeBlanc, A. R.: Design of ion-implanted MOSFETs with very small physical dimensions. IEEE J. Solid-State Circuits, **SC-9**, 256 (1974)
9. Davari, B.; Dennard, R. H.; and Shahidi, G. G.: CMOS scaling for high performance and low power-the next ten years. Proc. IEEE **83**, 595 (1995)
10. Baccarani, G.; Wordeman, M. R.; and Dennard, R. H.: Generalized scaling theory and its application to a 1/4 micrometer MOSFET design. IEEE Trans. Electron Devices **ED-31**, 452 (1984)
11. Frank, D. J.; Dennard, R. H.; Nowak, E.; Solomon, P. M.; Taur, Y.; and Wong, H.-S. Ph.: Device scaling limits of Si MOSFETs and their application dependencies. Proc. IEEE **89**, 259 (2001)
12. Taur, Y. and Ning, T. H.: Fundamentals of modern VLSI devices. Cambridge: Cambridge University Press, 271 (1998)
13. Gonzalez, R.; Gordon, B. M.; and Horowitz, M. A.: Supply and threshold voltage scaling for low power CMOS. IEEE J. Solid-State Circuits **32**, 1210 (1997)
14. Likharev, K. K.: Electronics below 10 nm. In: Nano and Giga, Challenges in Microelectronics. Amsterdam: Elsevier, 27 (2003)
15. Nowak, E. J.; Aller, I.; Ludwig, T.; Kim, K.; Joshi, R. V.; Ching-Te, C.; Bernstein, K.; and Puri, R.: Turning silicon on its edge [double gate CMOS/FinFET technology]. IEEE Circuits and Devices Magazine **20**, 20 (2004)
16. IMEC results (2005).
17. Nowak E. J.; Maintaining the benefits of CMOS scaling when scaling bogs down. IBM J. Res. & Dev. **46**, 169 (2002)

Chapter 3
Interconnects in ULSI Systems: Cu Interconnects Electrical Performance

Avinoam Kolodny

3.1 Introduction

Integrated electronic systems have advanced in complexity at an exponential rate during the last four decades, as measured by the number of transistors on a single silicon chip [1, 2]. This growth, which had major implications on economy and society, was enabled by continuous miniaturization of transistor devices and the metallic wire structures used for making interconnections among them. In recent technology generations, as a result of the scaling down of all device and wire dimensions, the interconnect structures have become dominant limiters of system performance, power, and cost [3].

This chapter provides a perspective on the role of interconnects in integrated electronic systems. The structure of integrated electronic systems and the required interconnections within such systems are described in Section 3.2. Metrics for evaluating the quality of signal interconnections are presented in Section 3.3 followed by simple electrical circuit models for interconnect wires (Section 3.4). In light of these models, the fundamental problem of interconnect scaling is presented in Section 3.5, and design approaches for addressing the problem are briefly surveyed in Section 3.6. The unique requirements of power distribution interconnect are mentioned in Section 3.7.

3.2 On-Chip Interconnect Requirements

An electronic system is typically represented by a block diagram model consisting of several functional units (a.k.a. *blocks* or cells), which communicate with each other by electrical signaling over wires (a.k.a. *nets* or nodes). Each block has input/output terminals (a.k.a. *ports* or pins), and a net typically connects an output

A. Kolodny (✉)
Faculty of EE, Technion IIT, Haifa 32000, Israel
e-mail: kolodny@ee.technion.ac.il

Y. Shacham-Diamand et al. (eds.), *Advanced Nanoscale ULSI Interconnects: Fundamentals and Applications*, DOI 10.1007/978-0-387-95868-2_3,

port of a block with input ports of one or more other blocks. Ideally, a net is an equipotential surface, such that the output voltage signal produced by the driver block can be sensed directly and immediately by all the receivers. The number of possible connections can be very large for systems with many blocks. System architects cope with this complexity by using hierarchy: the system consists of only a few top-level blocks, and each block is modeled internally by several interconnected sub-blocks. This principle is repeated recursively until low-level sub-blocks describe elementary structures such as logic gates, which are composed of interconnected transistors. This connectivity model is illustrated in Fig. 3.1. In physical implementation, transistors are laid out on the surface of a semiconductor wafer, and the nets are formed by patterning multiple layers of metal and connecting vias. The logical block boundaries do not appear in the silicon implementation, but high-level blocks typically occupy clearly distinguishable rectangular areas on the die surface in the so-called *floorplan* (Fig. 3.2).

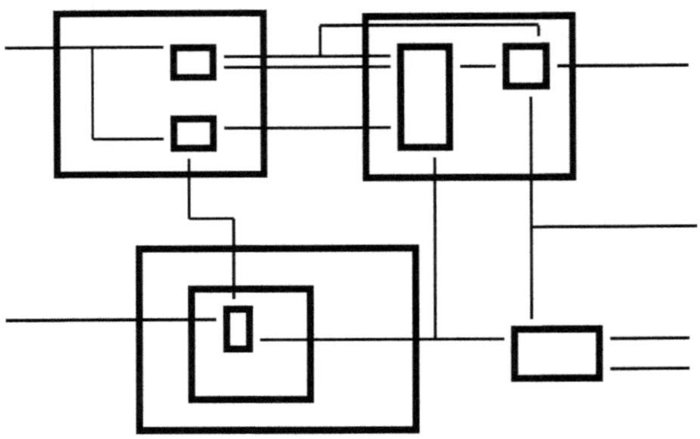

Fig. 3.1 Illustration of a system connectivity model. The structure is expanded to view several hierarchical levels of nested cells. Nets make logical connections among terminals of logical blocks, penetrating the entire hierarchy

Transistors within low-level blocks are laid out in close proximity to each other, such that the nets interconnecting them are very short. Hence, internal nets within low-level blocks are called *local wires*. Nets connecting blocks at the highest levels are typically very long and are called *global wires*. Global wires often reach several centimeters of length, as the die edge size is typically 1–2 cm.

Circuit architectures emphasize the importance of shared global wires as the main public transportation arteries for communications in the system. The shared wires are typically organized as *buses* composed of multiple parallel lines carrying control signals and data values among multiple functional units. Buses operate according to predefined protocols, enabling the connected functional blocks to request temporary control of the bus and perform transactions of information transfer without interfering with each other. A centralized arbitration mechanism is required to prevent collisions when simultaneous requests are issued for access to the bus. Another specialized circuit architecture, used in field-programmable devices, employs *configurable interconnect*. These circuits contain a fabric of wires and general logic

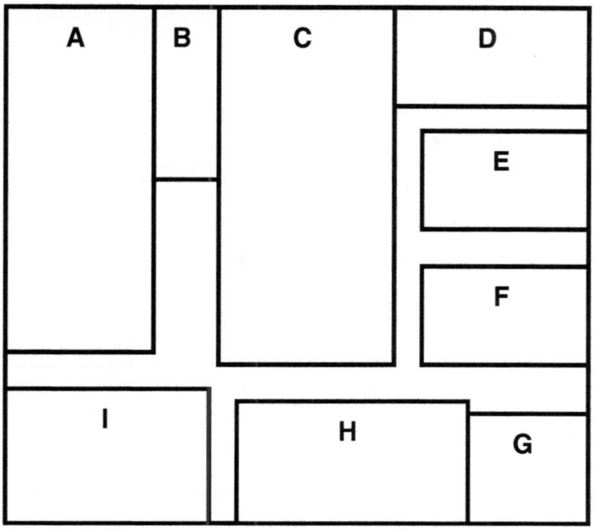

Fig. 3.2 Floorplan of a VLSI chip. IP blocks are labeled by A, B, C,. . .

elements. Special switches are used to define the desired logic functions and to assign specific wires for connections among them.

The number of nets and terminals at each level of the hierarchy depends on system architecture and design style. These parameters have been studied statistically in many practical systems and found to obey an empirical formula known as *Rent's rule* [4]:

$$T = kN^r \tag{3.1}$$

where T is the number of terminals of a block containing N sub-blocks and k is the average number of terminals of the sub-blocks with $r< 1$ (Rent's exponent). Practical circuits typically fit this expression with the exponent r in the range 0.5–0.75. Note that point-to-point connections among all N system elements could grow in proportion to N^2, so the low value of r is a result of using hierarchy and shared buses, as described above. A model for predicting the total number of wires in an integrated system and their length distribution has been developed on this basis [5]. Typical length distribution data (Fig. 3.2) exhibit a large number of short local nets and a continuum of bigger nets. The number of global nets is smaller by orders of magnitude, and their average length is larger by orders of magnitude. The total length of all wires in a state-of-the-art microprocessor is in the range of several kilometers. The mathematical model derived to represent such a distribution has been used to predict system level interconnect requirements, and to design appropriate process technologies with adequate number of metal layers having appropriate thicknesses and pitches (Fig. 3.4) [6].

Fig. 3.3 Interconnect length distribution in typical digital systems. Number of nets is arbitrarily normalized

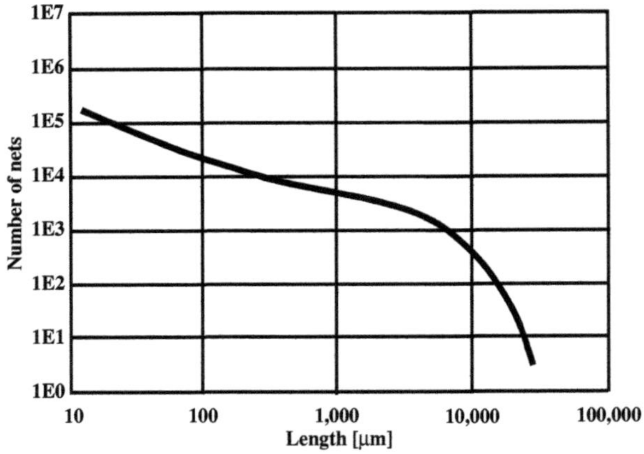

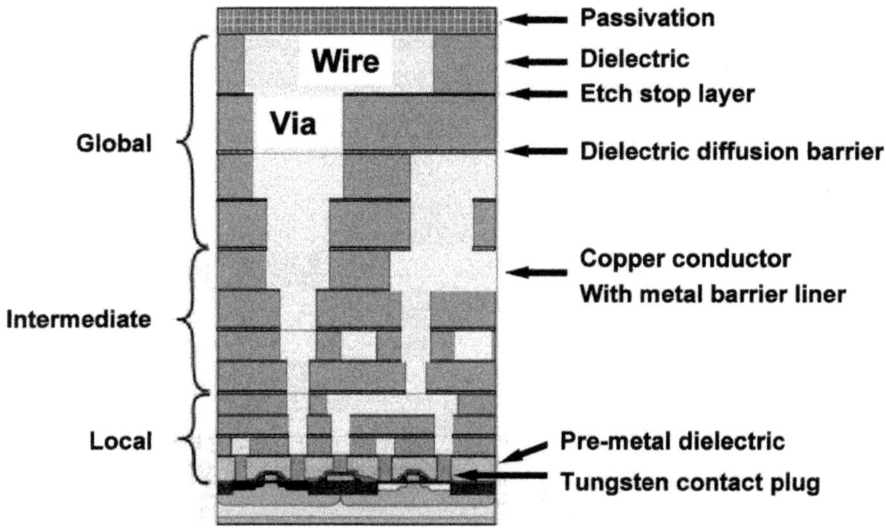

Fig. 3.4 Cross section of metallization stack in 130 nm technology
Source: ITRS 2005

A metallization stack is depicted in Fig. 3.4. The bottom metal layers use dense narrow wires, used only for local nets. Higher layers use progressively thicker metal and larger spaces, used for routing global nets. The top layers are also used for power distribution, conducting current supply to all the logic gates. (Power lines are often not shown in logical diagrams such as Fig. 3.1.) Usually, each metal layer employs a preferred routing direction, and successive layers use alternating orthogonal directions.

The physical design of actual integrated circuit layouts is a complex task, involving multiple computer-aided design (CAD) tools for operations such as floorplanning, cell generation, cell placement, and routing [7]. The primary tools for

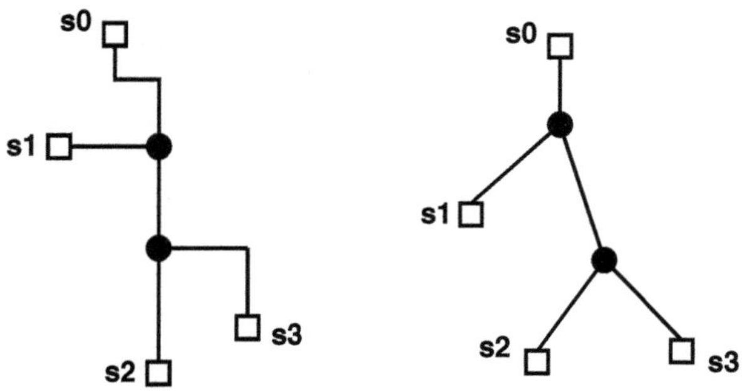

Fig. 3.5 Routing tree of a single net, with signal source terminal s0 and sinks s1, s2, s3. Abstract topology can be represented by a binary tree (*right*). The embedded form, with actual wire segment *center-lines*, is shown on the *left*. Wire segments may utilize different metal layers. Actual layout is typically restricted to rectilinear shapes ("Manhattan geometry")

interconnect design are *channel routers* and *area routers*. Channel routers are used for dedicated regions, pre-allocated as interconnect channels, using a well-defined set of parallel tracks, where wire segments are placed. Area routers (also called *maze* routers) freely route wires in the plane, bypassing obstacles. Typically, routers work net by net, performing metal layer allocation, and placing wire segments. Since each routed net becomes an obstacle for the following nets, the order of nets is of extreme importance. In the complete physical layout, each net is typically represented by a *routing tree* (Fig. 3.5), such that the root of the tree represents the driving point (signal *source*) and the leaves of the tree represent all signal receivers (signal *sinks*). Special signals (e.g., clock) may utilize non-tree topology such as a grid. The optimal routing problem is computationally intractable (NP-hard), and it becomes much harder as system complexity grows. Oftentimes, routers perform "rip-up and reroute," trying a different order of nets. Human intervention is sometimes required in the process of routing. Modern routers must become more and more sophisticated, since their task should involve optimizations and trade-offs among several objectives described below.

3.3 Interconnect Metrics

A real interconnect structure is quite different from the ideal abstraction of a logical net as defined above. In reality, signal propagation takes time, consumes power, and might be unreliable. Practical implementation involves trade-offs among these parameters as well as consideration of manufacturing costs and constraints. The following metrics are often used to evaluate and optimize real interconnect structures:

- *Geometrical metrics (circuit area and wire length)*: These parameters are easy-to-compute metrics which translate into manufacturing cost. They are also correlated with delay and power (smaller is faster and better). Therefore, early CAD

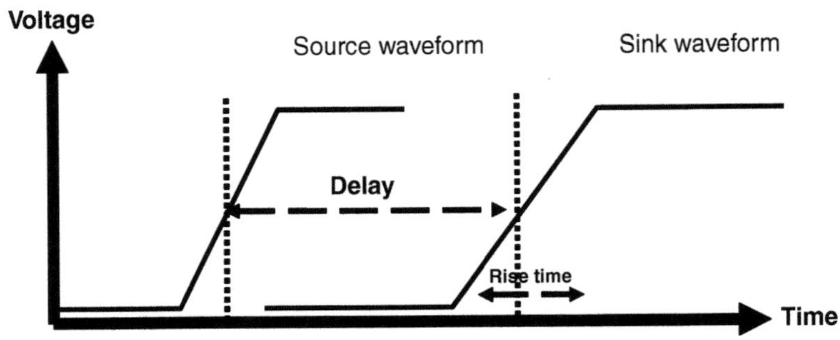

Fig. 3.6 Definition of interconnect delay by crossing 50% of the logic swing

tools have used silicon area and wire length estimates as cost functions for minimization in layout optimization.

- *Delay*: Interconnect delay is the primary performance metric for wires, since state-of-the-art circuit speed is limited by signal propagation over wires. Propagation delay of the signal from the source to the sink, defined for digital logic signals as the difference between time points where the voltage waveforms during logic transitions at the source and the sink cross their 50% amplitude values (Fig. 3.6). Each of these waveforms is characterized by its rise time (or transition time), defined as the time from 10 to 90% of the voltage transition. Electrical modeling of interconnect delay is discussed in more detail in the next sub-section.
- *Power*: Energy is dissipated while transmitting information over interconnect, releasing heat within the transistors which drive the interconnect [8]. The term "interconnect power" is often used to describe the power required for charging/discharging the interconnect in transitions between logic 0 and logic 1 [9], according to the formula for a switching net.

$$P = \alpha C_{\text{int}} V^2 f \tag{3.2}$$

where α is the average activity factor of the net ($0 < \alpha < 1$), C_{int} is the interconnect capacitance of the net, V is the logical signal amplitude (usually same as supply voltage), and f is the clock frequency of operation. Note that additional switching power is wasted in the driver and receiver circuits because of their gate and diffusion capacitances.

- *Signal integrity*: Quality of signaling over the interconnect is affected by capacitive and inductive crosstalk between wires, which might lead to errors in data transmission. The primary source of crosstalk is capacitive coupling between adjacent lines, as illustrated by Fig. 3.7. The model shows a voltage transition on a wire called the *aggressor*, which induces undesirable changes in the voltage at a neighboring wire called the *victim*. The noise is usually characterized in terms of the peak voltage amplitude induced on a "quiet" victim, and by the delay change associated with the noise effect on a transitioning victim. This latter delay uncertainty effect caused by noise is a primary concern, limiting the

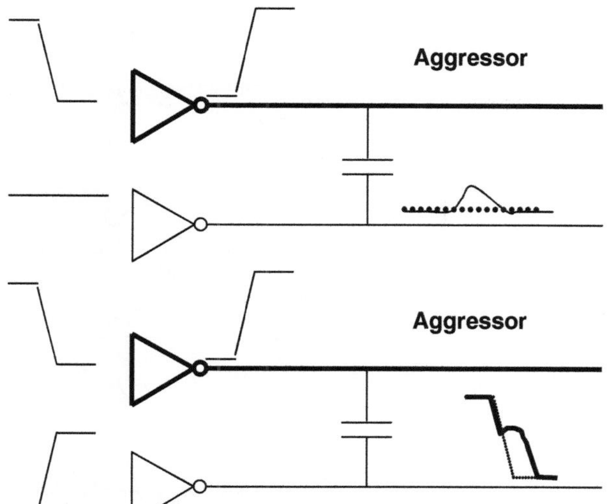

Fig. 3.7 Capacitive crosstalk model. (**a**): Transition of the aggressor signal induces a voltage spike on the victim through interwire capacitance, possibly changing its logic state momentarily. (**b**): Transition of the aggressor while the victim makes an opposite transition distorts the victim's waveform and moves the 50% delay point further out in time

maximal operating speed of synchronous circuits [8, 10]. The interconnect must be designed to keep these noise effects at an acceptable level.

- *Reliability*: Probability of permanent irreversible failure because of electromigration or self-heating [11]. These effects are sensitive to electrical current densities in the metal. Wire dimensions must be chosen according to expected currents and reliability design rules.

3.4 Circuit Models of Interconnect

Electrical modeling of interconnect has evolved over the years, in a progression that reflected the growing importance of interconnect effects on circuit speed, as a result of technology scaling. Figure 3.8 illustrates the types of interconnect models.

3.4.1 Ideal Interconnect

In the early days of VLSI, ideal interconnect was an adequate model, because high integration of logic functions was the primary task, circuit speed was not an issue (performance was dominated by properties of the gates and the active devices, so that interconnect effects were negligible), and power dissipation was not a significant problem. Hence, this electrically ideal model considers only geometrical metrics (occupied area and wire length).

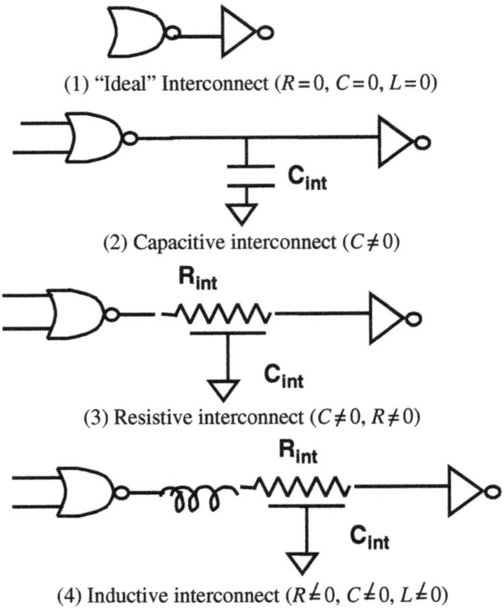

Fig. 3.8 Sequence of interconnect model types from top to bottom

3.4.2 Capacitive Interconnect

During the 1970s and 1980s, interconnect could be adequately modeled as "parasitic" capacitance which increased the loading on the driving gates and slowed the signal transitions somewhat. In the days of single layer metallization, interconnect capacitances were modeled and extracted as area capacitances to the substrate. Later on, fringing field capacitance has been added. In multilevel interconnects the adjacent layers are often considered as metal planes, and line-to-line cross-capacitances are included. Figure 3.9 depicts first-order capacitance calculations per unit length of a wire, where the C_a and C_x components per unit length of the wire are called *area*

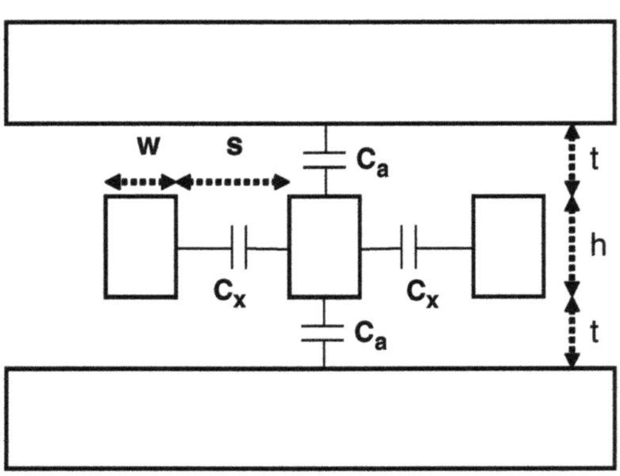

Fig. 3.9 First-order modeling of interconnect capacitances in a cross section of parallel wire segments

capacitance and *cross-capacitance*, respectively. Fringing field effects are ignored in this simple model, such that parallel-plate capacitor expressions can be used:

$$C_\mathrm{a} = \varepsilon_0 \varepsilon_r \frac{w}{t}, \tag{3.3}$$

$$C_\mathrm{x} = \varepsilon_0 \varepsilon_r \frac{h}{s}. \tag{3.4}$$

Interconnect capacitance modeling has now evolved to use 3D electrical field solvers and sophisticated approximations [12, 13]. Basically, capacitances can be reduced by using insulator materials with low dielectric coefficient ε_r (a.k.a "low-K materials").

A circuit delay model assuming capacitive interconnect is shown in Fig. 3.10 . In CMOS, the delay through a logic gate can be approximated as a linear function of the capacitive load it drives:

$$d_\mathrm{stage} = d_0 + R_\mathrm{eff} C_\mathrm{load} = d_0 + R_\mathrm{eff}(C_\mathrm{fanout} + C_\mathrm{int}) \tag{3.5}$$

where d_0 is a constant internal delay through the gate, R_eff is an effective output resistance of the gate, and the load capacitance is the sum of all input capacitances of fanout gates (the signal receivers) and the interconnect capacitance C_int is the sum of area and cross-capacitances. This model assumes an ideal step function (zero rise time) at the input. For an input waveform with nonzero rise time the delay is longer. Improved models contain additional terms to account for this effect [14].

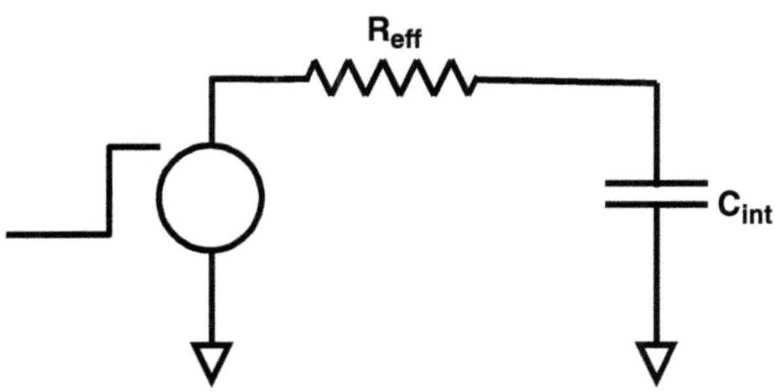

Fig. 3.10 Simplified circuit model of a logic gate driving a capacitive load. The gate is approximated by a linear Thevenin equivalent circuit with output resistance R_eff. The gate's internal delay is not shown. C_int represents the interconnect capacitance and includes the sum of input capacitances of any fanout gates driven by this gate

For a given C_int, the circuit may be speeded-up by "upsizing" the gate (using larger transistors, thus reducing R_eff). However, an upsized gate presents a heavier fanout load on the preceding logic stages, which may require further upsizing. Excessive gate sizing consumes larger silicon area and dissipates excessive power [15]. Circuit design methodologies were developed for initial gate sizing ignoring interconnect [16], followed by iterative adjustments considering extracted "parasitic capacitances" from layout.

3.4.3 Resistive Interconnect

Modeling became further complicated in the late 1980s and 1990s as technology scaling caused the resistance of metal wires to become non-negligible in comparison with the effective resistance of driving transistors. The basic model for interconnect resistance is depicted in Fig. 3.11, where ρ represents an effective specific resistivity of the wire material.

$$R = \rho \frac{l}{wh} \tag{3.6}$$

With resistive interconnect, logic nets must be partitioned into resistive segments. Considering the capacitance and resistance, a wire can be modeled as a *lumped* or *distributed RC* stage. Figure 3.12 compares the response of these models to voltage step function from an ideal (zero internal resistance) source. The delays are approximately 0.7*RC* for the lumped model but only 0.4*RC* for the distributed model,

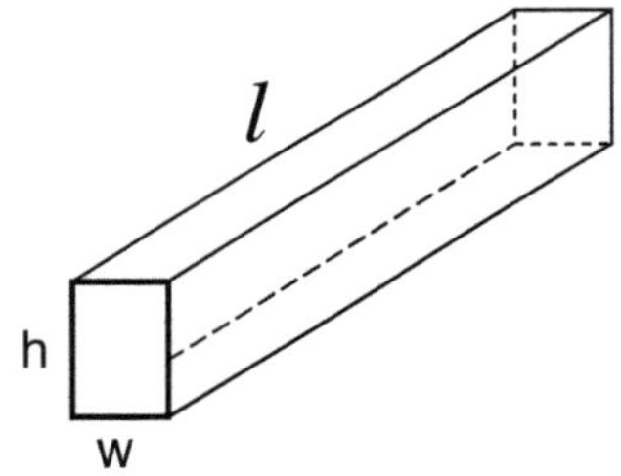

Fig. 3.11 Basic modeling of wire-segment resistance

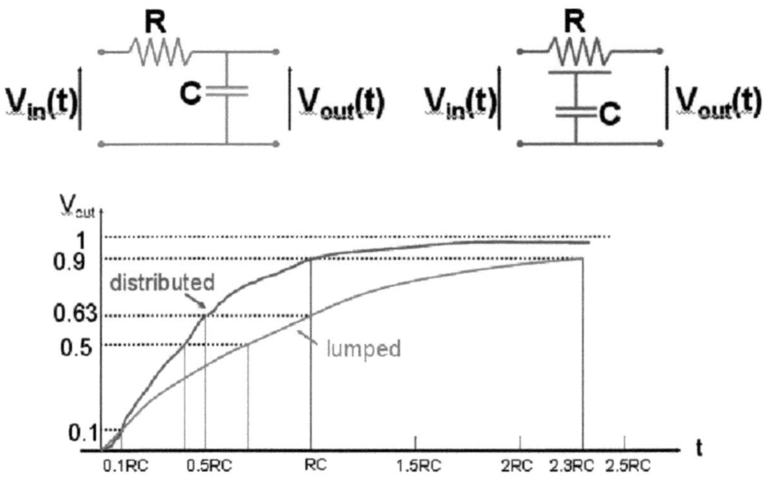

Fig. 3.12 Lumped (*top left*) and distributed (*top right*) *RC* stages, and their response to an ideal step voltage source. The 50% delay is 0.7*RC* for a lumped *RC* stage and 0.4*RC* for distributed *RC*

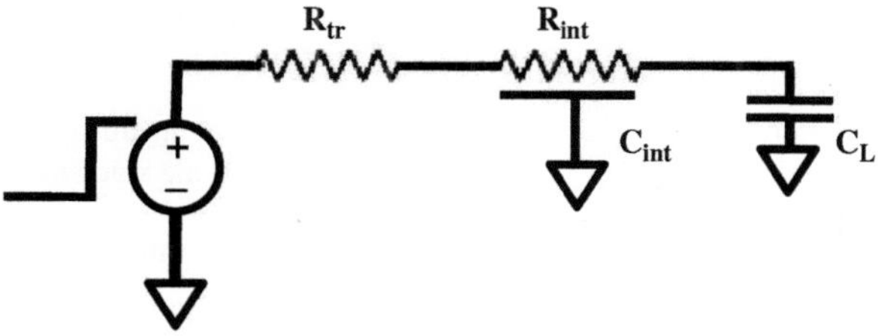

Fig. 3.13 Circuit model for a gate driving a distributed wire segment and sink capacitance C_L

because the lumped model assumes that current flows through the whole resistance to charge the capacitance, while in the distributed model some of the capacitance is charged through a fraction of the resistance only [17].

Considering a linearized gate model with effective resistance R_{eff} as a driver of a resistive wire segment, the circuit model shown in Fig. 3.13 is obtained.

This circuit model leads to a delay expression [17]

$$d_{stage} = d_0 + 0.7\mathrm{R_{eff}}C_{int} + 0.7\mathrm{R_{eff}}C_L + 0.7R_{int}C_L + 0.4R_{int}C_{int}. \quad (3.7)$$

Note that it is impossible to separate this expression into "gate delay" and "wire delay," as it includes combinations of parameters of the gate, wire, and fanout load. For long wires, $C_{int} >> C_L$ and the middle terms can be neglected, but still it is impossible to separate out a gate delay expression which is independent of interconnect parameters. Aggressive upsizing of the driver gate reduces the delay, but further upsizing yields a diminishing effect when the last term in the equation becomes dominant. This term contains the wire′s time constant $R_{int}C_{int}$, which is often (inaccurately) referred to as "wire delay". A practical method to extract the actual wire delay and to choose driver size for resistive wires is described in [18].

For a given metal layer in a given process technology, a wire's time constant $\tau = R_{int}C_{int}$ is independent of wire width (because making the wire wider to reduce R_{int} causes C_{int} to increase in the same proportion). However, both R_{int} and C_{int} are proportional to wire length l, such that τ is proportional to l^2.

$$\tau = R_{int}C_{int} = (R * l)(C * l) = RC * l^2. \quad (3.8)$$

This delay has been called diffusion delay [19], because signal propagation in distributed RC interconnect can be described by a diffusion equation.

A wire is just a simple particular case of resistive interconnect. For a more general resistive net corresponding to a routing tree (Fig. 3.5), the corresponding circuit model is an RC tree such as shown in Fig. 3.14. Computing delays in a general RC tree involves solution of a linear (but very large) system of differential equations. In order to deal with the complexity, sophisticated moment-based AWE methods have been developed for efficient computation of practical waveform approximations. However, a simple model due to Elmore [19], which is equivalent to the first

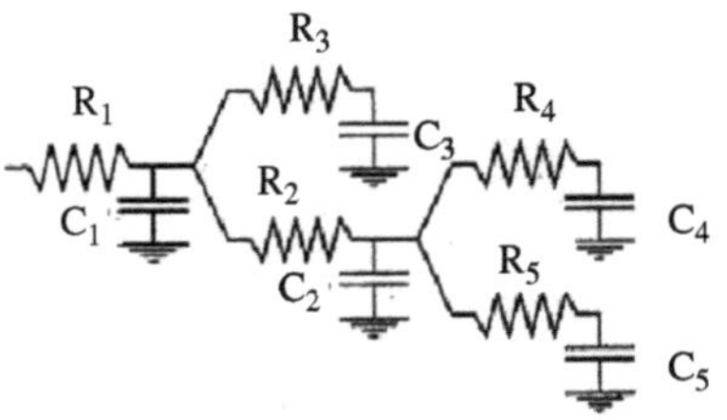

Upstream formula
$D_i = \sum_k C_k R_{ki}$
R_{kl} is the total upstream resistance common to node k and to node i.

Downstream formula
$D_i = \sum_{k \in \prod(i)} C_k \cdot L_k$
L_{ki} is the total downstream capacitive load driven by R_j.

Fig. 3.14 An *RC* tree example. Capacitance is denoted by node number

moment of the waveform, provides a useful easy-to-compute expression which has been proven as a high-fidelity objective function for interconnect optimization and a solid basis for physical CAD algorithms [20]. The Elmore delay from the source to node i in the tree can be expressed as

$$D_{\mathrm{i}} = \sum_{k \in \mathrm{path}(i)} R_{\mathrm{k}} L_{\mathrm{k}} \tag{3.9}$$

where k denotes all resistors on the path from the source to node i, and L_k is the total downstream capacitive load charged or discharged through the resistor R_k. The same delay can also be expressed as

$$D_i = \sum_k C_k R_{ki} \tag{3.10}$$

where C_k denotes any capacitor in the tree, and R_{ki} is the total upstream resistance common to node k and node i. The Elmore delay model overestimates the delay, and a factor of 0.7 is typically used as a rough calibration factor [21].

3.4.4 Inductive Interconnect

Most wires in ULSI systems are dominated by resistive delay as described above. The speed of signal propagation over resistive wires is much slower than the ultimate speed of electromagnetic wave propagation, which is [17]

$$v = \frac{c_0}{\sqrt{\varepsilon_{\mathrm{r}}}} = \frac{1}{\sqrt{LC}}, \tag{3.11}$$

where c_0 is the speed of light and ε_{r} is the relative dielectric constant of the insulating material surrounding the wires (v is about 15 cm/s for silicon dioxide). L and C are inductance and capacitance per unit length of the wire, and they are interrelated by this equation. Wires designed especially for very high-speed transmission, with

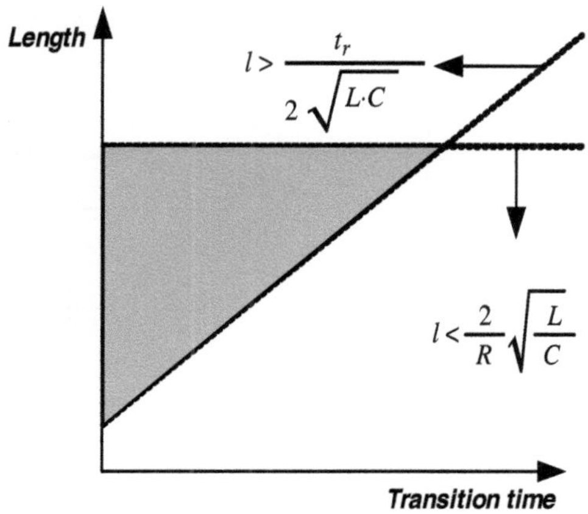

Fig. 3.15 Range of interconnect length where inductance is significant, marked in *grey* [23]. t_r is the source signal rise time, l is the wire length, and R, L, C are wire parameters per unit length. Inductance must be considered where the two indicated equalities are valid. Inductance is NOT significant in the *white* areas, where input transition time is long (relative to the ultimate time of flight through the wire) or the wire is long (such that resistive damping dominates delay)

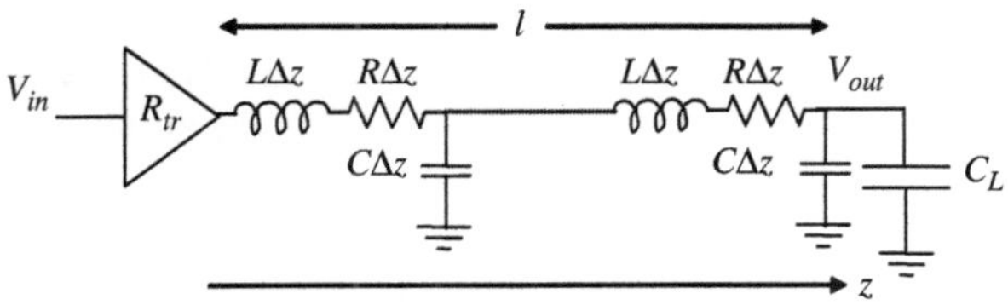

Fig. 3.16 *RLC* wire model using two *RLC* segments

minimal resistive effects, can approach this propagation speed. In such wires, the signal rise time is comparable to the ultimate time of flight l/v where l is the wire length. Such wires exhibit inductive characteristics and require an *RLC* model. The conditions which require *RLC* modeling are illustrated in Fig. 3.15 [22]. The indicated gray range is non-existent for most wires, but special signals (such as clocks) in the system indeed require an *RLC* model.

The behavior of an *RLC* wire model (Fig. 3.16) is illustrated in Fig. 3.17, including a comparison with an *RC* model which neglects the inductance. Inclusion of inductance in the model causes a slower initial response (reflecting signal time of flight over distance), a steeper slope in the waveform (representing an improvement of rise time), and some overshoot (with "ringing") before settling at the stable final value. The 50% delay is always underestimated by the *RC* model. Correction terms can be applied to *RC* delay models for representing the effects of inductance and reducing such errors [24]. Note that when resistive effects are negligible, the wire becomes a lossless transmission line which must be modeled as a circuit with many *LC* segments, while its delay can be calculated from (3.11).

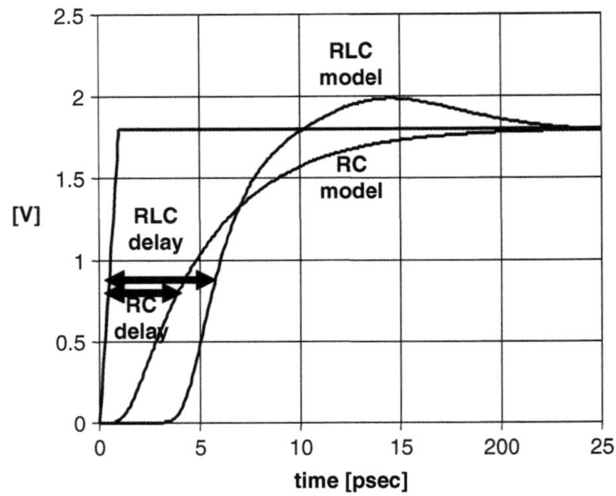

Fig. 3.17 Comparison of *RC* wire model and *RLC* wire model. Delay is measured between 50% points at the near end and far end of the wire

It is important to note that while capacitance is a local property of the wire in its near vicinity, inductance is not truly a property of a wire, but rather a property of a complete current loop. The current return path may involve quite remote wires, is generally not known in advance, and might vary with signal frequency. Therefore, a model such as Fig. 3.16 is misleading because the inductance depends on current return paths which are not shown and may not be known. To address this problem, highly complex inductance extraction methods must be used [25]. Alternatively, this complication can be bypassed by laying out fast signal link interconnects as transmission lines with built-in explicit return path [26]. An additional potential issue with interconnect inductance is related to skin and proximity effects, where current paths can vary with signal frequency, causing increased resistance and reduced inductance at high frequencies. However, these effects are usually of minor importance in wires of practical thickness and width [27].

3.5 The Interconnect Scaling Problem

The term scaling describes systematic miniaturization of device dimensions, along with reduction of supply voltage and introduction of new materials and fabrication techniques. Continued technology scaling during the last four decades has provided tremendous benefits by providing together cost reduction, speed improvement, and power saving per logic function. Over the years, interconnect scaling has become a problem of growing importance [28]. Issues of interconnect scaling have been surveyed in [3, 29, 30]. The primary issue with interconnect scaling is the global wire scaling problem. Put simply, the problem stems from the fact that system complexity is growing. At each generation transistors get smaller, but there are many more transistors per chip such that chip size stays fixed or grows slowly. Therefore, global wire length l roughly matches the die size, and it does not shrink. Assuming that all local dimensions shrink by a factor s and there is no change in materials. ($s= \sim 0.7$ per technology generation), capacitance per unit length $C = \varepsilon_0\varepsilon_r w/t_{ox}$ stays unchanged while wire resistance per unit length $R = \rho/wh$ grows by a factor $1/s^2$. Consequently, the delay $RCl^2$of a resistive local wire stays unchanged because local distances l scale by s, canceling the growth in resistance. However, since global distances do not scale, global delay grows in proportion to the wire resistance. Notably, gate delays become shorter as device scaling improves the speed of transistors, while local wire delay stays fixed and global wire delay grows longer. Advanced scaling techniques (e.g., weak scaling of wire thickness and usage of low dielectric constant insulator material) can improve the situation somewhat such that local wire delays almost track the improvement in gate delays. However, global wire delays become much worse than gate delays in every technology generation, since their length does not scale, thus becoming a major bottleneck.

The quadratic dependence of wire delay on distance was recognized early on as a major technology limiter, considering fixed-length global wires, which justified the move from aluminum to copper metallization [31]. Approximating (3.8) for the geometry shown in Fig. 3.18, the wire time constant is

$$\tau \approx \left(\rho\frac{1}{0.5P*T}\right)\left(2\varepsilon_r\varepsilon_0\left(\frac{l*0.5P}{T}+\frac{l*T}{0.5P}\right)\right) = 2\rho\varepsilon_r\varepsilon_0 l^2(1/T^2+4/P^2). \tag{3.12}$$

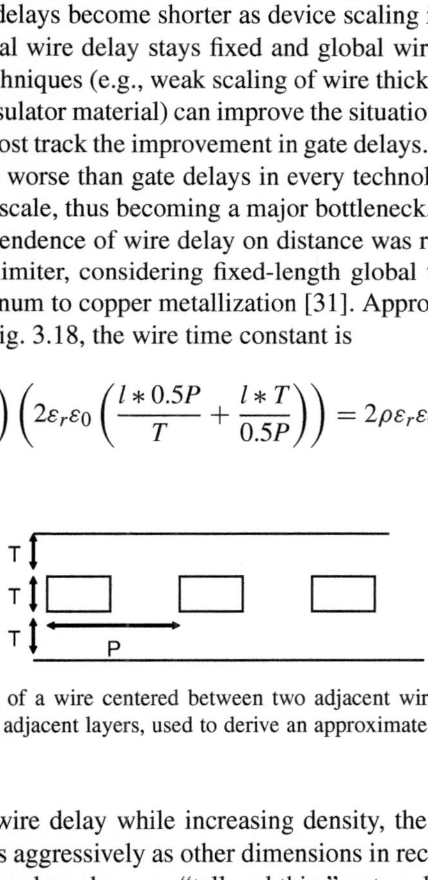

Fig. 3.18 Cross section of a wire centered between two adjacent wires at the same layer and assumed metal planes in adjacent layers, used to derive an approximate expression for the wire's time constant τ

To minimize the wire delay while increasing density, the vertical dimension T has not been scaled as aggressively as other dimensions in recent technology generations. Therefore, wires have become "tall and thin," not as drawn in the Fig. 3.18, with an *aspect ratio* $T/(0.5P)$ around 2.

Note, however, that while no metals better than the Cu are available, the effective specific resistivity of nanowires will actually get worse with scaling, as cladding layer conductivity, surface scattering, and other effects become more pronounced [28]. The metal resistivity effectively grows as wire width is scaled down Fig. 3.19.

A partial solution to the global wire delay problem, which reduces the delay dependence from quadratic to linear, is repeater insertion (described in some detail in the next sub-section). Basically, repeaters are line amplifiers which divide a long wire into segments which are driven separately. This solution is expensive in terms

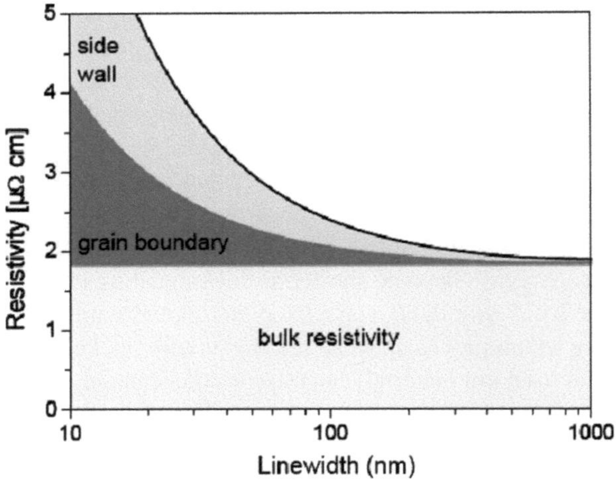

Fig. 3.19 Effective metal resistivity as a function of wire width
Source: ITRS 2005.

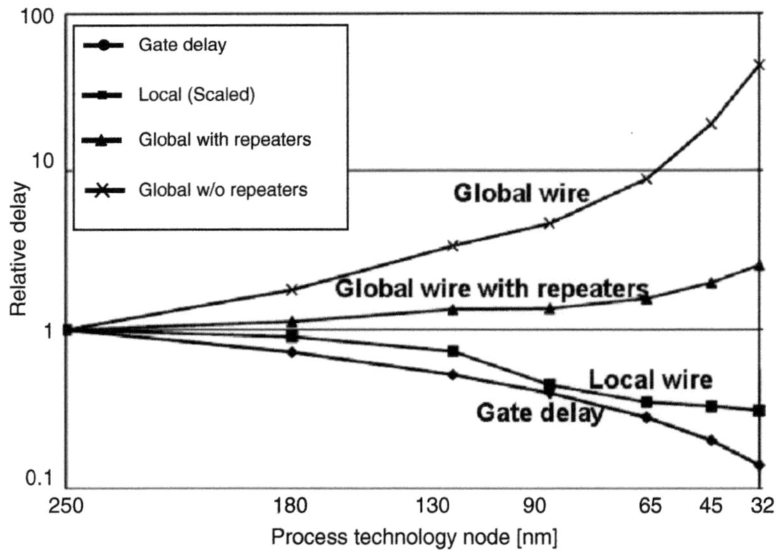

Fig. 3.20 The global wire scaling problem: Global wire delay grows with every technology generation (relative to gate delay and local wire delay)
Source: ITRS 2005

of area and power. As shown in Fig. 3.20, global wire delay with repeaters still becomes worse relative to gate delay by about a factor of 2 at every new technology generation.

The growing importance of interconnect in scaled technology is also due to other reasons in addition to wire delay: Interconnect power has become the major

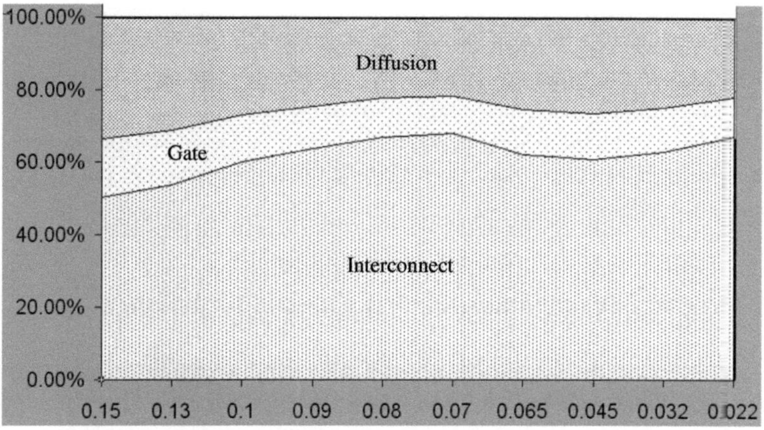

Fig. 3.21 Interconnect power is around 50% of dynamic switching power in microprocessors versus technology node [9]. It is expected to grow in future technology generations if the same circuit architectures will be used

component of dynamic power dissipation and is expected to keep growing Fig. 3.21 [9]. Interconnect noise because of crosstalk has also become a major design consideration and source of delay uncertainty, as aspect ratio (thickness/width) of wires has grown, and hence cross-capacitance to adjacent wires has become the biggest component of total interconnect capacitance. Many circuit layouts in modern technologies are constrained by routing resources (namely wire congestion and wire area) rather than by active device area and processing costs are affected by the growing number of required metal layers.

The significance of scaled interconnect issues has far-reaching implications on system architecture [32], design methodologies, and CAD tools [30]. In system architecture, locality becomes an important consideration since signals can no longer traverse the whole chip within a clock cycle. In synchronous systems, long wires must become pipelined and their delay must be explicitly modeled by architects. An interconnect-centric design methodology is required for addressing the various issues [33].

3.6 Design Approaches and Techniques for Addressing Interconnect Problems

While new materials and processing methods are developed to improve the characteristics of ULSI interconnect, there is a variety of design techniques which can help address the issues described above, for any given fabrication process. These can be classified into the four following categories.

3.6.1 Metallization Stack Design

Optimize the number of metal layers, their thicknesses, pitches, and design rules to enable CAD tools to make the required density of interconnections at appropriate signaling speeds for the system at hand. Sub-optimal metallization choices might lead to a waste (e.g., enable high density which is useless because the wires are too resistive and hence slow) [6, 34].

3.6.2 Repeater Insertion and Other Circuit Techniques

The quadratic growth of wire delay with wire length equation (3.8) is the root cause of the global wire problem. The classical approach to address this problem is repeater insertion (see Fig. 3.20) [23]. The global wire delay problem stems from the fact that current must flow from the driver through the entire wire resistance to reach capacitance at the receiver end. The solution idea is to drive each segment of the wire locally by a repeater, which is a line amplifier or buffer. In practice, logical inverters are used (a pair of logical inverters preserves signal polarity). The model of a wire after repeater insertion is shown in Fig. 3.22 .

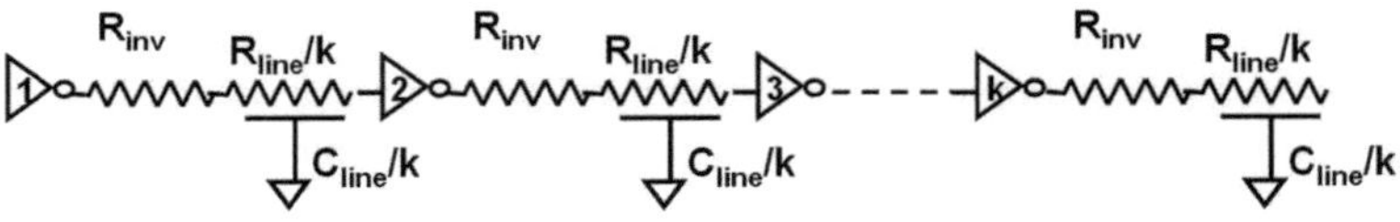

Fig. 3.22 A global wire divided into *k* segments driven by repeaters

The delay of the segmented wire can be calculated by applying (3.15) to each of the *k* segments, obtaining

$$T_{50\%} = k\left[0.7R_{inv}\left(\frac{C_{line}}{k}+C_{inv}\right)+\frac{R_{line}}{k}\left(0.4\frac{C_{line}}{k}+0.7C_{inv}\right)\right] \tag{3.13}$$

where R_{line} and C_{line} are resistance and capacitance of the full length of the wire, C_{inv} is a repeater's input capacitance, and R_{inv} is its effective output resistance. Each additional repeater adds some internal delay due to C_{inv}, but reduces the resistive delay of each segment. An optimal number of segments and repeaters can be found by differentiation with regard to k. The derivative becomes zero at

$$k_{opt} = \sqrt{\frac{0.4R_{line}C_{line}}{0.7R_{inv}C_{inv}}} \tag{3.14}$$

Since k must be an integer value and at least 2, one can calculate the minimal wire length which justifies the insertion of the first repeater. The misleading term *critical distance* is sometimes used for this length. It is quite short for low-level metal layers, and much longer for top-level metal layers. The product $R_{inv}C_{inv}$ is a technology constant representing the inherent speed of transistors in the given technology, and it is independent of inverter size [16]. Hence k_{opt} is independent of repeater size, which can be optimized separately. If a minimal size repeater is upsized by a factor x, its parameters become R_{inv}/x and $C_{inv}*x$. Differentiating by x and equating the derivative to zero leads to the optimal sizing factor

$$x_{opt} = \sqrt{\frac{R_{inv}C_{line}}{R_{line}C_{inv}}} \tag{3.15}$$

The optimal wire delay, using the optimal number of optimally upsized repeaters, is

$$T_{50\%\mathrm{opt}} = 2.5\sqrt{R_{\mathrm{inv}}C_{\mathrm{inv}}R_{\mathrm{line}}C_{\mathrm{line}}}. \tag{3.16}$$

This delay of a repeated wire grows linearly with wire length (recall from (3.8) that $R_{\mathrm{line}}C_{\mathrm{line}}$ is proportional to l^2). Delay-optimal repeater insertion, using (3.15) and (3.16), is not practical in terms of area and power. Note that optimal repeaters consume area and waste additional power. The total capacitance of the added repeaters is about the same as the total wire capacitance, doubling the dynamic power. Repeaters also suffer from relatively high leakage power. Major savings in power can be made by slightly increasing delay above the achievable minimum. This trade-off is described in [35, 36]. Fast wires, in which the inductance is significant, require fewer repeaters in comparison with purely resistive wires [37].

Specialized and innovative circuits such as boosters and advanced repeaters, rather than simple inverters, have been proposed [38, 39]. Specialized signaling methods, data encoding, and driver waveform shaping have been proposed too [40]. Methods for optimizing the sizes of logic gates in conjunction with the wires, using the gates driving the wires also as repeaters, are described in [50, 51, 54].

3.6.3 Layout Optimization Techniques

CAD tools and methods for physical design of VLSI layout are continuously being developed and improved to become interconnect centric [33]. While "classical" layout tools have been developed for minimizing geometrical metrics (chip area and wire length), new tools are geared toward interconnect performance. They make timing/power/noise trade-offs during placement and routing of cells and wires [30, 33, 41]. A major consideration in this context is *layer assignment*, where routing resources at higher metal layers (which are typically thicker with shorter delays) are assigned to the timing-critical signals. Routing algorithms and tools are modified to optimize delay, noise, or power rather than just minimize the total wire length. Wire sizing (width determination) and spacing are helpful techniques too, among other fine-tuning strategies for wires [42, 43, 49, 52, 53]. Increasing the spacing between adjacent wires (at the cost of reducing wire density) helps to reduce wire-to-wire cross-capacitances, which are major contributors to circuit delay, interconnect power, and crosstalk noise. Critical signals, such as clock distribution nets, are sometimes laid out with shield wires adjacent to them. Shield wires are connected to fixed voltages in order to screen their neighbor wires from noise [44].

3.6.4 Circuit Architecture

The highest leverage approach for addressing the issues of interconnect speed, power, and noise in ULSI systems is to modify the overall circuit architecture. The guiding principles should be *local computation* (minimizing the transfer of data across long distances) [32], and *wire sharing* (using the same physical wires for transmitting many different signals).

To exemplify the local computation, consider a large uniproprocessor such as in Fig. 3.2, where signals may need to traverse the whole chip among functional units, in order to execute an instruction. Instead, the same area could be divided among several small local computational cores, in which signals do not have to travel long distances. Although the performance of each small core is lower, they may work in parallel and provide higher performance in a more power-efficient manner. The principle of local computation is the basis for developing new chip multi-processors (CMP).

The idea of wire sharing is not new. Shared buses have been used in integrated systems for a long time. However, the scalability of traditional buses is limited, and modern circuit architectures tend toward interconnection fabrics with more parallelism, such as Network-on-Chip [45, 46]. In a NoC system, modules such as processor cores, memories, and specialized IP blocks exchange data encoded in packets of bits, using a network as a "public transportation" sub-system for the information traffic. A NoC is similar to a modern telecommunications network, using digital bit-packet switching over multiplexed links. The wires in the links of the NoC are shared by many signals. A high level of parallelism is achieved, because all links in the NoC can operate simultaneously on different data packets. Therefore, as the complexity of integrated systems keeps growing, an NoC provides enhanced performance and scalability in comparison with previous communication architectures (e.g., dedicated point-to-point signal wires, shared buses, or segmented buses with bridges) [47].

3.7 Power Supply Interconnect

The on-chip power distribution network is a major, critical sub-system of metallization interconnects structures, which is quite different from the logic signal wires described in the previous sections. Instead of propagating changing voltage waveforms, a power distribution system must maintain a constant voltage across all of its terminals. Any voltage deviation is termed power noise. State-of-the-art systems

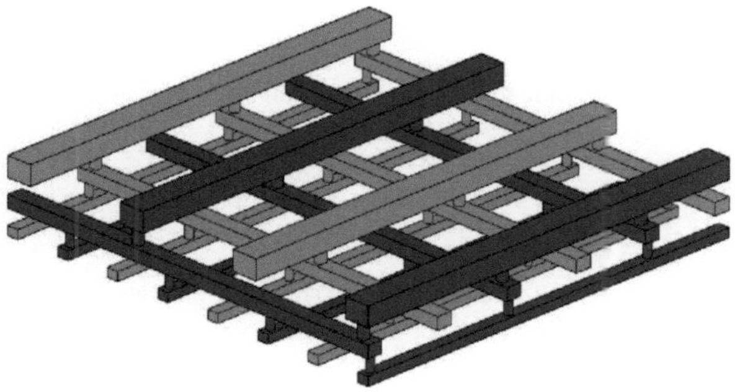

Fig. 3.23 Power distribution network employing grids for supply voltage and ground lines
Source: [48]

typically tolerate power noise up to about 5% of nominal supply voltage which is around 1 V. The purpose of a power distribution network is to provide high currents to the logic devices on the chip, while avoiding excessive *IR* voltage drops across its current paths. It must also avoid inductive voltage spikes $L(\mathrm{d}i/\mathrm{d}t)$ whenever changes occur in the supply current.

A typical power distribution system is laid out as a *grid*, as illustrated in Fig. 3.23. The loops in the grid structure complicate analysis, as current can be fed to any point from multiple directions, but it helps to maintain a stable voltage. Power distribution networks must be designed to have very low resistance and very low inductance to minimize supply voltage noise. Therefore, they use the thickest, top-level metal layers in the stack, and employ interleaved power (VCC) and ground (VSS) grids to minimize current loop areas and achieve minimal inductance. On-chip decoupling capacitors help stabilize the supply voltage. Future ULSI chips are expected to contain separate power distribution sub-systems with multiple voltage levels. The interested reader is referred to a monograph covering on-chip power distribution issues [48].

3.8 Summary

Interconnect has become a crucial element in the integration of complex electronic systems. Circuit speed is dominated by signal transmission delays through interconnects, because wire delays do not scale well (relative to gate delays) when all device dimensions are scaled down at each generation of silicon CMOS technology. This interconnect scaling problem has triggered the development of Cu interconnect technology, and has stimulated research and development efforts in diverse areas toward the future, ranging from low-*K* materials and innovative fabrication processes to physical layout design techniques, circuit design methods, and ULSI system architecture. Furthermore, interconnects dominate the dynamic power dissipation in microprocessors and other integrated systems, as well as coupling noise, reliability, and fabrication cost. Future design methodologies and tools must consider the complex trade-offs among these parameters.

References

1. E.:Cramming more components onto integrated circuits. Electronics **86(1)**, 1117–119 (1965). Reprinted in Proceedings of the IEEE, **82**, 85 (1998)
2. Borkar, S.: Electronics beyond nano-scale CMOS. In Proceedings of the 43rd Annual Conference on Design Automation. San Francisco, CA, USA (2006)
3. Davis, J. A. et al.: Interconnect limits on gigascale integration (GSI) in the 21st century. Proc. IEEE **89(3)**, 305 (2001)
4. Landman, B. S.; and Russo, R. L.: On a pin versus block relationship for partitions of logic graphs. IEEE Trans. Comput. **C-201**, 469 (1971)
5. Stroobandt, D.: A priori system-level interconnect prediction: Rent's rule and wire length distribution models. In Proceedings of System Level Interconnect Prediction (SLIP), 3 (2001)
6. Davis, J. A.; Meindl, J. D.; and Venkatesan, R.: Performance enhancement through optimal N-tier multilevel interconnect architectures. Proceedings of the 12th IEEE ASIC/SOC Conference, Washington D.C. 19 (1999)
7. Sait, S. M. and Youssef, H.: VLSI Physical Design Automation Theory and Practice. World Scientific, New Jersey (1999)
8. Pedram, M.: Power minimization in IC design: principles and applications. ACM Trans. Des. Autom. Electron. Syst. **1**, 3 (1996)
9. Magen, N.; Kolodny, A.; Weiser, U.; and Shamir, N.: Interconnect-power dissipation in a Microprocessor. International System Level Interconnect Prediction workshop (SLIP 2004), Paris (2004)
10. Sato, T.; Cao, Yu; Agarwal, K.; Sylvester, D.; and Hu, C.: Bidirectional closed-form transformation between on-chip coupling noise waveforms and interconnect delay-change curves. IEEE Trans. Computer-Aided Des. Integrated Circuits Syst. **l.22(5)**, 560 (2003)
11. Banerjee, K. and Mehrotra, A.: Global (interconnect) warming. IEEE Circuits Devices Mag. **17(5)**, 16 (2001)
12. Sakurai, T.; and Tamaru, K.: Simple formulas for two- and three-dimensional capacities. IEEE Trans. Electron Devices **ED-30(2)**, (1983)
13. Wong, S. C.; Lee, G. W.; and Ma, D. J.: Modeling of interconnect capacitance, delay and crosstalk in VLSI. IEEE Trans. Semiconductor Manuf. **13(1)**, (2000)
14. Sapatnekar, S. S.: Timing. Springer, New York (2004)
15. Rabaey, J. M.; Chandrakasan, A.; and Nikolic, B.: Digital Integrated Circuits (2nd Edition), Prentice Hall, Upper Saddle River, New Jersey (2003)
16. Sutherland, I.; Sproull, B.; Harris, D.: Logical Effort – Designing Fast CMOS Circuits. Morgan Kaufmann, San Fransisco, CA (1999)
17. Bakoglu, H. B.: Circuits, Interconnections and Packaging for VLSI. Addison-Wesley, Boston, MA, 194, 1990.
18. Sylvester, D. and Keutzer, K.: Getting to the bottom of deep submicron. In Proc. ICCAD, 203 (1998)
19. Kahng, A. B. and Muddu, S.: Delay analysis of VLSI interconnections using the diffusion equation model, 31st Conference on Design Automation, 563 (1994)
20. Elmore, W. C.: The transient response of damped linear networks with particular regard to wide band amplifiers. J. Appl. Phys., **19(1)** (1948).
21. Boese, K. D.; Kahng, A. B.; McCoy, B. A.; and Robins, G.: Fidelity and near-optimality of Elmore-based routing constructions. In Proceedings of 1993 IEEE International Conference on Computer Design (ICCD '93), 81 (1993)
22. Ismail, Y. I. Friedman, E. G.; and Neves, J. L.: Figures of Merit to characterize the importance of On-Chip Inductance. IEEE Trans. Very Large Scale Integr. (VLSI) Syst. **7(4)**, 442 (1999)
23. Bakoglu, H. B. and Meindl, J. D.: Optimal interconnection circuits for VLSI. IEEE Trans. Electron Devices, **ED-32**, 903 (1985)

24. Ismail, Y. I. and Friedman, E. G.: Effects of inductance on the propagation delay and repeater insertion in VLSI circuits. IEEE Trans. Very Large Scale Integr. (VLSI) Syst. **82(2)**, 195 (2000)
25. Gala, K.; Blaauw, D.; Wang, J.; Zolotov, V.; and Zhao, M.: Inductance 101: analysis and design issues. In Proceedings of the 38th Conference on Design Automation (DAC 2001). Las Vegas, Nevada, US (2001)
26. Goren, D. et al.: On-chip interconnect-aware design and modeling methodology, based on high bandwidth transmission line devices. In Proceedings of the 40th Conference on Design Automation, Anaheim, CA, USA (2003)
27. Barger, A.; Goren, D.; and Kolodny, A.: Design and modelling of network on chip interconnects using transmission lines. In Proceedings of the 2004 11th IEEE International Conference on Electronics, Circuits and Systems, (ICECS 2004), Tel-Aviv, Israel, 403 (2004)
28. http://public.itrs.net/
29. Dally, W. J.; and Poulton, J. W.: Digital Systems Engineering. Cambridge University Press, Cambridge (1998)
30. Ho, R.; Mai, K.; and Horowitz, M.: The future of wires. Proc. IEEE **89(4)**, (2001)
31. Bohr, M. T.: Interconnect scaling—the real limiter to high-performance ULSI. In Proc. IEDM., 241 (1995)
32. Dally, W. J.: Interconnect-limited VLSI architecture. IEEE International Conf. Interconnect Technol., 15 (1999)
33. Cong, J.: An interconnect-centric design flow for nanometer technologies. Proc. IEEE. **89(4)**, 505 (2001)
34. Gupta, P.; Kahng, A. B.; Kim, Y.; and Sylvester, D.: Investigation of performance metrics for interconnect stack architectures. In Proceedings of the 2004 International Workshop on System Level Interconnect Prediction. Paris, France (2004)
35. Kapur, P.; Chandra, G.; and Saraswat, K. C.: Power estimation in global interconnects and its reduction using a novel repeater optimization methodology. In Proceedings of the 39th Conference on Design Automation. New Orleans, Louisiana, USA (2002)
36. Chen, G. and Friedman, E. G.: Low-power repeaters driving RC and RLC interconnects with delay and bandwidth constraints. IEEE Trans. Very Large Scale Integrated Syst. **14(2)**, 161 (2006)
37. Ismail, Y. I. and Friedman, E. G.: Optimum repeater insertion based on a CMOS delay model for on-chip RLC interconnect. Proceedings of Eleventh Annual IEEE International ASIC Conference, 369 (1998)
38. Nalamalpu, A.; Srinivasan, S.; and Burleson, W. P.: Boosters for driving long on chip interconnects – design issues, interconnect synthesis, and comparison with repeaters. IEEE Transactions on Computer-Aided Design of Integrated Circuits and Systems. **21(1)**, (2002)
39. Kaul, H.; and Sylvester, D.: Low-power on-chip communication based on transition-aware global signaling (TAGS). IEEE Transactions on Very Large Scale Integrated Systems. **12(5)**, 464 (2004)
40. Svensson, C. and Caputa, P.: Well-behaved interconnect. IEEE Trans. Circuits Syst. – I, **52(2)**, (2005)
41. Chu, C. and Wong, D. F.: Closed form solution to simultaneous buffer insertion/sizing and wire sizing. ACM Trans. Design Automation of Electronic Systems **6(3)**, 343 (2001)
42. Kahng, A. B.; Muddu, S.; Sarto, E.; and Sharma, R.: Interconnect tuning strategies for high-performance ICs. Proceedings Design, Automation and Test in Europe. 471 (1998)
43. Wimer, S.; Michaely, S.; Moiseev, K.; and Kolodny, A.: Optimal bus sizing in migration of processor design. IEEE Transactions on Circuits and Systems I. **53(5)**, 1089 (2006)
44. Zhang, J. and Friedman, E. G.: Crosstalk modeling for coupled RLC interconnects with application to shield insertion. IEEE Transactions on Very Large Scale Integration (VLSI) Systems **14(6)**, 641 (2006)
45. De Micheli, G. and Benini, L.: Networks on Chips: Technology and Tools. Morgan Kaufmann, San Fransisco, CA (2006)

46. Jantsch, A. and Tenhunen, H. (Eds.): Networks on Chip. Kluwer Academic Publishers, Dordrecht (2003)
47. Bolotin, E.; Cidon, I.; Ginosar, R.; and Kolodny, A.: Cost considerations in network on chip integration. VLSI J **38(1)**, 19 (2004)
48. Popovich, M.; Mezhiba, A. V.; and Friedman, E. G.: Power Distribution Networks with On-Chip Decoupling Capacitors. Springer, New York (2007)
49. Barger, A.; Goren, D.; and Kolodny, A.: Simple criterion for maximizing data rate in NoC links. 10th IEEE Workshop on Signal Propagation on Interconnects, Berlin, (2006)
50. Otten, R. H.; and Brayton, R. K.: Planning for performance. In Proc. DAC-127. (1998)
51. Moreinis, M.; Morgenshtein, A.; Wagner, I. A.; and Kolodny, A.: Repeater insertion combined with LGR methodology for on-chip interconnect timing optimization. Proceedings of the 2004 11th IEEE International Conference on Electronics, Circuits and Systems, 125 (2004)
52. Moiseev, K.; Wimer, S.; and Kolodny, A.: On optimal irdering of signals in parallel wire bundles. Integration – the VLSI Journal. **41**, 253–268 (2008)
53. Moiseev, K.; Kolodny, A.; and Wimer, S.: Timing-aware power-optimal ordering of signals. ACM Transactions on Design Automation of Electronic Systems. **13**(4), Article 65, Sept (2008)
54. Morgenshtein, A.; Friedman, E.G.; Ginosar, R.; and Kolodny, A.: Timing optimization in logic with interconnect. Proceedings of the ACM/IEEE International Workshop on System Level Interconnect Prediction, pp. 19–26, April (2008)

Chapter 4
Electrodeposition

Madhav Datta

4.1 Introduction

Electrodeposition is the process of cathodic deposition of metals, alloys, and other conducting materials from an electrolyte using an external potential (electric current) for the cation reduction process to occur at the working substrate. The deposition process is also known as electrolytic plating, electroplating, or simply plating. Electrodeposition is widely employed in a variety of applications ranging from coatings for wear and corrosion resistance to nanoscale feature fabrication for ultra-large-scale integration (ULSI). The deposition thickness may vary from few angstroms of uniformly deposited compact films to electroformed structures that are millimeters thick. Compared to competing vacuum deposition processes, electrodeposition has emerged as more environmentally friendly and cost-effective micro/nanofabrication method. These features of electrodeposition make it an enabling technology for applications such as chip metallization and flip-chip solder bumping. Electrodeposition has thus become an integral part of wafer processing fabs and an enabling technology in many aspects of microelectronic packaging. Although some aspects of electrodeposition still remain empirical, the gap between fundamental understanding and manufacturing application is narrowing [1]. In the following text the terms electrodeposition, electroplating, and plating are used synonymously.

Advances in electrodeposition have played a major role in the phenomenal growth of storage, chip interconnects, microelectronic packaging, microelectromechanical systems (MEMS), and many other microelectronic and micromechanical components [2–10]. Some early examples of electrodeposition in the electronics industry include fabrication of printed circuit boards. Continued advances in plating processes for fine line wiring technology have contributed to the development of advanced boards and packages that are used today. Development of alloy plating

M. Datta (✉)
Cooligy Inc., 800 Maude Avenue, Mountain View, CA 94043, USA
e-mail: mdatta@cooligy.com

Y. Shacham-Diamand et al. (eds.), *Advanced Nanoscale ULSI Interconnects: Fundamentals and Applications*, DOI 10.1007/978-0-387-95868-2_4,

process, precision plating tool, and the application of through-mask plating for thin film heads laid the foundation for the advances in electrochemical technology in the micro- and nanoelectronics industry [2]. Continued efforts on the development of novel magnetic materials and their precision plating have led to the advanced storage devices [3, 4]. Phenomenal advances in electroplating occurred in the last decade when it enabled a paradigm shift in chip making with the introduction of plated flip-chip technology and Cu metallization for chip interconnects [6–8]. Indeed, the material change from aluminum to copper for chip metallization has been heralded as a major breakthrough that will enable extension of Moore's law beyond its earlier expectations. Other applications of electroplating in microelectronics include fabrication of connectors and interconnect, and metallization for multi-chip modules and other advanced packages [9]. These applications heavily relied on the advances in electroplating technologies which have been possible due to simultaneous progress in different areas. They include (i) continuous improvements and innovations in photolithography, (ii) fundamental understanding of the engineering principles that govern electrochemical micro- and nanofabrication processes and the ability to produce tailored materials and structures, and (iii) development of high yielding electrochemical processing tools that are compatible with ultra-clean semiconductor fabrications [10]. Some of these aspects are briefly discussed below.

4.2 Key Considerations

For electroplating on silicon wafers, one of the essential requirements is the presence of a continuous conducting layer (seed layer), which provides electric current path from the wafer edge contact to all points in the wafer where deposition is desired. Two different types of electroplating are applicable in the fabrication of a metallic structure: (i) through-mask plating (ii) and Damascene plating. While photoresist masks in through-mask plating are stripped to release the plated structure, in Damascene plating the patterned dielectric remains intact and forms a functional part of the structure. A continuous seed layer conformally covers the patterned dielectric. Plating occurs all over the surface thus creating challenges for void-free structure fabrication [7]. Chemical mechanical polishing (CMP) is used for planarization and removal of excess ''overburden'' metal and seed layers. Development of electroplating processes and tools for fabrication of micro- and nanostructures requires a thorough understanding of the underlying electrochemical engineering principles. This involves the understanding of the ability to produce tailored materials and structures and the application of the principles of mass transport and current distribution [11].

Small changes in process conditions can have an enormous effect on the microstructure and composition of the deposit and hence its properties. These small changes may be in the form of additives or complexing agents or in the deposition parameters. These aspects of electroplating offer the possibility of fabricating structures with tailored materials and properties. Precise control of the additives and process parameters is extremely critical to obtain reproducibility and uniformity of deposition. Internal stress that frequently develops in electrodeposits can cause cracking or loss of adhesion. Mismatch between substrate and deposit, grain coalescence during growth, and incorporation of additives or hydrogen may contribute to the internal stress. It is well known that electrodeposition under limiting current conditions leads to dendritic or powdery deposits [12, 13]. Therefore; the value of the operating current density with respect to the mass transport limited current density is a critical parameter for deposit morphology.

Mass transport conditions at the wafer limit the rate of electroplating and influences the current distribution and microstructure of the deposit. These criteria determine the thickness and uniformity of blanket deposited layers. In patterned plating, they also determine the shape evolution, leveling, and superfilling. The current distribution in patterned plating is considered on three different scales: substrate scale, repeating pattern scale, and feature scale [14]. At the substrate level, the current distribution is governed by the overall cell geometry and the uniformity of current distribution is generally achieved by using auxiliary electrodes or current shielding concepts. On the pattern scale, the current distribution depends on the feature geometry and the spacing. Current density on a feature is higher when it is spaced farther away from a neighboring feature. On a feature scale, the current distribution evolves with time due to the continuously changing shape of the feature. The current distribution within the features is influenced by the use of suitable additives [7, 14, 15].

Additives are widely employed in electroplating practice for grain refining, stress relieve, leveling, and brightening. Surfactants are also commonly added to plating electrolytes to facilitate evacuation of gas bubbles. A leveling agent is a suitably chosen additive that generally acts as an inhibitor for the metal deposition reaction. It is consumed at the cathode and its reaction rate is mass transport controlled [15–18]. Since peaks are more accessible than valleys, they are more strongly inhibited by the additives, leading to preferential metal deposition into recess. These concepts have been used to develop understanding of superfilling during dual-Damascene plating and to develop electroplating baths for Cu interconnects. Grain refining and brightening are related to inhibition which affects nucleation and growth. Early studies of the role of additives for the development of microstructure in electrodeposition were performed by Seiter and Fischer [19], who recognized the importance of inhibition for obtaining fine-grained deposits. Fisher's concepts were further refined by Winand [20] and the factors affecting microstructure of electrodeposits were also discussed by Landolt [12]. The use of additives in electrodepositon of copper is a widely studied topic. Schimdt et al. [21] found that in a sulfate solution without additives copper nucleation on gold was three-dimensional. The presence of BTA decreased the size of nuclei and increased their number, while the presence of thiourea led to the formation of small flat plates. Armstrong and Muller [22] found that BTA inhibits the growth of specific planes of copper but does not affect the number of nuclei. Kelly et al. [23] investigated the synergistic effects of adsorbed species using electrochemical methods and near field microscopy. These authors demonstrated a strong synergistic effect between the adsorbed plating additive and the chloride ion. Such studies of the interactions between different additive species and their role in the nucleation and growth of deposits in nanostructures are of great importance to understand the mechanisms involved in Damascene plating of copper interconnects.

4.3 Pulse Plating and Pulse Reverse Plating

Pulsating the current (or voltage) permits the plating process to be operated at higher average current density than dc plating without the formation of dendrites. Furthermore, the composition of alloy deposits can be varied by varying pulse parameters while keeping the average current density constant.

Mass transport in pulse plating has been studied by a number of authors [24–26]. In principle, pulsating current creates a combination of steady-state and non-steady-state diffusion processes such that by varying the pulse parameters, the relative contribution of each diffusion process can be controlled. Two different limiting currents must be distinguished, the steady-state limiting current density i_L which for a given electrolyte concentration depends only on hydrodynamic conditions and the pulse-limiting current density, i_{pL}. The pulse-limiting current density i_{pL} corresponds to that value of peak current, i_p, where the reactant concentration at the surface reaches zero just at the end of a pulse. For a applied pulse current density close to i_{pL} the current distribution is governed mostly by non-steady-state mass transport and therefore may become relatively uniform. Indeed, in metal deposition using pulse plating one usually works well below the pulse-limiting current density in order to avoid dendrite formation. In alloy deposition, on the other hand, when the nobler component is present at small concentration in the electrolyte its pulse-limiting current density is exceeded during deposition of the less noble component. This can lead to spatial uniformity of composition because one component is governed by tertiary current distribution and the other by secondary current distribution.

Depending on the mass transport conditions at the cathode, the current distribution in pulse plating can be less or more uniform than in dc plating. In the absence of significant mass transport effects the Wagner number, $Wa = \frac{\mathrm{d}\eta/\mathrm{d}i}{\rho_e L}$, which governs secondary current distribution in pulse-plating depends on the pulse current density rather than the average current density. Because the pulse current density is always higher than the average current density the Wagner number corresponding to a given deposition rate is smaller in pulse plating than in dc plating and the current distribution therefore is less uniform. In the presence of significant mass transport effects the situation is quite different, however, and the current distribution in pulse plating may be more uniform than in dc plating. Due to the high instantaneous current densities applied in pulse plating the reaction rate during the pulse on time may become limited by non-steady-state mass transport even under conditions where the average current density is well below the dc limiting current density. For an average current density corresponding to a secondary current distribution in dc plating a tertiary current distribution may therefore prevail in pulse plating under certain conditions leading to a more uniform current distribution.

Pulse reverse plating offers the possibility of achieving more uniform current distribution than in dc plating. In order to achieve this, the pulse parameters are so chosen that during the cathodic deposition cycle the current distribution is more uniform than during the anodic dissolution cycle. For example, one can apply a cathodic pulse corresponding to a relatively uniform secondary or tertiary current distribution, followed by a short high current density anodic pulse corresponding to a primary current distribution. The application of pulse reverse current is through-hole plating was studied by Pesco and Cheh [27]. Although the use of pc leads to a less uniform current distribution than the use of dc, pulse reverse plating was considered by the authors as a promising technique for improving current distribution. Wan et al. [28] investigated the applicability of pulse plating and pulse reverse plating in through-hole plating taking into account both the potential distribution and the prevailing mass transport conditions. From their theoretical analysis the authors concluded that throwing power is not improved by pulse plating but the use of pulse reverse plating may lead to more uniform deposits, in agreement with Pesco and Cheh [27]. Yung et al. [29] showed that the critical parameter characterizing the deposition rate in through-hole plating is proportional to L^2/r where L is the printed circuit board thickness and r is the hole radius. The current distribution therefore depends on the absolute dimension of the hole in addition to the aspect ratio.

4.4 Electrodeposition of Copper

Copper is the most commonly plated material for its varying decorative, functional, and engineering applications. Several different types of copper plating systems that have been reported in the literature include alkaline cyanide, fluoborate, pyrophosphate, and acid sulfate baths [30]. Due to toxicity and waste treatment issues, cyanide baths are getting replaced by non-cyanide baths. High speed plating fluoborate baths and pyrophosphate baths used in plated through-hole PCBs have been mostly replaced by acid sulfate baths that have become the most commonly used, cost-effective copper plating systems in the micro- and nanoelectronics.

Copper sulfate and sulfuric acid are the primary constituents of the acid sulfate baths. The sulfate bath generally contains chloride ions in varying range of concentrations to influence the deposit properties such as microstructure, microhardness, crystallographic orientation, internal stress, and surface appearance. Several organic additives are also frequently added to the bath to influence surface finish, grain refining, and suppression of dendritic deposition. A list of these additives and their functions in electrodeposition of copper is presented in [30] and of applied current density on the deposit morphology has been reported in the literature [13, 31–33]. Increasing inhibition and/or increasing current density produces fine-grained structures. Kinetically limited growth tends to favor compact columnar or equiaxial growth in copper deposits while mass transport limited growth favors formation of loose dendritic deposits. During electroplating of micro-features, the use of proper additives is essential for leveling and superfilling [7, 34].

Additives that act as inhibitors tend to promote formation of fine equiaxial grains, which may lead to increased internal stress. Electrodeposited copper may contain non-equilibrium grain structures, which spontaneously recrystallize even at room temperature [35]. As a consequence, structure-dependent properties such as sheet resistance and internal stress of deposits may change slowly with time after deposition.

4.4.1 Electrodeposition Process for Copper Interconnects

Cu interconnects are fabricated by dual-Damascene process which is referred to a metallization patterning process by which two insulator (dielectric) levels are patterned, filled, and planarized to create a metal layer consisting of vias and lines [7]. A sandwich of two levels of insulator and etch stop layers is patterned as holes for vias and troughs for lines. They are then filled with a single metallization step. Finally, the excess material is removed, and the wafer planarized by CMP. Electroplating enables deposition of Cu in via holes and overlying trenches in a single step thus eliminating via/line interface and significantly reducing the cycle time. Due to these reasons and due to relatively less expensive tooling, electroplating is a cost-effective and efficient process for Cu interconnects fabrication.

The Damascene copper plating process is typically based on a chloride-containing acid copper sulfate solution [7, 36]. In such a system formation of voids is very common, particularly in narrow trenches, where top ridges build up first thus creating a void in the middle of the structure. Formation of seams is commonly observed during conformal plating as well. Such seams can be as disastrous as the voids. So formed voids and seams lead to electromigration problems in submicrometer interconnect structures. These defects can also entrap electrolyte that can lead to serious corrosion issues. Elimination of voids and seams is therefore very critical to minimize electromigration and other reliability issues.

In dual-Damascene plating, defect-free filling of trenches can be achieved by creating a condition where higher deposition rate is achieved in less accessible bottom of the trenches, compared with the easily accessible flat top surface and upper side walls. Small amounts of organic additives added in right concentrations can increase the plating rate inside trenches and vias relative to the planar surface. This differential bottom-up plating rate leads to superfilling. For defect-free, superfilling deposition, several organic additives are added to the bath. Based on their function, these additives can be broadly categorized into two types: an accelerator and an inhibitor. The accelerator is essentially an organic sulfur-containing compound (a mercapto species) that preferentially adsorbs at the bottom of a via. Inhibitor additive (s) in the bath may contain one or more components. The main inhibitor is generally a glycol, which acts to suppress the electrodeposition rate especially in the presence of chloride ions. Some baths also contain a nitrogen-containing molecule that acts as a leveling agent. Competitive adsorption of these inhibitor molecules together with the fact that one or more of these co-adsorbed species may be mass transport controlled lead to a condition whereby via/trench bottoms are less inhibited.

Mathematical modeling work by Dukovic et al. [7, 14] assumed that the additives react under diffusion control while the metal deposition reaction is under activation controlled. Thus by optimizing the reaction rates (by varying additive concentration and adjusting the current density of Cu deposition), it is possible to obtain higher deposition rate at the bottom of via/trench. Moffat et al. [34, 37–39] and West et al. [23, 40] explained the superconformal electrodeposition of copper by a curvature enhanced accelerator coverage model. The model is based on the assumptions that the local growth velocity is proportional to the surface coverage of the accelerator and the accelerator remains segregated at the metal/electrolyte interface during copper deposition. During deposit growth on nonplanar geometries, this leads to enrichment of the accelerator on the evolving concave surfaces and depletion on convex surfaces. This phenomenon gives rise to bottom-up superfilling of submicrometer trenches and vias.

In addition to the superfilling property, these additives also influence the deposit structure (hence stress) and roughness. It is evident that a delicate balance of respective additives is needed in the bath to obtain precise fabrication of void-free, nanoscale Cu interconnects that are to be uniformly laid on hundreds to thousands of chips in a 300 mm wafer.

4.5 Concluding Remarks

This chapter presented a brief description of electrodeposition process and a focused discussion of copper electrodeposition for chip interconnects. It must be emphasized that different chip manufacturers have their unique combination of electrodeposition tools, proprietary tailored bath, and integration scheme for copper chip interconnects. Successful implementation of the copper electrodeposition process in high volume chip manufacturing involves equal attention to metrology, process integration, and reliability issues. Besides void-free deposition of interconnect structures, selection of a reliable CMP process, excellent adherence of copper lines/vias to the dielectric, low resistivity, and resistance to electromigration are some of the key requirements for a high yielding, reliable interconnect electrodeposition process. All these aspects are addressed in greater detail in different chapters of this book.

References

1. Schlesinger, M. and Paunovic, M.: Modern Electroplating (eds.), 4th edition, Wiley Interscience, New York (2000)
2. Romankiw, L. T.; Croll, I.; and Hatzakis, M.: Batch fabricated thin film. Magnetic recording head. IEEE Trans. Magn. **6(4)**, 729 (1970)
3. Osaka T.: Electrochemical formation and microstructure in thin films for high functional devices. Electrochim. Acta. **42**, 3015 (1997)
4. Romankiw, L. T.: A path: from electroplating through lithographic masks in electronics to LIGA in MEMS. Electrochim. Acta. **42**, 2985 (1997)

5. Romankiw, L. T. and Turner, D. (eds.): Electrodeposition Technology: Theory and Practice, PV 86–17, Electrochemical Society Proceedings, New Jersey (1987)
6. Datta, M.; Shenoy, R. V.; Jahnes, C.; Andricacos, P. C.; Horkans, J.; Dukovic, J. O.; Romankiw, L. T.; Roeder, J.; Deligianni, H.; Nye, H.; Agarwala, B.; Tong, H. M.; and Totta, P. A.: Electrochemical fabrication of mechanically robust C4s. J. Electrochem. Soc. **142**, 3779 (1995)
7. Andricacos, P. C.; Uzoh, C.; Dukovic, J. O.; Horkans, J.; and Deligianni, H.; Damascene copper electroplating for chip interconnections. IBM J. Res. Dev. **42**, 567 (1998)
8. Datta, M.: Electrochemical processing technologies in chip fabrication: Challenges and opportunities. Electrochim. Acta. **48**, 2975 (2003)
9. Seraphim, D. P.; Barr, D. E.; Chen, W. T.; Schmitt, G. P.; and Tummala, R. R.: In Microelectronic Packaging Handbook, Part III, 2nd edition, Tummala, R. R.; Rymaszewski, E. J.; and Klopfenstein, A. G. (eds.), Chapman and Hall, New York (1997)
10. Datta, M.: In New Trends in Electrochemical Technology, Microelectronic Packaging, Datta, M.; Osaka, T.; and Schultze, J. W. (eds.), CRC Press, New York, **3**, 3 (2005)
11. Datta, M. and Landolt, D.: Fundamental aspects and applications of electrochemical microfabrication. Electrochim. Acta. **45**, 2535 (2000)
12. Landolt D.: Electrochemical and materials science aspects of alloy deposition. Electrochim. Acta. **39**, 1075 (1994)
13. Ibl, N.: In Comprehensive Treatise of Electrochemistry. Yeager, E.; Bockris, J. O'M.; and Conway; B. (eds.), Plenum Press, New York **6(1)**, 133, 239, (1982)
14. Dukovic, J. O.: Feature-scale simulation of resist-patterned electrodeposition. IBM J. Res. Dev. **37(2)**, 125 (1993)
15. Madore, C.; Matlosz, M.; and Landolt, D.: Blocking inhibitors in cathodic leveling. I. Theoretical analysis. J. Electrochem. Soc. **143(12)**, 3927 (1996)
16. Kardos, O.: Current distribution on microprofiles, Part I, II, III. Plating, **61**, 129, 229, 316 (1974)
17. Kruglikov, S. S.; Kudriavtsev, N. T.; Vorobiova, G. F.; and Antonov, A. Ya.: On the mechanism of levelling by addition agents in electrodeposition of metals. Electrochim. Acta. **10(3)**, 253 (1965)
18. Dukovic, J. and Tobias, C. W.: Simulation of leveling in electrodeposition. J. Electrochem. Soc. **137**, 3748 (1990)
19. Seiter, H. and Fischer, H.: Electrocrystallization of metals. Z. Elektrochemie. **63**, 249 (1959)
20. Winand, R.: Electrodeposition of metals and alloys-new results and perspectives. Electrochim. Acta. **39(8-9)**, 1109 (1994)
21. Schimdt, W. U.; Alkire, R. C.; and Gewirth, A.: Mechanic [sic] study of copper deposition onto gold surfaces by scaling and spectral analysis of in situ atomic force microscopic images. J. Electrochem. Soc. **143(10)**, 3122 (1996)
22. Armstrong, M. J. and Muller, R. H.: *In situ* scanning tunneling microscopy of copper deposition with Benzotriazole. J. Electrochem. Soc. **138(8)**, 2303 (1991)
23. Kelly, J. J.; Tian, C.; and West, A. C.: Leveling and microstructural effects of additives for copper electrodeposition. J. Electrochem. Soc. **146**, 2540 (1999)
24. Ibl, N.: Some theoretical aspects of pulse electrolysis. Surface Technology. **10**, 81 (1980)
25. Chin, D. T.: Mass transfer and current-potential relation in pulse electrolysis. J. Electrochem. Soc. **130**, 1657 (1983)
26. Datta, M. and Landolt, D.: Experimental investigation of mass transport in pulse plating. Surface Technol. **25**, 97 (1985)
27. Pesco, A. M. and Cheh, H. Y.: The current distribution within plated through-holes. J. Electrochem. Soc. **136(2)**, 408 (1989)
28. Wan, H. H.; Chang, R. Y.; and Yang, W. L.: Current distribution in a jet through-hole system during periodic electrolysis. J. Electrochem. Soc. **140(5)**, 1380 (1993)
29. Yung, E. K.; Romankiw, L. T.; and Alkire, R. C.: Plating of copper into through-holes and vias. J. Electrochem. Soc. **136(1)**, 206 (1989)

30. Dini, J. W.: In Modern Electroplating. Schlesinger, M.; and Paunovic, M.: (eds.), 4th edition, Wiley Interscience, New York, 61 (2000)
31. Winnad, R.: Electrodeposition of metals and alloys-new results and perspectives. Electrochim. Acta. **39**, 1091 (1994)
32. Donepudi, V. S.; Venkatachalapathy, R.; Ozemoyah, P. O.; Johnson, C. S.; and Prakash, J.: Electrodeposition of copper from sulfate electrolytes: Effects of Thiourea on resistivity and electrodeposition mechanism of copper. Electrochem. Solid-State Lett. **4(2)**, C13 (2001)
33. Landolt, D.: Electrodeposition science and technology in the last quarter of the twentieth century. J. Electrochem. Soc. **149(3)**, S, 9 (2002)
34. Moffat, T. P.; Bonewich, J. E.; Huber, W. H.; Stanishevsky, A.; Kelly, D. R.; Stafford, G. R.; and Josell, D.: Superconformal electrodeposition of copper in 500–90 nm features. J. Electrochem. Soc. **147(12)**, 4524 (2000)
35. Cabral, C.; Andricacos, P. C.; Cignac, L. M.; and Noyan, I. C.: Room temperature annealing of damascene plated Cu chip metallization. Adv. Metallization Conf. Proc., ULSI XIV, 81 (1998)
36. Dubin, V. M.; Simka, H. S.; Shankar, S.; Moon, P.; Marieb, T.; and Datta, M.: In New trends in electrochemical technology, microelectronic packaging. Datta, M.; Osaka, T.; and Schultze, J. W. (eds.), CRC Press, New York. **3**, 31 (2005)
37. Moffat, T. P.; Wheeler, D.; Huber, W. H.; and Josell, D.: Superconformal electrodeposition of copper. Electrochem. & Solid-State Lett. **4**, C26 (2001)
38. Josell, D.; Wheeler, D.; Huber, W. H.; Bonevich, J. E.; and Moffat, T. P.: A simple equation for predicting superconformal electrodeposition in submicrometer trenches. J. Electrochem. Soc. **148**, C767 (2001)
39. Wheeler, D.; Josell, D.; and Moffat, T. P.: Modeling of superconformal electrodeposition using the level set method. J. Electrochem. Soc. **150**, C302 (2003)
40. West, A. C.; Mayer, S.; and Reid, J.: A superfilling model that predicts bump formation. Electrochem. & Solid-State Lett. **4**, C50 (2001)

Chapter 5
Electrophoretic Deposition

David Brandon

5.1 Electrophoresis and Electrophoretic Deposition

Electrophoresis is the migration under the influence of an electric field of charged particles held in a dispersion. Electrophoretic deposition (EPD) is then the deposition of these migrating particles onto a deposition electrode or porous membrane [1]. At a deposition electrode the charge on the particles is neutralized by the current flowing through the conducting electrode, while in a porous membrane it is the electrolyte in the part of the cell that is separated by the membrane from the dispersion that carries the neutralizing current. The two geometries are shown schematically in Fig. 5.1. Deposition membranes are formed from a microporous non-conductor, while deposition electrodes have to have sufficient conductivity to carry the deposition current. In either case, the deposit closely follows the surface topology of the deposition electrode or membrane, and quite complex shapes have been produced which may either be detached from the substrate, as a freestanding component, or form an adherent surface coating.

5.1.1 Process Limitations

The thickness of the deposit may be closely controlled, from a few micrometers up to many millimeters, with excellent surface finish. The composition of the deposit mirrors that of the particles in the dispersion, and metals, polymers, semi-metals, and ceramics have all been successfully deposited by EPD to yield components with a wide variety of electrical, optical, and magnetic properties. The major limitations of this manufacturing technology arise from the discrete nature of the particulate deposit, which limits the packing density so that the deposit usually has to be consolidated, typically by sintering, in order to improve the strength of the product and reduce or eliminate the porosity in the green, as-deposited mate-

D. Brandon (✉)
Faculty of MSE, Technion IIT, Haifa 32000, Israel
e-mail: brandon@technion.ac.il

Y. Shacham-Diamand et al. (eds.), *Advanced Nanoscale ULSI Interconnects: Fundamentals and Applications*, DOI 10.1007/978-0-387-95868-2_5,

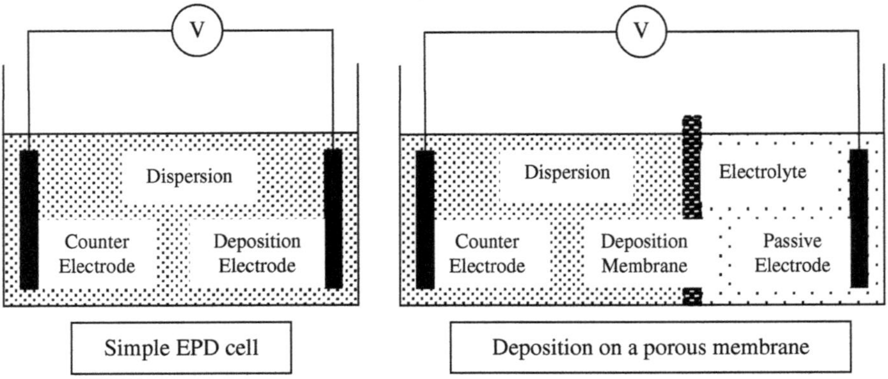

Fig. 5.1 Charged particles in dispersion migrate away from an electrode of the same sign. They deposit either on a shaped deposition electrode of opposite sign (*left*) or on a microporous, non-conducting membrane (*right*). In the case of deposition on a porous membrane, an electrolyte completes the electric circuit

rial. The limited range of particle sizes that can be successfully dispersed without sedimentation may also limit possible applications. Coarse particles, larger than a few micrometers, cannot be retained in dispersion, while very small particles, less than 0.1 μm, tend to aggregate in the presence of an electric field. In general, EPD is expected to have advantages for the production of microcomponents from sub-micrometer or nanoparticulate materials, where the stability of the dispersion that constitutes the feedstock is not limited by sedimentation.

5.1.2 Theoretical Basis

The basic equation for electrophoretic mobility is $v_{\mathrm{EPD}} = E\mu = \frac{2E\varepsilon\xi}{3\eta}\left[1+f(kr)\right]$ [2]. In this equation, v_{EPD} is the particle velocity under the influence of an electric field E, μ is the mobility, ε is the dielectric constant of the dispersion medium, η is the viscosity of the dispersion, ξ is the "zeta-potential" (the potential at the shear boundary between the particle together with its associated liquid double layer and the liquid medium). The factor $\left[1+f(kr)\right]$ is included to indicate that there is a weak dependence of mobility on r, the particle radius. By selecting suitable surface-active additives it has proved possible to control the zeta potential, and hence electrophoretic deposition, in a wide range of aqueous and non-aqueous dispersion media [3]. Hydrogen evolution at the deposition electrode in aqueous media is often a problem, and most reported results have therefore been obtained using non-aqueous dispersion media.

5.2 Potential Applications

The application of EPD to the production of microelectronic components and devices is proving to be a difficult challenge, but four specific areas of application have been identified and prototype products have been demonstrated. The four applications are

Fuel cell technology (which will not be discussed here, other than to note that the controlled porosity of the EPD product plays a critical role in determining the efficiency of the fuel cell).
Conducting lines on LTCC tapes.
Embedded passive components and thermal vias in LTCC tape.
High-performance, low-profile solid electrolyte capacitors.

5.2.1 Conducting Lines

It has proved possible to adapt EPD to the production of silver interconnects by first depositing the interconnect pattern onto a substrate and then transfer-printing onto tape [4, 5]. The connectivity and edge definition are excellent (Fig. 5.2) but many technological problems remain to be resolved if this technology is to become commercially viable. To ensure that the deposit is free of large aggregates and other defects, deposition is performed in a special cell in which the patterned deposition electrode is positioned vertically at the top of the deposition cell (Fig. 5.3).

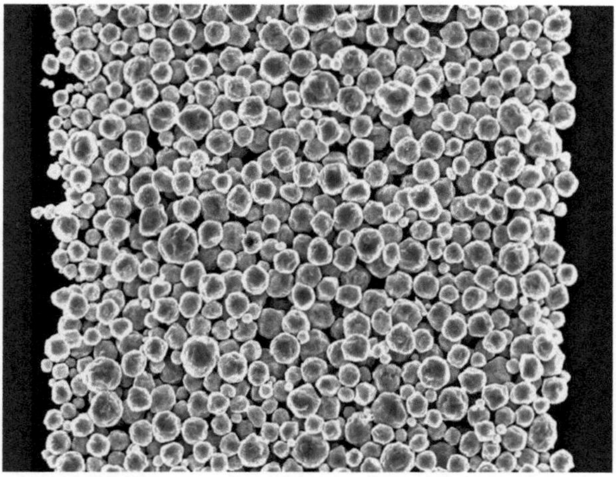

Fig. 5.2 Unsintered Ag–Pd alloy conduction line demonstrating excellent edge definition and uniformity of the 1 μm particle size

5.2.2 Embedded Passive Components

A very promising long-term application of EPD is in the production of embedded passive components and thermal vias in LTCC tape assemblies. By tailoring the sintering characteristics of the powders used for the passive components to that of the LTCC tapes it is possible to ensure the integrity of the passive component. Additions

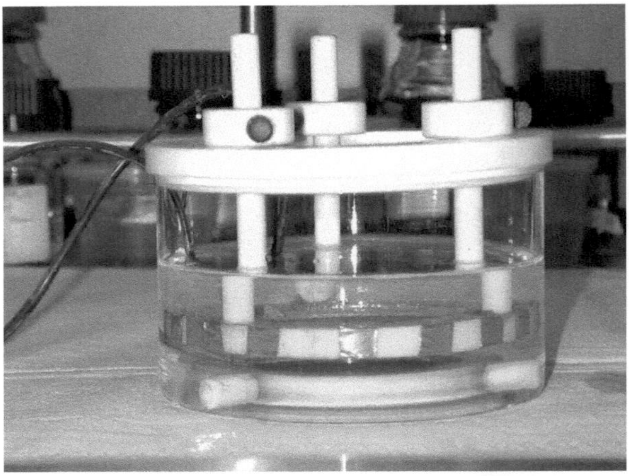

Fig. 5.3 EPD cell for vertical deposition of interconnects and conduction lines on a patterned electrode

of small volume fractions of a suitable glass powder to a standard, X7R capacitor powder, together with the constraint imposed within the multilayer tape sandwich, have made it possible to produce demonstration products which meet all the engineering requirements and are free of processing defects associated with cavitation, distortion, decohesion, or shrinkage cracking (Fig. 5.4). Similar low volume fraction glass additions to silver powders have been found suitable for the EPD formation of thermal vias in thick, multilayer tape assemblies. However, this technology is also a

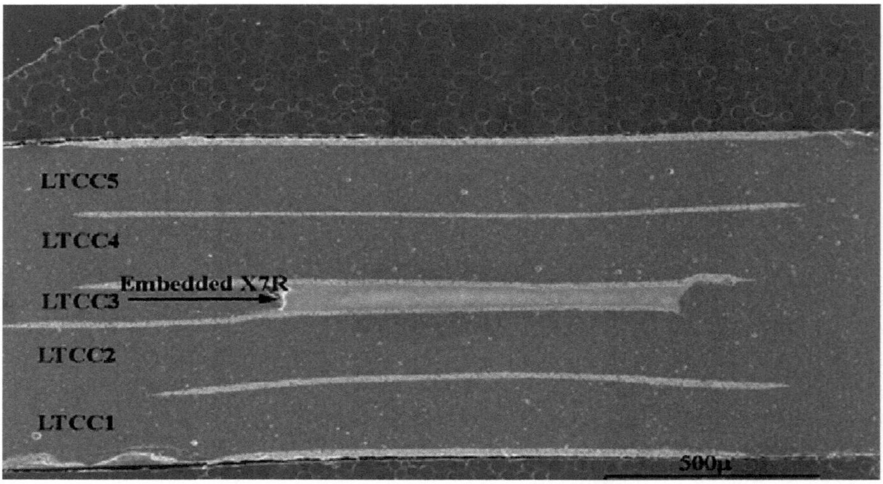

Fig. 5.4 Embedded X7R capacitor in a sintered tape assembly. The capacitor powder has been modified by the addition of a few percent of a glass composition in order to match the sintering characteristics of the LTCC tape

long way from the demonstration of commercial feasibility, which will require the prototype production of simple circuits that include embedded capacitor, resistor, and other elements.

5.2.3 Solid Electrolyte Capacitors

Perhaps the most successful feasibility demonstration at this time is the use of EPD to manufacture high-performance, solid electrolyte capacitors. These capacitors have an exceptionally high capacitance per unit volume, with minimal direct current leakage and equivalent series resistance at high frequencies. They rely on a partially sintered porous anode which is separated from an impregnated cathode by a dielectric oxide layer formed by anodizing the anode. EPD formation of the anodes (Fig. 5.5) is possible using a modular assembly and yields anodes with very uniform, controlled porosity and highly reproducible dimensions. The compacts are

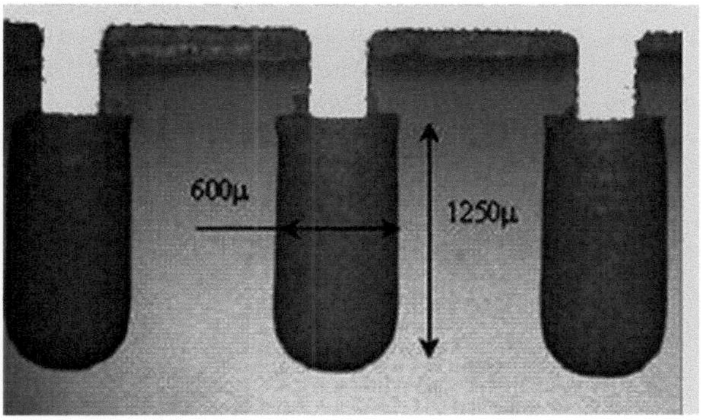

Fig. 5.5 Green anode blanks formed for pilot plant production of EPD solid electrolyte capacitors. Note the freedom from process defects

Fig. 5.6 Prototype low-profile, high-performance capacitors. The thickness of the capacitors is 0.3 mm

free of processing defects and can be formed in dimensions well below those that are feasible using the conventional, cold-pressing technology. Moreover, the EPD anodes can make excellent use of the nanopowders that are now available and can be sintered at lower temperatures to yield much higher surface to volume ratios, leading to exceptional capacitance per unit volume. A prototype product complete with termination is shown in Fig. 5.6.

Acknowledgements This contribution was made possible through the efforts of the development team at Cerel – Ceramic Technologies Ltd., especially Martin Zarbov and Assaf Thon, whose efforts, over an extended period, have turned tentative speculation into practical demonstrations of technical and commercial feasibility.

References

1. Randall, C. and Van Tassel, J.: Encyclopedia of Mat. Sci. & Technol., 2733 (2001)
2. Hunter, R. J.: Foundations of Colloid Science (2nd edition) Oxford University Press, Oxford (2001)
3. Sarkar, P. and Nicholson, P. S.: Electrophoretic deposition (EPD): mechanisms, kinetics, and application to ceramics. J. Am. Ceram. Soc. **79(8)**, 1987 (1996)
4. Tassel, J. V.; Daga, A.; and Randall, C. A.: Electrophoretic deposition for fabrication of ultra-thin multi-component electroceramic tapes. IMAPS 3906, 647 (1999)
5. Tassel, J. V. and Randall, C. A.: Electrophoretic deposition and sintering of thin/thick PZT films. J. Eur. Ceram. Soc. **19**, 955 (1999)

Chapter 6
Wafer-Level 3D Integration for ULSI Interconnects

Ronald J. Gutmann and Jian-Qiang Lu

6.1 Introduction

Three-dimensional (3D) integration in a system-in-a-package (SiP) implementation (packaging-based 3D) is becoming increasingly used in consumer, computer, and communication applications where form factor is critical. In particular, the hand-held market for a growing myriad of voice, data, messaging, and imaging products is enabled by packaging-based 3D integration (i.e., stacking and connecting individual chips). The key drivers are for increased memory capacity and for heterogeneous integration of different IC technologies and functions.

Wafer-level 3D integration (i.e., 3D stacking prior to singulation of wafers into individual chips) has become an increasingly active research topic, without any present (early in 2007) large-scale IC manufacturing. While wafer-level 3D technology is appreciably more complex than packaging-based 3D, significant advantages are obtained compared to packaging-based 3D integration, namely [1–3],

- higher density of inter-chip interconnects;
- lower electrical parasitics of inter-chip interconnects (therefore, higher interconnect electrical performance);
- lower high-volume manufacturing cost, since monolithic, wafer-level interconnectivity is extended to multiple device levels.

In addition, the form factor and heterogeneous integration advantages of packaging-based 3D integration are maintained. Wafer-level 3D integration is the next significant technology for integrated electronics with ever high performance, density, and functionality after copper/low-k interconnect, complementing both the traditional printed circuit board (PCB) and related packaging techniques and advanced 2D IC interconnects introduced in the past decade.

R.J. Gutmann and J.-Q. Lu (✉)
RPI, Low Center for Industrial Innovation, Troy, New York 12180, USA
e-mail: gutmar@rpi.edu, luj@rpi.edu

Y. Shacham-Diamand et al. (eds.), *Advanced Nanoscale ULSI Interconnects: Fundamentals and Applications*, DOI 10.1007/978-0-387-95868-2_6,

For wafer-level 3D integration, a number of inherent issues need to be either accommodated or solved prior to wide-scale use. These include

- establishment of integration architecture and design tools;
- die yield and impact on wafer-level stacking;
- common die size requirement for full silicon area utilization;
- thermal constraints with high power consumption;
- wafer-level processing equipment qualified for 24/7 manufacturing.

While these technical and infrastructure considerations for wafer-level 3D integration appear daunting, the performance advantage of through-chip micrometer-sized inter-chip vias for high-speed multi-core processors, high memory capacity with reduced processor-memory latency, heterogeneous integration of mixed signal ICs with high-performance interconnects and many other leading-edge products with 3D-enabled integration are driving research in academic, government, consortia, and individual company laboratories.

6.2 Types of 3D Integration

Three-dimensional (3D) integration can be classified into three categories as follows:

- post-singulation or packaging based;
- pre-bottom-wafer singulation or die-on-wafer;
- pre-singulation or wafer to wafer.

While the research emphasis is on wafer-to-wafer 3D integration, packaging-based 3D and die-on-wafer 3D are summarized to provide a technology context for, and technology roadmap to, wafer-to-wafer 3D.

The packaging industry has already adapted chip stacks for form-factor considerations, including stacking of single-chip packages (see Fig. 6.1a), multi-level thinned die (see Fig. 6.1a, c) [4–6]. Current manufacturing is limited to wire bonding of peripheral input/output (I/O) connections to a printed circuit board (PCB) or within a chip-scale package (CSP). Appreciable industrial R&D is devoted to through-Si vias (TSVs) compatible with areal flip-chip connections, although these TSVs are tens of micrometers in diameter [7–9]; while lower electrical parasitics (particularly inductance) is obtained with packaging-based TSVs, the vertical interconnect density is at least two orders of magnitude lower than obtained with micrometer-sized vertical interconnects available with wafer-level 3D [1–3, 8–15]. The impact of this technology on hand-held consumer products has been, and will continue to be, significant.

Pre-bottom wafer singulation 3D integration (or die-on-wafer) is a technology intermediate between post-singulation 3D described in the previous paragraph and

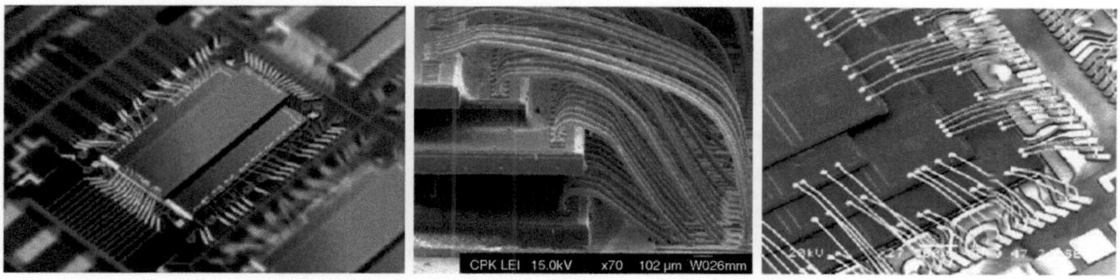

(a) (b) (c)

Fig. 6.1 Chip stacks indicative of packaging-based 3D integration: (**a**) chip scale (courtesy of Amkor) [4], (**b**) and (**c**) multi-level thinned die (courtesy of ChipPac Inc. [5], courtesy of Amkor [4, 6])

pre-singulation (i.e., wafer-to-wafer or wafer-level) 3D. Post-singulated thinned chips with TSVs that are pre-tested can be mounted on a wafer using pick-and-place equipment. Only known-good-die (KGD) are aligned and mounted on good die in the substrate wafer, thereby alleviating one major disadvantage of wafer-to-wafer 3D (i.e., the yield penalty of having good die with bad die in a non-repairable stack). The key disadvantage of this approach is that the die-to-wafer alignment with pick-and-place equipment is limited to 10–20 μm, better than die-to-die assembly but still two orders of magnitude worse in TSV areal density than with wafer-to-wafer alignment. Examples are shown in Fig. 6.2.

(a) (b)

Fig. 6.2 Die-on-wafer 3D integration (courtesy of Ziptronix) [16]: (**a**) assembly equipment; (**b**) typical die-stack cross section

One of the pioneering 3D integration techniques by Irvine Sensors is a modification of die-on-wafer [17, 18]. As depicted in Fig. 6.3, a pseudo-wafer is used, eliminating bad die at the substrate level as well (assuming sufficient electrical performance testing is feasible prior to assembly). The individual die, at least in the original Irvine Sensors implementation, has beam-lead type I/O connections and form a reasonably planar surface for stacking of subsequent die. Originally developed for high-capacity memory, up to 100 pseudo-wafers have been stacked,

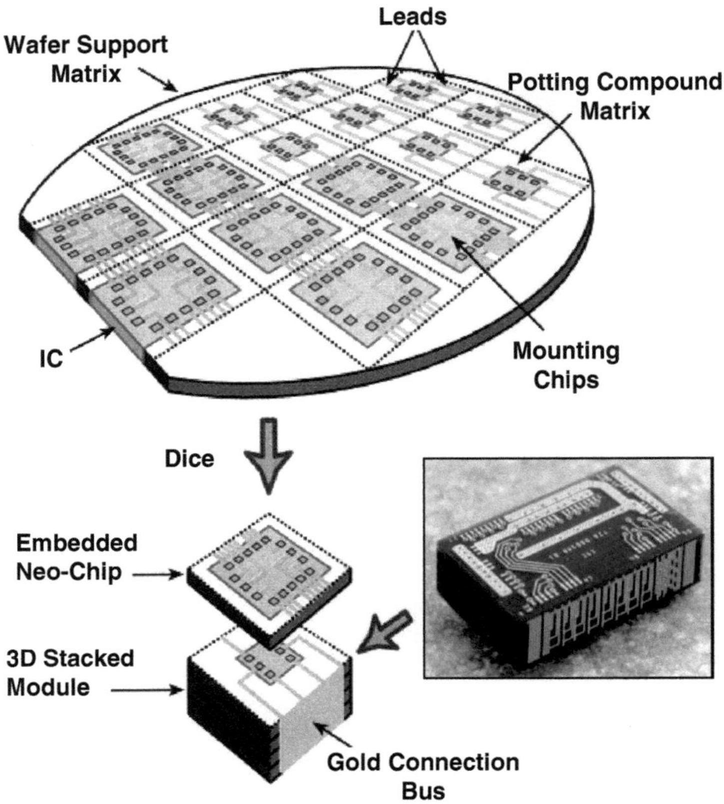

Fig. 6.3 Pseudo-wafer 3D integration (courtesy of Irvine Sensors), adapted from [17, 18]

followed by singulation into chip stacks, polishing of the edges of the chip stack, and subsequent thin-film wiring at the edge(s) of the stack as shown in Fig. 6.3. The relatively expensive vertical interconnectivity on the edge(s) of the stack is not generally pursued today, replaced by TSVs using deep trench etching, conductor fill, and wafer thinning.

Wafer-to-wafer (or wafer-level) 3D integration can be further classified as either front-end-of-the-line (FEOL) based or back-end-of-the-line (BEOL) based as the technologies pursued are appreciably different. Wafer-level FEOL-based integration incorporates multiple, parallel active semiconductor layers which can be of single crystal structure by using either epitaxial lateral overgrowth techniques or recrystallization of polycrystalline [19, 20]. A simpler FEOL process flow is feasible if polycrystalline silicon can be used for active devices; however, a major difficulty is to obtain high-quality electrical devices and interconnects. While obtaining single-crystal device layers in a generic IC technology remains in the research stage, polycrystalline devices suitable for non-volatile memory (NVM) have not only been demonstrated but have been commercialized; a cross-section of such a NVM memory stack from Matrix Semiconductor is shown in Fig. 6.4.

Fig. 6.4 Three-dimensional non-volatile memory (NVM) with polysilicon memory devices (courtesy of Matrix Semiconductor) [19]

Most of the R&D on wafer-level 3D integration is BEOL based, where fully processed and interconnected wafers are wafer-to-wafer aligned, bonded, thinned, and interconnected [1–3, 8–15]. An interesting approach with the interconnectivity of FEOL-based 3D but with the processing approach and constraints of BEOL-based 3D is to bond two fully processed wafers with oxide-to-oxide bonding after full FEOL processing and local-interconnect processing of each wafer. This local-interconnect-based 3D approach limits the high aspect ratio (HAR) requirements to form a high density of inter-wafer vias in BEOL-based 3D platforms, thereby providing close to the active layer interconnectivity obtained with FEOL-based 3D. The cross-sectional diagrams in Fig. 6.5 depict the technology under development at IBM. One difficulty with this approach is the limited die-yield mapping capability prior to wafer-to-wafer bonding; only active devices and test structures can be mapped prior to bonding.

The numerous BEOL-based wafer-level 3D integration technology platforms are not as easy to classify [1–3, 8–15]. The key differentiators are as follows:

- type of wafer bonding, either metal to metal or polymer to polymer;
- type of inter-wafer interconnection, with copper (Cu), tungsten (W)
- or heavily doped polysilicon (poly) the most dominant;
- via-first or via-last process flow, with or without a handling wafer.

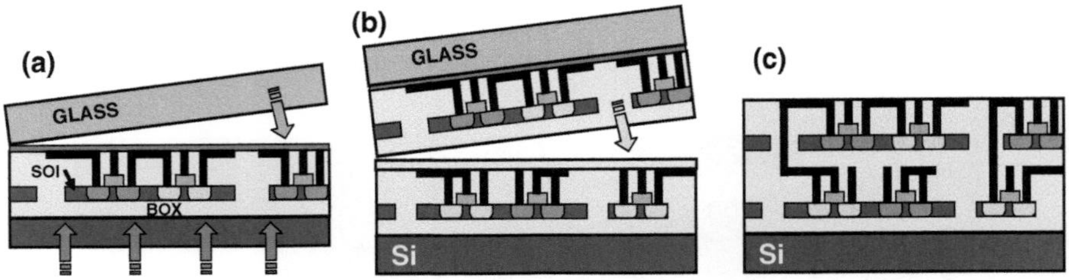

Fig. 6.5 Local interconnect-based 3D, with BEOL processing constraints and near-FEOL interconnectivity capabilities (courtesy of IBM) [12]: (**a**) device wafer attached to glass handle wafer and Si substrate removed; (**b**) top device wafer aligned and bonded to bottom device wafer; and (**c**) handle wafer removed and vias formed between device wafers

The selection of the optimum technology platform is subject to ongoing development and debate. Cu-to-Cu bonding with a via-first process flow has significant advantages for highest interwafer interconnectivity [3, 8, 9, 11], comparable to oxide-to-oxide bonding after local interconnectivity within each wafer as described in the previous paragraph; as a result, this approach is desirable for microprocessors and digitally based system-on-a-chip (SoC) technologies. Polymer-to-polymer bonding with a via-last process flow with high bonding strength is attractive when heterogeneous integration of diverse technologies is the driver and the interwafer interconnect density is more relaxed; benzocyclobutene (BCB) is the polymer most widely investigated [1, 10].

6.3 BEOL-Based Wafer-Level 3D Processing Considerations

The types of wafer bonding potentially suitable for wafer-level 3D integration are depicted in Fig. 6.6 [1]. Both the oxide-to-oxide bonding [12] and the polymer-to-polymer bonding [1] are inherently via-last process flows; the interwafer vias are formed after wafer-to-wafer alignment and bonding (by HAR etching, metal fill, and chemical–mechanical planarization (CMP)). Any metal-to-metal bonding

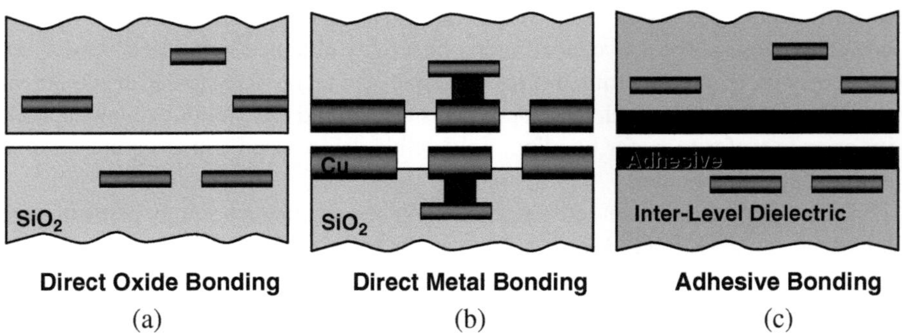

Fig. 6.6 Wafer bonding techniques for wafer-level 3D integration: (**a**) oxide to oxide; (**b**) metal to metal; and (**c**) polymer to polymer

can be used in a via-first process flow, where the interwafer vias are formed in the bonding process [3, 8, 9, 11]; note that appropriate interconnect processing within each wafer is required to enable 3D interconnectivity.

Oxide-to-oxide bonding of fully processed IC wafers requires an atomic-scale smoothness, which is difficult to achieve; in addition, wafer distortions introduced by FEOL and BEOL processing introduce sufficient wafer bowing and warping to decrease bonding strength, possibly to levels that inhibit subsequent processing requirements. While oxide-to-oxide bonding after FEOL and local-interconnect processing as depicted in Fig. 6.5 has been shown to be promising, the increased wafer distortion and oxide roughness after multi-level interconnect processing does not allow a robust process window.

Both copper-to-copper bonding using the via-first process flow described earlier and polymer-to-polymer bonding with a via-last process flow are the most viable approaches with fully interconnected wafers (after FEOL and BEOL processing); these are depicted in Figs. 6.7 and 6.8, respectively. In both cases, all process conditions are BEOL-compatible (e.g., $T \leq 400°C$ and with limited bonding pressure).

These wafer-level 3D platforms do not utilize handling wafers, but use face-to-face alignment and bonding with all required I/Os brought to the thinned backside of the top wafer (which becomes the face of the two-wafer stack). Another approach is to bond the top wafer to a handling wafer, after which the device wafer is thinned (to expose interwafer interconnects for metal-to-metal via-first bonding or a silicon or oxide surface for subsequent polymer-to-polymer bonding) and bonded to the full-thickness bottom wafer; after this bonding the handling wafer is removed with a compatible energy source such as ultraviolet (UV) light.

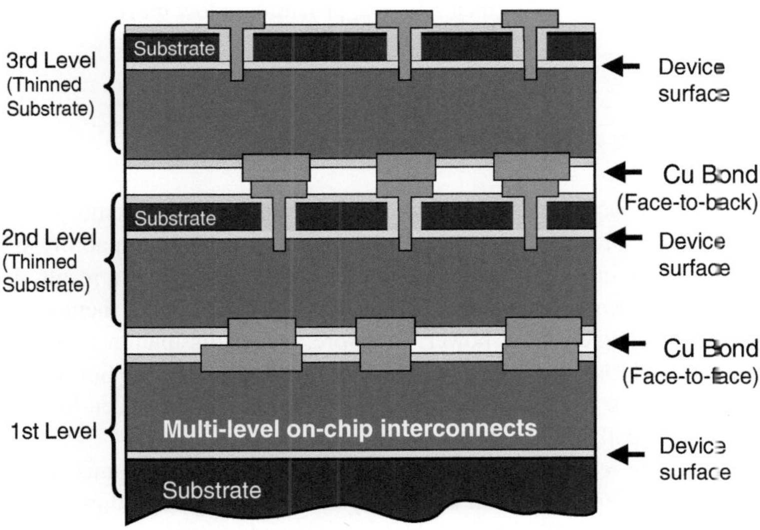

Fig. 6.7 Copper-to-copper via-first cross section (3-wafer stack illustrating face-to-face bonding and process flow of first two wafers)

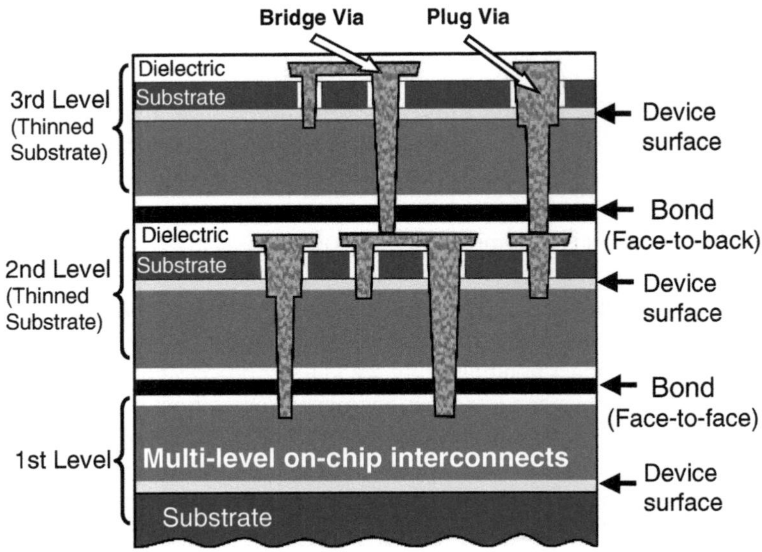

Fig. 6.8 Polymer-to-polymer via-last schematic cross section (3-wafer stack illustrating face-to-face bonding and process flow of first two wafers) [10]

A recent addition to the wafer-level 3D technology platform portfolio combines the bonding strength of polymer-to-polymer (in particular, BCB to BCB) bonding with the process flow advantages of metal to metal (in particular Cu to Cu), as shown Fig. 6.9. This platform also provides a Cu–BCB redistribution layer between strata in a 3D stack, a significant advantage in using functional wafers for different 3D configurations. Moreover, the fabrication requirements appear compatible with wafer-level packaging (WLP) technologies, providing a low barrier to manufacturing acceptance.

6.4 Wafer-Level 3D Design Opportunities

A key driver for wafer-level 3D technology is high-capacity memory with short access time [15, 21, 22]. With the redundancy and reprogrammability of cache memory, 100% memory cell yield across a die is not required; as a result, die yield issues inherent in wafer-level stacking are alleviated. In addition, memory technology dissipates relatively little power; therefore, power dissipation issues are relatively minor. Both the pioneering pseudo-wafer stacking of Irvine Sensors [17] and the initial prototype products introduced by Tezzaron [9] have been high-capacity memories, principally due to these two factors.

Another strong technology driver for wafer-level 3D is the promise of incorporating more active silicon within a clock cycle for leading-edge digital processors [2, 9, 18]. Wafer-level 3D for such high-speed, multi-core processors are significantly more complex, partially because conventional 2D IC design and physical implementation are less established and partially because the layer partitioning and

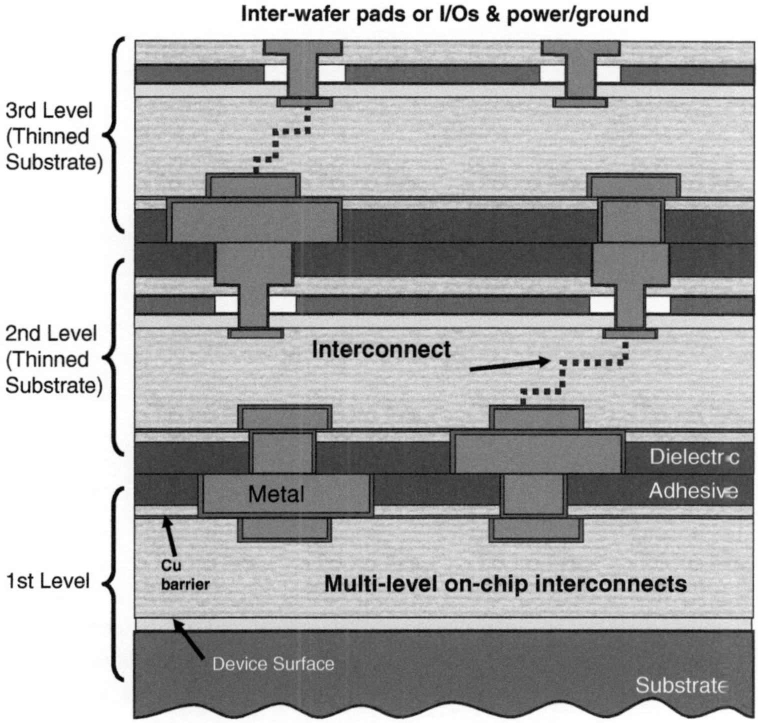

Fig. 6.9 Schematic of Cu/BCB redistribution layer bonding for via-first 3D integration (Cu–Cu bonds provide inter-wafer interconnects and BCB–BCB adhesive bonds provide mechanical wafer attachment) [11]

vertical interconnectivity are more difficult to design and fabricate. High-density interwafer interconnectivity is needed for design flexibility in such products. Die yield impacts can only be limited by keeping die size small and/or introducing redundancy.

Besides these two digital areas, complex analog-mixed signal designs, software-defined and software radios, and prime power delivery approaches are other electronic application areas for 3D integration [21–25]. Multi-technology design opportunities include micro-electrical mechanical systems (MEMS), sensors, imagers with pixel-by-pixel processing, optical and electro-optical applications, and the micro-nano interface (use of 3D to interface nanotechnology-based IC components with scaled microelectronics-based IC components).

Imagers with pixel-by-pixel processing is the first multi-technology design product that challenges technology capabilities of wafer-level 3D technology [26, 27]. A possible implementation is indicated in Fig. 6.10 [26], illustrating that signal processing electronics are contained in a 3D stack "behind" each pixel element. As the signal processing electronics shrink with each IC generation, either improved pixel

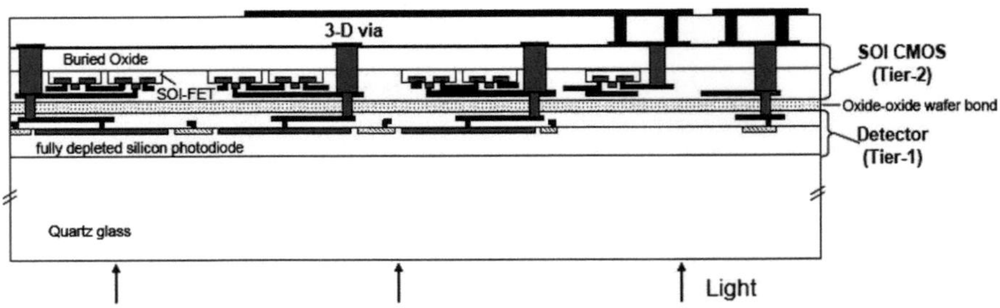

Fig. 6.10 Schematic of 3D stacked imager circuit with a Si photodetector wafer (Tier-1) and an inverted FDSOI CMOS wafer (Tier-2) [26]

resolution and/or additional signal processing can be introduced. Such imagers can be a wafer-level 3D technology driver.

6.5 Future Projections

While the timing of the future of wafer-level 3D integration is uncertain, the performance and integration advantages become compelling since scaled CMOS becomes increasingly expensive as the 16 nm technology node is reached. As recent FEOL device advances are incorporated in the 45nm and 32nm technology nodes, BEOL limitations will become more severe. New ULSI interconnect technologies described elsewhere in this book offer promising alternatives, but are not expected to be manufacturing worthy by the 32 nm node. It is therefore expected that another evolutionary transition will occur during the transition from the 45 to 32 nm node. Wafer-level 3D integration using scaled copper Damascene patterning for inter-wafer interconnects is the most promising breakthrough.

Memory stacks offer a near-term IC driver for 3D process development and establishment of 3D design tools. Redundancy is more easily incorporated in memory ICs, and die yields are higher than in microprocessors or ASICs. While smart imagers as described in Section 6.4 (see Fig. 6.10) are a promising application, a large-scale manufacturing capability will be most readily established for a proven large-scale market, such as multifunctional portable electronics, networks, and games. Memory stacks will become an important component in these and other technology-aggressive applications.

References

1. Lu, J.-Q.; Cale, T. S.; and Gutmann, R. J.: Wafer-level three-dimensional hyper-integration technology using dielectric adhesive wafer bonding. In Materials for Information Technology: Devices, Interconnects and Packaging. Zschech, E.; Whelan, C.; and Mikolajick, T. (eds.), Springer-Verlag, 386 (2005)
2. Meindl, J. D.; Venkatesan, R.; Davis, J. A.; Joyner, J. W.; Naeemi, A.; Zarkesh-Ha, P.; Bakir, M.; Mulé, T.; Kohl, P. A.; and K. P. Martin: Interconnecting Device Opportunities for

Gigascale Integration (GSI). IEEE 2001 International Electronic Devices Meeting (2001 IEDM), 525 (2001)
3. Das, S.; Fan, A.; Chen, K.-N.; Tan, C. S.; Checka, N.; and Reif, R.: Technology, performance, and computer-aided design of three-dimensional integrated circuits. In Proceedings of the International Symposium on Physical Design (ISPD'04), 108 (2004)
4. Walker, J.: Market transition to 3d integration and packaging: density, design, and decisions. In International Conference on 3D Architectures for Semiconductor Integration and Packaging, Tempe, AZ (2005)
5. Carson, F.: 3D SiP: established and emerging trends. In Preconference Symposium of the 3rd International Conference on 3D Architectures for Semiconductor Integration and Packaging, Burlingame, CA (2006)
6. Tessler, T.: Recent development in wafer level packaging. In 3rd IC Packaging Technology Exposition and Conference, Japan (2002)
7. Savastiouk, S., Siniaguine, O., and Korczynski, E.: 3D Wafer level packaging. In 2000 International Conference on High-Density Interconnect and Systems Packaging (2000)
8. Morrow, P.; Park, C.-M.; Ramanathan, S.; Kobrinsky, M. J.; and Harmes, M.: Three-dimensional wafer stacking Via Cu-Cu bonding integrated with 65-nm Strained-Si/Low-k CMOS technology. IEEE Electron Device Lett. **27(5)**, 335 (2006)
9. Patti, R.: Three-dimensional integrated circuits and the future of system-on-chip designs. Proc. IEEE **94(6)**, 1214 (2006)
10. Lu, J.-Q.; Jindal, A.; Kwon, Y.; McMahon, J. J.; Lee, K.-W.; Kraft, R. P.; Altemus, B.; Cheng, D.; Eisenbraun, E.; Cale, T. S.; and Gutmann, R. J.; 3D System-on-a-chip using dielectric glue bonding and Cu Damascene inter-wafer interconnects. In Thin Film Materials, Processes, and Reliability, Mathad, S. et al. (Eds.), ECS Proc. **PV 2003–13**, 381–389 (2003)
11. McMahon, J. J.; Lu, J.-Q.; and Gutmann, R. J.: Wafer bonding of damascene-patterned metal/adhesive redistribution layers for via-first three-dimensional (3D) interconnect. In IEEE 55th Electronic Components and Technology Conference (ECTC 2005), 331 (2005)
12. Guarini, K. W.; Topol, A. W.; Ieong, M. R.; Yu, Shi, L.; Newport, M. R.; Frank, D. J.; Singh, D. V.; Cohen, G. M.; Nitta, S. V.; Boyd, D. C.; O'Neil, P. A.; Tempest, S. L.; Pogge, H. B.; Purushothaman, S.; and Haensch, W. E.: Electrical integrity of state-of-the-art 0.13 μm SOI CMOS devices and circuits transferred for three-dimensional (3D) integrated circuit (IC) fabrication. In Digest of International Electron Device Meeting, 943 (2002)
13. Burns, J.; Aull, B.; Chen, C.; Chen, C.-L.; Keast, C.; Knecht, C. J.; Suntharalingam, V.; Warner, K.; Wyatt, W. P.; and Yost, D.-R.: A wafer-scale 3-D circuit integration technology. IEEE Trans. Electron Devices **53(10)**, 2507 (2006)
14. Ramm, P.; Bonfert, D.; Gieser, Haufe, H. J.; Iberl, F.; Klumpp, A.; Kux, A.; Wieland, R.: InterChip via technology for vertical system integration. 2001 IEEE International Interconnect Technology Conference (IITC), 160 (2001)
15. Lee, K. W.; Nakamura, T.; One, T.; Yamada, Y.; Mizukusa, T.; Hasimoto, H.; Park, K. T.; Kurino, H.; and Koyanagi, M.: Three-dimensional shared memory fabricated using wafer stacking technology. In IEEE 2000 International Electron Devices Meeting (2000 IEDM), 165 (2000)
16. Markunas, B.: 3D Architectures for semiconductor integration and packaging. In International Conference on 3D Architectures for Semiconductor Integration and Packaging, Burlingame, CA (2004)
17. Gann, K.: Neo-stacking technology. High Density Interconnect Magazine **2** (1999)
18. Goldstein, H.: Packages go vertical. IEEE Spectr. **38(8)**, 46 (2001)
19. Lowton, G.: The lowdown on high-rise chips. In Computer magazine, IEEE Comput. Soc., 24 (2004)
20. Souri, S. J. and Saraswat, K. C.: Interconnect performance modeling for 3D integrated circuits with multiple Si layers. IEEE International Interconnect Technology Conference (IITC 1999), 24 (1999)

21. Zeng, A. Y.; Lu,J.-Q.; Rose, K.; and Gutmann, R. J.: First-order performance prediction of cache memory with wafer-level 3D Integration. IEEE Design Test Comput. **22(6)**, 548 (2005)
22. Gutmann, R. J.; Zeng, A. Y.; Devarajan, S.; Lu J.-Q.; and Rose, K.: Wafer-level three-dimensional monolithic integration for intelligent wireless terminals. J. Semiconductor Technol. Sci. **4(3)**, 196–203 (2004)
23. Lu, J.-Q.; Devarajan, S.; Zeng, A. Y.; Rose, K.; and Gutmann, R. J.: Die-on-wafer and wafer level three-dimensional (3D) integration of heterogeneous IC technologies for RF-microwave-millimeter applications. In Materials, Integration and Packaging Issues for High-Frequency Devices II. Cho, Y. S.; Shiffler, D.; Ranall, C. A.; Tilmans, H. A. C.; and Tsurumi T. (eds.) MRS Proc. **833**, 229–234 (2004)
24. Lu, J.-Q.; Sun, J.; Giuliano D.; and Gutmann, R. J.: 3D architecture for power delivery to microprocessors and ASICs. In Proceedings CD of the 3rd International Conference on 3D Architectures for Semiconductor Integration and Packaging, Burlingame, CA (2006)
25. Sun, J.; Lu, J.-Q.; Giuliano, D.; Chow, T. P.; Gutmann, R. J.: 3D power delivery for microprocessors and high-performance ASICs. 22nd Annual IEEE Applied Power Electronics Conference and Exposition (APEC 2007), 127 (2007)
26. Suntharalingam, V.; Berger, R.; Burns, J.; Chen, C.; Keast, C.; Knecht, J.; Lambert, R.; Newcomb, K.; O'Mara, D.; Rathman, D.; Shaver, D.; Soares, A.; Stevenson, C.; Tyrell, B.; Warner, K.; Wheeler, B.; Yost, D.; and Young D.: Megapixel CMOS image sensor fabricated in three-dimensional integrated circuit technology. In Proceedings of the IEEE International Solid-State Circuits Conference, 356 (2005)
27. Temple, D.; Bower, C. A.; Malta, D.; Robinson, J. E.; Coffman, P. R.; Skokan M. R.; and Welch, T. B.: High density 3-D integration technology for massively parallel signal processing in advanced infrared focal plane array sensors. In Digest of 2006 IEEE International Electron Device Meeting (2006 IEDM), 143 (2006)

Part III
Interconnect Materials

This part is about interconnect materials. Chapter 7 presents about the materials for diffusion barriers for ULSI copper metallization. The main challenge is to achieve a significant reduction of the barrier thickness while maintaining high conductivity of this ultra-thin layer. In addition, the deposition process must be highly conformal in order to maintain uniform barrier properties throughout the high-aspect trenches and vias of the dual-Damascene process. The leading candidates for achieving this are atomic layer deposition of TaN, Ru, and WNC/Ru bi-layers. Self-assembled molecular layers and self-forming barriers are also possible deposition methodologies for reaching this goal. In addition, electroless deposition, in particular that of CoWP, is the leading candidate for use as a capping layer, although the industry is concerned with electrical shorts, which may be created by this metallic barrier. The motivation for the introduction of copper metallization and difficulties of implementing copper metallization have been discussed in one of the section of this chapter. Types of diffusion barriers and some of the deposition methods for diffusion barriers have also been discussed in detail. Evaluation of diffusion barriers for copper metallization is presented in detail. Chapter 8 gives the introductory idea about silicides used as contact materials; mainly TiS_2, $CoSi_2$, NiSi, SiGe into silicide/Si contacts have been discussed briefly. Some important properties of these materials are summarized. The formation process, advantages, and disadvantages of these silicides are separately mentioned. Additionally, several technologies to control crystalline and electrical properties of silicide contacts are also discussed. Chapter 9 gives the overview of electrical properties of the materials used in ULSI metallization. Large-grained Cu interconnects and ultrathin barrier layers are essential to realize low-resistance nano-scale Cu interconnects in the future ULSI devices. For development of a fabrication technique of large-grained Cu interconnects, grain growth mechanism of Cu thin films was understood. A new grain growth model for Cu thin films was proposed based on the strain energy criterion model at temperatures where dislocation glide was the dominant strain relaxation mechanism. The fabrication technique to self-form the thin barrier layers in Cu(Ti) alloy films by annealing at relatively low temperatures ($\sim$400°C) was demonstrated. The selection rule of the Cu alloy elements required for the barrier self-formation and low film resistivity was proposed based on understanding of the self-formation mechanism of the barrier layer. Chapter 10 reviews low-κ materials and proposes future development. Though low-κ materials have been actively investigated by many researchers, target values of ITRS (International Technology Roadmap for Semiconductor) have not been achieved. To reduce RC delay, both lower resistance wiring material, copper, and low-κ material are necessary. Especially for recent IC chips, ratio of BEP (back-end process) has rapidly increased with increasing wiring layers. Chapter 11 discussed about the electrical and mechanical characteristics of air-bridge copper interconnects. Multilevel air-bridge Cu interconnects with air gaps in the M1 and M2 metal line level were analyzed to evaluate their electrical and stress characteristics. As variations of the low-*k* dielectrics, three materials porous MSQ, SiCOH, and TEOS were considered. The combination of fully dense SiCOH dielectric in the M2 level and air-bridging down to the via 1 level was found to provide an effective dielectric constant of 2.27. This meets the Technology Roadmap requirement for the 45 nm technology node and is extendable to the 32 nm node. Simulation showed that the stress levels induced during processing in the air-bridge Cu lines and vias using SiCOH as a dielectric were intermediate between those with TEOS and p-MSQ. The effective modulus B was calculated to evaluate the effect of air-bridge structure on mechanical integrity. The result showed that SiCOH and p-MSQ provide a similar degree of confinement to Cu lines, compared with TEOS which provides more confinement due to its mechanical rigidity. Chapter 12 discussed about theatomic layer deposition (ALD) seed layers for plating and electroless plating. Metal ALD is very challenging due to limited precursor selection and the necessity of an organic reducing agent. Pd ALD is a significant improvement over the acid-bath solution route to Pd activation. Ultra-thin films 12–50 Å that are highly conformal can be deposited via $Pd^{II}(hfac)_2$ and hydrogen via a remote hydrogen plasma or glyoxylic acid. Improvement is certainly possible with these ultra-thin metallic layers deposited via ALD. Electroless copper was deposited with ease on these Pd PA-ALD layers. Electroless copper can either be deposited in "conformal" mode or "trench fill" mode depending on the experimental process parameters employed.

Chapter 7
Diffusion Barriers for Ultra-Large-Scale Integrated Copper Metallization

A. Kohn and M. Eizenberg

7.1 The Motivation for the Introduction of Copper Metallization

The microelectronics industry strives to continuously improve the speed and functionality of its integrated circuits. A significant contribution toward achieving this goal is the miniaturization of the semiconductor devices, in particular the reduction of the length of the gate of the metal-oxide-semiconductor (MOS) transistor. The typical size of this feature defines a term called "technology generation." The device miniaturization also requires to reduce the lateral dimensions of the conducting interconnects and via-contacts (termed "Vias").

The interconnect metallization is defined as the electrical conduction lines that connect the individual semiconductor devices. Figure 7.1 presents a schematic cross section of an integrated circuit, which shows that the metallic interconnects are mechanically supported and electrically insulated by an inter-level dielectric (ILD). Despite the advantages of Cu interconnects, various difficulties have hindered its introduction. Consequently, effective diffusion barriers are crucial in order to apply Cu interconnects.

Despite the miniaturization of the semiconductor devices, their number has increased and additional functional elements are added to the integrated circuits such as memory cells. Thus, the length of the metal interconnects increases as well as their density.

After the 0.25 μm generation, the speed limitation of the integrated circuit was no longer determined mainly by the device dimensions, but also by the metallic interconnects. The reason for this speed limitation is that the metallic interconnects and the ILDs form a large array of resistors and capacitors, which introduce a so-called *RC* delay time [2].

The possibilities for reducing this delay time are

- Reducing the resistivity, ρ, of the interconnect by replacing the previously used Al–Cu 4 wt% alloy (effective $\rho = 3.5$ μΩ cm) with Cu (effective $\rho =$ 1.8 μΩ cm).

A. Kohn (✉)
Department of Materials, University of Oxford, Parks Road, Oxford OX1 3PH United Kingdom
e-mail: amit.kohn@materials.ox.ac.uk

Y. Shacham-Diamand et al. (eds.), *Advanced Nanoscale ULSI Interconnects: Fundamentals and Applications*, DOI 10.1007/978-0-387-95868-2_7,

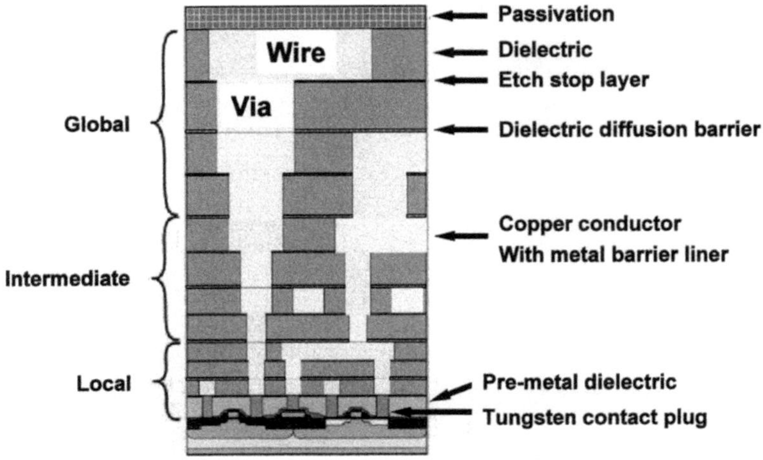

Fig. 7.1 Schematic cross section of an integrated circuit metallization (From [1])

- Reducing the dielectric constant (*K*) by replacing the SiO_2 interlayer dielectric (ILD, with $K = 4$) with a low-*K* one.
- Change of the geometric design of the interconnect layout.

The transition to Cu and a low-*K* dielectric in the interconnect scheme was mandatory from the 0.18 μm generation in order to increase the speed of the integrated circuits [3].

Additional motivation for replacing Al interconnects with Cu ones is the latter's improved resistance to electromigration [2]. Electromigration is an increasing problem with the miniaturization trend as the current density increases resulting in enough momentum transfer from electrons to atoms to induce atom movement. Copper is more resistant than Al to thermally activated diffusion processes of electromigration as well as stress-induced voiding because its melting temperature (1086°C) is significantly higher than that of Al (660°C) [2].

Functional evaluations have shown that the mean time to failure of Cu interconnect lines is at least 1–2 orders of magnitude longer at a temperature of 275°C and current density of 1.5×10^7 A/cm^2 [4].

Additional advantages of Cu for interconnect metallization are its higher thermal conductivity (Cu: 401 W m^{-1} K^{-1}, Al: 238 W m^{-1} K^{-1}), lower coefficient of thermal expansion (Cu: 16.5×10^{-6} K^{-1}, Al: 23.5×10^{-6} K^{-1}), and better mechanical properties (e.g., Young's modulus at 20°C – Cu: 129.8 GPa, Al: 70.6 GPa) [5]. Thus, Cu interconnects are more reliable than Al ones because they are not as sensitive to Joule heating and thermal stress.

7.2 Difficulties of Implementing Cu Metallization

Despite the advantages of Cu interconnects, their use in devices faces several substantial difficulties, which may be overcome by using an effective diffusion barrier and capping (or cladding) layer.

7.2.1 Copper Diffusion Through and Degradation of the Dielectric

Copper diffuses through ILDs mainly as Cu^+ ions, enhanced when an electric field is applied as a result of the fast mobility of these ions [6, 7]. Copper is ionized in the presence of humidity, impurity particles on the surface, and in the bulk of the dielectric [4]. The degree of the ionization is also dependent on the ambient [6]. Thus, ionization and subsequent Cu^+ diffusion is enhanced in an atmospheric or N_2 ambient compared to high vacuum.

The dielectric is degraded by Cu^+ ions due to the introduction of positive charge, creation of traps in the bulk, and creation of Cu-rich precipitates at the interface with Si. From the device perspective, failure occurs already when the positively charged Cu ions diffuse into the dielectric, degrading it by increasing the leakage current, and reducing the breakdown voltage. At later stages, Wendt et al. [8] showed that when a Si wafer is contaminated by Cu, the oxide fails due to precipitation of Cu-rich silicide particles at the SiO_2/Si interface.

The oxide is penetrated by these particles growing out of the substrate, opening possible conducting paths, or the breakdown strength is reduced by local thinning of the oxide. Thermal diffusion parameters of Cu in thermally grown SiO_2 have not been reported in the literature. Gupta reports on Cu diffusivity values for various amorphous dielectrics, such as 4%-phosphorus-silicate glass, hydrogenated SiN-16%H, and several polyimide thin films [9]. Typical diffusivity values at the temperatures of 200°C and 500°C are $\sim 5 \times 10^{-16}$ cm^2/s and $\sim 5 \times 10^{-14}$ cm^2/s, respectively. The values of Cu^+ion mobility in thermally grown SiO_2 vary between reports in the literature.

McBrayer et al. [6] measured an activation energy of 1.82 eV for Cu diffusion under a bias thermal stress (BTS) of 1 MV/cm and temperatures of 350–450°C in an ambient of H_2/N_2. Shacham-Diamand et al. [10] reported on Cu ions mobility in thermally oxidized SiO_2 under bias of 1 MV/cm in the temperature range of 250–350°C (the ambient was not reported). The measured activation energy was 0.92 ± 0.20 eV and mobility values of approximately 7×10^{-16} cm^2/V·s and 2×10^{-14} cm^2/V·s were reported at temperatures of 250°C and 350°C, respectively.

Raghavan et al. [11] detected diffusion of Cu ions in thermally oxidized SiO_2 already at 100°C as a result of applying a large bias of 2.5–6 MV/cm. The activation energy of electric field aided Cu diffusion in thermal oxide was determined to be 1.2 eV. By thermally annealing these MOS structures at 150°C and subsequently applying BTS, it was demonstrated that neutral copper atoms also have high diffusivity in thermally grown SiO_2 and can significantly diffuse through this dielectric even in the absence of an electric field.

Loke et al. [12] determined that the Cu^+ ion drift rate in plasma enhanced chemical vapor deposited SiO_2 is approximately $1.57 \times 10^{20} \exp(-1.13\ \mathrm{eV}/k_B T)$ ions/(cm^{2}s).

Willis et al. [13] show that at temperatures below 450°C and electric fields lower than 1 MV/cm or simultaneous bias conditions of 300°C and 0.5 MV/cm only oxidized copper is transported in the SiO_2dielectric with no diffusion of metallic copper. However, copper electrodes oxidized by ambient gases or pure oxygen show measurable copper ion diffusion in the temperature range of 200–300°C and bias fields of 0.1–1 MV/cm. The authors suggest a mechanism where oxidized copper is a source of copper ions. These ions are transported through the dielectric layer via diffusion and drift at an apparent activation energy of 1.8 eV. However, the model suggested reveals that unique activation energy cannot be assigned due to uncertainty about the chemical and physical processes at the Cu/SiO_2 interface.

Kohn et al. [14, 15] reported on an intermediate Cu-related degradation of SiO_2 in metal-oxide-semiconductor capacitors subjected to bias thermal stress (BTS). This degradation occurs between the commonly reported flat-band voltage shifts and the decrease in the breakdown voltage of the oxide. The commonly used stress of bias larger than 1 MV/cm at 300°C results in increased oxide conductivity, which is large enough to enable the leakage of minority carrier charge generated at the inversion layer, while the oxide capacitance and high-frequency deep depletion capacitance can still be measured accurately. This phenomenon termed as 'No Inversion Layer' (NIL) does not occur following BTS when the metallization is Al, or following thermal stress, even in Cu. This effect is also important when using metal-oxide-semiconductor capacitors for evaluating the quality of diffusion barriers, and will therefore be referred to later (Section 7.4).

7.2.2 Copper Adhesion to the Dielectric

Copper adhesion to SiO_2 is poor because the free energy for formation of a copper oxide is considerably smaller than that of silicon oxide [2]. As a result, Cu does not reduce the silicon oxide, namely the adhesion of Cu is only mechanical. Therefore, an adhesion layer between the Cu metallization and the ILD is required.

7.2.3 Copper Passivation

Copper oxides, Cu_2O and CuO, are readily formed at temperatures above 100°C on the exposed surfaces of the interconnect. Hu et al. [16] studied the oxidation rates in either dry or wet oxygen ambient. The chemical composition is Cu_2O at oxidation temperatures less than 400°C and becomes a mixture of Cu_2O and CuO at higher temperatures. The activation energies for Cu film oxidation were measured at 0.68 eV for dry oxidation and 0.43 eV for wet oxidation. These copper oxides do not form a protective layer as opposed to Al oxides. As a result, a passivation layer is required to protect the Cu interconnect from corrosion [4].

7.2.4 Copper Diffusion in and Reaction with Si

7.2.4.1 Diffusion

Copper is characterized by high solubility in Si in interstitial sites, e.g., the solubility at 500°C is $1.2 \times 10^{14} cm^{-3}$ [17]. Copper is the fastest diffusing transition metal in Si and diffuses through interstitial sites. The unique feature of the interstitial Cu is its positive charge state at both room and elevated temperatures [18] so that Cu diffusivity is dependent not only on temperature but also on the conductivity type and the doping level of the sample.

In the case of p-type Si, immobile acceptors, such as B^{-1}, reduce the intrinsic Cu diffusivity by trapping the Cu_i^+ donors [19, 20]. The intrinsic and effective Cu diffusion parameters in B-doped Si up to doping levels of 10^{17} cm^{-3} have been reported by Istratov et al. [21], which show that Cu can diffuse at room temperature through a 700 μm thick intrinsic Si wafer in under 3 h, and through a moderately B-doped Si wafer (10^{15} cm^{-3}) in about 15 h.

During Cu diffusion, a fraction of 10^{-4} to 10^{-3} of the Cu contamination becomes electrically active deep level states in the forbidden energy bandgap of Si [22]. These traps act as effective generation recombination centers for electron–hole pairs, causing the electrical characteristics of devices such as MOS capacitors, Schottky diodes, and p–n junctions to be significantly altered [18].

7.2.4.2 Precipitation

Precipitation of Cu depends on the type and concentration of the shallow dopants of the Si wafer. Copper precipitates are electrically amphoteric and change their charge state from positive to negative when the Fermi level crosses their electro-neutrality level at 0.2 eV below the conduction band [23]. The electrostatic repulsion between the positively charged nuclei of Cu precipitates and Cu_i^+ suppresses the precipitation of Cu in p-Si [24, 25]. In n-Si, the Cu precipitates are neutral or negatively charged, and there is either no significant electrostatic repulsion between them or there is an electrostatic attraction that enhances precipitation. These precipitates create localized electronic states [26], which impact on semiconductor devices, as discussed in Section 7.4.

7.2.4.3 Silicides

A solid state reaction between Cu and Si occurs already at 200°C in order to form the orthorhombic η"-Cu_3Si silicide, which is stable at room temperature [27]. The silicide formation consumes Si, increases the volume of the material, and may cause spiking through the space charge region resulting in the change of the electrical characteristics.

In addition, η"-Cu_3Si particles catalyze the room temperature oxidation of Si [27]. Hong et al. [28] also found that Cu_3Si is the dominant phase in the Cu–Si system. The growth of the silicide is a diffusion controlled process with an activation energy of 0.95 eV in the temperature range of 200–260°C, where Cu appears to be the dominant diffusing species.

To summarize Section 7.2.4: The relatively low concentration of point defects (a fraction of 10^{-3} of the total Cu concentration) is a result of a lack of stable traps for Cu in a perfect Si crystal. Even ionized acceptors only retard Cu diffusion to stable sinks by several hours at room temperature since their dissociation energy is low, e.g., 0.61 eV for CuB pairs [18]. Additionally, bulk precipitation is suppressed in p-Si. Therefore, Cu does not affect bulk lifetime in p-Si up to contamination levels of 10^{14} cm^{-3} to 10^{16} cm^{-3}, while in n-Si the bulk recombination activity is considerably more sensitive comparable to Fe.

Low Cu contamination in p-Si results in its diffusion out of the bulk to the wafer surfaces or precipitates in the near-surface region, without forming detectable electrically active defects in the bulk. However, this behavior is dangerous for the electrical properties of devices, which are fabricated in the near-surface layer, and contain areas favorable for Cu precipitation such as local tensile strain fields. The influence on Cu detection using electrical methods as a means to evaluate diffusion barriers is discussed in Section 7.4.

7.2.5 Processing

Reactive ion etching (RIE) is not applicable for Cu metallization since the etching by-product, Cu halides, are not volatile at room temperature [29]. The industry solution is the so-called dual-Damascene process in which the trenches of a metal layer and vias in the prior ILD are etched at once, followed by deposition of the metallization [4]. The wafers are exposed to lithographic patterning using a photoresist. Anisotropic dry etch then cuts through a surface hard mask such as SiN and the low-*K* dielectric. Etching is stopped by another SiN layer. The photoresist is removed, leaving behind either a trench or in the currently preferred approach, a via-contact. The lithography process is again applied allowing for the next layer, the via-contact or the trench to be defined. At this stage, the diffusion barrier is deposited, which lines the dual-Damascene structure. A Cu seed layer is next deposited using PVD in order to enable electrochemical deposition of the Cu interconnects. The excess metal is polished back and planarized using chemical mechanical polishing. Finally, a thin SiN barrier is deposited on top and the dual-Damascene structure is completed [30]. In order to reduce the *RC* delay time, the depth of the trench is not reduced at the same rate as its lateral width. Consequently, the electrochemically deposited Cu and in particular the deposition of the diffusion barrier must be highly conformal and ideally serve as the conductive seed layer for the electrochemical deposition.

7.3 Diffusion Barriers

The difficulties of implementing Cu metallization discussed in Section 7.2 mean that diffusion barriers and capping layers are mandatory in order to apply this metallization for ULSI devices. The diffusion barrier is a technological requirement to which numerous demands are presented. These requirements are not necessarily related and in some cases are even contradictory. Nicolet [31] listed the properties of an ideal diffusion barrier during the operational time of the device subjected to the working temperature, bias, and ambient: The transport and loss rate between Cu, the ILD, and Si substrate should be small. The barrier should be thermodynamically stable and provide good adhesion between the Cu and the ILD. The thickness of the barrier should be small and uniform (~6.5 nm for the 80 nm technology generation) [32] with high electrical and thermal conductivity in order to retain the improvement in resistivity reduction. The barrier also needs to resist mechanical and thermal stress. The necessity for a diffusion barrier is mainly during the processing. This stability requirement is quantified using a 'thermal budget,' for example, in the 0.18 μm generation, the barrier must function at approximately 400°C for up to an hour.

Former reviews of diffusion barriers tested for Cu metallization can be found in [33, 34]. In this overview, we present the main approaches for choosing the materials, structures, and the deposition process. These approaches are demonstrated by examples of diffusion barriers investigated for Cu metallization. The examples outline trends and comparisons and not absolute criteria for barrier integrity due to the variety of methods employed and thicknesses of the barriers.

7.3.1 Approaches and Examples

7.3.1.1 Passive Metallic Thin Films

The first approach is to use a passive, thermodynamically stable metal, which is immiscible with Cu, namely a high segregation factor. The choice of a metal is due to its low resistivity, which is important for reducing the overall resistivity, and improving electromigration reliability.

However, many metal thin films, which comply with these requirements, fail as diffusion barriers because of their microstructure. The microstructure of thin metallic films is usually polycrystalline, characterized by relatively small grains and high density of defects such as dislocations [35]. The volume fraction of grain boundaries in nanocrystalline thin films is considerable. For example, in a thin film in which the grain size is approximately 10 nm, the grain boundaries comprise an estimated 5–10 vol.% of the film. As a result, rapid Cu diffusion occurs through these defect paths causing barrier failure.

The failure time of the barrier is determined by the diffusion kinetics, which is the result of the bulk material properties (e.g., binding energy, crystal structure), the microstructure of the thin film, and the boundary conditions. This can be quantitatively determined by the interrelation of the following four characteristic lengths: the product of the segregation factor, s, and the grain boundary width, δ; the characteristic penetration depth of the diffuser in the volume, L; the characteristic penetration depth of the diffuser along the grain boundary, L_b; and the grain size, d.

The three main categories of diffusion kinetics are [36]

- Type-C kinetics – Volume diffusion is negligible, and diffusion takes place only along the grain boundaries, without any essential leakage to the volume.
- Type-B kinetics – Grain boundary diffusion takes place simultaneously with volume diffusion around the grain boundaries. The volume diffusion fields

of neighboring grain boundaries do not overlap each other. Individual grain boundaries are isolated, and the solutions are similar to the isolated grain boundary case.

- Type-A kinetics – The volume diffusion length is greater than the grain size so that the volume diffusion fields overlap extensively.

The main types of diffusion kinetics, including several sub-divisions, are presented in Fig. 7.2 along with the interrelation of the four criteria, which determine

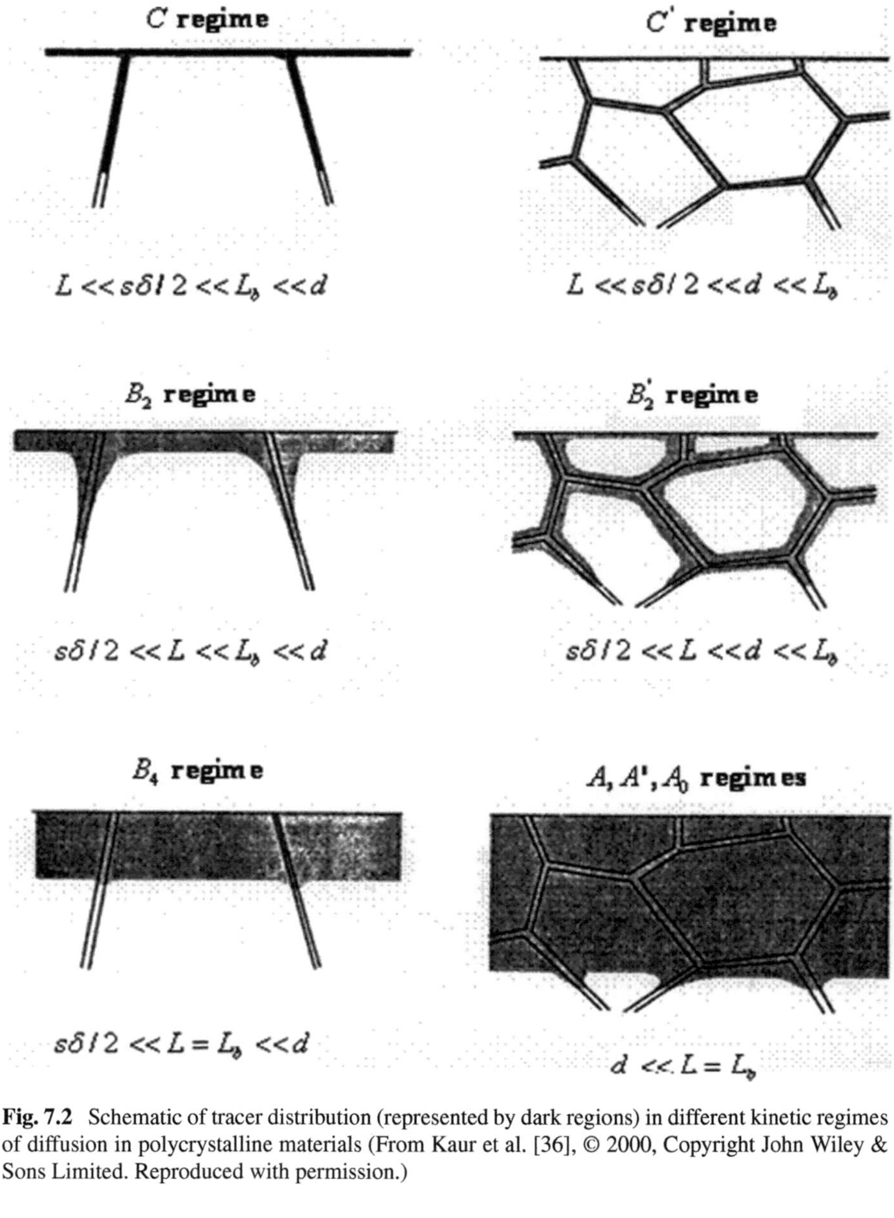

Fig. 7.2 Schematic of tracer distribution (represented by dark regions) in different kinetic regimes of diffusion in polycrystalline materials (From Kaur et al. [36], © 2000, Copyright John Wiley & Sons Limited. Reproduced with permission.)

the type of diffusion [36]. Since L, L_b, and d depend on the temperature and time, several diffusion kinetic models can exist during the functioning of the barrier. In general, both bulk and grain boundary diffusion, which determine L and L_t, respectively, are related to the melting temperature [36]. The relative diffusivity is reduced as the melting temperature of the material is increased. This can be qualitatively understood due to the increase of atomic binding energy, which retards the diffusion.

An example of this approach, the use of passive metallic thin films for Cu is the Cu/Co pair. These two metals are immiscible and do not react with each other [37]. However, O'Sullivan et al. [38] measured significant Cu diffusion into a 100 nm thick evaporated Co film after annealing at 400°C in forming gas for several hours. This significant Cu diffusion at a relatively low temperature is a result of grain boundary diffusion. O'Sullivan et al. [38] also reported significant Cu diffusion into a 50 nm thick evaporated Cr film, which is also a passive metal with respect to Cu. This rapid Cu diffusion was attributed to voids in the film.

Ono et al. [39] compared the barrier properties of transition metals against Cu diffusion. Most of these transition metals fit this category of passive metals with respect to Cu. The barrier integrity was ranked from Cr (worst) through Mo, Ta, to W (best). Ono et al. noted that this ranking coincides with the relative self diffusion coefficient of the barrier materials, which also coincides with their relative melting temperature. Consequently, polycrystalline pure Cr and Co thin films are ineffective barriers against Cu diffusion, while Ta and W may be effective barriers because of their high melting temperature, and despite their polycrystalline structure. Similar interdiffusion of passive metallic films coupled with Cu was reported for Pb, Bi, and Cr [31].

7.3.1.2 Thermodynamically Stable Barriers

In a thermodynamically stable diffusion barrier, the free energy required to form a compound with the Cu or dielectric layers is positive. Materials which are characterized by a large negative free energy of formation may be suitable, such as oxides, nitrides, borides, and silicides of transition metals.

The most widespread of these barriers are TaN and TiN deposited by PVD and chemical vapor deposition (CVD). The failure mechanism of these barriers is usually by grain boundaries diffusion. A recent survey of the numerous studies related to these barriers is presented in [34], which shows that the TaN-based barriers are currently the most widely used for Cu metallization. In this respect, the engineering approach is to solve one difficulty at a time. Therefore, a practical barrier may be comprised of several layers, each of which serves some function. This is the currently used method in industry for obtaining diffusion barriers for Cu metallization. The barrier is comprised of an hcp/fcc TaN film, which provides adhesion to the dielectric, followed by a bcc Ta (α-Ta) layer, which maximizes conduction because of its relatively low resistivity of 15–30 $\mu\Omega \cdot$ cm [40]. The tetragonal phase of Ta (β-Ta) is avoided as its resistivity is larger than 150 $\mu\Omega \cdot$ cm. For example, when Ta and TaN_x are deposited by PVD using a N_2/Ar gas mixture, the increase of N_2 flow enables the transition from β-Ta to the low resistivity α-Ta. Min et al. [41]

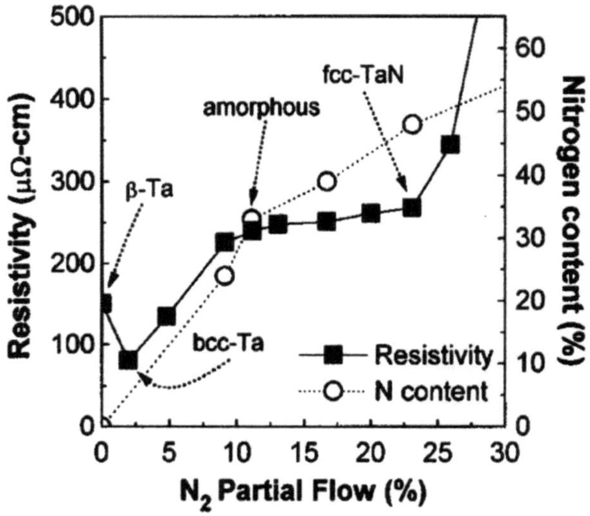

Fig. 7.3 The electrical resistivity and the nitrogen content of TaNx films deposited at various N_2/Ar gas ratios (Reused with permission from Min et al. [41], © 1996, American Institute of Physics.)

demonstrated this transition along with the subsequent formation of amorphous Ta_2N and crystalline fcc-TaN. Figure 7.3 demonstrates how the increase of N_2 flow determines these transitions along with the change in electrical resistivity.

7.3.1.3 "Stuffed" Barriers

Grain boundaries in nanocrystalline ultra-thin films, which are the reason for barrier failure in many cases, can be enriched by several atomic percent of an alloying or impurity element. These elements, such as oxygen, nitrogen, and carbon, may hinder copper diffusion in the grain boundaries. It should be noted that this concept, termed "stuffed" grain boundary, is still under debate and lacks an accepted quantitative model. This is because proving that an element, especially a light one, is present in the grain boundary of a nanometer-sized grain is experimentally challenging as is the distinction between an element and an intermetallic compound in the grain boundary. Consequently, the number of reported cases in this category is limited.

An example, which reports on this approach, is of CVD W that was exposed to nitrogen [42]. The barrier integrity significantly improved in comparison with pure W films. An XPS study of the WN_x film found that the measured W binding energy was characteristic of metallic bonding. This is an indirect proof that the N accumulated in the grain boundaries without the formation of a compound with W. An additional example was reported by Nam et al. [43] on a multi-layered diffusion barrier of TiN/Al/TiN for Cu metallization. In this case, Al was used as a stuffing material of TiN grain boundaries by forming Al_2O_3, which is possible due to oxygen saturated in the grain boundaries. Thus, the diffusion barrier performance was significantly improved compared to a TiN barrier for Cu metallization

An example of an electroless deposited Co-based diffusion barrier in which phosphorus and tungsten segregate to the grain boundaries was demonstrated by Kohn et al. [44–46]. Copper grain boundary diffusivity was measured to be two to three orders of magnitude smaller than in PVD pure Co films making this film an effective diffusion barrier above 450°C. The reduced diffusivity in the electroless films was a result of the significantly lower value of the pre-exponential diffusion factor. The reduction of this factor was explained by grain boundary 'stuffing' because the Cu concentration in the grain boundaries of electroless deposited Co-based films was found to be considerably smaller than in pure Co [47]. This barrier is currently a leading candidate for application as a capping layer because the electroless deposition process is self-aligning on the Cu interconnects [32].

TiN is used in many cases as the diffusion barrier for Al metallization. The good barrier properties are attributed to exposing the barrier to air thus enabling a 'stuffing' mechanism of oxygen at the grain boundaries [48]. A modification of the TiN microstructure and the grain boundaries composition, which improves the barrier integrity against Al diffusion, has been achieved by H_2/N_2 plasma treatment [49, 50]. However, Park et al. [51] demonstrated that stuffing the grain boundaries with oxygen is not effective for blocking Cu diffusion because its reaction with oxygen is thermodynamically less favorable. Attempts to retard Cu diffusion through these barriers with Si were attempted by exposing H_2/N_2 plasma-treated TiN films to silane [52, 53]. Improvement of the barrier properties was attributed to the formation of Si_3N_4.

In the case of e-beam deposited Ta films, the research of Clevenger et al. [54] does suggest that oxygen at the grain boundaries reduces Cu diffusion. Tantalum films were deposited in high-vacuum (HV) and ultra-high-vacuum (UHV), where in the latter case, in situoxygen dosing was introduced between the deposition of the Ta and Cu layers. The best diffusion barriers were HV deposited films without oxygen dosing (failure temperature, T_f ~600–630°C), followed by dosed HV (T_f ~570–630°C), with the UHV without oxygen dosing being the worse (T_f ~320–630°C). No significant microstructural differences between the films were observed. However, care should be exercised by concluding that this is indirect proof of grain boundary 'stuffing' because increasing the oxygen pressure during the UHV deposition resulted in a deterioration of the barrier properties.

7.3.1.4 Sacrificial Barriers

In the case of sacrificial barriers, the thin film does react with the metallization and the dielectric or Si. However, the kinetic mechanisms of these reactions are known. If these reactions are laterally uniform, it is possible to predict the expected lifetime. The thickness requirement for Cu metallization, namely less than 10 nm for the 90 nm generation, most probably render this approach impractical as the expected lifetime would be too short for such thin films.

Castoldi et al. [55] investigated this approach for 50 nm thick Ti layers in order to prevent Cu diffusion. The formation of a laterally non-uniform layer of $TiCu_4$ was measured, which is presumably preceded by the formation of Cu–Ti. The growth of the compound is kinetically controlled by means of a diffusion coefficient having activation energy of 1.7 eV and a 5×10^{-2} cm^2/s pre-exponential factor. The authors suggest that the formation of a titanium copper compound ensures a reliable and low resistance electrical contact especially at the vias. This 50 nm thick Ti layer in conjunction with a TiN barrier was found to be an efficient Cu diffusion barrier following a 20 min 500°C heat treatment.

7.3.1.5 Self-forming Barriers

Murarka et al. [2, 56] introduced the concept of 'self-forming' barriers by alloying copper with an element that has low solubility. Following thermal anneal, this element will segregate to the interface with the ILD or Si. If this element has a strong tendency to oxidize, such as Al and Mg, the formation of a thin passivating layer is expected. This approach was studied for Cu–Mg (supersaturated at 5–12 at.%) on SiO_2 by Frederick et al. [57] and was recently applied to Cu–Mn (7–9 at.% Mn) for which an amorphous 3–4 nm thick Mn layer formed an effective barrier between Cu and silicon oxide [58]. Following exposure to a 450°C heat treatment, no interdiffusion was detected between Cu and SiO_2 by X-ray energy dispersive spectroscopy.

In addition, Cu–Ti (0.02 wt.%) alloys, which form a TiO_x interface with SiO_2 have also shown promising results by maintaining low current leakage in MOS devices after exposure to a 700°C vacuum anneal [59]. However, alloying of Cu bears a fundamental problem, which is the increase of resistivity in the Cu alloy compared with pure Cu interconnects.

7.3.1.6 Single Crystalline Barriers

In a single crystalline thin film, the Cu diffusivity may be significantly lower as the fast diffusion paths (i.e., grain boundaries and defects) are absent. This approach is not practical due to technological and economical considerations. Few attempts have been made to investigate this structure as a barrier for Cu metallization.

Wang et al. [60] demonstrated the influence of structure on barrier properties by comparing single crystal NaCl-structured TaN and polycrystalline hexagonal ε-TaN thin films. By obtaining z-contrast images using scanning transmission electron microscopy, a planar diffusion front in the single crystal was observed, as opposed to rapid diffusion along grain boundaries in the polycrystalline film. Copper diffusivity in the single crystal was estimated to have an activation energy of 3.27 ± 0.10 eV, which is approximately twice as large as that for grain boundary, and results in low diffusivity values.

The influence of preferred orientation was demonstrated in the case of PVD Cu/W (110)/Si [61]. The W (110) barrier was more stable than a thin W film without preferred orientation, which was attributed to a decrease of fast grain boundary diffusion paths.

7.3.1.7 Amorphous Barriers

An alterative method to remove the fast grain boundaries diffusion paths is by using an amorphous structure. Differentiating between polycrystalline films with nanometer-sized grains and films with random long range order is not always conclusive. However, the substantial property of these films is the lack of defects the length of which is comparable to the film thickness. Even though Cu diffusivity is expected to be higher than in a single crystalline material, it can be significantly lower than grain boundary diffusion.

Attempts to deposit amorphous thin films, predominately by PVD, have mainly concentrated on the ternary systems of TM–Si–(N or O), where TM is a transition metal such as Ta, Ti, and W. These systems, which were initially proposed by Nicolet [62], demonstrate good barrier properties at relatively high temperatures due to their thermodynamic stability.

Recently, reactive sputter depositions of amorphous ternary W–Ge–N thin films, which require 800°C to crystallize, have also shown promising barrier properties [63]. Distinct interfaces with Cu and SiO_2 were maintained after a 600°C thermal treatment, as determined by compositional depth profile analysis performed by Auger electron spectrometry. Nicolet, who termed these highly metastable amorphous alloys as 'mictamict,' explained that the resulting structure is due to the dissimilar crystalline structure of the constituting quasi-binary compounds with one common species, usually the nitrides or oxides. In addition, since the elements are of differing atomic sizes, and possibly different bonding character, crystallization temperatures of up to 1000°C have been reported [64]. However, the highly stable amorphous structure results in the main drawback of these barriers, which is high electrical resistivity. Resistivity values are at best several hundred $\mu\Omega \cdot cm$ so that the improvement in interconnect conductivity obtained by the move to copper can be significantly reduced.

For less stable amorphous barriers, where the resistivity may be lower, the failure mechanism can be due to crystallization and grain boundary diffusion. An example of this is electroless deposited CoWP films, which upon deposition have a significant amorphous component. These films crystallize at approximately 300°C, as demonstrated by the sharp reduction in resistivity shown in Fig. 7.4 [45]. At this stage, the polycrystalline film functions as an effective 'stuffed' barrier, the details of which were discussed in Section 7.3.1.3.

7.3.1.8 Self-assembled Molecular Layers

A major challenge for the sub-65 nm generation is to obtain effective barriers while maintaining conformality in high aspect ratio features. This is because barrier thicknesses below 5 nm will be required. An approach to obtain monolayer thick diffusion barriers using self-assembled molecular (SAM) deposition has been demonstrated by Ramanath et al. [65, 66] in which one end of a bifunctional organosilane ligand is adsorbed on the ILD while the other end is anchored to the Cu interconnect.

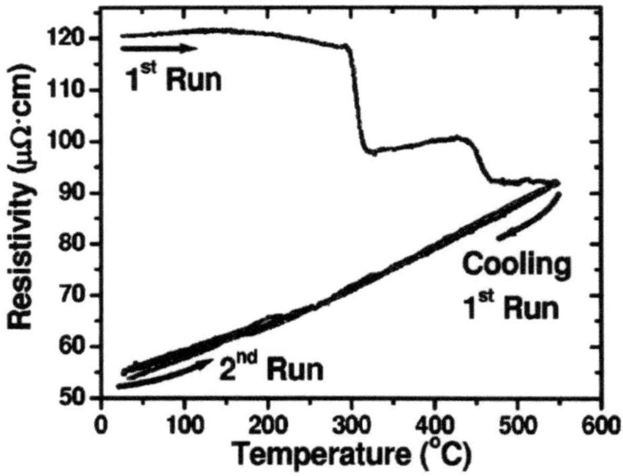

Fig. 7.4 In situ resistivity as a function of temperature of a $Co_{0.9}W_{0.02}P_{0.08}$ film. Heating at a constant rate of 2.8°C/min. After the specimen was cooled to room temperature, a second in situ resistivity measurement was performed, which demonstrates that the phase transition is irreversible (Reused with permission from Kohn et al. [45], © 2003, American Institute of Physics.)

SAMs can be assembled either by vapor or wet-chemical methods, and molecules as short as 2 nm were found to be effective diffusion barriers in device applications.

Such layers were evaluated as diffusion barriers for the Cu/SiO_2 interface by annealing MOS devices at 200°C in a 2 MV/cm electrical field. When 3-[2-(trimethoxysilyl)ethyl]pyridine was used, the leakage current was more than four orders of magnitude lower and the time to failure increased by a factor of 4 compared with Cu/SiO_2 reference samples. Krishnamoorthy et al. [65] proposed that steric hindrance caused by the terminal groups in the SAMs is responsible for the improved barrier properties.

This deposition method demonstrates good step coverage in high aspect ratio features because of the high probability of one end of the molecules attaching to the substrate, and the low tendency for multi-layer formation. In addition, the process can be tailored to improve the adhesion of the Cu layer to the dielectric [67]. However, these layers are not conductive, resulting in an increase of the via series resistance, which is an increasingly important parameter as device size shrinks.

7.3.2 Deposition Methods

As was just demonstrated in the case of SAM, the properties and type of diffusion barrier are determined in many cases by the deposition method itself. The choice of the deposition method becomes increasingly important because highly conformal, sub-10 nm thick barriers are required for future ULSI generations. In the following paragraph, the leading deposition methods are surveyed.

A leading method for depositing diffusion barriers is by physical vapor deposition, which enables a wide choice of materials with high purity. However, PVD methods do not result in conformal layers, even when attempting to improve the directionality of the depositing atoms by ionized metal plasma (IMP). Alternatively, chemical vapor deposition does enable conformal films, and the derivative of this method, atomic layer deposition (ALD) enables to deposit ultra-thin layers [68, 69]. In ALD, the films are grown in a cyclic process usually consisting of exposure of the first precursor, purge of the reaction chamber, exposure of the second precursor, and a further purge of the reaction chamber. The growth cycles are repeated as many times as required for the desired film thickness. The precursor molecules chemisorb or react with the surface groups up to saturation, so that no further adsorption is possible after the formation of that layer. Therefore, film growth is self-limiting enabling excellent control of film thickness at the monolayer level. The suitability of the ALD process has been studied for the sub-90 nm Cu/low-*K* metallization scheme [70], with the leading material candidates being Ta [71], and cubic TaN [72–74]. Recent work has demonstrated that good barrier properties can be obtained for even 0.6 nm thick films, in which the structure was nanocrystalline or amorphous [75]. Good barrier properties were maintained up to 700°C as determined by synchrotron X-ray diffraction analysis. The limitations of CVD methods are material choice, impurities, in particular oxygen and carbon, high deposition temperature, poor adhesion to the substrate, and roughness of the film (compared to ALD).

Tsai et al. [76] compared between the barrier properties of metal-organic CVD and PVD TaN films between Cu and Si. The barrier properties of PVD TaN ($Tf \sim 600°C$) were better than the CVD films ($Tf \sim 550°C$) as determined by the leakage current in shallow p–n junction diodes. This result was attributed to the (111) preferred orientation and 20 nm grain size of the PVD films compared with the (200) preferred orientation and 60 nm grain size of the CVD films. These results indicated that the PVD TaN film can act as a better diffusion barrier than the CVD TaN film.

Electroless deposition for ULSI metallization has the advantages of high conformality, low deposition temperature, and ease of integration with the currently used electrochemical deposition of the Cu metallization [77]. Electroless deposition is a process in which a catalytic substrate is introduced into a solution that is comprised of metal ion complexes and a reducing agent. This reducing agent enables the deposition of neutral metal atoms on the substrate, which forms a continuous film. The process is self-initiating and auto-catalytic and therefore is interesting for applying as a capping layer [38].

7.4 Evaluation of Diffusion Barriers for Cu Metallization

The transition from Al to Cu metallization also requires the adaptation of the methods used for evaluating the quality of diffusion barriers, because their detrimental effects on the dielectric and the Si semiconductor are different. Aluminum is considered a stable metallization for SiO_2 as opposed to Cu [4], although it should be noted that in polymeric ILDs, such as a hybrid organosiloxane polymer, Al ions diffuse more rapidly than Cu [78]. The interaction with the Si semiconductor, which is the major reliability issue, is considerably different. The failure of a barrier for Al metallization is primarily a result of a reaction with and accumulation of Al at the interface between Si and the diffusion barrier. This Al accumulation can affect the electrical contact properties, and additionally damage the devices through the formation of Al spikes at the interface, which may penetrate up to several micrometers deep into the Si. Diffusion of Si into the Al metallization also results in device failure. Aluminum diffusion in Si is many orders of magnitude lower than in Cu, e.g., less than 10^{-20} cm^2/s at 500°C [79]. Consequently, using a metallurgical parameter such as the formation of a copper silicide at the interface as an indication of barrier failure may be misleading, as it occurs only after the bulk of the wafer is saturated with Cu up to its equilibrium solubility. The electrical characteristics of the devices are in fact degraded at an earlier stage. Rha et al. [80] demonstrated that the failure temperature for the same diffusion barrier can vary by as much as 100–200°C depending on the evaluation method. However, the trends were similar, namely the conclusion regarding the relative ranking of different barriers against Cu diffusion was the same irrespective of the evaluation method. Choosing the most sensitive methods to detect Cu penetration is important in order to determine the correct failure mechanism. The electrical evaluation methods are the most sensitive in detecting Cu penetration. An exception to this statement is the four-point probe method, which is sensitive to compound formation such as Cu_3Si. In addition, within the same evaluation method, e.g., the MOS capacitor, different temperatures of failure can be determined depending on the chosen measured parameter.

The evaluation techniques can be broadly divided into the following groups:

- Analytical techniques based on depth profiles of surface analysis methods such as Auger electron spectroscopy, Rutherford backscattering spectroscopy, X-ray photoelectron spectroscopy, and secondary ion mass spectrometry. The detection limit for Cu in the first three techniques is in the range of 0.1–1.0 at.%. The detection limit of SIMS is considerably better, in the ppm range, approximately 10^{16}–10^{17} cm^{-3}.
- Microscopic-based techniques, namely transmission and scanning electron microscopy. These techniques usually detect phase formation such as silicides. SECCO etch-pitting test (aqueous solution of $K_2Cr_2O_7$ and HF) is used to highlight the formation of Cu silicides in the silicon [53].

An example, which compares the SIMS analytical technique with the etch-pitting microscopy method for evaluating a TiN diffusion barrier for Cu is presented in Fig. 7.5 from [53]. The penetration of Cu through the barrier is clearly observed in the SIMS depth profiles already following the 450°C treatment. However, the etch pit SECCO test indicated the failure of the barrier only after the 600°C treatment for 1 h.

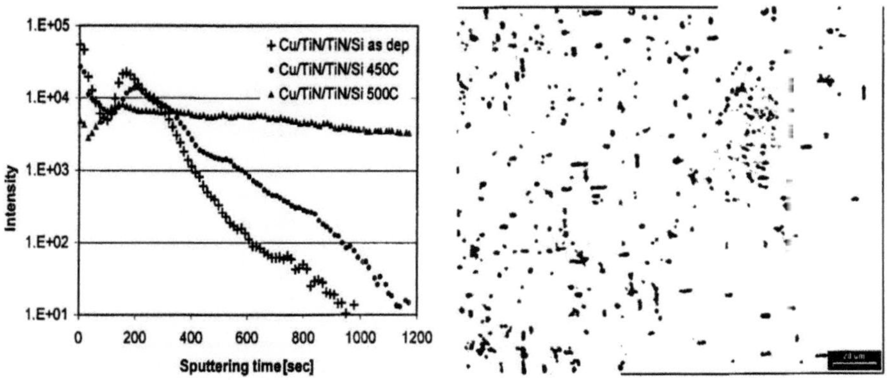

Fig. 7.5 SIMS depth profile of Cu in a 2 × 7.5 nm TiN sample treated by H_2/N_2 plasma after thermal treatments up to 500°C (*left*). Etch pits obtained from the Cu/TiN/TiN/Si sample after annealing at 600°C for 1 h and subjecting to the SECCO etch-pitting test (*right*) (Reused with permission from Joseph et al. [53], © 2002, American Institute of Physics.)

- Electrical evaluation, which is a functional method, offers the potential of the most sensitive techniques for detecting Cu. However, significant differences in the sensitivity of Cu detection exist between p- and n-type Si, as well as between the various devices, namely p–n junction, Schottky diode, and the MOS capacitor.
- In Section 7.2.4, it was demonstrated that Cu does not affect bulk lifetime in p-Si up to contamination levels of 10^{14} cm^{-3} to 10^{16} cm^{-3}, while in n-Si the bulk recombination activity is considerably more sensitive, comparable to Fe. However, low Cu contamination in p-Si results in its diffusion out of the bulk to the wafer surfaces or precipitates in the near-surface region. This behavior is dangerous for electrical properties of devices, which are fabricated in the near-surface layer, and contains area favorable for Cu precipitation such as local tensile strain fields.

The following electrical devices are commonly used for evaluation of diffusion barriers:

- Evaluation of leakage current of reverse-biased shallow p–n junctions (e.g., [81, 82]). The increase of leakage current following exposure to thermal stress is associated with Cu penetration. An example of this evaluation technique for Ta barriers along with a schematic of the test structure as demonstrated by Wang et al. [83] is presented in Fig. 7.6. Istratov et al. suggest that Cu does not usually form deep level traps at concentrations sufficient to affect the leakage currents of these junctions through the generation of minority carriers [18]. This method is not sensitive because leakage current increases only when Cu forms precipitates in the junction area, starting from the n-side where its precipitation is facilitated by the Fermi level affects discussed in Section 7.2. This suggestion agrees with

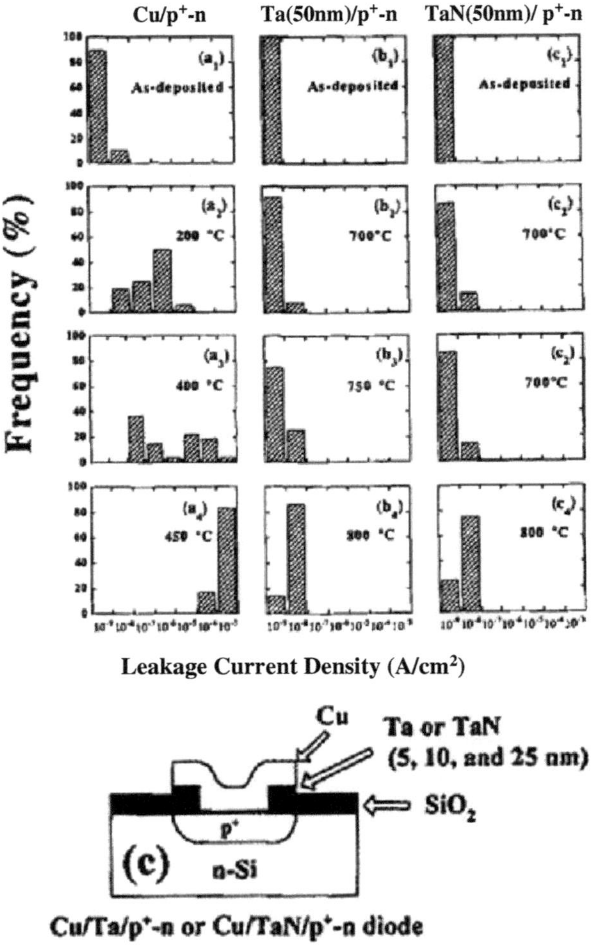

Fig. 7.6 *Top*: Histograms showing the distributions of reverse bias leakage current density for (**a**) Cu/Ta(25 nm)/p–n, (**b**) Cu/Ta (10 nm)/p–n, and (**c**) Cu/Ta(5 nm)/p–n junction diodes annealed at various temperatures. *Bottom*: Schematic cross sections of the Cu/barrier/p–n junction diode use to evaluate the barrier quality (From Wang et al. [33], © 1998, Reproduced by permission of ECS – The Electrochemical Society)

results in the literature [82]. However, no quantitative data are available as to what dissolved Cu concentration is required to affect the properties of p–n junctions.

- Measurements of current–voltage characteristics of Schottky diodes formed by a conductive, non-reacting diffusion barrier on p-Si. Following thermal anneal, the presence of Cu at the contact interface has a strong influence on the capacitance of a reverse-biased Schottky diode and its current–voltage characteristics. This method is highly sensitive to Cu penetration in the form of barrier height, leakage current, and ideality factor (e.g., [84]), an example of which is presented in Fig. 7.7 for TiN barriers.

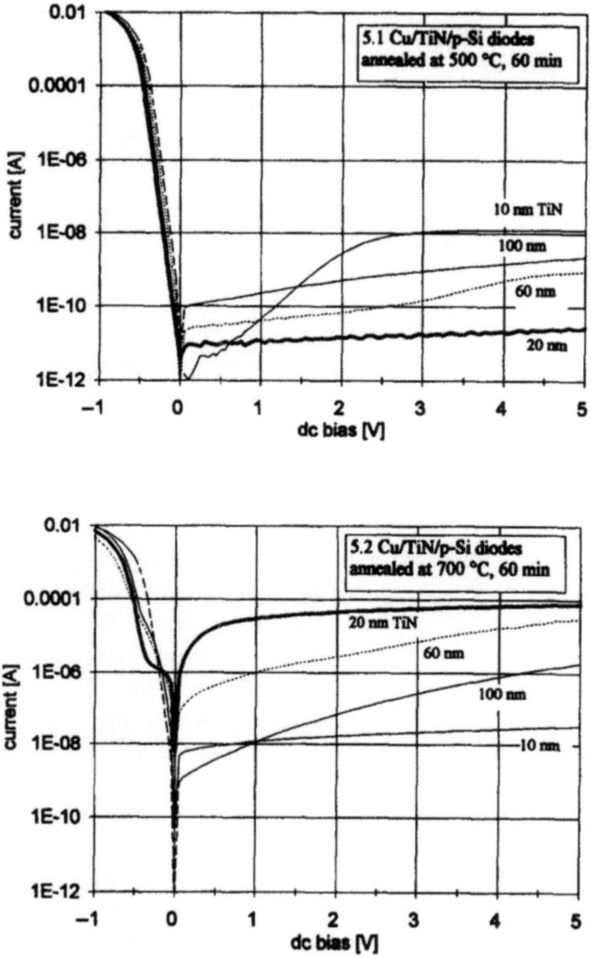

Fig. 7.7 Current–voltage plots of Cu/TiN/p–Si Schottky diodes for different TiN thicknesses after annealing at 500°C (*Top*) and 700°C (*Bottom*) (Reprinted from Ahrens et al. [84], with permission from Elsevier.)

- Measurements of leakage current, capacitance – voltage, and generation lifetime of metal-oxide-semiconductor capacitors subjected to thermal and bias stress. This method is the most functional evaluation in terms of the end use of the diffusion barrier between the metallization and the ILD. The metal-oxide-semiconductor (MOS) capacitor is a two-terminal device composed of a thin dielectric layer, usually SiO_2, sandwiched between a Si substrate and a metallic field plate.

For evaluation of diffusion barriers, the field plate materials are Cu and the tested diffusion barrier. The application of bias and thermal stress (BTS) is a means of

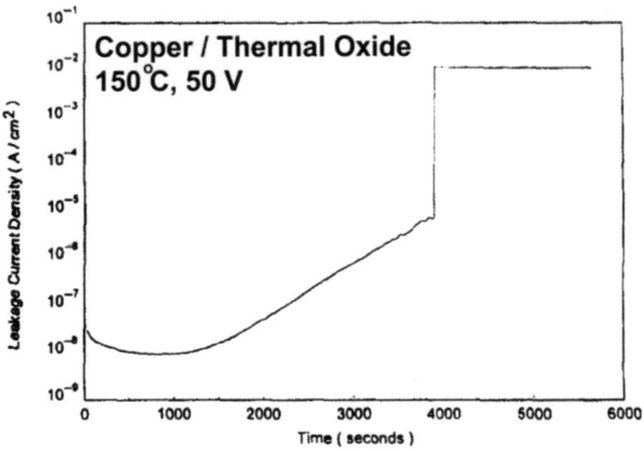

Fig. 7.8 Leakage current as a function of time for copper on thermal oxide: 150°C and 50 V (Reprinted from Raghavan et al. [11], with permission from Elsevier.)

increasing Cu ion mobility and inducing thermal diffusion, as well as degrading the insulating properties of the dielectric. Electric fields of 0.5–2 MV/cm applied at 150–300°C on oxide thicknesses of approximately 100 nm are typically chosen because these cause Cu-related degradation within minutes to several hours. Measuring the leakage current during the applied bias and thermal stress offers the advantage of in situtracking of the degradation caused by the Cu ions [11]. A typical example is shown in Fig. 7.8 from [11] of a Cu/SiO_2/Si MOS capacitor subjected to BTS conditions of 150°C and approximately 4.1 MV/cm. The current–time (*I*–*t*) curve has an initial decrease in the leakage current, attributed to the injection of copper ions into the dielectric, which sets up an opposing space charge. This is followed by a second region of low and constant current. Finally, in the third region, the applied potential across the "un-diffused" part of the oxide increases results in an increase of the leakage current, which leads to oxide breakdown. The current is then observed to saturate at the compliance limit of the measuring instrument, which can be referred to as the failure point.

Alternatively, shifts in the flat-band voltage of equilibrium capacitance–voltage characteristics can be used to track the overall change in concentration of Cu ions in the dielectric [10, 85].

An example of such a measurement by Loke et al. [12] of MOS capacitors with an oxide deposited by plasma enhanced CVD (PECVD) is presented in Fig. 7.9. An example of how this shift appears in a capacitance–voltage measurement is presented in Fig. 7.10.

However, it is difficult to evaluate how much Cu has penetrated into the dielectric, its location in the dielectric, and the minimum amount of Cu necessary to affect a 100 nm thick oxide in a MOS capacitor is expected to be higher than that for 5–10 nm thick oxides. In addition, it is not known how much Cu is required to significantly change the shape of the C–V curve [18]. However, if it is assumed that all the ions are concentrated at the ILD/Si interface, concentrations as low as

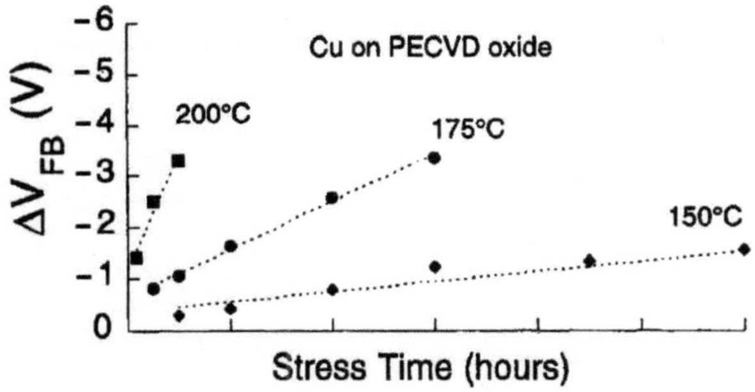

Fig. 7.9 The shift in the flat-band voltage of MOS capacitors with a PECVD oxide dielectric stressed at temperatures of 150, 175, and 200°C and 1 MV/cm (From Loke et al. [12], © 1996, IEEE. Reprinted with permission.)

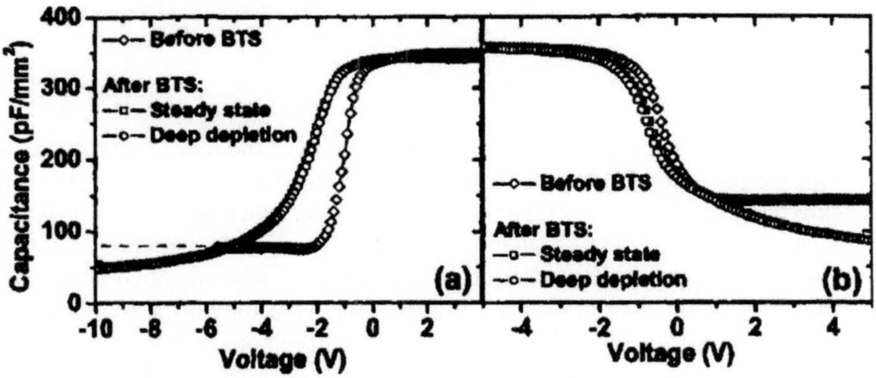

Fig. 7.10 Capacitance–voltage curves of a (**a**) n-Si and (**b**) p-Si MOS capacitor with a Cu metallization after initial anneal and bias thermal stress demonstrating the NIL phenomenon (Reused with permission from Kohn et al. [14], © 2004, American Institute of Physics.)

10^{10} cm^{-2} can be detected [86]. Distortions of the equilibrium C–V curve, which are caused by Cu-induced surface interface states in the semiconductor (Si), are also used to detect failure of the diffusion barrier [10]. In addition, Kohn et al. [14, 15] show that applying a large bias (1 MV/cm) in the presence of Cu degrades relatively thin dielectrics (~100 nm) by increasing the leakage current through it. Even though the oxide conductivity is large enough to enable the leakage of minority carrier charge generated at the inversion layer, the oxide capacitance and high-frequency deep depletion capacitance can still be accurately measured. This is an intermediate degradation, termed by the authors as the "No Inversion Layer" (NIL) effect, which occurs between the commonly reported flat-band voltage shifts and the decrease in the breakdown voltage of the oxide. This phenomenon does not occur following BTS when the metallization is Al, or following thermal stress, even in Cu. The NIL phenomenon is demonstrated in Fig. 7.10 for both p- and n-Si, where following BTS, the inversion characteristic of the capacitance–voltage curve disappears even

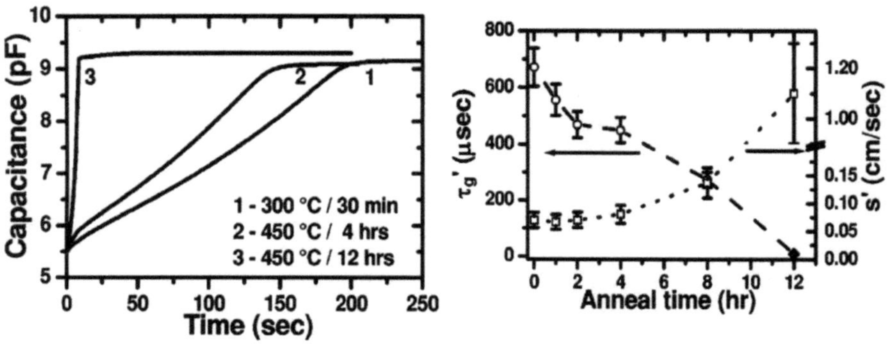

Fig. 7.11 C–t measurements of MOS capacitors with CoPW diffusion barriers after thermal anneals (*left*). Effective generation lifetime (τ') and effective surface generation velocity (s') derived from the C–t measurements of these MOS capacitors after thermal annealing (*right*) (Reused with permission from Kohn et al. [44], © 2003, American Institute of Physics.)

when measuring under equilibrium conditions. However, the remaining C–V curve can still be measured accurately. Note the shift in the flat-band voltage is to more negative voltages, regardless of the type of Si substrate, which is due to the diffusion of positively charged Cu ions into the dielectric.

The generation statistics of minority charge carriers can be measured by examining the transient behavior of the charge in the semiconductor when a single voltage step is applied to switch the MOS capacitor from accumulation to inversion [87]. The reduction of minority carrier lifetime is related to Cu-induced deep level state in the Si. Examples of this approach for evaluating Cu barriers are presented in [44, 88]. Figure 7.11 shows such a capacitance vs. time (C–t) measurement from [44], demonstrating the reduction in the minority carrier lifetime due to Cu penetration into the Si substrate.

It should be noted that generation statistics and interface states can also be characterized by quasi-static C–V measurements [87]. Finally, a method of detecting the type of diffusing ion and quantifying its concentration is by the triangular-voltage sweep method [89]. This is a quasi-static C–V measurement performed at a low sweep rate, and high temperature (150–300°C), which enables ion diffusion during the measurement. The additional measured current can be quantitatively related to the type of diffusing ion. For example, Cohen et al. [90] and Ganesan et al. [91] attributed the –1 MV/cm and –0.6 MV/cm peaks, respectively, to Cu ion diffusion following a 0.2 V/s sweep at 200°C. An example of such a measurement is presented in Fig. 7.12 from [92] for an MOS capacitor where Cu was deposited on a nominally 5 nm thick TiN barrier exposed to a 30 s exposure of H_2/N_2 plasma. The wide peak located at –4.0 ±0.2 V is attributed to Cu ions because this peak does not appear in the reference Al metallization MOS capacitors. The sharp peak at 0.40 ±0.05 V, which also appeared in the reference MOS capacitors, is attributed to high mobility ions such as an alkali ion contamination.

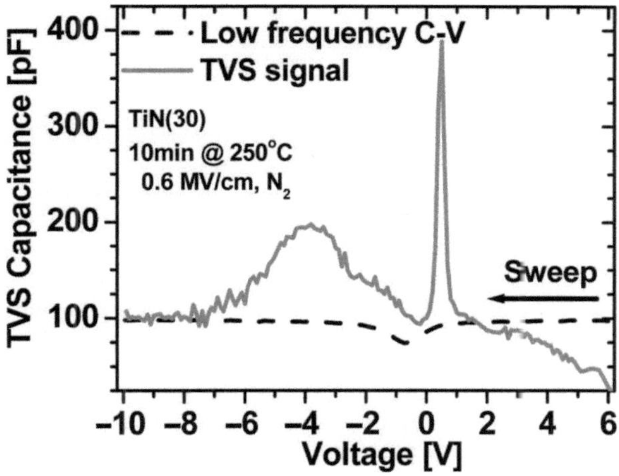

Fig. 7.12 TVS measurement of a capacitor with a TiN barrier (exposed to a 30 s H_2/N_2 plasma) after BTS at 250°C, 0.6 MV/cm. Measurement conducted at 250°C, sweep rate of 0.05 V/s (from [92])

7.5 Summary and Future Trends

The 2005 ITRS roadmap predicts that in the 45 nm technology generation, expected in 2010 (where the wiring pitch of the first metal layer is 90 nm), the thickness of the barrier or cladding layer should be 3.3 nm. At present, a manufacturable solution for this barrier thickness is not known [32].

Therefore, the main challenge is to achieve a significant reduction of the barrier thickness while maintaining high conductivity of this ultra-thin layer. In addition, the deposition process must be highly conformal in order to maintain uniform barrier properties throughout the high-aspect trenches and vias of the dual-Damascene process. The leading candidates for achieving this are atomic layer deposition of TaN, Ru, and WNC/Ru bi-layers [32]. Self-assembled molecular layers and self-forming barriers are also possible deposition methodologies for reaching this goal. In addition, electroless deposition, in particular that of CoWP, is the leading candidate for use as a capping layer, although the industry is concerned by electrical shorts, which may be created by this metallic barrier.

The various deposition methodologies must maintain a uniform morphology of the barrier, because it also forms the seed layer for the electrochemical deposition of Cu. A non-uniform seed layer may cause difficulties in Cu filling of the trenches and vias. Also, a controlled interface between the barrier and the Cu metal is important for reducing electron scattering, thus maintaining high electrical conductivity of the interconnects.

For both the ALD and the SAM deposition methodologies, the precursor materials may penetrate into the porous low-*K* dielectrics being developed for future technology generations. This can result in an increase of the dielectric constant and degradation of the ILD's mechanical properties. Thus, a modification of the deposition process may be required or adjustment of the ILD material.

References

1. SIA International Technology Roadmap for Semiconductors (Semiconductor Industry Association, San Jose, CA (2001)
2. Murarka, S. P.: Multilevel interconnections for ULSI and GSI era. Mater. Sci. Eng. **R19(3-4)**, 87 (1997)
3. Bohr, M. T.: Interconnect Scaling – The Real Limiter to High Performance ULSI 1995 International Electron Devices Meeting Technical Digest, 241 (1995)
4. Murarka, S. P.; Verner, I. V.; and Gutmann, R. J.: Copper – Fundamental Mechanisms for Microelectronic Applications. John Wiley and Sons, New York. (2000)
5. Smithells, C. J. (Ed.): Metals Reference Book. 5th edition, Butterworth, London & Boston (1976)
6. McBrayer, J. D.; Swanson, R. M.; and Sigmon, T. W.: Diffusion of metals in silicon dioxide. J. Electrochem. Soc. **133(6)**, 1242 (1986)
7. Atkinson, A.: Diffusion Phenomena in Thin Films and Microelectronic Materials. Gupta, D. and Ho, P. S. (Eds.) Noyes Publication, Berkshire (1988)
8. Wendt, H.; Cerva, H.; Lehmann, V.; and Pamler, W.: Impact of copper contamination on the quality of silicon oxides. J. Appl. Phys. **65(6)**, 2402 (1989)
9. Gupta, D.: Diffusion in several materials relevant to Cu interconnection technology. Mater. Chem. Phys. **41(3)**, 199 (1995)
10. Diamand, Y. S.; Dedhia, A.; Hoffstetter, D.; and Oldham, W. G.: Copper transport in thermal SiO_2. J. Electrochem. Soc. **140(8)**, 2427 (1993)
11. Raghavan, G.; Chiang, C.; Anders, P. B.; Tzeng, S.-M.; Villasol, R.; Bai, G.; Bohr, M.; and Fraser, D. B.: Diffusion of copper through dielectric films under bias temperature stress. Thin Solid Films **262(1–2)**, 168 (1995)
12. Loke, A. L. S.; Ryu, C.; Yue, C. P.; Cho, J. S. H.; and Wong, S. S.: Kinetics of copper drift in PECVD dielectrics. IEEE Electron Device Lett. **17**, 549 (1996)
13. Willis, B. G.; and Lang, D. V.: Oxidation mechanism of ionic transport of copper in SiO_2 dielectrics. Thin Solid Films **467(1–2)**, 284 (2004)
14. Kohn, A.; Lipp, E.; Eizenberg, M.; and Shacham, Y.: Copper-related degradation of SiO2 in metal–oxide–semiconductor capacitors subjected to bias thermal stress: Leakage of the minority charge carriers in the inversion layer. Appl. Phys. Lett. **85(4)**, 627 (2004)
15. Lipp, E.; Kohn, A.; and Eizenberg, M.: Lifetime-limited current in Cu-gate metal-oxide-semiconductor capacitors subjected to bias thermal stress. J. Appl. Phys. **99(3)**, 034504 (2006)
16. Hu, Y. Z.; Sharangpani, R.; and Tay, S. -P.: In situ rapid thermal oxidation and reduction of copper thin films and their applications in ultralarge scale integration. J. Electrochem. Soc. **148(12)**, G669 (2001)
17. Weber, E. R.: Properties of Silicon. Section 14.15 Solubility of Copper in Silicon. (INSPEC, the Institution of Electrical Engineers) (1988)
18. Istratov, A. A.; Flink, C.; and Weber, E. R.: Impact of the unique physical properties of copper in silicon on characterization of copper diffusion barriers. Phys. Stat. Sol. (b) **222**, 261 (2000)
19. Reiss, H. C.; Fuller, C. S.; and Morin, F. J.: Chemical interactions among defects in germanium and silicon. Bell Syst. Tech. J. **35**, 535 (1956)
20. Frank, F. C.; and Turnbull, D.: Mechanism of diffusion of copper in germanium. Phys. Rev. **104(3)**, 617 (1956)
21. Istratov, A. A.; Flink, C.; Hieslmair, H.; Weber, E. R.; and Heiser, T.: Intrinsic diffusion coefficient of interstitial copper in silicon. Phys. Rev. Lett. **81(6)**, 1243 (1998)
22. Istratov, A. A.; and Weber, E. R.: Electrical properties and recombination activity of copper, nickel and cobalt in silicon. Appl. Phys. A. **66**, 123 (1998)
23. Istratov, A. A.; Hedemann, H.; Seibt, M.; Vyvenko, O. F.; Schröter, W.; Heiser, T.; Flink, C.; Hieslmair, H.; and Weber, E. R.: Electrical and recombination properties of copper-silicide precipitates in silicon. J. Electrochem. Soc. **145(11)**, 3889 (1998)

24. Istratov, A. A.; Flink, C.; Hieslmair, H.; McHugo, S. A.; and Weber, E. R.: Diffusion, solubility and gettering of copper in silicon. Mater. Sci. Eng. B **72(2)**, 99 (2000)
25. Flink, C.; Feick, H.; McHugo, S. A.; Mohammed, A.; Seifert, W.; Hieslmair, H.; Heiser, T.; Istratov, A. A.; and Weber, E. R.: Out-diffusion and precipitation of copper in silicon: an electrostatic model. Phys. Rev. Lett. **85(23)**, 4900 (2000)
26. Broniatowski, A.: Multicarrier trapping by copper microprecipitates in silicon. Phys. Rev. Lett. **62(26)**, 3074 (1989)
27. Stolt, L.; Charai, A.; D'Heurle, F. M.; Fryer, P. M.; and Harper, J. M. E.: Formation of Cu_3Si and its catalytic effect on silicon oxidation at room temperature. J. Vac. Sci. Technol. A **9(3)**, 1501 (1991)
28. Hong, S. Q.; Comrie, C. M.; Russel, S. W.; and Mayer, J. W.: Phase formation in Cu-Si and Cu-Ge. J. Appl. Phys. **70(7)**, 3655 (1991)
29. Li, J.; Diamand, Y. S.; and Mayer, J. W.: Copper deposition and thermal stability issues in copper-based metallization for ULSI Technology. Mater. Sci. Rep. **9**, 1 (1992)
30. Stanley, W.: Silicon Processing for the VLSI Era, Volume **4**: Deep Submicron Process Technology, Lattice Press, Sunset Beach, CA (2002)
31. Nicolet, M.-A.: Diffusion barriers in thin films. Thin Solid Films **52(3)**, 415 (1978)
32. SIA International Technology Roadmap for Semiconductors, Semiconductor Industry Association, San Jose, CA (2005)
33. Wang, S. Q.: Barriers against copper diffusion into silicon and drift through silicon dioxide. MRS Bull. **19(8)**, 30 (1994)
34. Ganesan, P. G.; and Eizenberg, M.: Diffusion barriers for copper metallization. Internal report (2002)
35. Chopra, K. L.: Thin film phenomena, McGraw-Hill, New York (1969)
36. Kaur, I.; Mishin, Y.; and Gust, W.: Fundamentals of Grain and Interphase Boundary Diffusion, Wiley, Chichester, UK (2000)
37. Nishizawa, T.; and Ishida K.: The Co-Cu (Cobalt-Copper) system. Bull. Alloy Phase Diagrams **5**, 161 (1984)
38. O'Sullivan, E. J. A.; Schrott, G.; Paunovic, M.; Sambucetti, C. J.; Marino, J. R.; Baily, P. J.; Kaja, S.; and Semkow, K.W.: Electrolessly deposited diffusion barriers for microelectronics. IBM. J. Res. Develop. **42**, 607 (1998)
39. Ono, H. ; Nakano, T.; and Ohta, T.: Diffusion barrier effects of transition metals for Cu/M/Si multilayers (M=Cr, Ti, Nb, Mo, Ta, W). Appl. Phys. Lett. **64(12)**, 1511 (1994)
40. Edelstein, D.; Uzoh, C.; Cabral Jr.C.; DeHaven, P.; Buchwalter, P.; Simon, A.; Cooney III, E.; Malhotra, S.; Klaus, D.; Rathore, H.; Ararwala, B.; and Nguyen, D.: An optimal liner for copper damascene interconnects. Proc. Adv. Metal. Confer., 541 (2001)
41. Min, K. H.; Chun, K. C.; and Kim, K. B.: Comparative study of tantalum and tantalum nitrides (Ta_2N and TaN) as a diffusion barrier for Cu metallization. J. Vac. Sci. Tech. B **14(5)**, 3263 (1996)
42. Chang, K.-M.; Yeh, T.-H.; Deng, I.-C.; and Shih, C.-W.: Amorphous like chemical vapor deposited tungsten diffusion barrier for copper metallization and effects of nitrogen addition. J. Appl. Phys. **82(3)**, 1469 (1997)
43. Nam, K. T.; Datta, A.; Kim, S.-H.; and Kim, K.-B.: Improved diffusion barrier by stuffing the grain boundaries of TiN with a thin Al interlayer for Cu metallization. Appl. Phys. Lett. **79(16)**, 2549 (2001)
44. Kohn, A.; Eizenberg, M.; and Diamand, Y. S.: Copper grain boundary diffusion in electroless deposited cobalt based films and its influence on diffusion barrier integrity for copper metallization. J. Appl. Phys. **94(5)**, 3015 (2003)
45. Kohn, A.; Eizenberg, M.; and Diamand, Y. S.: Structure of electroless deposited $Co_{0.9}W_{0.02}P_{0.08}$ thin films and their evolution with thermal annealing. J. Appl. Phys. **94(6)**, 3810 (2003).
46. Kohn, A.; Eizenberg, M.; Diamand, Y. S.; and Sverdlov, Y.: Characterization of electroless deposited Co (W, P) thin films for encapsulation of copper metallization. Mat. Sci. Eng. A **302(1)**, 18 (2001)

47. Kohn, A.; Eizenberg, M.; and Y Diamand, Y. S.: Improved diffusion barriers for copper metallization obtained by passivation of grain boundaries in electroless deposited cobalt-based films. J. Appl. Phys. **92(9)**, 5508 (2002)
48. Sinke, W.; Frijlink, G. P.A.; and Saris, F. W.: Oxygen in titanium nitride diffusion barriers. Appl. Phys. Lett. **47(5)**, 471 (1985)
49. Danek, M.; Liao, Tseng, M. J.; Littau, K.; Saigal, D.; Zhang, H.; Mosely, R.; and Eizenberg, M.: Resistivity reduction and chemical stabilization of organometallic chemical vapor deposited titanium nitride by nitrogen rf plasma. Appl. Phys. Lett. **68(7)**, 1015 (1996)
50. Kröger, R.; Eizenberg, M.; Marcadal, C.; and Chen, L.: Plasma induced microstructural, compositional, and resistivity changes in ultrathin chemical vapor deposited titanium nitride films. J. Appl. Phys. **91(8)**, 5149 (2002)
51. Park, K. C.; and Kim, K. B.: Effect of annealing of titanium nitride on the diffusion barrier property in Cu metallization. J. Electrochem. Soc. **142(9)**, 3109 (1995)
52. Marcadal, C.; Eizenberg, M.; Yoon, A.; and Chen, L.: Metallorganic chemical vapor deposited TiN barrier enhancement with SiH_4 treatment. J. Electrochem. Soc. **149(1)**, C, 52 (2002)
53. Joseph, S.; Eizenberg, M.; Marcadal, C.; and Chen, L.: TiSiN films produced by chemical vapor deposition as diffusion barriers for Cu metallization. J. Vac. Sci. Technol. B **20**, 1471 (2002)
54. Clevenger, L. A.; Bojarczuk, N. A.; Holloway, K.; Harper, J. M. E.; Cabral, C. Jr.; Schad, R. G.; Cardone, F.; and Stolt, L.: Comparison of high vacuum and ultra-high-vacuum tantalum diffusion barrier performance against copper penetration. J. Appl. Phys. **73(1)**, 300 (1993)
55. Castoldi, L.; Visalli, G.; Morin, S.; Ferrari, P.; Alberici, S.; Ottaviani, G.; Corni, F.; Tonini, R.; Nobili, C.; and Bersani, M.: Copper–titanium thin film interaction. Microelec. Eng. **76**, 153 (2004)
56. Ding, P. J.; Lanford, W. A.; Hymes, S.; and Murarka, S. P.: Oxidation resistant high conductivity copper films. Appl. Phys. Lett. **64(21)**, 2897 (1994)
57. Frederick, M. J.; Goswami, R.; and Ramanath, G.: Sequence of Mg segregation, grain growth, and interfacial MgO formation in Cu–Mg alloy films on SiO_2 during vacuum annealing. J. Appl. Phys. **93(10)**, 5966 (2003)
58. Koike, J.; and Wada, M.: Self-forming diffusion barrier layer in Cu–Mn alloy metallization. Appl. Phys. Lett. **87(4)**, 041911 (2005)
59. Liu, C. J.; and Chen, J. S.: Low leakage current Cu(Ti)/SiO_2 interconnection scheme with a self-formed TiOx diffusion barrier. Appl. Phys. Lett. **80**, 2678 (2002)
60. Wang, H.; Tiwari, A.; Zhang, X.; Kvit, A.; and Narayan, J.: Copper diffusion characteristics in single-crystal and polycrystalline TaN. Appl. Phys. Lett. **81(8)**, 1453 (2002)
61. Takeyama, M.; Noya, A.; and Fukuda, T.: Thermal stability of Cu/W/Si contact systems using layers of Cu(111) and W(110) preferred orientations. J. Vac. Sci. Tech. A **15(2)**, 415 (1997)
62. Kolawa, E.; Chen, J. S.; Reid, J. S.; Pokela, P. J.; and Nicolet, M.-A.: Tantalum-based diffusion barriers in Si/Cu VLSI metallizations. J. Appl. Phys. **70(3)**, 1369 (1991)
63. Rawal, S.; Norton, D. P.; Anderson, T. J.; and McElwee-White, L.: Properties of W–Ge–N as a diffusion barrier material for Cu. Appl. Phys. Lett. **87(11)**, 111902 (2005)
64. Nicolet, M.-A.; and Giauque, P. H.: Highly metastable amorphous or near-amorphous ternary films (mictamict alloys). Microelec. Eng. **55(1–4)**, 357 (2001)
65. Krishnamoorthy, A.; Chanda, K.; Murarka, S. P.; Ramanath, G.; and Ryan, J. G.: Self-assembled near-zero-thickness molecular layers as diffusion barriers for Cu metallization. Appl. Phys. Lett. **78(17)**, 2467 (2001)
66. Ramanath, G.; Cui, G.; Ganesan, P. G.; Guo, X.; Ellis, A. V.; Stukowski, M.; Vijayamohanan, K.; and Doppelt, P.: Self-assembled subnanolayers as interfacial adhesion enhancers and diffusion barriers for integrated circuits. Appl. Phys. Lett. **83(2)**, 383 (2003)
67. Ganesan, P. G.; Cui, G.; Vijayamohanan, K.; Lane, M.; and Ramanath,G.: Effects of amine- and pyridine-terminated molecular nanolayers on adhesion at Cu-SiO_2 interfaces. J. Vac. Sci. Technol. B **23(1)**, 327 (2005)

68. Ritala, M.; and Leskel, M.: Handbook of Thin Film Materials. Deposition and Processing of Thin Films, Vol. 1, Nalwa, H. S. (Ed.), Academic Press, 103 (2002)
69. Leskel, M.; and Ritala, M.: Atomic layer deposition chemistry: Recent. developments and future challenges. Angew. Chem. Int. Ed. **42**, 5548 (2003)
70. Bayer, G.; Satta, A.; Schuhmacher, J.; Maex, K.; Besling, W.; Kilpela, O.; Sprey, H.; and Tempel, G.: Development of sub-10-nm atomic layer deposition barriers for Cu/low-k interconnects. Microelec. Eng. **64**, 233 (2002)
71. Kim, H.; Cabral, C.; Lavoie, C.; and Rossnagel, S. M.: Diffusion barrier properties of transition metal thin films grown by plasma-enhanced atomic-layer deposition. J. Vac. Sci. Technol. B **20(4)**, 1321 (2002)
72. Kim, H.; Kelloch, A. J.; and Rossnagel, S. M.: Growth of cubic-TaN thin films by plasma-enhanced atomic layer deposition. J. Appl. Phys. **92(12)**, 7080 (2002)
73. Kim, H.; Lavoie, C.; Copel, M.; Narayanan, V.; Park, D.-G.; and Rossnagel, S. M.: The physical properties of cubic plasma-enhanced atomic layer deposition TaN films. J. Appl. Phys. **95(10)**, 5848 (2004)
74. Wu, Y. Y.; Kohn, A.; and Eizenberg, M.: Structures of ultra-thin atomic-layer-deposited TaN_x films. J. Appl. Phys. **95(11)**, 6167 (2004)
75. Kim, H.; Detavenier, C.; van der Straten, O.; Rossnagel, S. M.; Kellock, A. J.; and Park, D. G.: Robust TaN_x diffusion barrier for Cu-interconnect technology with subnanometer thickness by metal-organic plasma-enhanced atomic layer deposition. J. Appl. Phys. **98(1)**, 014308 (2005)
76. Tsai, M. H.; Sun, S. C.; Tsai, C. E.; Chuang, S. H.; and Chiu, H. T.: Comparison of the diffusion barrier properties of chemical-vapor-deposited TaN and sputtered TaN between Cu and Si. J. Appl. Phys. **79(9)**, 6932 (1996)
77. Diamand, Y. S.; Dubin V.; and Angyal, M.: Electroless copper deposition for ULSI. Thin Sold Films **262(10)**, 93 (1995)
78. Mallikarjunan, A.; Murarka, S. P.; and Lu, T.-M.: Metal drift behavior in low dielectric constant organosiloxane polymer. Appl. Phys. Lett. **79(12)**, 1855 (2001)
79. De Cogan, D.; Haddara, Y. M.; and Jones, K.: Properties of Crystalline Silicon. Hull, R. (Ed.), Inspec, London (1999)
80. Rha, S. K.; Lee, W. J.; Lee, S. Y.; Hwang, Y. S.; Lee, Y. J.; Kim, D. I.; Kim, D. W.; Chun, S. S.; and Park, C. O.: Improved TiN film as a diffusion barrier between copper and silicon. Thin Solid Films **320(1)**, 134 (1998)
81. Reid, J. S.; Sun, X.; Kolawa, E.; and Nicolet, M.-A.: Ti-Si-N diffusion barriers between silicon and copper. IEEE Electron Device Lett. **12(8)**, 298 (1994)
82. Baumann, J.; Kaufmann, C.; Rennau, M.; Werner, T.; and Gessner, T.: Investigation of copper metallization induced failure of diode structures with and without a barrier layer. Microelect. Eng. **33(1–4)**, 283 (1997)
83. Wang, M. T.; Lin, Y. C.; and Chen, M. C.: Barrier properties of very thin Ta and TaN layers against copper diffusion. J. Electrochem. Soc. **145(7)**, 2538 (1998)
84. Ahrens, C.; Ferretti, R.; Friese, G.; and Weidner, J. O.: Thermal stress effects on capacitance and current characteristics of Cu/Si and Cu/TiN/Si Schottky-diodes. Microelect. Eng. **37/38**, 211 (1997)
85. Angyal, M. S.; Diamand, Y. S.; Ried, J. S.; and Nicolet, M.-A.: Performance of tantalum-silicon-nitride diffusion barriers between copper and silicon dioxide. Appl. Phys. Lett. **67(15)**, 2152 (1995)
86. Barbottin G.; and Vapaille, A.: Instabilities in Silicon Devices. North-Holland, Amsterdam (1986)
87. Schröder, D. K.: Advanced MOS devices. Addison-Wesley Publishing Company, Boston, MA (1987).
88. Kohn, A.; Eizenberg, M.; Diamand, Y. S.; Israel, B.; and Sverdlov, Y.: Evaluation of electroless deposited Co(W, P) thin films as diffusion barriers for copper metallization. Microelec. Eng. **55(1–4)**, 297 (2001)

89. Kuhn, M.; and Silversmith, D. J.: Ionic contamination and transport of mobile ions in MOS structures. J. Electrochem. Soc. **118(6)**, 966 (1971)
90. Cohen, S. A.; Liu, J.; Gignac, L.; Ivers, T.; Armbrust, D.; Rodbell, K. P.; and Gates, S. M.: Proc. Adv. Interconnects and Contacts Confer. Edelstein, C. (Ed.): Materials Research Society, Warrendale Pa, **564**, 551 (1999)
91. Ganesan, P. G.; Gamba, J.; Ellis, A.; Kane, R. S.; and Ramanath, G.: Polyelectrolyte nanolayers as diffusion barriers for Cu metallization. Appl. Phys. Lett. **83(16)**, 3302 (2003)
92. Lipp, E.: M.Sc. Thesis, Technion – Israel Institute of Technology, Haifa, Israel (2004)

Chapter 8
Silicides

Osamu Nakatsuka and Shigeaki Zaima

8.1 Introduction

Silicides are intermetallic compounds with Si and many metals form stable silicides. The interfacial reactions between various metals and Si have been investigated for many years and some silicides have been developed for contact and gate materials in Si ultra-large-scale-integrated (ULSI) applications [1–7].

A self-aligned silicide (salicide) process is one of the most important technologies for the shrinkage of metal-oxide-semiconductor field effect transistors (MOSFETs). Figure 8.1 shows a schematic diagram of the salicide formation in MOSFETs.

Silicide is formed by the deposition of a metal thin film on a Si substrate and the post-deposition annealing. When this formation process is carried out on the patterned structure, silicide is formed with the reaction between metal and Si only on poly-silicon gates and source/drain (S/D) contacts where metal directly contacts with Si. Metal on insulators such as SiO_2 or Si_3N_4 does not react or hardly reacts. Unreacted metals on these insulator areas can be selectively removed by wet chemical solutions after the silicidation, and self-aligned silicide remains only in the gate and contact regions. This salicide process prevents the shortening of the gate to the S/D electrodes without any lithography and patterning processes and minimizes the overlaps of these electrodes.

The application of metal silicide for MOSFETs provides the low series resistance of the poly-silicon gate and the reduction of contact resistances in the S/D regions. The requirements for silicides in ULSI applications are as follows: low resistivity for low sheet resistance, low contact resistivity with Si for reduction of the parasitic resistance, high thermal robustness during front- and back-end processes, and low reactivity to prevent the undesired reaction between interconnect materials and Si. According to the International Technology Roadmap for Semiconductors

O. Nakatsuka (✉)
Department of Crystalline Materials Science, Graduate School of Engineering, Nagoya University, Furo-cho, Chikusa-ku, Nagoya 464-8603, Japan
e-mail: nakatuka@alice.xtal.nagoya-u.ac.jp

Y. Shacham-Diamand et al. (eds.), *Advanced Nanoscale ULSI Interconnects: Fundamentals and Applications*, DOI 10.1007/978-0-387-95868-2_8,

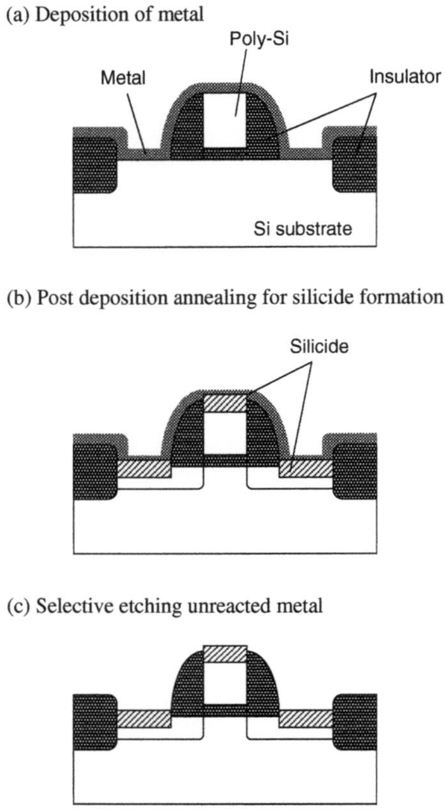

Fig. 8.1 The schematic diagram of the salicide formation in the MOSFET

as shown in Fig. 8.2 [8], it is required to develop ultra-thin silicide layers below 20 nm on ultra-shallow S/D junctions and ultra-low contact resistivity as low as $10^{-8}\ \Omega\ cm^2$.

The contact resistivity of an ohmic metal/semiconductor contact is essentially dependent on the Schottky barrier height (SBH) at the metal/Si interface and the impurity concentration in semiconductors. The theoretical estimation of contact resistivity has been given in some reports [9, 10]. Figure 8.3 shows the contact resistivities of $CoSi_2$/Si contacts as a function of the impurity concentration [11]. Contact resistivities numerically calculated for SBHs of 0.3, 0.4, and 0.5 eV are also shown in Fig. 8.3. As shown by this figure, lower SBH and higher impurity concentration are required to reduce the contact resistivity. However, there are some limitation factors, such as the energy bandgap, the electron affinity, and the solid solubility of impurity atoms in Si, to achieve much lower contact resistivity. In particular, it is difficult to minimize the contact resistivities for both n^+- and p^+-Si simultaneously by using the same contact material in CMOS devices because the sum of SBHs for n^+- and p^+-Si is equal to the bandgap of Si, a material with a low contact resistivity for

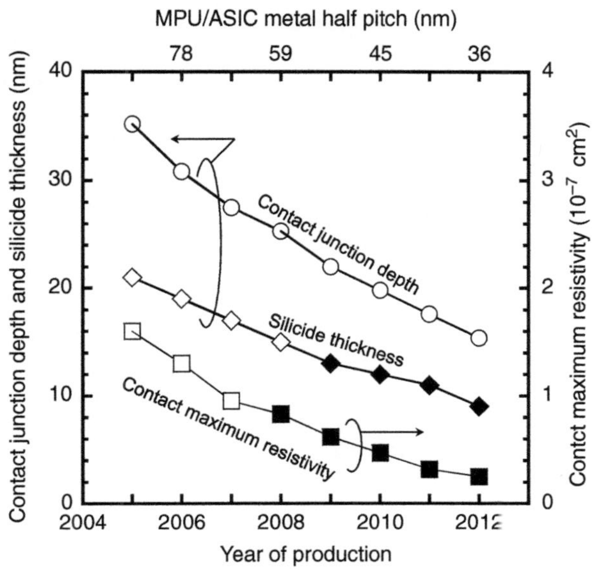

Fig. 8.2 Contact junction depth, silicide thickness, and maximum contact resistivity for bulk MPU/ASIC as predicted in ITRS2005 [8]. Filled points correspond to "Manufacturable solutions are NOT known"

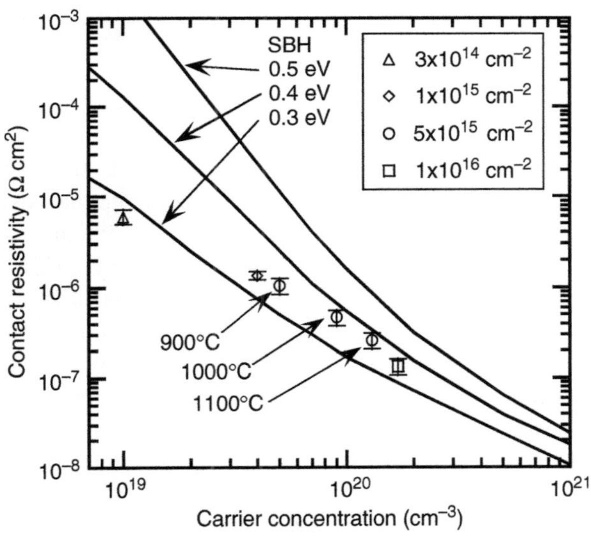

Fig. 8.3 Contact resistivities of $CoSi_2/p^+$-Si samples with implantation doses from 3×10^{14} to 1×10^{16} cm^{-2} [11]. The samples with a dose of 5×10^{15} cm^{-2} are post-annealed after the implantation at temperatures of 900, 1000, and 1100°C, and others are annealed at 1100°C. *Solid lines* are the calculated results as a function of carrier concentration for SBHs of 0.3, 0.4, and 0.5 eV. (Reprinted from Nakatsuka et al. [11], with permission from Elsevier.)

one type of Si is disadvantageous for another type. The incorporation of a SiGe layer into the silicide/Si interface is attractive to solve this dilemma, as described later.

From viewpoints as mentioned above, $TiSi_2$ and $CoSi_2$, and NiSi have been extensively investigated for CMOS applications. Some important properties of these materials are summarized in Table. 8.1 [1, 2, 12]. The formation process, advantages, and disadvantages of these silicides are separately mentioned below. Additionally, several technologies to control crystalline and electrical properties of silicide contacts are also discussed.

Table 8.1 Important properties of C54-$TiSi_2$, $CoSi_2$, and NiSi [1, 2, 12]

	C54-$TiSi_2$	$CoSi_2$	NiSi
Crystalline structure	Face-centered orthorhombic	Cubic	Orthorhombic
Formation temperature (°C)	700	700	350
Melting point (°C)	1500	1326	992
Si consumption per nm of metal (nm)	2.27	3.64	1.83
Thin film resistivity (Ω μm)	13–20	14–20	14–20
Dominant diffusion species for silicide formation	Si	Co and Si	Ni
SBH for n-type Si (eV)	0.49	0.61	0.68
SBH for p-type Si (eV)	0.51	0.43	0.38

8.2 $TiSi_2$

Ti disilicide is one of the most popular contact materials for ULSI applications in refractory metal silicides, because $TiSi_2$ has high thermal stability and low resistivity. Ti disilicide has two types of the crystalline structure, a base-centered orthorhombic C49-structure and a face-centered orthorhombic C54-structure. C49- and C54-$TiSi_2$ are the high-resistivity (~60 μΩ cm) and low-resistivity phases (13–20 μΩ cm), respectively [13–16]. A C54-$TiSi_2$ layer for contacts is generally formed by the salicide process with two-step annealing. In the first step, C49-phase is formed by annealing in N_2 atmosphere at the temperature range between 600 and 700°C. This annealing suppresses the lateral diffusion of Si due to TiN formed on the surface and prevents the bridging of silicide between the gate and S/D regions. After the selective etching of TiN and unreacted Ti, the C49-phase is transformed to the C54-phase by the second step annealing above 800°C.

The narrow-line effect of $TiSi_2$ restricts the application of this silicide to nanoscale devices. The phase transformation from C49 to C54 phases does not occur for narrower poly-Si gate lines or smaller S/D contact areas below 500 nm and 100 μm^2, respectively [17, 18]. The reason considered is that the nucleation mode of the C54-phase is dependent on the size of $TiSi_2$ layers [13, 19]. The energy barrier for the nucleation increases with decreasing the layer thickness because disk-shaped C54-nucleus in a very thin layer has the larger surface-to-volume ratio than the spherical C54-nucleus in a thick layer. Hence the nucleation preferentially occurs at the triple points of C49-grain boundaries for silicide layers thinner than 55 nm, while the nucleation of the C54-phase predominantly occurs at the grain boundary of the two C49-grains for layers thicker than 100 nm. As a result, the transformation of C54-$TiSi_2$ is limited depending on the size of C49-$TiSi_2$ layers and the resistance of $TiSi_2$ lines increases with narrowing the line width below 1 μm [13, 19, 20].

8.3 $CoSi_2$

$CoSi_2$ is an alternative material to C54-$TiSi_2$ for ULSI applications [17, 21]. $CoSi_2$ has low resistivity and high thermal stability as well as $TiSi_2$. $CoSi_2$ is also formed by salicide process with a two-step annealing like C54-$TiSi_2$. The first-step annealing at a low temperature ranging between 400 and 500°C provides a Co-rich phase silicide such as Co_2Si or CoSi. After the chemical-selective etching of unreacted Co on insulators, $CoSi_2$, low-resistivity phase, is formed by the second-step annealing at a high temperature ranging between 600 and 800°C.

It is well known that the formation mechanism of Co silicide is quite complicated, because diffusion species are dependent on formed phases. In the first stage of silicidation, the Co_2Si layer grows predominantly by the Co diffusion during the Co/Si interfacial reaction, and then the CoSi layer grows predominantly by the Si diffusion [22–26]. In the stage of $CoSi_2$ formation, the nucleation of $CoSi_2$ occurs at triple points between CoSi grains and a Si substrate and the $CoSi_2$ layer grows predominantly by Co diffusion.

At first, the advantage of $CoSi_2$ was its insensitivity to decreasing line width compared to C54-$TiSi_2$. However, it is recently reported that $CoSi_2$ thin layers also exhibit a scaling issue: the resistance of $CoSi_2$ lines increases with narrowing below 200 nm [27, 28].

The junction leakage is also a considerable issue when using $CoSi_2$ for shallow S/D junctions in ULSI applications. The formation of $CoSi_x$ and $CoSi_2$ spikes in $CoSi_2$/Si contacts causes the junction leakage [29]. Additionally, Co atoms dissolve in Si substrate by post-annealing of very thin $CoSi_2$ islands on Si(001) over 600°C [30]. In-diffusion of Co atoms into a Si substrate during the formation of $CoSi_2$/Si contacts and the formation of generation-recombination centers are suspected to be the causes of the intrinsic leakage current [31].

8.4 NiSi

NiSi is one of the promising candidates for silicide/Si contact materials in next-generation ULSI devices [32–35]. This is because NiSi/Si contacts have several advantages as follows: (1) low resistivity of NiSi as low as that of C54-$TiSi_2$ and $CoSi_2$; (2) low consumption of Si during silicidation compared to disilicides like $TiSi_2$ and $CoSi_2$, which is suitable for ultra-shallow junctions; (3) low formation

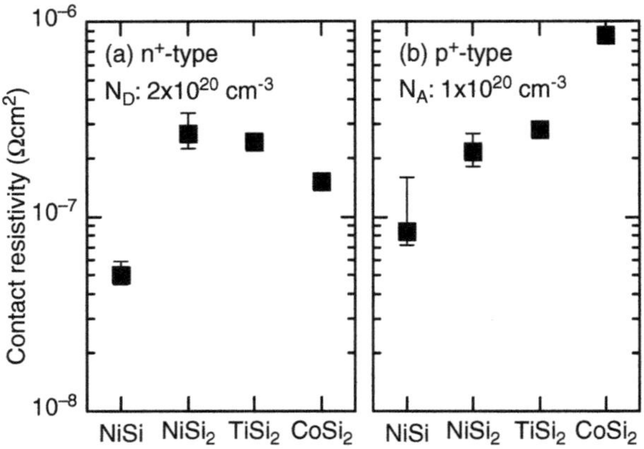

Fig. 8.4 Contact resistivities of NiSi, $NiSi_2$, C54-$TiSi_2$, and $CoSi_2$/Si contacts for (**a**) n^+- and (**b**) p^+-types. Doping concentrations of P and B in the Si region are 2×10^{20} and 1×10^{20} cm^{-3}, respectively (Reprinted from Zaima [12], with permission from Elsevier.)

temperature of NiSi below 400ºC; and (4) low contact resistivity as low as 10^{-8} $\Omega\ cm^2$. Figure 8.4a, b shows contact resistivities of NiSi, $NiSi_2$, C54-$TiSi_2$, and $CoSi_2$/Si contacts for n^+- and p^+-types, respectively [12]. The contact resistivity of NiSi shows the lowest values for both n^+- and p^+-type contacts in these silicides. This is mainly attributed to the pile-up of the dopant impurity, P at the NiSi/n^+-Si interface during the silicide formation, and the low SBH for the NiSi/p-type contact.

However, the poor thermal stability of NiSi layers is a severe issue for ULSI applications and it causes the degradation of the electrical properties after the annealing processes [32–35]. Indeed, a NiSi thin layer easily agglomerates by annealing over 650°C, and NiSi is transformed to the high-resistivity $NiSi_2$ phase by annealing over 750°C. A $NiSi_2$ layer is formed epitaxially on a Si(001) substrate by the solid-phase reaction of a Ni/Si contact. However, the interface morphology is generally not superior for contacts due to the inhomogeneous formation of $NiSi_2$ and {111} facets at the epitaxial $NiSi_2$/Si interface because the nucleation of $NiSi_2$ preferentially occurs in the vicinity of the Si-exposed region and agglomerated NiSi grains [32, 36].

The incorporation of the third elements into a silicide/Si system is one of the effective solutions to improve the properties of silicide films; controlling the solid-phase reaction, raising the thermal stability, and lowering the film and the contact resistance. Several elements into Ni/Si system to improve the properties of NiSi/Si contacts have been reported, such as Pt [37–39], Pd [40], Ge [12, 41–43], C [12, 44], N [45, 46] and H [47]. Some examples are introduced below.

The incorporation of Pt into NiSi is effective to suppress the agglomeration of a polycrystalline NiSi layer. Pt in NiSi affects the texture structure of a NiSi layer on a Si substrate [37, 48]. Grains in polycrystalline NiSi layers exhibit five different types of preferential orientation on a Si(001) substrate. These texture components changes and the epitaxial alignment of NiSi on Si becomes preferable due to the expansion of unit cell caused by the addition of Pt [48]. The expansion of the NiSi lattice with Pt also causes the transformation of MnP-type structure (orthorhombic) into more stable NiAs-type structure (hexagonal). As a result, NiSi layers with Pt on a Si substrate become more stable over 800°C.

Moreover, the incorporation of Pt into NiSi raises the thermal stability of the low-resistivity NiSi phase. Since the crystalline structure of PtSi is MnP type as well as NiSi and the lattice mismatch between NiSi and PtSi is less than 15%, NiSi and PtSi form a solid solution in bilayer and concentrated alloy thin layers. Pt–Si compounds do not form a disilicide phase in contrast to Ni–Si system and the solid solubility of Pt in $NiSi_2$ may be low. Hence, the formation of $NiSi_2$ phase is effectively suppressed by the incorporation of Pt because of required expulsion of Pt from NiSi for the $NiSi_2$ nucleation.

Pd also plays a role in raising the phase transition temperature from NiSi to $NiSi_2$ [40]. PdSi has the same crystalline structure (MnP type) as that of NiSi and it is generally formed above 700°C, which is much higher than the formation temperature of NiSi. There is no Si-rich silicide in the Pd–Si system. Hence, the phase transformation temperature to $NiSi_2$ phase is raised due to the introduction of the PdSi phase into the NiSi layer as well as the Ni–Pt–Si system.

The incorporation of C into Si substrates effectively improves on the thermal stability and the electrical properties of NiSi/Si contacts [12, 44]. Increase in sheet resistance of a NiSi layers on Si is suppressed by the C implantation into Si substrates before the silicidation, which is due to preventing the agglomeration of polycrystalline NiSi grains. The contact resistance of NiSi/p^+-Si contacts with the C implantation is also reduced due to the pile-up of B atoms at the NiSi/Si interface, which is enhanced with the C incorporation.

Recently, metal gates are expected to replace poly-Si gates in MOSFETs in order to reduce the effective gate oxide thickness by eliminating the poly-Si depletion region. A fully silicided (FUSI) gate using Ni silicides is one of promising candidates for metal gates [49, 50]. Silicides having appropriate work functions for metal gates must be selected for the optimum operation of each n- and p-type MOSFET. The effective work function of Ni silicides strongly depends on silicide phases, kinds of gate dielectrics, and dopant impurities in a poly-Si before silicidation [49, 51, 52]. The Fermi level pinning at the metal/dielectric interface is also a serious issue. It is reported that using Ni-rich silicide like Ni_3Si on HfSiON provides unpinning of Fermi level [52]. The formation of controlled silicide/dielectric interfaces in nano-scale gate regions with the appropriate work functions for CMOS applications remains a critical issue.

8.5 SiGe Incorporation into Silicide/Si Contacts

The application of $Si_{1-x}Ge_x$ in a metal/Si ohmic contact has some advantages such as (1) controllability of the energy bandgap, (2) the possibility of realizing higher doping concentrations of impurities than those in Si, (3) the heteroepitaxial growth of $Si_{1-x}Ge_x$ on Si(001) substrates, and (4) the good compatibility with Si processes. Since the energy band gap of $Si_{1-x}Ge_x$ can be changed from 1.12 to 0.66 eV by increasing the Ge fraction, x, the band offset is mainly formed at the valence band edge interfaces [53],we can expect the smaller SBHs at the metal/p–$Si_{1-x}Ge_x$ interface than the metal/Si interface. And smaller SBH at the metal/semiconductor interface promises the lower contact resistivity. The smaller SBH of the $TiSi_2$/SiGe/Si contact than the $TiSi_2$/Si contact has been reported [54].

The incorporation of Ge into silicide/Si system also controls the stability of silicide phase. The Ni monosilicide phase is thermally stabilized by the Ge introduction [12, 41–43]. While the NiSi phase transforms to the $NiSi_2$ phase after annealing above 800°C, a mono-germanosilicide phase formed on $Si_{1-x}Ge_x$ layers remains even after the annealing at 850°C [12]. In a Ni–Ge system, germanium-rich phase is not formed at any temperatures contrary to the Ni–Si system. Hence, when Ni mono-germanosilicide phase transforms to $NiSi_2$ phase in a Ni–Si–Ge system, an additional thermally activated process is required, i.e., the segregation of Ge atoms from the germanosilicide. This leads to raising the phase transformation temperature compared to the case without Ge.

References

1. Murarka, S. P.: Silicides for VLSI Applications. Academic Press, New York (1983)
2. Nicolet M.-A. and Lau, S. S.: VLSI Electronics, Microstructure Science. Eispruch, N. and Larrabee, G. (Eds.), Academic, New York, **6**, 329 (1983)
3. Murarka, S. P.: Silicide thin films and their applications in microelectronics. Intermetallics **3**, 173 (1995)
4. Maex, K.: Silicides for Integrated Circuits: $TiSi_2$ and $CoSi_2$. Mater. Sci. Eng. Rev. **R11**, 53 (1993)
5. Maex, K. and van Rossum, M. (Eds.): Properties of Metal Silicides, INSPEC, London (1995)
6. Gambino J. P. and Colgan, E. G.: Silicides and ohmic contents. Mater. Chem. Phys. **52**, 99 (1998)
7. Zhang, S.-L. and Östling, M.: Metal silicides in CMOS technology: Past, present, and future trends. Critical Rev. Solid State. Mater. Sci. **28(1)**, 1 (2003)
8. International Technology Roadmap for Semiconductors 2005 Edition, Semiconductor Industry Association (see the web page, http://www.itrs.net/).
9. Yu. A. Y. C.: Electron tunneling and contact resistance of metalsilicon contact barriers. Solid-State Electron. **13**, 239 (1970)
10. Chang, C. Y.; Fang, Y. K.; and Sze. S. M.: Specific contact resistance of metal-semiconductors barriers. Solid-State Electron. **14**, 541(1971)
11. Nakatsuka, O.; Ashizawa, T.; Nakai, K.; Tobioka, A.; Sakai, A.; Zaima, S.; and Yasuda, Y.: Dependence of contact resistivity on impurity concentration in Co/Si systems. Appl. Surf. Sci. **159–160**, 149 (2000)
12. Zaima, S.; Nakatsuka, O.; Sakai, A.; Murota, J.; and Yasuda, Y.: Interfacial reaction and electrical properties in Ni/Si and Ni/SiGe(C) contacts. Appl. Surf. Sci. **224 (1–4)**, 215 (2004)
13. Ma, Z. and Allen, L. H.: Kinetic mechanisms of the C49-to-C54 polymorphic transformation in titanium disilicide thin films: A microstructure-scaled nucleation-mode transition. Phys. Rev. B **49(19)**, 13501 (1994)
14. Mann R. W. and Clevenger, L. A.: The C49 to C54 phase transformation in $TiSi_2$ thin films. J. Electrochem. Soc. **141(5)**, 1347 (1994)

15. Clevenger, L. A.; Harper, J. M. E.; Cabral, Jr., C.; Nobili C.; Ottaviani, G.; Mann, E.: Kinetic analysis of C49-$TiSi_2$ and C54-$TiSi_2$ formation at rapid thermal annealing rates. J. Appl. Phys. **72(10)**, 4978 (1992)
16. van Houtum, H. J. W.; Raaijmakers, I. J. M. M.; and Menting, T. J. M.: Influence of grain size on the transformation temperature of C49 $TiSi_2$ to C54 $TiSi_2$. J. Appl. Phys. **61(8)**, 3116 (1987)
17. Lasky, J. B.; Nakos, J. S.; Cain, O. J.; and Geiss, P. J.: Comparison of transformation to low-resistivity phase and agglomeration of $TiSi_2$ and $CoSi_2$. IEEE Trans. Electron. Dev. **38(2)**, 262 (1991)
18. DiGregorio J. F. and Wall, R. N.: Small area versus narrow line width effects. On the C49 to C54 transformation of TiSi. IEEE Trans. Electron Dev. **47(2)**, 313 (2000).
19. Ma, Z.; Allen L. H.; and Allman, D. D. J.: Effect of dimension scaling on the nucleation of C54 $TiSi_2$. Thin Solid Films **253 (1–2)**, 451 (1994)
20. Clevenger, L. A.; Roy, R. A.; Cabral, C. Jr.; Saenger, K. L.; Brauer, S.; Morales, G ; Ludwig, K. F. Jr.; Gifford, G.; Bucchignano, J.; Jordan-Sweet, J.; Dehaven, P.; and Stephenson, G. B.: A comparison of C54-$TiSi_2$ formation in blanket and submicron gate structures using *in situ* x-ray diffraction during rapid thermal annealing. J. Mater. Res. **10(9)**, 2355 (1995)
21. Maex, K.; Lauwers, A.; Besser, P.; Kondoh, E.; de Potter, M.; and Steegen, A.: Self-aligned $CoSi_2$ for 0.18 mm and below. IEEE Trans. Electron Dev. **46(7)**, 1545 (1999)
22. Lau, S. S.; Mayer, J. W.; and Tu, K. N.: Interactions in the Co/Si thin-film system. I. Kinetics. J. Appl. Phys. **49(7)**, 4005 (1978)
23. van Gurp, G. J.; van der Weg, W. F.; and Sigurd, D.: Interactions in the Co/Si thin-film system. II. Diffusion-marker experiments. J. Appl. Phys. **49(7)**, 4011 (1978)
24. d'Heurle F. M. and Petersson, C. S.: Formation of thin films of $CoSi_2$: Nucleation and diffusion mechanisms. Thin Solid Films **128**, 283 (1985)
25. Appelbaum, A. R.; Knoell, V.; and Murarka, S. P.: Study of cobalt-disilicide formation from cobalt monosilicide. J. Appl. Phys. **57(6)**, 1880 (1985)
26. van Ommen, A. H.; Bulle-Lieuwma, C. W. T.; and Langereis, C.: Properties of $CoSi_2$ formed on (001) Si. J. Appl. Phys. **64(5)**, 2706 (1988)
27. Ohguro, T.; Saito, M.; Morifuji, E.; Yoshitomi, T.; Morimoto, T. T.; Momose, H. S.; Katsumata,Y.; and Iwai, H.: Thermal stability of $CoSi_2$ film for CMOS salicide. IEEE Trans. Electron Dev. **47(11)**, 2208 (2000)
28. Lauwers, A.; de Potter, M.; Chamirian, O.; Lindsay, R.; Demeurisse, C.; Vrancken, C.; and Maex, K.: Silicides for the 100-nm node and beyond: Co-silicide, Co(Ni)-silicide and Ni-silicide. Microelectron. Eng. **60**, 221 (2002)
29. Goto, K.-I.; Fushida, A.; Watanabe, J.; Sukegawa, T.; Tada, Y.; Nakamura, T.; Yamazaki, T.; and Sugii, T.: A new leakage mechanism of Co salicide and optimized process conditions [for CMOS]. IEEE Trans. Electron Dev. **46(1)**, 117 (1999)
30. Ikegami, H.; Ikeda, H.; Zaima, S.; and Yasuda, Y.: Thermal stability of ultra-thin $CoSi_2$ films on Si(100)-2 $\times$ 1 surfaces. Appl. Surf. Sci. **117–, 118 (2)**, 275 (1997)
31. Tsuchiaki, M.; Hongo, C.; Takashima, A.; and Ohuchi, K.: Intrinsic junction leakage generated by cobalt in-diffusion during $CoSi_2$ formation. Jpn. J. Appl. Phys. **41(4B)** , 2437 (2002)
32. Morimoto, T.; Ohguro, T.; Sasaki, H.; Iimura, T.; Kunishima, I.; Suguro, K.; Katakabe, I.; Nakajima, H.; Tsuchiaki, M.; Ono, M.; Katsumata, Y.; and Iwai, H.: Self-aligned nickel-mono-silicide technology for high-speed deep submicrometer logic CMOS ULSI. IEEE Trans. Electron Dev. **42(5)**, 915 (1995)
33. Lauwers, A.; Steegen, A.; de Potter, M.; Lindsay, R.; Satta, A.; Bender, H.; and Maex, K.: Materials aspects, electrical performance, and scalability of Ni silicide towards sub-0.13 mm technologies. J. Vac. Sci. Technol. B **19(6)**, 2026 (2001)
34. Tsuchiya, Y.; Tobioka, A.; Nakatsuka, O.; Ikeda, H.; Sakai, A.; Zaima, S.; and Yasuda, Y.: Electrical properties and solid-phase reactions in Ni/Si(100) contacts. Jpn. J. Appl. Phys. **41(4B)**, 2450 (2002)

35. Lavoie, C.; d'Heurle, F. M.; Detavernier, C.; and Cabral, C.: Towards implementation of a nickel silicide process for CMOS technologies. Microelectron. Eng. **70(2–4)**, 144 (2003)
36. Okubo, K.; Tsuchiya, Y.; Nakatsuka, O.; Sakai, A.; Zaima, S.; and Yasuda, Y.: Influence of structural variation of Ni Silicide thin films on electrical property for contact materials. Jpn. J. Appl. Phys. **43(4B)**, 1896 (2004)
37. Mangelinck, D.; Dai, J. Y.; Pan, J. S.; and Lahiri, S. K.: Enhancement of thermal stability of NiSi films on (100)Si and (111)Si by Pt addition. Appl. Phys. Lett. **75(12)**, 1736 (1999)
38. Liu, J. F.; Chen, H. B.; Feng, J. Y.; and Zhu, J.: Improvement of the thermal stability of NiSi films by using a thin Pt interlayer. Appl. Phys. Lett. **77(14)**, 2177 (2000)
39. Lee, P. S.; Pey, K. L.; Mangelinck, D.; Ding, J.; Chi, D. Z.; and Chan, L.: New salicidation technology with Ni(Pt) alloy for MOSFETs. IEEE Electron Dev. Lett. **22(12)**, 568 (2001)
40. Tsai, C. J.; Cheng, P. L.; and Yu, K. H.: Stress evolution of Ni/Pd/Si reaction system under isochronal annealing. Thin Solid Films **365(1)**, 72 (2000)
41. Thompson, R. D.; Tu, K. N.; Angillelo, J.; Delage, S.; and Iyer, S. S.: Interfacial reaction between Ni and MBE grown SiGe alloy. J. Electrochem. Soc. **135(12)**, 3161–, 3163 (1988)
42. Luo, J.-S.; Lin, W.-T.; Cheng, C. Y.; and Tsai, W. C.: Pulsed KrF laser annealing of Ni/$Si_{0.76}Ge_{0.24}$ films. J. Appl. Phys. **82(7)**, 3621(1997).
43. Jarmar, T.; Seger, J.; Ericson, F.; Mangelinck, D.; Smith, U.; and Zhang, S.-L.: Morphological and phase stability of nickel–germanosilicide on $Si_{1-x}Ge_x$ under thermal stress. J. Appl. Phys. **92(12)**, 7193 (2002)
44. Nakatsuka, O.; Okubo, K.; Sakai, A.; Ogawa, M.; Yasuda, Y.; and Zaima, S.: Improvement in NiSi/Si contact properties with C-implantation. Microelectron. Eng. **82(3–4)**, 479 (2005)
45. Cheng, L. W.; Cheng, S. L.; Chen, J. Y.; Chen, L. J.; and Tsui, B. Y.: Effects of nitrogen ion implantation on the formation of nickel silicide contacts on shallow junctions. Thin Solid Films **355–356**, 412 (1999)
46. Lee, P. S.; Pey, K. L.; Mangelinck, D.; Ding, J.; Wee, A. T. S.; and Chan, L.: New salicidation technology with Ni(Pt) alloy for MOSFETs. IEEE Electron Dev. Lett. **21(11)**, 568 (2001)
47. Choi, C.-J.; Ok, Y.-W.; Hullavarad, S. S.; Seong, T.-Y.; Lee, K.-M.; Lee, J.-H.; and Park, Y.-J.: Effects of hydrogen implantation on the structural and electrical properties of nickel silicide. J. Electronchem. Soc. **149(9)**, G517 (2002)
48. Detaverniera, C. and Lavoie, C.: Influence of Pt addition on the texture of NiSi on Si(001). Appl. Phys. Lett. **84(18)**, 3549 (2004)
49. Qin, M.; Poon, V. M. C.; and Ho, S. C. H.: Investigation of polycrystalline nickel silicide films as a gate material. J. Electrochem. Soc. **148(5)**, G271 (2001)
50. Maszara, W. P.: Fully silicided metal gates for high-performance CMOS technology: A review. J. Electrochem. Soc., **152(7)**, G550 (2005)
51. Sim, J. H.; Wen, H. C.; Lu, J. P.; and Kwong, D. L.: Dual work function metal gates using full nickel silicidation of doped poly-Si. IEEE Trans. Electron Dev. Lett. **24(10)**, 631 (2003)
52. Kittl, J. A.; Lauwers, A.; Pawlak, M. A.; van Dal, M. J. H.; Veloso, A.; Anil, K. G.; Pourtois, G.; Demeurisse, C.; Schram, T.; Brijs, B.; de Potter, M.; Vrancken, C.; and Maex, K.: Ni fully silicided gates for 45 nm CMOS applications. Microelectron. Eng. **82(3–4)**, 441 (2005)
53. Van deWalle C.G. and Martin, R.: Theoretical calculations of heterojunction discontinuities in the Si/Ge system. Phys. Rev. B **34(8)**, 5621 (1986)
54. Zaima, S. and Yasuda, Y.: Study of reaction and electrical properties at Ti/SiGe/Si(100) contacts for ultralarge scale integrated applications. J. Vac. Sci. Tech. B **16(5)**, 2623 (1998)

Chapter 9
Materials for ULSI metallization – Overview of Electrical Properties

S. Tsukimoto, K. Ito, and M. Murakami

9.1 Introduction

Since performance of ULSI Si devices was found to be controlled by *RC* delay (where *R* is the electrical resistance at the interconnects and *C* is the capacitance of the insulators), efforts have been continued to reduce the wiring resistance and the insulator capacitance. Replacement of aluminum alloy interconnect materials by copper (which has about 40% lower resistivity compared with the aluminum alloy) reduced not only the device switching times but also the fabrication cost. However, the resistivity of the Cu wires was demonstrated experimentally to increase rapidly [1, 2] when the line width approached to the mean free path (~39 nm) of the conducting electrons as predicted by theories [3–5]. Figure 9.1 shows theoretical resistivity calculated as a function of line widths with two average grain sizes (*D*) using Mayadas and Shatzkes (MS) model [5] with $P = 0$ and $R = 0.5$, which were experimentally determined [2]. Significant increase in the theoretical resistivity is predicted by reducing the line width less than 70 nm and the resistivity increment becomes large for the wires with small average grain sizes. This suggested that the conducting electrons primarily scattered at the grain boundaries in Cu thin films [1], causing an increase in the electrical resistivity of sub-micron scale Cu wires with small grains. Thus, large-grained Cu interconnects are essential to realize high-performance ULSI Si devices.

The diffusion barriers are needed to prevent intermixing of Cu with the surrounding dielectric materials in manufacturing Cu interconnects with Damascene structure. The barrier layers reduce the cross-sectional area of the interconnects, and the effective resistivity of the Cu wires becomes larger than the barrier-free Cu wires. The resistivity increase due to the barrier layers becomes significantly large upon reducing the line width of the interconnects as shown in Fig. 9.2a, b. The effective resistivity in nano-scale Cu interconnects with barrier layers was calculated as a function of the barrier layer thickness using the determined scattering coefficients

M. Murakami (✉)
Department of Materials Science and Engineering, Kyoto University, Sakyo-Ku, Kyoto, Japan
e-mail: masanori.murakami@materials.mbox.media.kyoto-u.ac.jp

Y. Shacham-Diamand et al. (eds.), *Advanced Nanoscale ULSI Interconnects: Fundamentals and Applications*, DOI 10.1007/978-0-387-95868-2_9,

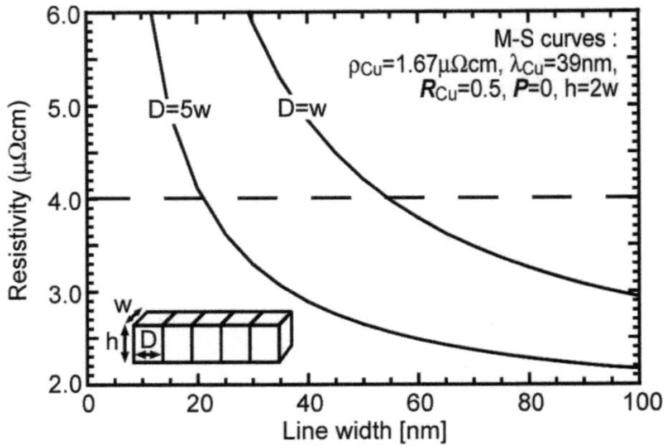

Fig. 9.1 Theoretical resistivities calculated using the MS model with $P = 0$ and $R = 0.5$ for the Cu interconnects with two grain sizes of $D = w$ and $5w$, where w is line width

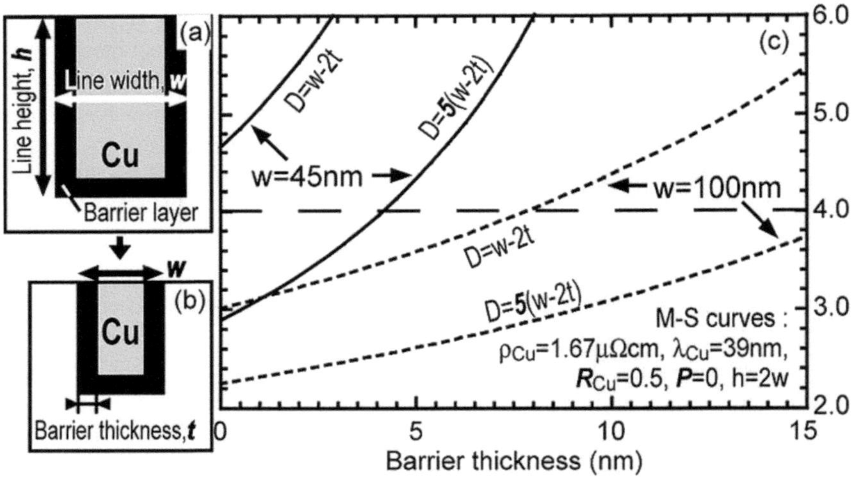

Fig. 9.2 Schematic illustration of the (**a**) wide and (**b**) narrow Cu interconnects with a diffusion barrier layer. (**c**) Effective resistivites of Cu interconnects with the line widths (w) 45 (*solid curves*) and 100 (*broken curves*) nm for two grain sizes (D). They were calculated using the MS model with $P = 0$ and $R = 0.5$ as a function of barrier thickness (t). The effective line width w^* is defined as w–$2t$

P and R. The results of the effective resistivity for the Cu interconnect with the line widths of 45 and 100 nm are shown by solid and broken curves in Fig. 9.2c, respectively [2]. Significant increase in the theoretical resistivity is observed by increasing the barrier thickness and the increment becomes large for the interconnects with the small line width of 45 nm. The resistivity of the Cu interconnect with line width of 45 nm and grain size 5 times larger than the film thickness is close to 4 μΩ cm (when the barrier thickness (t) is 4 nm), which is required by the device designer for high-speed ULSI devices with 70-nm-wide Cu interconnects. Thus, the thickness of the barrier layer should be less than about 4 nm for the Cu interconnects with large average grain sizes to suppress the effective resistivity less than 4 μΩ cm. This suggested that the fabrication technique to prepare nano-scale Cu interconnects with a very large grain size and an ultrathin barrier layer should be developed to satisfy the designer's requirements.

To develop the fabrication technique of large-grained Cu interconnects, understanding of a grain growth mechanism in Cu thin films is essential. Although several parameters (such as surface energy, grain boundary energy, impurities, etc.) to control the grain growth of the Cu thin films were proposed [6–9], we demonstrated experimentally [10, 11] that the strain (or stress) introduced into the thin films was the primary factor to enhance the grain growth specially at low temperatures in addition to surface/grain boundary energy. In the next section, the experiments [10, 11] are briefly reviewed which demonstrated importance of strain (or stress) in thin Cu films to enhance grain growth. Then a mechanism why the strain (or stress) in the films enhanced growth of Cu grains was explained based on the strain energy criterion model [12].

A fabrication technique to form the barrier layers in Cu alloy films by annealing at elevated temperatures was extensively studied [13–18]. This technique is called "self-formation of barrier layer" and is applied to prepare the self-passivation layers on the Cu film surface in order to increase the resistance of the surface corrosion [19, 20]. The Cu alloy films with various solutes (Mg, Ti, Al, Cr, etc.) were extensively investigated for the possibility of the self-formation of the passivation (interfacial barrier) layers [13–21]. Titanium is a good candidate as an alloy element to Cu for the self-formation of the thin barrier layer. In the third section, Ti segregation to film/substrate interface and the barrier self-formation in the Cu(Ti) alloy films after annealing at relatively low temperatures (~400°C) are reviewed [22]. Then, the selection rule of the Cu alloy elements required for the barrier self-formation is proposed based on our understanding of the self-formation mechanism of the barrier layer.

9.2 Large-Grained Cu Interconnects

9.2.1 Abnormal Grain Growth of Cu Thin Films

When the Cu films were annealed at elevated temperatures after film deposition, a few grains grew locally at the expense of the smaller neighboring grains. This bimodal grain growth is conventionally related to an abnormal grain growth [23]. In electroplated Cu films that are widely used in manufacturing Si ULSI devices, abnormal grain growth was observed even during room temperature storage [24–33]. Although several mechanisms for grain growth in the Cu films were proposed by several authors, a key factor that induced abnormal grain growth at room temperature was not identified [25–31, 33, 34].

In order to study the effect of intrinsic and extrinsic strains on the grain growth of the Cu thin films, 100 nm-thick Cu films were prepared on different substrates by a magnetron sputter-deposition technique with base pressure of 2×10^{-8} Torr [10, 11]. The cross sections of the samples were schematically shown in Fig. 9.3. Since the thermal expansion coefficients of Si are smaller than those of Cu, the compressive strain was introduced into the Cu film upon heating sample A. The choice of the rock salt substrate (sample B) was that the thermal expansion coefficients of the rock salt are larger than those of Cu and thus the tensile strain was introduced into the Cu film upon heating sample B. Another reason to use the rock salt substrate was that a free-standing Cu film (sample C) could be prepared simply by dissolving the rock salt of sample B in water.

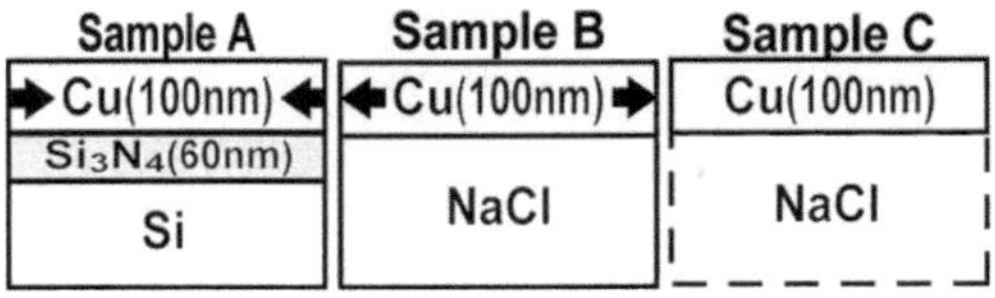

Fig. 9.3 Cross sections of samples used for a grain growth study of sputtered Cu films deposited on different substrates

Grain growth of the Cu films was strongly enhanced by existence of the substrates during storage of the samples at room temperature. Significant grain growth with bimodal grain size distribution was observed in both samples A and B, while no grain growth was observed in the sample C, as shown in Fig. 9.4. Grain growth of samples A and B was enhanced upon annealing at elevated temperatures, which was believed to be due to introduction of thermal strains into the Cu films. Although both the compressive (sample A) and tensile (sample B) strains are introduced into the Cu films, the grain growth behaviors were not influenced by the sign of the strains. Grain growth of sample A was initiated at locations close to the substrate surface where the large strain was observed in the Cu film [11]. Carbral et al. [35] observed simultaneously relaxation of strain in the film and grain growth during room temperature storage.

Grain growth during room temperature storage was found to be more significant for the Cu films deposited on the (100)-oriented Si wafers covered sequentially with

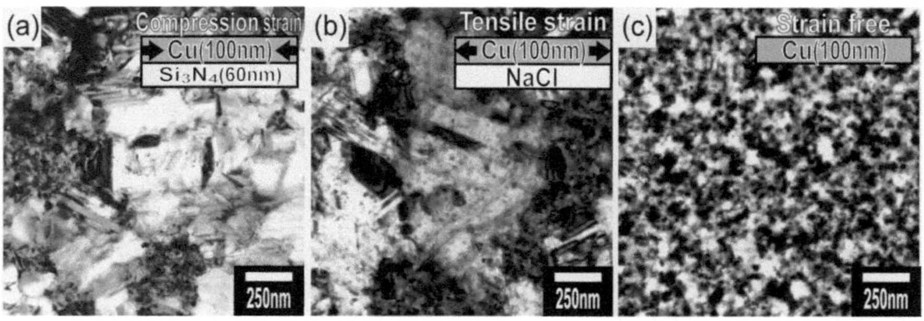

Fig. 9.4 Plan-view TEM images of the 100-nm thick sputtered Cu films annealed at room temperature (**a**) for 73 h with a Si_3N_4 substrate, (**b**) for 264 h with a rock salt substrate, and (**c**) 192 h without a substrate

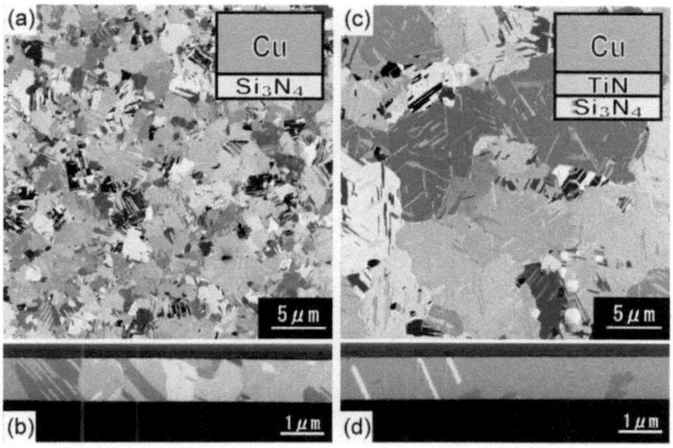

Fig. 9.5 Plan-view and cross-sectional SIM images of 100-nm thick sputtered Cu films annealed at room temperature (**a**) and (**b**) for 25 h with a Si_3N_4 substrate and (**c**) and (**d**) for 18 h with a TiN substrate

TiN and Si_3N_4 layers than that on the (100)-oriented Si wafers covered with the Si_3N_4 layers (sample A), as shown in Fig. 9.5.

This indicates that larger strain was introduced into the Cu films by the TiN under-layers than by the Si_3N_4 layers. Since the TiN layer is one of the best barrier layers for Cu interconnects, large-grained Cu interconnects with good diffusion barriers would be realized. The Cu interconnects were fabricated by patterning selectively a giant single grain in the blanket Cu films by using FIB techniques. These interconnects could be treated as "single crystal" Cu interconnects and provided relatively low resistivities of about 1.85 $\mu\Omega\,cm^2$. (111)-oriented Cu grains were found to grow at the expense of the small grains with other orientations in both Cu films in Fig. 9.5, and many twins and dislocations introduced in large grains were observed after grain growth was completed.

9.3 Grain Growth Mechanism of Cu Thin Films

The strain energy criterion model [12] (which was originally developed to determine the maximum thermal strain that could be supported elastically in Pb thin films [36] deposited on the Si substrates) was applied here to understand the grain growth mechanism of the Cu thin films [10, 11]. This model is briefly reviewed here. The total strain energy E of a film (which is strained intrinsically or extrinsically and the strain is partially relaxed by dislocation glide) is divided into two parts:

$$E = E_{\text{elast}} + E'_{\text{disloc}} \tag{9.1}$$

where E_{elast} is the energy associated with a uniform elastic strain and E'_{disloc} is the energy of the dislocation strain field. These energies are calculated in a grain with rectangular volume ($D \times D \times h$) with grain size D and thickness h. Assuming N dislocations glide across a grain and relaxed the intrinsic (or thermal) strain level from ε_0 to an average level ε, the E_{elast} term is given by

$$E_{\mathrm{elast}} = \frac{1}{2}\Upsilon\varepsilon^2 D^2 h \tag{9.2}$$

where Y is the Young's modulus of the film. The E'_{disloc} is given by

$$E'_{\mathrm{disloc}} = E_{\mathrm{disloc}} N \tag{9.3}$$

where E_{disloc} is the energy associated with a dislocation loop present in the film, and N is the number of dislocations given by

$$N = (\varepsilon_0 - \varepsilon) / \Delta\varepsilon \tag{9.4}$$

where $\Delta\varepsilon$ is the amount of strain each dislocation can relax and is given by $\Delta\varepsilon = fb'/D$ (where b' is the projection of the edge component of the Burgers vector on the film surface and f is the average displacement in units of the Burgers vector).

The total energy E^0 (per unit volume) of the film is given by

$$E^0 = \frac{1}{2}\Upsilon\varepsilon^2 + \frac{(\varepsilon_0 - \varepsilon)\, E_{disloc}}{b'fD \cdot h} \tag{9.5}$$

The total energy E^0 given by Eq. (9.5) is schematically shown in Fig. 9.6 as a function of ε together with the elastic energy term (the first term of Eq. (9.5)) and the energy associated with dislocations (the second term). Figure 9.6 indicates that when the strain ε_0 is applied to the film with grain size D and thickness h, the strain is (plastically) relaxed at the amount of $(\varepsilon_0 - \varepsilon^*)$ and the film is elastically supported by the amount of the critical strain ε^* at which E^0 is minimum. The calculated critical

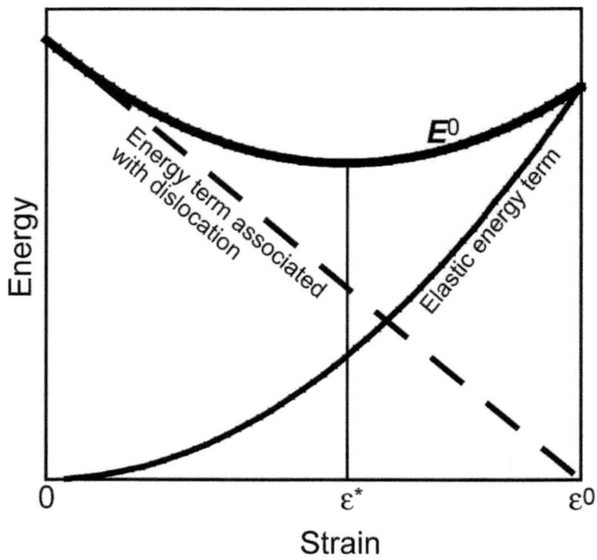

Fig. 9.6 Schematic illustration of total energy (E^0), elastic energy (E_{elast}), and energy associated with dislocations (E'_{dislo}) (normalized by unit volume) as a function of strain ε

strain was in an excellent agreement with the residual elastic strains measured in Pb thin films which were deposited onto the Si substrates and cooled to 4.2 K was observed [36], indicating the validity of this model to calculate the elastic strain supported by the film under tensile or compressive strain.

When a polycrystalline film with various grain size D_i (as shown schematically in Fig. 9.7) is strained intrinsically or extrinsically to the value of ε_0, the residual elastic strain in the grain is ε.

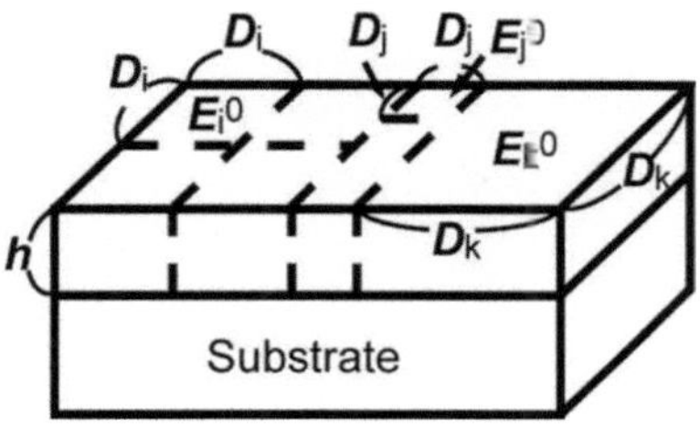

Fig. 9.7 Schematic illustration of a film with thickness *h* deposited on the substrate: D_i and E_1^0 indicate grain size and energy stored in a grain *i*, respectively

The energies E_1^0 and E_2^0 are schematically shown in Fig. 9.8 for the films with grain sizes D_1 and D_2, respectively, where $D_1 > D_2$ [37]). The E_1^* and E_2^* represent the minimum elastic energies stored in the grains with sizes D_1 and D_2, respectively. The grain with small size D_2 stores the strain energy (per unit volume) larger than that of the grain with large size D_1. Therefore, the best way to reduce the strain energy stored in the grain with small size is to grow the grain size which enhances the strain relaxation by dislocation glide. Therefore, grain growth was observed in samples A and B which introduces strain into the films from the substrate and not in free-standing sample C [10]. Also, based on the present grain growth model, a higher density of dislocations or stacking faults are expected to be introduced into the grains after grain growth [37]. These dislocations and stacking faults were observed experimentally in the previous TEM experiment [10].

Although Eq. (9.5) gives the strain energy in a film with isotropic elastic constants, metallic films such as Cu have, in general, anisotropic elastic constants. The elastic energies (E_{hkl}) of the Cu film which has various (*hkl*) grain orientations were calculated [37]. The maximum and minimum values in E_{hkl} are observed in (111)- and (100)-oriented grains, respectively, and the E_{hkl} value of the (111)-oriented grains is more than a factor of two larger than that of (100)-oriented grains. This result indicates that at a given strain a (111)-oriented grain stores an elastic strain energy larger than that of a (100)-oriented grain with the same grain size. Therefore, to reduce the strain energy efficiently, relaxation of the elastic strain in the (111)-oriented grain by increasing the grain growth is more favorable. Therefore, growth of (111)-oriented grains was observed in the Cu films upon annealing at elevated temperatures.

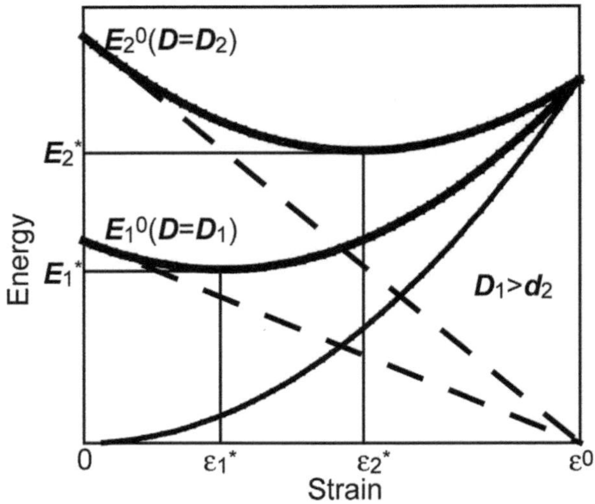

Fig. 9.8 Schematic illustration of total energies and for grains with grain size D_1 and D_2, respectively, as a function strain ε, where $D_1 > D_2$

9.3.1 Self-formation of Nano-scale Ti Diffusion Barrier Layers in Cu(Ti) Alloy Films

Figure 9.9a, b shows schematic illustrations of the barrier formation in the Cu(Ti) alloy films after annealing at elevated temperatures. Alloy films with supersaturated Ti solutes beyond the limit of the thermodynamically equilibrium phase boundary can be prepared by the evaporation and sputter-deposition techniques, when the vapor atoms are quenched on the cold substrates (Fig. 9.9a). The supersaturated Cu(Ti) alloy films are thermally unstable and the annealing of the Cu(Ti) alloy films at an elevated temperature facilitates transformation of the thermodynamically stable two phases of Ti-rich layers and the Cu-rich film with the equilibrium Ti concentrations. Figure 9.9c shows cross-sectional TEM-EDS elemental mapping image of the Cu(2.9 at. % Ti) alloy film after annealing at 400°C in an Ar gas ambient for

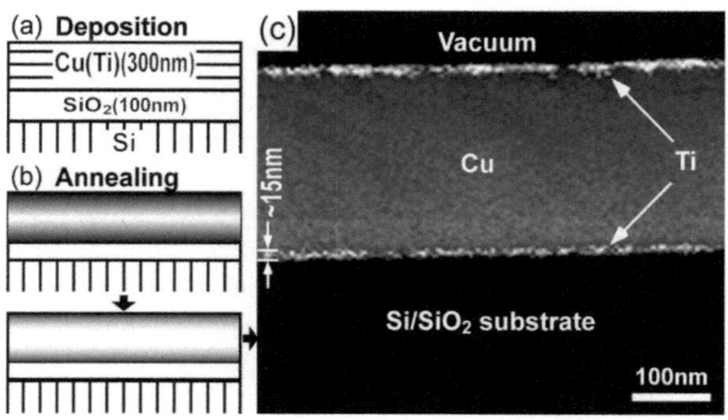

Fig. 9.9 (**a**) and (**b**) Schematic illustration of the barrier self-formation of sputtered Cu alloy films after annealing at elevated temperatures. (**c**) A cross-sectional TEM-EDS elemental mapping image of the Cu (2.9 at. % Ti) alloy film after annealing at 400°C for 2 h

2 h [22]. The Ti element is mapped by bright color. The Ti atoms were observed clearly to segregate both at the film surface and at the film/substrate interface, while they are not observed inside the Cu film. The Ti segregation formed uniformly thin Ti-rich layers with thicknesses of about 10–20 nm at the surface and interface.

Rutherford backscattering spectrometry and the TEM-EDS elemental mapping image in the oxygen element showed O segregation at the Ti-rich layers at the film surface and at the film/substrate interface, indicating that the Ti-rich layers consisted of Ti oxides. The Ti atoms segregated at the film surfaces are found to react to oxygen contained in the Ar gas ambient, yielding Ti oxides. The Ti atoms segregated to the film/substrate interface were found to react with topmost layers on the substrates, forming Ti oxides in addition to Ti silicides. Liu et al. [38] reported that the TiO_x layers formed at the interface between the Cu(Ti) films and the SiO_2 substrates were demonstrated to have the good diffusion barrier property.

The electrical resistivities of the Cu(Ti) alloy films were measured during isothermal annealing at 400°C and the resistivities of the 300-nm thick Cu(1.3 at. % Ti) and Cu(2.9 at. % Ti) alloy films after the annealing for various times are shown in Fig. 9.10. The resistivities of Cu(2.9 at. % Ti) alloy films as-deposited and after annealing for 30 min are higher than those of Cu(1.3 at. % Ti) alloy films, but the annealing for 2 h resulted in almost the same resistivity of about 3.5 μΩ cm for both films. The reduction of the resistivities in the Cu(Ti) alloy films is explained partially by segregation of the Ti atoms due to the low solubility limit (~0.3 at. % Ti) of Ti in Cu at 400°C [39]. The annealing times required to reduce the resistivity are found to increase with increasing the Ti concentration in the films, although the final resistivity value is the same. The resistivity increase in the Cu(Ti) alloy films may be due to grain boundary scattering, because the alloy films have smaller grains. The annealing time for Ti atoms in the films to diffuse out from the films becomes

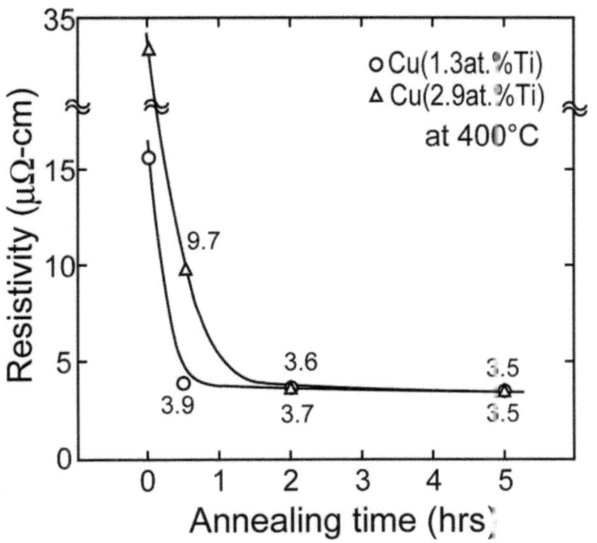

Fig. 9.10 Resistivity changes of 300-nm thick Cu(1.3 at. % Ti) and Cu(2.9 at. % Ti) alloy films after isothermal annealing at 400°C for various storage times

long with increasing the Ti concentration in the films. The excess Ti atoms at the boundary suppress grain growth of the Cu(Ti) alloy films and remain fine grains.

Self-formation of thin barrier layers at low-temperature annealing and low film resistivity in the Cu(Ti) alloy films would be applicable to metal wiring fabrication processes of ULSI devices. Therefore, the following two processes are essential: (1) oxygen contained in the Ar gas ambient facilitates Ti diffusion to the film surface, reducing Ti concentration in the Cu(Ti) alloy films and (2) low Ti concentration in the as-deposited Cu(Ti) alloy films reduces annealing time for the Ti atoms to diffuse to the film surface and the film/substrate interface, forming a thin barrier layer.

Selection of the proper elements solved in the Cu alloy films is essential for success in the device application. The Cu(Ti) alloy decomposes to two phases at low temperatures and the solubility limit with about 0.3 at. % Ti at 400°C is small enough for low resistivity of Cu(Ti) alloy films as shown in Fig. 9.11.

The reduction of the melting point of Cu facilitates segregation of the Ti atoms by diffusion at low temperatures. Thus, Ti is one of the best candidates in the Cu alloy elements. Based on the Cu(Ti) alloy films, an element (M) added to the Cu films should have the following four properties: (1) the Cu(M) alloy film should decompose to two phases that are Cu-rich and M-rich and the solubility limit of M in Cu should be small at low temperatures; (2) the Cu(M) should have a low melting point and the melting point of Cu should decrease with increasing the M concentration; (3) the element M should react with the substrates to form the thin, uniform interface layers by annealing at elevated temperatures; and (4) the M-rich layer formed at the interface should act as a diffusion barrier of Cu into the substrate.

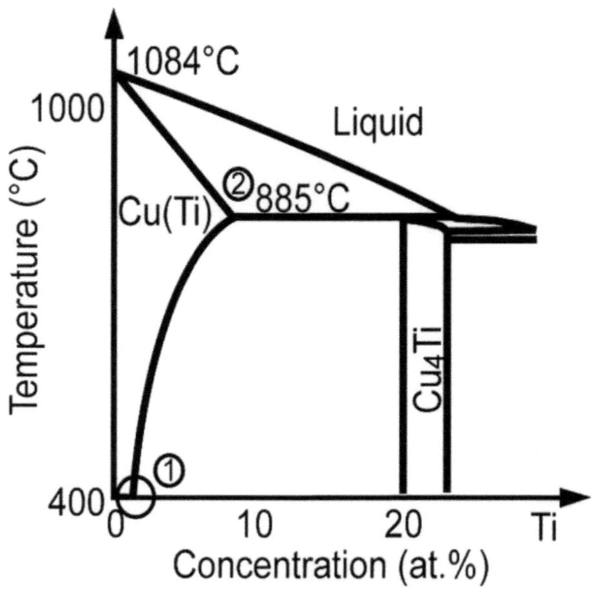

Fig. 9.11 Selection rules of an alloy element added to the Cu films for the barrier self-formation based on a portion of the Cu–Ti binary phase diagram

9.4 Summary

The grain boundary scattering primarily increases the resistivity of the Cu interconnects and resistivity increase due to the barrier layers becomes significantly large upon reducing the line width of the sub-100-nm Cu interconnects. Thus, large-grained Cu interconnects and ultrathin barrier layers are essential to realize low-resistance nano-scale Cu interconnects in the future ULSI devices.

For development of a fabrication technique of large-grained Cu interconnects, grain growth mechanism of Cu thin films was understood. Rapid grain growth of Cu films was observed in the strained films bonded to the rigid substrates during room temperature storage, but no grain growth was observed in the strain-free (free-standing) films. The grain growth was enhanced upon annealing at elevated temperatures, which was believed to be due to introduction of thermal strains into the Cu films. Although both the compressive and tensile strains are introduced into the Cu films, the grain growth behaviors were not influenced by the sign of the strains. A new grain growth model for Cu thin films was proposed based on the strain energy criterion model at temperatures where dislocation glide was the dominant strain relaxation mechanism. Based on this model, the grains of the Cu films under tensile or compressive strain were explained to grow primarily to reduce the elastic strain energy by dislocation glide, because the dislocations are easily introduced into large grains upon introduction of strain into the films.

The fabrication technique to self-form the thin barrier layers in Cu(Ti) alloy films by annealing at relatively low temperatures (~400°C) was demonstrated. Oxygen contained in the Ar gas ambient and initial low Ti concentration in the Cu(Ti) alloy films was essential for fabrication of the Cu(Ti) alloy films with thin barrier layers and low film resistivity. Also Ti showed to be one of the best candidates in the Cu alloy elements, because of the low solubility limit at low temperatures and reducing the melting point of Cu. The selection rule of the Cu alloy elements required for the barrier self-formation and low film resistivity was proposed based on our understanding of the self-formation mechanism of the barrier layer.

References

1. Moriyama, M.; Shimada, M.; Masuda, H.; and Murakami, M.: Determination of parameters to control electrical resistivities of nano-scale copper interconnects. Trans. Mater Res. Soc. Jpn. **29**, 51 (2004)
2. Shimada, M.; Moriyama, M.; Ito; K., Tsukimoto, S.; and Murakami, M.: Electrical resistivity of polycrystalline Cu interconnects with nano-scale linewidth. J. Vac. Sci. Technol B **24**, 190 (2006)
3. Fuchs, K.: The conductivity of thin metallic films according to the electron theory of metals. Proc. Camb. Phil. Soc. **34(8)**, 100 (1938)
4. Sondheimer, E. H.: The mean free path of electrons in metals. Adv. Phys. **1(1)**, 1 (1952)
5. Mayadas, A. F. and Shatzkes, M.: Electrical-resistivity model for polycrystalline films: the case of arbitrary reflection at external surfaces. Phys. Rev. B **1**, 1382 (1970)
6. Lingk, C. and Gross, M. E.: Recrystallization kinetics of electroplated Cu in damascene trenches at room temperature. J. Appl. Phys. **84(10)**, 5547 (1998)

7. Harper, J. M. E.; Cabral, Jr., C.; Andricacos, P. C.; Gignac, L.; Noyan, I. C.; Rodbell, K. P.; and Hu, C. K.: Mechanisms for microstructure evolution in electroplated copper thin films near room temperature. J. Appl. Phys. **86(5)**, 2516 (1999)
8. Brongersma, S. H.; Richard, E.; Vervoot, I.; Bender, H.; Vandervorst, W.; Lagrange, S.; Beyer, G.; and Maex, K.: Two-step room temperature grain growth in electroplated copper. J. Appl. Phys. **86**, 3642 (1999)
9. Chaudhari, P.: Grain growth and stress relief in thin films. J. Vac. Sci. Technol. **9(1)**, 520 (1972)
10. Moriyama, M.; Matsunaga, K.; and Murakami, M.: The effect of strain on abnormal grain growth in Cu thin films. J. Electron. Mater. **32**, 261 (2003)
11. Moriyama, M.; Matsunaga, K.; Morita, T.; Tsukimoto, S.; and Murakami, M.: The effect of strain distribution on abnormal grain growth in Cu thin films. Mater. Trans. **45**, 3033 (2004)
12. Murakami, M.; Kuan, T-S.; and Blech, I. A.: Mechanical Properties of Thin Films on Substrates in Treatize on Mater. Sci. Technol., Preparation and Properties of Thin Films, Tu, K. N. and Rosenberg, R. (Ed.) (Academic Press, Inc., New York, NY) **24**, 163 (1982)
13. Ding, P. J.; Lanford, W. A.; Hymes, S.; and Murarka, S. P.: Effects of the addition of small amounts of Al to copper: Corrosion, resistivity, adhesion, morphology, and diffusion. J. Appl. Phys. **75(7)**, 3627 (1994)
14. Adams, D.; Alford, T. L.; Theodore, N. D.; Russell, S. W.; Spreitzer, R. L.; and Mayer, J. W.: Passivation of Cu via refractory metal nitridation in an ammonia ambient. Thin Solid Films **262**, 199 (1995)
15. Liu, C. J. and Chen, J. S.: Effects of the addition of small amounts of Al to copper: Corrosion, resistivity, adhesion, morphology, and diffusion. Appl. Phys. Lett. **80(15)**, 2678 (2002)
16. Liu, C. J.; Jeng, J. S.; Chen, J. S.; and Lin, Y. K.: Effects of Ti addition on the morphology, interfacial reaction, and diffusion of Cu on SiO_2. J. Vac. Sci. Technol. B **20(6)**, 2361 (2002)
17. Frederick, M. J.; Goswami, R.; and Ramanath, G.: Sequence of Mg segregation, grain growth, and interfacial MgO formation in Cu–Mg alloy films on SiO_2 during vacuum. J. Appl. Phys. **93(10)**, 5966 (2003)
18. Frederick, M. J. and Ramanath, G.: Interfacial phase formation in Cu–Mg alloy films on SiO_2. J. Appl. Phys. **95(6)**, 3202 (2004)
19. Hoshino, K.; Yagi, H.; and Tsuchikawa, H.: Effect of titanium addition to copper interconnect on electromigration open circuit failure. Proc. 7th Int. VLSI Multilevel Interconnection Conf. Piscataway, NJ; IEEE. , 357 (1990)
20. Li, J.; Mayer, J. W.; and Colgan, E. G.: Oxidation and protection in copper and copper alloy thin films. J. Appl. Phys. **70(5)**, 2820 (1991)
21. Hu, C.-K.; Luther, B.; Kaufman, F. B.; Hummel, J.; Uzoh, C.; and Pearson, D. J.: Copper interconnection integration and reliability. Thin Solid Films **262(1–2)**, 84 (1995)
22. Tsukimoto, S.; Morita, T.; Moriyama, M.; Ito, K.; and Murakami, M.: Formation of Ti diffusion barrier layers in thin Cu(Ti) alloy films. J. Electron. Mater. **34(5)**, 592 (2005)
23. Smith, C. S.: Trans. AIME **188**, 1021 (1950)
24. Ritzdorf, T.; Graham, L.; Jin, S.; Mu, C.; and Fraser, D. B.: Self-annealing of electrochemically deposited copper films in advanced interconnect applications. Proc. Int. Interconnect Technology Conf. (New York: IEEE), 166 (1998)
25. Gross, M. E.; Takahashi, K.; Lingk, C.; Ritzdorf, T.; and Gibbons, K.: The role of additives in electroplating of void-free Cu in sub-micron damascene features. Advanced Metallization Conf. 1998, Sandhu, G. S., et al. (Ed.) MRS, Pittsburgh, PA, 51 (1999)
26. Lingk, C. and Gross, M. E.: Recrystallization kinetics of electroplated Cu in damascene trenches at room temperature. J. Appl. Phys. **84(10)**, 5547 (1998)
27. Lingk, C.; Brown, M. E.; Lai, W. Y. -C.; Miner, J. F.; Ritzdorf, T.; Turner, J.; Gibbons, K.; Klawuhn, E.; and Zhang, F.: Room temperature recrystallization of electroplated Cu in damascene trenches: kinetics and mechanisms. Advanced Metallization Conf. 1998, Sandhu, G. S., et al. (Ed.) MRS, Pittsburgh, PA, 89 (1999)

28. Walther, D.; Gross, M. E.; Evans-Lutterodt, K.; Brown, W. L.; Oh, M.; Merchant, S.; and Naresh, P.: Room temperature recrystallization of electroplated copper thin films: methods and mechanisms. Mater. Res. Soc. Symp. Proc. **612**, D. 10.1., 1 (2000)
29. Cabral, C. Jr. et al.: Room temperature evolution of microstructure and resistivity in electroplated copper films. Advanced Metallization Conf. 1998, Sandhu, G. S., et al. (Ed.), MRS, Pittsburgh, PA, 81 (1999)
30. Jiang, Q.-T. and Smekalin, K.: Variations in Cu CMP removal rate due to Cu film self-annealing. Advanced Metallization Conf. 1998, Sandhu, G. S., et al. (Ed.), MRS, Pittsburgh, PA, 209 (1999)
31. Harper, J. M. E.; Cabral, C. Jr.; Andricacos, P. C.; Gignac, L.; Noyan, I. C.; Rodbell, K. P.; and Hu, C. K.: Mechanisms for microstructure evolution in electroplated copper thin films near room temperature. J. Appl. Phys. **86(5)**, 2516 (1999)
32. Ueno, K.; Ritzdorf, T.; and Grace, S.: Seed effect on self-annealing of electroplated copper films. Advanced Metallization Conf. 1998, Sandhu, G. S., et al. (Eds.), MRS, Pittsburgh, PA, 95 (1999)
33. Brongersma, S. H.; Richard, E.; Vervoot, I.; Bender, H.; Vandervorst, W.; Lagrange, S.; Beyer, G.; and Maex, K.: Two-step room temperature grain growth in electroplated copper. J. Appl. Phys. **86(7)**, 3642 (1999)
34. Chaudhari, P.: Grain growth and stress relief in thin films. J. Vac. Sci. Technol. **9(1)**, 520 (1972)
35. Cabral, C. Jr.; Andricacos, P. C.; Gignac, L.; Noyan, I. C.; Rodbell, K. P.; Shaw, T. M.; Rosenberg, R.; Harper, J. M. E.; DeHaven, P. W.; Locke, P. S.; Malhotra, S.; Uzoh, C.; and Klepeis, S. J.: Room temperature evolution of microstructure and resistivity in electroplated copper films. MRS Conf. Proc. ULSI XIV. 81(1999)
36. Murakami, M. and Wook, R. W.: Strain relaxation mechanisms of thin deposited films. CRC Critical Review in Sol. Stat. Mater. Sci. **11**, 317 (1983)
37. Murakami, M.; Moriyama, M.; Tsukimoto, S.; and Ito, K.: Grain growth mechanism of Cu thin films. Mater. Trans. **46(7)**, 1737 (2005)
38. Liu, C. J.; Jeng, J. S.; Chen, J. S.; and Lin, Y. K.: Effects of Ti addition on the morphology, interfacial reaction, and diffusion of Cu on SiO_2. J. Vac. Sci. Technol. B **20(6)**, 2361 (2002)
39. Subramanian, P. R.; Chakrabarti, D. J.; and Laughlin, D. E.: Phase diagrams of binary copper alloys, ASM International, Materials Park, OH, 447 (1994)

Chapter 10
Low-κ Materials and Development Trends

Akira Hashimoto and Ichiro Koiwa

10.1 Introduction

Though low-κ materials have been actively investigated by many researchers, target values of ITRS (International Technology Roadmap for Semiconductor) have not achieved. To reduce *RC* delay, both lower resistance wiring material, copper, and low-κ material are necessary. Especially for recent IC chips, ratio of BEP (Back End Process) has rapidly increased with increasing wiring layers. This chapter reviews low-κ materials and proposes future development.

10.2 Change of Insulator Between Wiring

Change of insulator materials between wiring was listed Table 10.1. In the first stage, only inorganic materials were used. These materials were mainly prepared by chemical vapor deposition (CVD) method and partially by spin on glass (SOG) method. Recently some people use spin on dielectric (SOD) as SOG with a little different meaning. In this section, we use SOG for both SOG and SOD. The types of CVD were atmospheric pressure, low pressure, and plasma and source of CVD was mainly O_3-TEOS (tetraetoxysilane). On the other hand, the SOG method used SiO_2 and phosphorus-doped SiO_2 glass (PSG). In the second stage, an organic SOG was first used and an etch back step of photoresist was also used. From the third stage, research and development of low-κ materials became active. Since the design rule of this stage was less than 0.25μm, a delay due to wiring became serious. The new sources of CVD were used; oxygen atom was displaced by fluorine (F), carbon (C), and hydrogen. In the fourth stage, the low-κ materials were actively investigated considering combination with copper wiring.

I. Koiwa (✉)
Institute of Science and Technology, Kanto Gakuin University, 1-50-1 Mutsuurahigashi, Kanazawa-ku, Yokohama, Japan
e-mail: koiwa@kanto-gakuin.ac.jp

Y. Shacham-Diamand et al. (eds.), *Advanced Nanoscale ULSI Interconnects: Fundamentals and Applications*, DOI 10.1007/978-0-387-95868-2_10,

Table 10.1 Change of insulator materials between wiring [1]

First stage (only inorganic material)	Method	Chemical vapor deposition (CVD) Spin on glass (SOG)
	Material	Non-doped SiO_2 glass (NSG) Phosphorus-doped SiO_2 glass (PSG) SiO_2
Second stage (partially used organic SOG)	Method	Chemical vapor deposition (CVD) Spin on glass (SOG) or CVD + etching
	Material	SiO_2 Boro-phospho-silicate-glass (BPSG) Phospho-silicate-glass (PSG) Inorganic SOG Organic SOG (MSQ, HSQ)
Third stage (design rule less than 0.25 μm)	Method	Chemical vapor deposition (CVD) + spin on glass (SOG), Chemical vapor deposition (CVD) + chemical mechanical polisher (CMP)
	Material	SiO_2 SiOF SiOC SiOC (black diamond) $(HSiO_{3/2})$n-poly-hydrogen silsesquioxane (HSQ) $(HSiO_{3/2})$n-poly-methyl-silsesquioxane (MSQ)
Fourth stage (combination with Cu wiring)	Method	Spin on glass (SOG) + chemical mechanical polisher (CMP) Chemical vapor deposition (CVD) + chemical mechanical polisher (CMP)
	Material	Porous SiOC Porous SiOC (black diamond) SOG (containing polyallylether)

The dielectric constant value of SiO_2 which was used from first to third stage is 4.2. The dielectric constant values of $(HSiO_{3/2})$ n-poly-hydrogen-silsesquioxane (HSQ) and $(HSiO_{3/2})$ n-poly-methyl-silsesquioxane (MSQ) which was developed in the fourth stage were 3.0 and 2.4–2.7, respectively. The dielectric constant value decreased with scaling down the technology.

10.3 Low-κ Materials Trends and Their Condition for Practical Usage

For semiconductor manufacturing companies, there are many technical issues to overcome for practical usage of low-κ materials. They need significant capital investment, long time, and many engineers; therefore, companies hardly construct production lines using only low-κ. Many semiconductor companies made partial development for equipment companies and low-κ material companies. However, the above-mentioned method increases license fees and encourages strong competition. Recently, many semiconductor manufacturing companies collaborate not only with equipment and material companies but also with public research center and research groups of the same business companies.

Table 10.2 presents a summary of newly developed low-κ materials for CVD, and Table 10.3 presents similar data for SOG.

Now the main low-κ material which is used till 95 nm technology node is the black diamond which is prepared by plasma-CVD. It is easily manageable but its dielectric constant value is 3.0–3.5 as shown in Table 10.2. From the viewpoint of low-κ, porous materials have much advantage, their dielectric constant values are less than 2.5 [2, 3].

However, the porous materials are fragile and they need much attention to put them into production line. The dielectric constant value of final target of porous material is less than 1.5 and their pore ratio is more than 80%.

Table 10.2 Newly developed low-κ materials for CVD [1]

Materials system	Brand name	Relative dielectric constant (k)	Heat resistance (°C)	Manufacturer
SiO_2 system	–	4.2		–
SiOF system	–	3.7~3.5	–	–
SiOC system	Black diamond (BD)	2.7–2.4	450	Applied materials
	Corral	2.7–2.4	500	Novellus systems
	Aurora	2.7	450	Japan ASM

Table 10.3 Newly developed low-κ materials for SOG

Materials system	Brand name	Relative dielectric constant (k)	Heat resistance (°C)	Manufacture
HSQ system	FOX	2.9	–	Dow Corning
	OCD-12	3.4~2.9	450	Tokyo Ohka Kogyo
	OCD-32	2.5	450	Tokyo Ohka Kogyo
MSQ system	HSG-R7	2.8	650	Hitachi Chemical
	OCD T-7	2.7	600	Tokyo Ohka Kogyo
	LKDT400	2.2~2.7	400	JSR
Porous MSQ	HSG-6211X	2.4	650	Hitachi Chemical
	HSG-6210X	2.1	650	Hitachi Chemical
	ALCAP-S	2.3~1.8	450	AsahiKasei Co
	OCLT-77	2.2~1.9	600	Tokyo Ohka Kogyo
	Aerogel	1.4~1.11	>500	Kobe Steel Co
BCB	Cyclotene	2.65	>350	Dow Chemical
Organic (porous glass)	HOSP	2.5	550	Honeywell
	SiLK	2.65	>490	Dow Chemical
	PolyELK	<2	490	Air Products

10.4 Technical Issues of Porous Low-κ Materials and Their Provisions

There are four main issues to use porous low-κ materials: mechanical properties, adsorption, heat conduction, and relationship with other materials. In this section, we review each issue.

10.4.1 Mechanical Properties

Since porous materials have many pores, mechanical properties must be deteriorated. There are many processes that cause damage, such as chemical mechanical polishing (CMP), scribing, bumping, and back grind. For the CMP process, many provisions such as low pressure CMP, soft pad, mini-electron beam, selected-area UV irradiation are proposed. For scribing, laser scribing is proposed. To reduce residual stress, stress control layer is proposed.

10.4.2 Adsorption

The adsorption problems lead to other problems such as gas evolution of adsorption material (degas). There are many provisions such as silane-coupling treatment, decrease of pore size, capping layer, and closed pore.

10.4.3 Heat Conduction

Adding unnecessary metal wiring is proposed. Moreover, dielectric pole with high heat conduction and capping layer with high heat conduction are also proposed.

10.4.4 Relationship with Other Materials

There are important layers that use low-κ materials in different layers such as adhesion layer, barrier layer, seed layer, Cu wiring, capping layer, and etch-stopping layer. All layers are fine-pattern formed with high aspect ratio and they should be thinner. Main materials and their formation methods are listed in Table 10.4.

Table 10.4 Important materials that use low-κ materials

Layer	Material	Method
Adhesion layer	Ta	Sputtering
Barrier layer	TiN, TaN, TiW, WN	Sputtering
Seed layer	Cu	Sputtering
Cu wiring	Cu	Electroplating
Capping layer	SiC, SiO_2, CoWP	CVD, electroless plating
Etch-stopping layer	SiN, SiC	CVD, SOD

10.5 Ultimate Porous Low-κ Material

As mentioned above, the pore ratio in low-κ materials will increase to decrease the dielectric constant; the ultimate low-κ material is air whose dielectric constant is 1.0. Wiring without low-κ materials is ultimate and this structure makes the above technical issues unnecessary. Some researchers proposed the ultimate structure to eliminate insulator after formation of wiring process. Figure 10.1 shows the schematic model of air-gap formation method which was proposed by Toshiba [4–6]. First, the copper wiring was formed by conventional method. Second, the graphite layer was formed by sputtering method and this layer works as insulator. Third, the surface

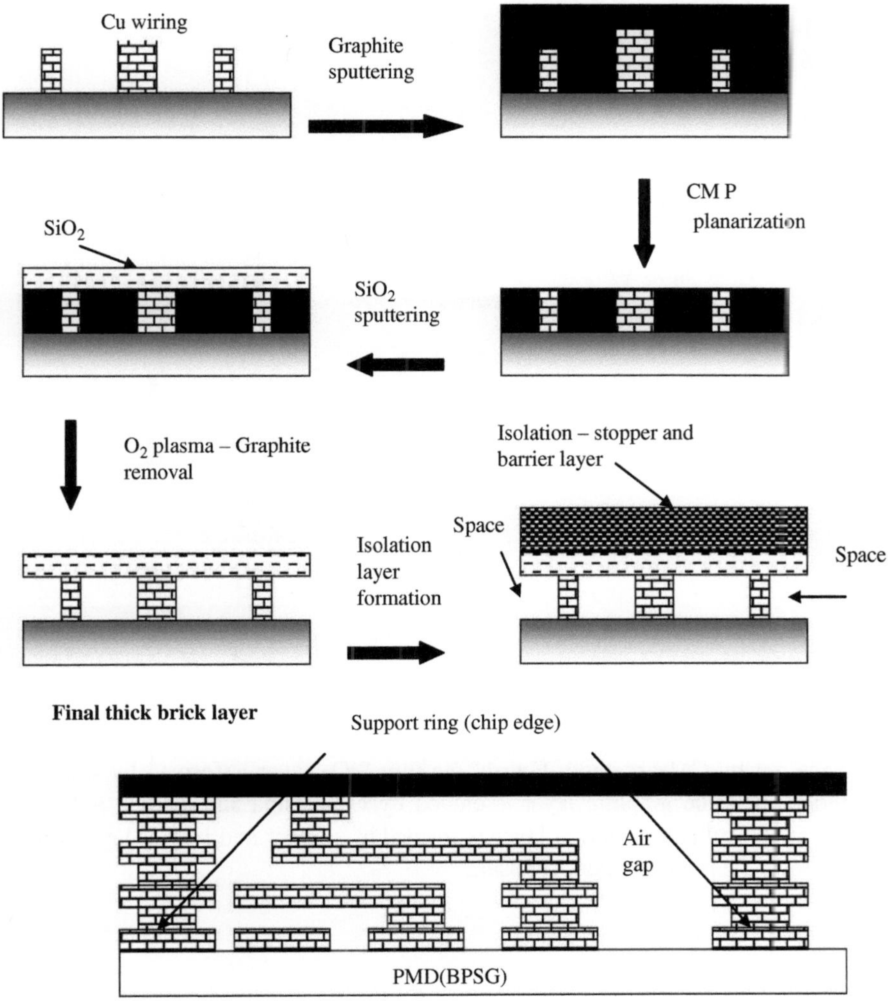

Fig. 10.1 Schematic model of air gap

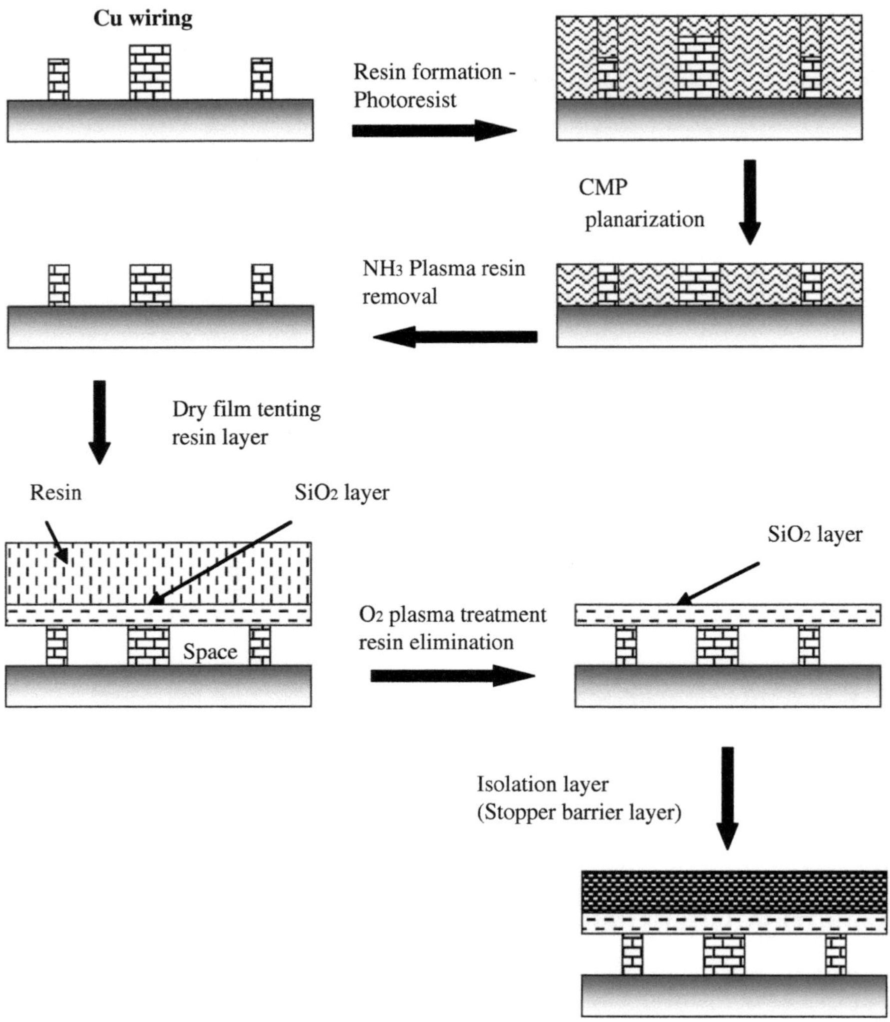

Fig. 10.2 Schematic model of our proposal

is made flat by CMP method. Fourth, the thin SiO_2 layer is formed by sputtering method. Fifth, the graphite layer is etched by O_2 plasma and air gap is formed. Sixth, a high reliable isolation layer is formed as stopper and barrier layer. Figure 10.2 shows the model of our proposal.

References

1. Nikkei Microdevices, **8**, 175 (2000)
2. Electronic Journal, **12**, 8 (2003)
3. Provisional Publication No.2002-25223, 2522 4 (Toshiba)
4. Provisional Publication No.9-237831(Toshiba)
5. Provisional Publication No. 2000-269327 (Toshiba)
6. Nikkei Microdevices, **4**, 130 (1997)

Chapter 11
Electrical and Mechanical Characteristics of Air-Bridge Cu Interconnects

Hyun Park, Matthias Kraatz, Jay Im, Bernd Kastenmeier, and Paul S. Ho

11.1 Introduction

As Cu/low-*k* technology continues to advance, extensive efforts are being focused to implement ultra-lower *k* dielectrics for future interconnects. For the 45 nm node, the interconnect structure is required to have an effective dielectric constant of 2.3–2.6 [1], which will require the dielectric material to have a bulk dielectric constant of about 2.1 if no air-gap design is considered.

Air-bridge Cu interconnects is a promising approach to satisfy the Technology Roadmap requirement for future chips, as air may be considered as the ultimate insulating material with *k* being unity. Implementation of air-bridge structures has been reported using non-conformal chemical vapor deposition into patterned trenches [2–4], or by removing a sacrificial material [5–7]. Efforts so far have been focused on building prototype air-bridge structures; little systematic investigation of the electrical or mechanical characteristics of air-bridge structures has been reported. Such information can be useful for design of feasible air-bridge interconnects.

The mechanical stability of air-bridge Cu interconnects is an important concern. Fabrication of interconnect structures involves multitudes of thermal loading during processing, resulting in thermal stresses and residual stresses in the finished structure [8–10]. Thermal stresses originated from mismatch in coefficient of thermal expansion (CTE) can cause failure at weak links such as vias or interfaces where different materials join and processing-related defects can easily be formed. Residual stresses can have a significant impact on the mechanical performance and reliability of interconnect structures, which may cause stress-induced voiding and electromigration failure in Cu lines and vias, and interfacial delamination and cracking. The magnitude and the nature of thermal stresses in a interconnect structure are greatly influenced by the thermo-mechanical properties of the materials used and might dictate the failure behavior at vias and interfaces. The confinement effect on Cu line by

P.S. Ho (✉)
Microelectronics Research Center, The University of Texas at Austin,
TX 78712-1100, USA
e-mail: paulho@mail.utexas.edu

Y. Shacham-Diamand et al. (eds.), *Advanced Nanoscale ULSI Interconnects: Fundamentals and Applications*, DOI 10.1007/978-0-387-95868-2_11,

surrounding materials can influence material transport along the interconnect and affect the overall electromigration characteristics [11–13]. When an interconnect is strongly confined by the surrounding materials, a relatively high stress-induced counterflux by the accumulated material develops, resulting in an improvement of electromigration resistance [11, 12]. The degree of confinement can be estimated by the effective elastic modulus, *B*, which is determined by the mechanical properties of the surrounding materials. When air-bridge elements are incorporated into the low-*k* structure, basic questions arise regarding the structural integrity of the Cu interconnects and which dielectric material would be most suitable for constructing the air-bridge structure.

In the present study, computer simulation was employed to evaluate the electrical and mechanical characteristics of air-bridge Cu interconnects. Several via level interconnect structures were modeled, which consisted of varying degrees of air gap/dielectric proportions. For these model structures, the effective dielectric constants, the volume-averaged stresses, and the effective elastic moduli were calculated using finite element analysis. The effects of the dielectric material and air-gap configuration on the dielectric and mechanical characteristics are examined.

11.2 Methods of Simulation

11.2.1 Model Structure

Two types of via level air-bridge structures were analyzed, as shown in Figs. 11.1 and 11.2. The first was an ideal case consisting of a single-via level model structure with one via level and two metal line levels (Fig. 11.1). In this structure, air was assumed to replace the solid dielectric between the inter-metal Cu lines in the M1 and M2 levels, while there was no air gap present in the via level. The processing

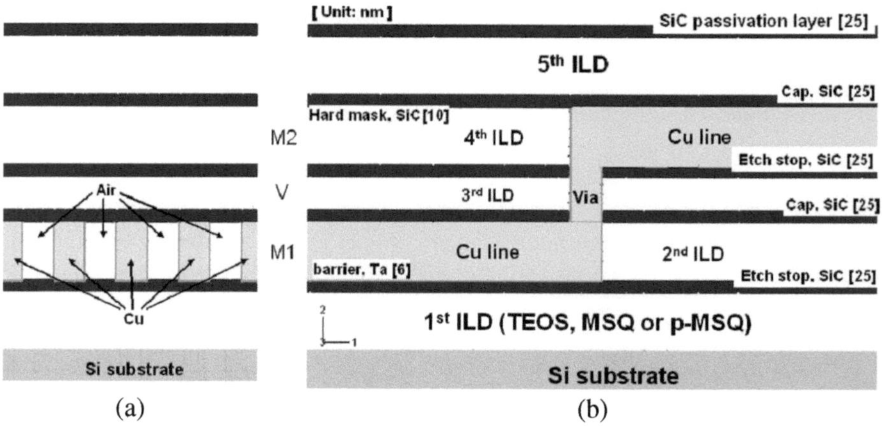

Fig. 11.1 Single-via level model structure: (**a**) front view and (**b**) longitudinal cross-section view

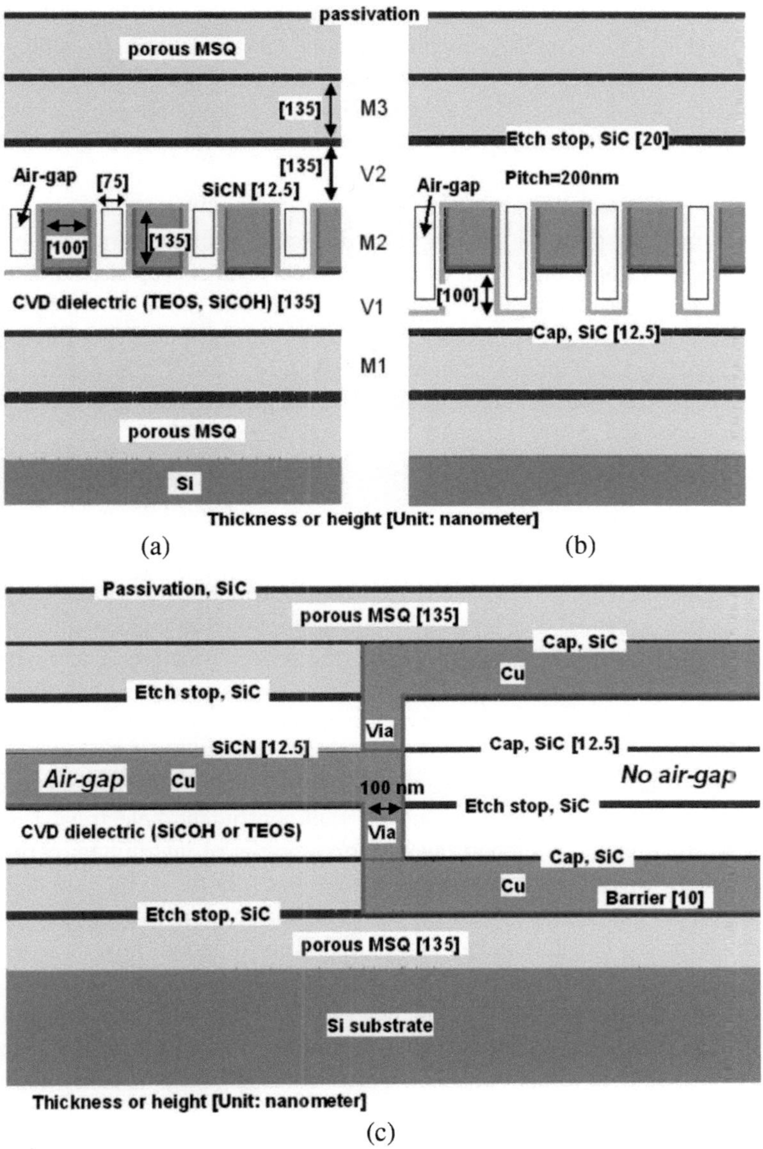

Fig. 11.2 Multi-via level model structures: front views of (**a**) trench air gap, (**b**) trench + via air gap, and (**c**) longitudinal cross-section

feasibility for such air-gap structures was not considered. Three dielectric materials: porous methylsilsesquioxane (MSQ) as a porous CVD low-*k* dielectric, MSQ as a fully dense CVD low-*k* dielectric, and TEOS as a control were investigated in this study. The Cu line width, height, and pitch were fixed at 60, 120, and 120 nm, respectively.

The second was a multi-via level structure with two via levels and three metal line levels (Fig. 11.2a, b). The structure might be fabricated through etching of the dielectric material in the M2 or M2 + V1 level, followed by deposition of the SiCN liner and a non-conformal CVD dielectric. Here, two kinds of air-gap configurations were considered: trench air gap and trench + via air gap (Fig. 11.2b). In the trench air-gap structure, we assumed that the air gap was present only in the M2 level. Furthermore, we assumed that the air bridge was partially (~40%) implemented, with the remainder (~60%) made up of a dielectric insulator surrounding the air gap at the side walls of the M2 lines. In the trench + via air-gap structure, air gaps were extended by an etching process through the M2 then the V1 levels by as much as 100 nm. Porous MSQ, SiCOH as a fully dense dielectric, and TEOS as a control were included as CVD dielectric materials in M2, V1, and V2. In the M1 and M3 levels, porous MSQ was assumed to have been deposited. The line width, height, and pitch were fixed at 100, 135, and 200 nm, respectively.

11.2.2 Electrical Simulation

The air-bridge structure was modeled as a two-dimensional cross-section perpendicular to the line direction. The level of detail was provided to include all constituent parts with their individual permittivity into the model structure. The commercial software tool TCAD-Raphael by Synopsys Inc. was used to calculate the capacitance of the model structure. The core computation engine used was a solver for the Poisson problem [14].

The goal was to extract the effective permittivity (or effective dielectric constant, k_{eff}), which is a composite property of all constituent dielectrics. The low-k regions with the higher-k support layers form a region of intermittent permittivity that ultimately affects signal propagation in the lines. In the worst case, the effective permittivity can be almost as high as the highest k in the structure, eliminating all benefits of low-k materials and air bridges. This situation must be avoided by carefully choosing the placement of the air bridges, support layers, and their thicknesses. In reverse, k_{eff} can never be as low as the lowest k in the structure, but it can be considerably minimized. The structure considered comprised of all the elements including Si substrate within one pitch, which coincided with the definition used in the ITRS.

The effective permittivity was calculated as follows: we first assumed that k of all constituent dielectrics was the same at one specific k and calculated the corresponding capacitance. There exists a linear relationship between this globally homogeneous k and the calculated capacitance of the homogenized structure. Thus, by changing the global k to a second specific value and calculating the resulting capacitance a model-specific linear calibration line was obtained. Using the previously calculated capacitance base on the individual permittivities of all constituent materials, the effective permittivity was determined as the corresponding k-value in the calibration curve. The flow chart diagram in Fig. 11.3 summarized the procedure described here.

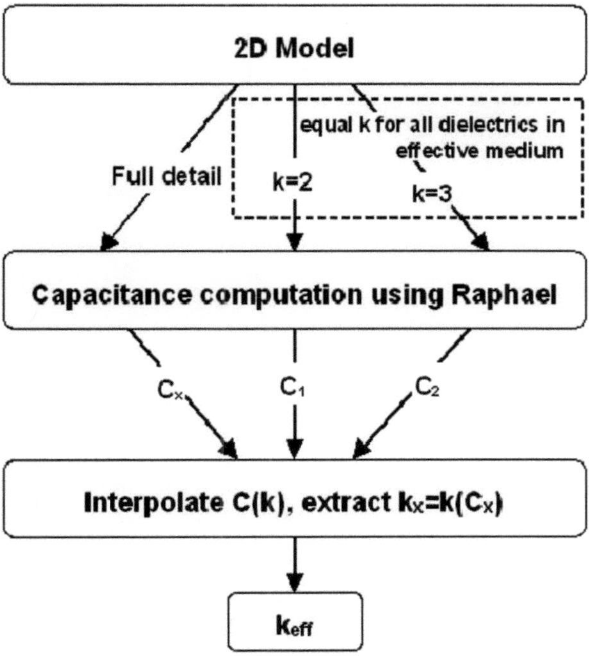

Fig. 11.3 Flow chart for extraction of effective dielectric constant. An effective medium is applied for two of the three computations. The capacitance is linearly interpolated between the two results to determine the value of k that corresponds to the third, fully detailed capacitance computation. The determined value of k is equivalent to the effective relative dielectric constant k_{eff}

11.2.3 Stress Simulation

Finite element analysis (FEA) was used for mechanical simulation of the model structures. We followed a series of wafer process steps by using an "Element Removal and Reactivation" technique [15]. Most FEA studies in the past assumed that a structure possessed a zero stress temperature, and then calculated the stresses developed after one thermal loading. In this study, however, we simulated the stress build up at each sub-process step and temperature. This allowed us to identify the critical processing steps in generating residual stresses in the interconnect structure, which have to be optimized for improving the structural integrity for air-bridge interconnects. The coordinate system for stress analyses of Cu lines and vias in (a) single-via structure and (b) multi-via structure is shown in Fig. 11.4.

M1 – lower level metal lines
M2 – intermediate level metal lines
M3 – top layer metal lines
Via – via contacts (V1 – between M1 and M2, V2 – between M2 and M3)

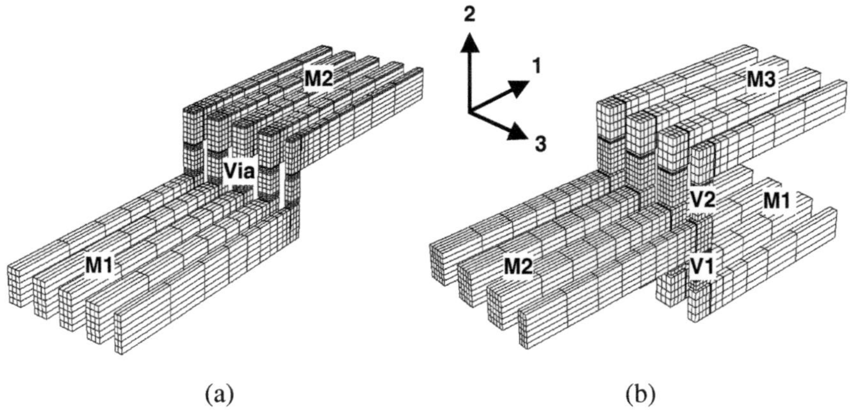

Fig. 11.4 Coordinate system for stress analyses of Cu lines and vias in (**a**) single-via structure and (**b**) multi-via structure

The aim of the stress simulation in this chapter was to calculate the volume-averaged stresses of Cu line and via and to estimate the general trend of stresses as a function of air-gap configurations and dielectric materials. They were calculated along three reference directions: the line direction (σ_{11}), vertical direction to the top plane (σ_{22}), and transverse direction (σ_{33}). Volume-averaged hydrostatic and von Mises stresses were determined from the principal stress components. The hydrostatic stress is expressed as

$$\sigma_H = \frac{(\sigma_x + \sigma_y + \sigma_z)}{3} \tag{11.1}$$

where σ_x, σ_y, and σ_z are three principal stresses.

The von Mises stress is expressed as

$$\sigma_{\text{von-Mises}} = \frac{1}{\sqrt{2}}\left[(\sigma_x - \sigma_y)^2 + (\sigma_y - \sigma_z)^2 + (\sigma_z - \sigma_x)^2\right]^{\frac{1}{2}}$$

The hydrostatic stress provides an estimate of the stress level that can lead to stress voiding, which is a reliability concern. The von Mises stress represents the maximum shear stress and is responsible for yielding or plastic deformation of the interconnect structure. All materials were assumed to behave elastically during thermal loading and unloading, and their properties are listed in Table 11.1. Air was assumed to be a tangible material with a negligibly small elastic modulus, and perfect interfacial adhesion was assumed between different materials.

The processing steps considered for the multi-via level model structure are described in Table 11.2. Each layer was assumed to be stress free during its

Table 11.1 Material properties for model structures

	Elastic modulus (GPa)	Poisson's ratio	CTE (ppm/°C)	Dielectric constant
Si	130	0.28	2.3	11.5
SiC (cap, etch stop)	440	0.17	4.5	5.1
SiCN	440	0.17	4.5	5.0
Ta (barrier)	185	0.34	6.5	–
Cu	104	0.35	17.0	–
Air	1.0×10^{-8}	0.3	100	1.0
Porous MSQ	3	0.3	60	2.0
MSQ	–	–	–	2.8
SiCOH	9.5	0.3	30	2.8
TEOS (SiO_2)	72	0.2	1.4	4.4

Table 11.2 Process procedure of multi-via level air-bridge structure for processing stress calculation

Step	Sub-process and temperature	Step	Sub-process and temperature
1	ILD, p-MSQ dep. (400°C)	14	Cap dep. (400°C)
2	Etch stop dep. (400°C)	15	SiCN liner dep. (400°C)
3	ILD, p-MSQ dep. (400°C)	16	CVD dielectric dep. and air-gap formation (400°C)
4	Barrier dep. (–20°C)	17	Etch stop dep. (400°C)
5	Cu dep. of M1 (25°C)	18	ILD, p-MSQ dep. (400°C)
6	Post-dep. anneal (150°C)	19	Barrier dep. (–20°C)
7	Cap dep. (400°C)	20	Cu dep. of V2 and M3 (25°C)
8	CVD dielectric dep. of V1 (400°C)	21	Anneal (150°C)
9	Etch stop dep. (400°C)	22	Cap dep. (400°C)
10	CVD dielectric of M2 (400°C)	23	ILD, p-MSQ dep. (400°C)
11	Barrier dep. (–20°C)	24	Passivation dep. (400°C)
12	Cu dep. of V1 and M2 (25°C)	25	Cooling to room temperature
13	Anneal (150°C)		

deposition. It was also assumed that the stresses in the Cu line and via were fully relaxed through post-deposition annealing at 150°C.

As a boundary condition, a mirror symmetry was applied to the front (at $x_1 = 0$) and right (at $x_3 = 0$) planes of the model structure. Since the dimension of the rigid Si substrate is several orders of magnitude larger than those of the fine structures on top, the back (at $x_1 = t$) and left ($x_3 = p$) planes were assumed to be constrained by the Si substrate, which remained flat at all temperatures.

11.2.4 Calculation of Effective Elastic Modulus, B

The effective elastic modulus, *B* is defined as

$$\frac{dC}{C} = -\frac{d\sigma}{B} \tag{11.3}$$

where *dC* is the change of the available lattice sites per unit volume due to mass transport in Cu, *dC/C* is the volumetric strain in the metal lines, and $d\sigma$ is the change of hydrostatic stress [11]. In this study, stress induced by material accumulation was simulated based on electromigration-induced mass transport using FEM [12, 13]. The CTE components in the 1-direction and 3-direction of the Cu were set to be zero but that in the 2-direction or axial direction was not and the CTEs for the other materials were set to be zero in order to simulate the change of lattice sites, which is determined by Cu transport along the top interface between the Cu line and the cap layer. The change of the lattice density can be calculated using the expansion of the Cu lines.

11.3 Simulation Results and Discussion

11.3.1 Effective Dielectric Constant, $\mathbf{k_{eff}}$

Figure 11.5a shows the k_{eff} with respect to air gap for various dielectric materials for the single-via level model structure. The figure shows that a large decrease of k_{eff} can be realized with introduction of air gaps. The reduction in k_{eff} was larger for the higher dielectric constant material. The combination of ~65% air gap with TEOS or ~35% air gap with p-MSQ can satisfy the ITRS need for the 45 nm technology node. It was found that the k_{eff} did not depend much on the choice of dielectric material at the via level when 100% air gap was assumed at the line level. For example, the k_{eff} for TEOS and porous MSQ are 2.1 and 2.0, respectively (Fig. 11.5b).

These results suggest that TEOS can be used as a dielectric material at the metal line levels. TEOS is mechanically stronger than MSQ or p-MSQ and should provide more support for Cu lines during thermal loading. However, it should be pointed out that 100% air gap is unrealistic from processing point of view. More realistically, the air gap will be surrounded by the residual solid dielectric left over from processing. Furthermore, no sub-process for air-gap formation was considered in our simulation. Therefore, the results based on the multi-via level model structure seem more relevant as discussed below.

Figure 11.6 shows the k_{eff} for the multi-via level model structure with respect to air-gap extension for various dielectric materials and SiCN liner thicknesses. We found that the combination of SiCOH, a fully dense CVD dielectric, and air-gap extension into the via level can drive k_{eff} down as low as 2.27. Therefore, the fully dense SiCOH meets the ITRS need for the 45 nm technology node for a wide range of air-gap extension and can be extended to the 32 nm node with a deeper air gap.

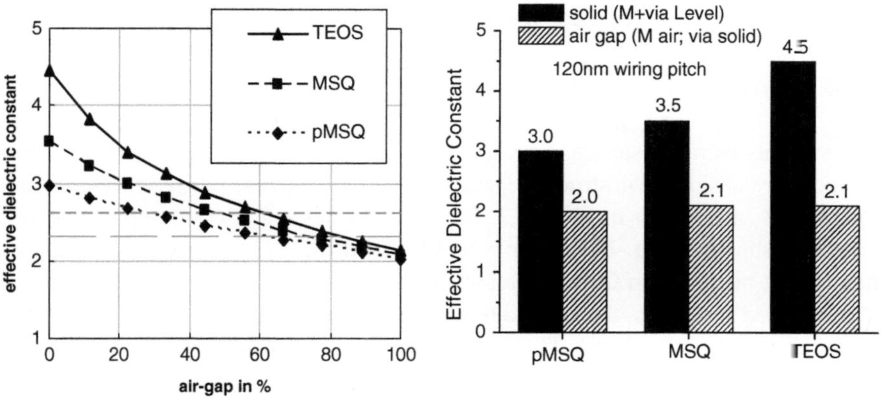

(a) (b)

Fig. 11.5 Calculated effective dielectric constants for single-via level model structure: (**a**) with air gap in % and (**b**) at 100% air gap

Fig. 11.6 Calculated effective dielectric constants for multi-via level model structure

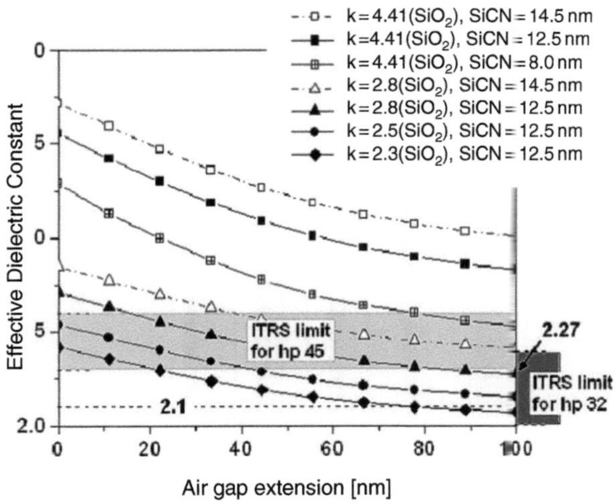

If a CVD dielectric with a dielectric constant of 2.3 is employed, k_{eff} can be reduced to less than 2.1.

For the next generation ICs, air-bridge Cu interconnects need to satisfy not only the electrical and mechanical requirements but also the manufacturability. Ease of processing and processing cost are important factors for choosing a particular air-bridge structure. On this basis, the fully dense SiCOH seems to be more versatile than TEOS from the standpoint of air-gap processing. Although the porous p-MSQ is more desirable due to the lower k_{eff} than SiCOH, the technology for porous low-k dielectrics is not as mature and processing would be more difficult.

11.4 Volume-Averaged Stresses

11.4.1 Stresses in Single-Via Structure

The volume-averaged stresses for Cu line and Cu via after the final cooling to room temperature in single-via structure, where air gaps occupy the entire spaces between metal lines, are shown in Fig. 11.7. For the Cu lines, all stress components were tensile as shown in Fig. 11.7a. For p-MSQ structures, the incorporation of air gaps had almost no effect on Cu line stresses although the effect was larger for σ_{22}. For TEOS-based structures, the air bridge was employed, the k_{eff} can be reduced to less than 2.1. On the other hand, TEOS can meet the ITRS need for only 45 nm node by both extending air gap almost 100% and reducing the liner thickness to 8 nm. Structure reduced significantly the line stresses, bringing them down to the levels similar to that of the p-MSQ structure. Therefore, the Cu line stresses in either TEOS or p-MSQ did not depend much on the dielectric material used to construct the air-gap structure. This is because the dielectric material was located only at the via level away from the metal level. The k_{eff} for the single-via level structure behaved similarly due to the same reason (Fig. 11.5b). Between M1 and M2 stresses, the difference was not significant.

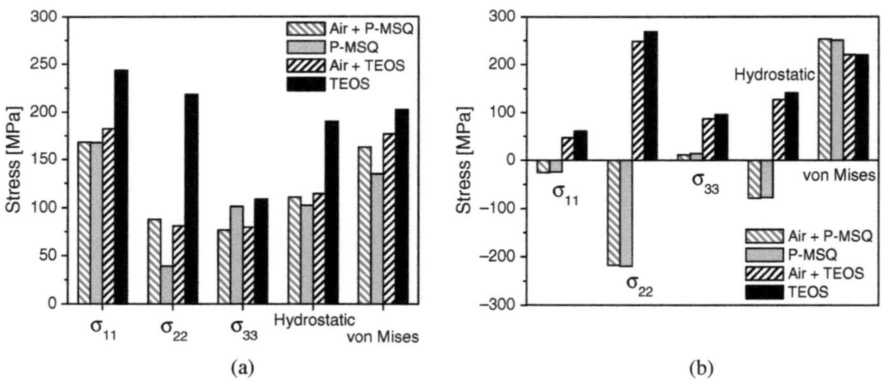

Fig. 11.7 Volume-averaged stresses in single-via level structure: (**a**) Cu line stresses and (**b**) Cu via stresses

For the Cu via stresses (Fig. 11.7b), the air-bridging effect is negligible, because air gaps are incorporated only in the metal line levels away from the via level. The via stresses does depend on the dielectric material surrounding the via. For air-gap structure with p-MSQ, the axial and hydrostatic stresses are compressive, opposite from the case of air-gap structure with TEOS. The axial stress difference between p-MSQ and TEOS structures can be attributed to the CTE mismatch between the Cu via and the surrounding dielectric. Compared with Cu, the higher CTE of p-MSQ leads to a compressive axial stress in the Cu via, while the lower CTE of TEOS leads to a tensile stress.

The magnitudes of the stresses shown in Fig. 11.7a, b are generally less than ~250 MPa, which may be deemed too small to cause mechanical failure of Cu lines or vias. However, caution should be exercised in interpreting the simulation results because they were calculated under several assumptions. For example, we assumed that stresses of the Cu lines and vias were fully relaxed by annealing at 150°C. If we assumed a higher stress relaxation temperature, the magnitude of Cu stresses at room temperature would have increased. The assumptions on the material properties and the interfaces also affect the calculated stress level. More than anything else, the stresses calculated are volume averaged, making them lower than the local stresses at stress concentration points. Therefore, the stresses in an actual structure will be higher than the calculated volume-averaged stresses.

The via stress results provide some insights into the mechanical stability of vias in air-bridge Cu interconnects. Vias are weak links in the interconnect structure, where various materials adjoin and structural defects can be easily developed during processing. Therefore, vias are expected to be more sensitive to thermal and residual stresses than the Cu lines. The compressive axial stress and von Mises stress for porous low-*k* dielectric such as p-MSQ, if high enough, can lead to via collapse or plastic deformation, while via pulling may occur for TEOS if a high enough tensile axial stress develops. A more localized stress calculation than the volume-averaged stresses as employed here will be necessary for a more realistic assessment of possible failures as will be further discussed in Section 11.5.

11.4.2 Stresses in Multi-Level Structure

The volume-averaged stresses of Cu line and via with various dielectrics for multi-level structures with air gaps in the trench level and extending into the via level are shown in Figs. 11.8 and 11.9. The M2 Cu line stresses for SiCOH and p-MSQ are similar, while the stresses for TEOS are generally higher (Fig. 11.8). It is also seen that the stress levels for the air gap in trench structure in general are similar to those for the air gap extending into the via structure. This indicates that extending the air gap into the via level has negligible effect on Cu line stresses.

Compared with the Cu line stresses, the Cu via stresses depend on both dielectric and air-gap extension. As shown in Fig. 11.9a, a high CTE dielectric leads to compressive axial and compressive hydrostatic via stresses in the SiCOH and p-MSQ air gap in trench structure. On the other hand, TEOS gives rise to tensile stresses. It is also seen that von Mises stress for p-MSQ was highest. The via stresses are similar to those from the single-via level structure, as no air-gap is present at the via level in either structure.

Air-gap extension into the via level increases the via stress levels or turns the axial and hydrostatic stresses to tensile (Fig. 11.9b). As a result, the stress levels for TEOS, particularly σ_{22} and σ_{33}, become highest. The stress increase toward the tensile direction can be explained by the confinement effect of the SiCN liner. When air gaps are extended in the via level, the SiCN liner, having a much higher elastic

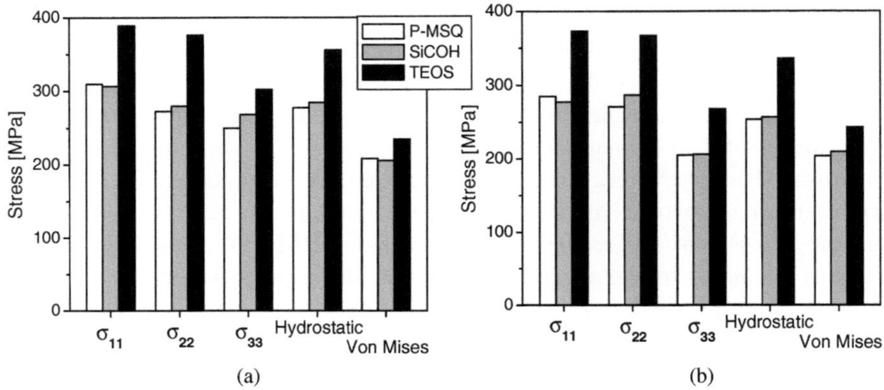

Fig. 11.8 Volume-averaged Cu line stresses in multi-via level structures: (**a**) trench and (**b**) trench + via r-gap structure

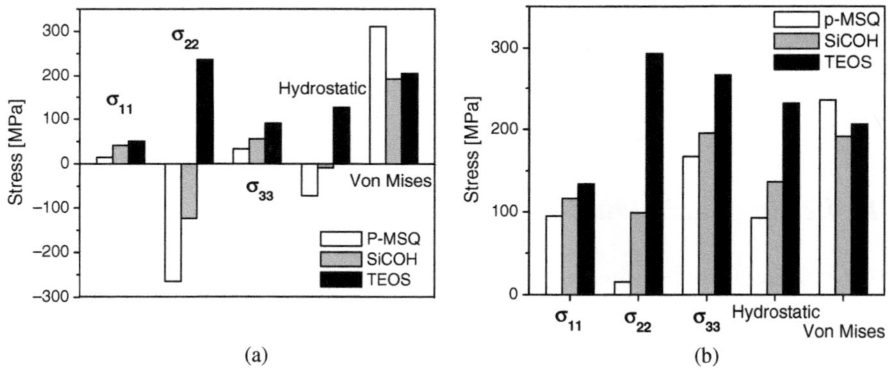

Fig. 11.9 Volume-averaged Cu via stresses in multi-via level structures: (**a**) trench and (**b**) trench + via air-gap structure

modulus and lower CTE than Cu, is also extended into the via level and confines the Cu via more tightly than a dielectric does, resulting in higher tensile stresses in the via.

In comparison with TEOS and p-MSQ, SiCOH led to a moderate stress level in both Cu lines and Cu vias. There is no stress component highly developed to cause an early alarm for SiCOH. This is due to its thermo-mechanical properties, which are intermediate between TEOS and p-MSQ (Table 11.1). It is interesting that the difference in Cu stresses for various dielectric materials is less significant than the difference in their elastic properties. For example, the CTE of SiCOH is ~ 21 times higher than TEOS and elastic modulus 8 times lower. Much of the CTE and modulus differences seem to be canceled out by each other with the resulting Cu line and via stress levels for the two materials within 3 times.

Although our stress results provided some valuable information, it should be pointed out that volume-averaged stress simulation alone is not sufficient to predict the mechanical stability of air-bridge Cu interconnect structures. For example, local stress distribution and stress concentration at specific locations are important [16, 17]. Even for a low average stress, a high local stress can develop to cause failure at specific locations. Interfacial fracture characterization is also important and should be considered. Interconnect structures consist of various materials and multitudes of interfaces. The interfaces work as a mechanically weak link where defects can form during processing, leading to cracking or delamination. Of particular interest is the stress imposed onto the interconnect structure during packaging assembly of the low-*k* chip. The impact of such chip–package interaction [18] on the structural integrity of air-bridge structure can be of important reliability concern. This problem is being investigated by calculating the strain energy release rate for the air-bridge Cu interconnect structures.

11.4.3 Effective Elastic Moduli, B

Figure 11.10 shows the calculated *B* for the multi-via level model structures. As expected, *B* increases as the elastic modulus of dielectric materials increases; thus, *B* is the highest for TEOS. The effect, however, is relatively small and independent of the low-*k* material. This is because the confinement on the Cu lines from the hard SiCN liner and the barrier layer is considerably larger than that from the ILD and IMD dielectrics. Air-gap extension into the via level leads to a slight decrease of *B* although the change was negligible for p-MSQ and SiCOH. The results imply that Cu mass transport and electromigration are not significantly influenced by the low-*k* dielectric material for the air-gap structure.

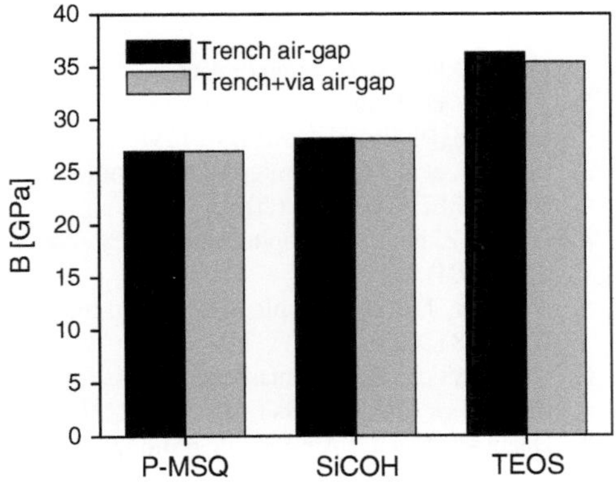

Fig. 11.10 Effective elastic moduli for multi-via level structures

11.5 Conclusions

Two kinds of air-bridge models were modeled in this study. One was the single-via level model structure where 100% air gap was assumed in the metal level but no air gap in the via level. The other were for two multi-via level structures, one with ~40% air-gap area was assumed in the metal level (the air gap in trench structure), and the other with the air gap extending into the via level (the air gap in the via structure). Simulation methods were carried out to evaluate their electrical and mechanical characteristics, including the effective dielectric constants, volume-averaged stresses, and effective elastic moduli of the Cu lines and Cu vias.

For the multi-via level structures, the results of k_{eff} showed that the combination of fully dense CVD low-*k* dielectric and air-gap extension into the via level is promising, in that k_{eff} can satisfy the ITRS need for 45 nm node and extendable to 32 nm node. In the multi-via level structures, the process-induced stress levels in Cu lines and vias for SiCOH were intermediate in between those for TEOS and p-MSQ. Compared with the difference in thermo-elastic properties among the dielectric materials considered, the Cu stress difference among these dielectric materials was relatively insignificant. In the multi-via level structures with the trench air gap, high compressive axial stress and high von Mises stress were evolved in the via when p-MSQ was used as a dielectric, while high tensile axial stress developed in the via for TEOS. For the trench + via air-gap structure, the use of TEOS caused a higher axial tensile stress than with other dielectrics. The results of *B* calculation indicate that EM characteristics are expected to be only moderately influenced by the dielectric material and air-gap configuration for the Cu/low-*k* air-bridge interconnects structures.

Acknowledgments This work was supported by SEMATECH through the Advanced Materials Research Center at the University of Texas at Austin and the Korea Research Foundation Grant funded by Korea Government (MOEHRD, Basic Research Promotion Fund) (KRF-2004-214-D00312).

References

1. ITRS 2004 Update: Interconnect.
2. Arnal, V. et al.: Integration of a 3 level Cu-SiO2 air gap interconnect for sub 0.1 micron CMOS technologies (Proc., IEEE IITC) 298 (2001)
3. Shieh, P. et al.: Electromigration reliability of low capacitance air-gap interconnect structures (Proc. IEEE IITC) 203 (2002)
4. Gabric, Z. et al.: Air gap technology by selective ozone/TEOS deposition (Proc. IEEE IITC) 151 (2004)
5. Noguchi, J. et al.: Simple self-aligned air-gap process with Cu/FSG structure (Proc., IEEE IITC) 68 (2003)
6. Uno, S. et al.: Dual-Damascene process for air-gap Cu interconnects using conventional CVD films as sacrificial layers (Proc., IEEE IITC) 174 (2005)
7. Daamen, R. et al.: Air gap integration for the 45 nm node and beyond (Proc., IEEE IITC) 240 (2005)

8. Yang, K.-H. et al.: Sequential process modeling for determining process-induced thermal stress in advanced Cu/Low-*k* interconnects (Mater. Res. Soc. Sym Proc.) Vol. 766, 39 (2003)
9. Rhee, S.-H.; Du, Y.; and Ho, P. S.: Thermal stress characteristics of Cu/oxide and Cu/low-k submicron interconnect structures. J. Appl. Phys. **93**, 3926 (2003)
10. Wang, G.: Thermal deformation of electronic packages and packaging effect on reliability for Copper/low-k interconnect structures (Ph.D. Dissertation, The University of Texas at Austin) 56, 60 (2004)
11. Korhonen, M. A. et al.: Stress evolution due to electromigration in confined metal lines. J. Appl. Phys. **73**, 3790 (1993).
12. Hau-Riege, S. P. and Thompson, C. V.: The effects of the mechanical properties of the confinement material on electromigration in metallic interconnects. J. Mater. Res. **15**, 1797 (2000)
13. Gan, D. W.: Thermal stress and stress relaxation in Cu metallization for ULSI interconnects (Ph.D. Dissertation, The University of Texas at Austin, USA) 190, 193 (2005)
14. Kastenmeier, B.; Pfeifer, K.; and Knorr, A.: Porous low-k materials and effective k. Semiconductor Int., **27**, 87 (2004)
15. ABAQUS/Standard, User's Manual, Vol. **I**, Hibbitt, Karlsson & Sorensen, Inc., 7.4.2
16. Auersperg, J.; Vogel, D.; and Michel; B.: Crack and delamination risk evaluation of thin silicon applications based on fracture mechanics approaches. 5th Int. Conf. on Thermal and Mechanical Simulation and Experiments in Micro-Electronics and Micro-Systems, 169 (2004)
17. Baik, J.-M.; Park, H.; Joo, Y. C.; and Park, K.-C.: Effect of dielectric materials on stress-induced damage modes in Damascene Cu lines. J. Appl. Phys. **97**, 104513 (2005)
18. Wang, G.; Groothuis, S.; and Ho, P. S.: Packaging effect on reliability for Cu/low k structures. IEEE ECTC, 727 (2003)

Chapter 12
ALD Seed Layers for Plating and Electroless Plating

Jay J. Senkevich

12.1 Introduction

Atomic layer deposition (ALD) in many ways is a logical extension of chemical vapor deposition (CVD) with close scrutiny over precursor deliver and one other aspect, namely, the process and chemistry defined by its self-limiting nature. During each pulse of precursor, no more than one chemical monolayer, which is often a metallorganic, is chemisorbed onto the substrate surface. The discussion here is focused on metal ALD; however, much of the discussion can be applied to other ALD systems. The precursor delivery with ALD is modulated unlike CVD where the precursors are all delivered together. In both cases a carrier and process gas is used with low pressure conditions (~1 mTorr to ~10 Torr regime). CVD can be conceptually written out according to Equation (12.1).

$$A + B \rightarrow C + \text{by-products} \uparrow \quad (12.1)$$

where A can be considered a metallorganic and B a reducing agent for metal ALD. C is the metal and the by-products are volatiles not incorporated into the deposited film. With ALD the reaction is separated into two half reactions shown in Equations (12.2), (12.3), and (12.4).

$$A \rightarrow A^{*}\text{-Sub} \quad (12.2)$$

First, the metallorganic A is dosed to the substrate and A*-Sub is the chemisorbed metallorganic on the substrate [1].

$$B + A^{*}\text{-Sub} \rightarrow \text{C-Sub} + \text{by-products} \uparrow \quad (12.3)$$

J.J. Senkevich (✉)
Brewer Science Inc., 2401 Brewers drive, Rolla, MO 65401 USA
e-mail: jsenkevich@brewerscience.com

Y. Shacham-Diamand et al. (eds.), *Advanced Nanoscale ULSI Interconnects: Fundamentals and Applications*, DOI 10.1007/978-0-387-95868-2_12,

The reducing agent B is then dosed to the chemisorbed metallorganic A*-Sub. These results with a metal on the substrate and the volatile by-products are purged away [2].

$$A \rightarrow A^{*}\text{-C-Sub} \tag{12.4}$$

The second cycle is characterized by dosing the metallorganic to the metal substrate A*-C-Sub.

This metallorganic could be interacting with the substrate directly or with the metal directly. The reason for this is that only one monolayer of metallorganic is deposited at best. The resulting metal is always less than a monolayer.

This can be schematically shown in Fig. 12.1 with palladium (II) hexafluoroacetylacetonate, an ALD precursor. $Pd^{II}(hfac)_2$ is dosed to a surface that has appropriate bonding sites X. The reducing agent [Red] or B is dosed to the surface and the hfac ligands are protonated and leave the surface. What is left is less than a monolayer of Pd. The cycle is repeated and in some cases the $Pd^{II}(hfac)_2$ interacts with X and in some cases with elemental Pd. The interaction between $Pd^{II}(hfac)_2$ and the substrate is critical. If the interaction is too weak then no deposition will ever occur. This interaction follows Langmuir kinetics [3] and will be discussed in the context of dosing next. Further, the reducing agent needs to liberate electrons to the metallorganic to reduce it to the elemental metal. It also needs to liberate the metallorganic ligand. In the case of coordination compounds the reducing agent protonates the ligand thus volatilizing it.

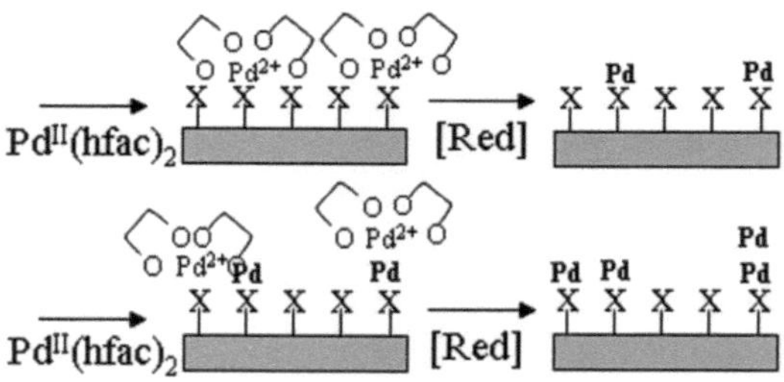

Fig. 12.1 Metal ALD process with $Pd^{II}(hfac)_2$. There is always less than a monolayer of Pd atoms per ALD cycle

The precursors are modulated with ALD. Intermixing the precursors can be considered parasitic CVD and is not desired for chemistry control and film thickness uniformity. The precursors are added to the vacuum system by the generalized procedure [4].

$$A \rightarrow \text{Purge} \rightarrow B \rightarrow \text{Purge} \tag{12.5}$$

where A is the metallorganic and B is the reducing agent as discussed above. From Equation (12.2), it was discussed that the metallorganic has to chemisorb to the substrate or metal C. The gas A is dosed to the substrate. When the dose is stopped there is a desorption isotherm that exists. How fast A desorbs from the surface is determined by the interaction energy (Langmuir kinetics). Low temperatures favor less desorption during the purge cycle. The purge cycle should be kept as short as possible to purge out any metallorganic but if it is too short parasitic CVD can result. For high throughput manufacturing processing parasitic CVD might be desired; however, the film conformality and chemistry will often be between ALD and traditional thermal CVD [5].

Low temperatures favor slow desorption kinetics and also favor the chemisorption of an intact hfac ligand in the case of $Pd^{II}(hfac)_2$ and other hfac-based coordination compounds such as $Cu^{II}(hfac)_2$. It is not clear if tmhd-based ligands have the same instability issues as the hfac ligand since there is no molecular tag. The hfac ligand has a $-CF_3$ group that can be analyzed readily by surface science techniques such as X-ray photoelectron spectroscopy (XPS) [6]. If this was the whole story, then dosing and deposition should always be undertaken at lower temperatures. However, the reducing agent has to be thermally or catalytically cleaved to liberate its electrons. Hydrogen is not a very effective reducing agent since the H–H bond is very stable. It will cleave with relative ease on an elemental metallic surface [7]. However, the metal surface has to be oxide free. This can be possible with plasma clean of tantalum as part of the TaNx/Ta stack [8]. However, if the ALD metal is deposited on a non-metallic substrate, hydrogen is not the best choice.

Finally, there are two temperature limits for the deposition of any ALD material. The lower limit is determined by the condensation of the metallorganic (or other reactants). The upper limit is determined by the thermal decomposition of the metallorganic. If this temperature is exceeded, self-limiting chemistry will not occur and a "dirty" film will result. Thermal decomposition breaks carbon–carbon, carbon–oxygen, and carbon–fluorine bonds that result in carbon in the metal deposit. At lower temperatures than the decomposition temperature, self-limiting chemistry can be evident; however, dissociation can be evident. Dissociation is where a few bonds are broken, which leaves residue in the metallic film but no metal is deposited. In the case of $Pd^{II}(hfac)_2$, the $-CF_3$ group can dissociate from the hfac ligand. A few percent impurities in a metallic film can severely disrupt its texture and therefore the texture of the electrochemically deposited Cu film.

12.2 Thermal and Plasma-Enhanced ALD

The discussion of metal ALD will be focused on palladium. However, the discussion here can be applied to nearly any transition metal that possesses the analogous coordination compound. These coordination compounds have the pseudo-six member ring resonance structure. The structure of $Pd^{II}(hfac)_2$ is shown in Fig. 12.2. Beyond the hfac ligand found in $Pd^{II}(hfac)_2$, there are two other common ligands tetramethylheptanedionate (tmhd) and acetylacetonate (acac). Tmhd has t-butyl and acac has just protons, respectively, instead of $-CF_3$ (trifluoromethyl) groups with the hfac ligand. Typically, the acac coordination compounds are not very volatile because of the strong molecule–molecule interactions. Low polarizable and bulky substituent groups like hfac and tmhd, respectively, allow the coordination complex to be easily

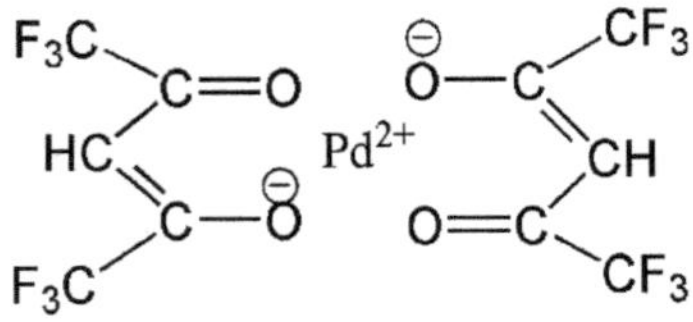

Fig. 12.2 Structure of $Pd^{II}(hfac)_2$

sublimable. In general if the solid melts it will vaporize or sublime and could be an appropriate ALD precursor.

The choice of reducing agents has the same requirements as the metallorganic. Namely, the reducing agent has to be able to "fly" without decomposing. The reducing agent needs to yield electrons to the metallorganic and should, in the case of coordination compounds, yield protons to remove the organic ligand. Hydrogen is the logical choice but it is fairly inert. The H–H bond does not cleave until ~560°C (autoignition temperature). This temperature is prohibitive for most ALD processes. As previously discussed hydrogen works well with catalytic surfaces but often it is desired to deposit the catalytic layer on a non-metallic surface like a metal oxide or functionalized organic. In these cases an alternate reducing agent is needed. Some work has been undertaken with glyoxylic acid [2] and formalin (37 w/w% formaldehyde in water) [9]. However, the initial work with formalin was undertaken above the precursor's, $Cu^{II}(hfac)_2$, thermal stability. Further, since formalin is not pure formaldehyde, the redox reaction at the substrate surface can be rather complicated, especially since water can be an oxidizing agent. Both of these reducing agents have been used in electroless deposition [10–11]. Metal ALD has many analogies with electroless metal deposition since they are both redox processes. The last option that exists for delivering a reducing agent to the chemisorbed metallorganic at the substrate surface is the use of a remote plasma source, which is not available for electroless processes. The plasma source can cleave molecular hydrogen to atomic hydrogen. The drawback of this method is the lack of conformality of the atomic hydrogen. However, it may be suitable for ~10:1 aspect ratio trenches used in the back end of the line for future technology nodes (45 nm and beyond).

The first consideration to make for an ALD process is the "window" of potential operation as discussed above. The upper bound is determined by the thermal decomposition of the precursor. Figure 12.3 shows two Cu ALD precursors: $Cu^{II}(hfac)_2$ that readily forms a hydrate and $Cu^{II}(tmhd)_2$. $Cu^{II}(hfac)_2$ and $Pd^{II}(hfac)_2$ are nearly identical except that $Cu^{II}(hfac)_2$ forms a hydrate because copper has a stronger tendency to interact with oxygen-based Lewis bases.

In Fig. 12.3, the precursors were dosed for 2 min each. The $Cu^{II}(hfac)_2$ precursor thermally decomposes at 230°C much like $Pd^{II}(hfac)_2$ [12]. However, due to the presence of water $Cu^{II}(hfac)_2$ shows non-self-limiting growth above ~110°C. Water has been known to "chew" up on the hfac ligand and has been used as a nucleating agent for ALD processes [13]. $Cu^{II}(tmhd)_2$ is known to decompose at ~260°C but there is no dramatic jump at this temperature because of the different thermal decomposition mechanisms between $Cu^{II}(hfac)_2$ and $Cu^{II}(tmhd)_2$. Therefore, just

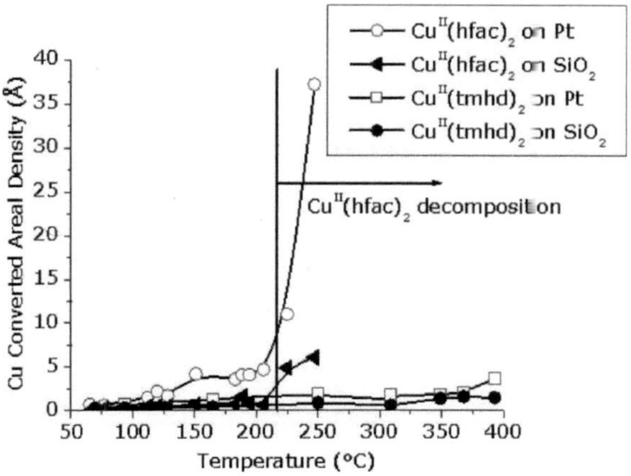

Fig. 12.3 Cu precursor thermal decomposition at $SiO_2$and Pt surfaces as a function of substrate temperature as measured by Rutherford backscattering (RBS)

because "self-limiting" behavior is observed other chemistry might be occurring that leads to much contamination in the films. Also, it is important to note, irregardless of the precursor or whether the ALD process is thermal or plasma assisted, the ALD window needs to be established by looking at the thermal decomposition of the precursor. Investigating the chemistry on the monolayer scale is also important since it can translate to high quality metal deposits with good texture.

12.3 Palladium on Noble Metal

Depositing a noble metal on a noble metal might seem strange and of little interest; however, a noble metal substrate is well defined and it allows the catalytic cleavage of molecular hydrogen. Figure 12.4 shows two deposition curves for Pd on iridium. The upper limit for deposition is 230°C for $Pd^{II}(hfac)_2$. The deposition curves were generated by a 5/10/20/10 s for $Pd^{II}(hfac)_2$/purge/hydrogen/purge. The long purge time was undertaken to ensure that no parasitic CVD occurred. The difference in deposition rates between the films deposited at 80°C and at 130°C was due to desorption of $Pd^{II}(hfac)_2$ from the Ir or Pd surfaces. As discussed earlier this follows Langmuir kinetics. The deposition at 130°C is particularly interesting. Ir (7.6 $\pm$ 1.9 $Å^3$) (G. D. Doolen, Los Alamos National Laboratory, Unpublished) possesses a significantly greater electronic polarizability compared to Pd (7.6 $\pm$ 1.9 $Å^3$) [14] due to the presence of the f orbitals. This greater polarizability allows for stronger interaction between $Pd^{II}(hfac)_2$ and Ir compared to $Pd^{II}(hfac)_2$ and the deposited Pd. The initial sluggish deposition at 80°C was attributed to adventitious carbon that needs to be removed by hydrogen/argon purge before a consistent deposition rate existed.

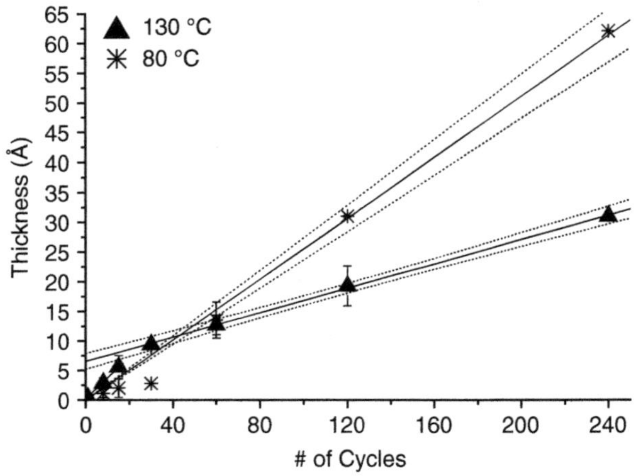

Fig. 12.4 The sequential growth of Pd ALD films as a function of the number of cycles for $Pd^{II}(hfac)_2$ pulsed separately with H_2. The films at 130°C show less growth with more cycles due to the desorption of $Pd^{II}(hfac)_2$ on Pd vs. Ir

12.4 Palladium on Tetrasulfide Silane

If $Pd^{II}(hfac)_2$ is dosed on hydroxylated SiO_2, no palladium is evident but fluorine can be observed with X-ray photoelectron spectroscopy (XPS) [6]. Longer doses and higher substrate temperatures result in higher fluorine levels at the surface.

Researchers have exploited the strong interaction between H_2O and the hfac ligand to undertake metal deposition on metal oxide surfaces. However, interfacial carbon and fluorine can cause serious issues with the quality of the ultra-thin (<50 Å) metal deposit. Palladium has a strong affinity for sulfur and the other group VIII catalytic metals. Bonding bis[3-(triethoxysilyl)propyl]-tetrasulfide to a hydroxylated SiO_2 surface results in a surface with a high affinity for $Pd^{II}(hfac)_2$. The next hurdle to overcome is the lack of catalytic activity of the tetrasulfide silane or SiO_2 toward the cleavage of molecular hydrogen. Instead of hydrogen, glyoxylic acid was used as a reducing agent [2]. Above 200°C glyoxylic acid thermally cleaves liberating electrons and other by-products [15].

The experimental procedure is shown in Fig. 12.5. The current pulse sequence does not allow very high deposition rates because the strong desorption is at 210°C. However, it is the quality of the film that is more of a concern than the deposition rate since electroless/electrolytic copper will be grown off of the catalytic surface. The chemical quality of the palladium or copper seed layer is related to the texture of the film. The texture of the ALD deposit in turn is related to the texture of the copper or other metal deposit.

Figure 12.6 shows the reflection high energy electron diffraction (RHEED) spectra for ultra-thin ALD Pd films deposited on Ir and on the tetrasulfide silane. The Ir substrate exhibits preferential (111) orientation. After Pd is deposited on the Ir substrate the film is highly (111) textured. The metal deposit, albeit ALD, e-beam,

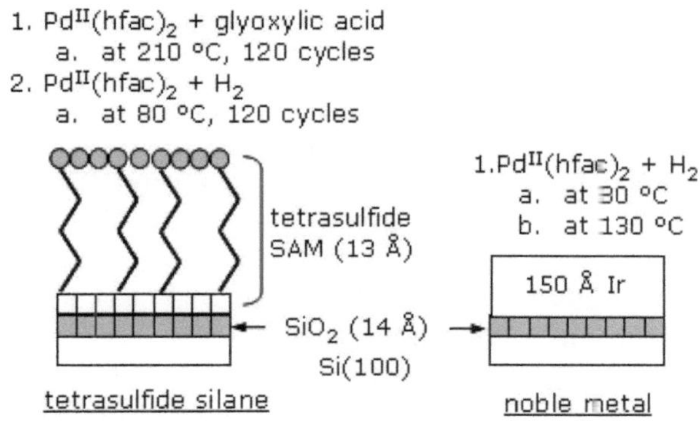

Fig. 12.5. The experimental procedure

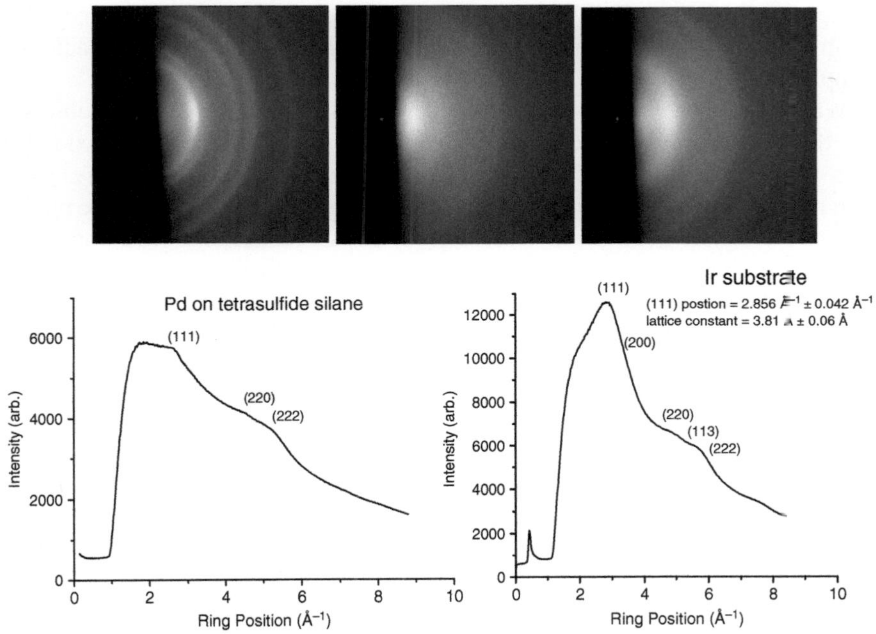

Fig. 12.6 Reflection high energy electron diffraction (RHEED) spectra for Pd ALD films on tetrasulfide silane (scheme in Fig. 12.5) (*top*) and on Ir (*middle*). The bottom spectrum and image are from the e-beam deposited Ir film on hydrogen-terminated Si. The texture of the Pd ALD film on the tetrasulfide silane was disrupted from the carbon and fluorine impurities and the amorphous substrate

etc., will become more textured compared to the substrate, especially if both metals have good lattice match. The ALD Pd on the tetrasulfide silane has almost no preferential texture and is in contrast to Pd deposited on Ir. This lack of texture is the result of Pd deposited on an amorphous substrate and the carbon and fluorine contaminations at the interface. Higher deposition temperatures favor dissociation reactions, which yield carbon and fluorine fragments, and especially the thermally unstable $-CF_3$ group.

12.5 Electroless Deposition of Cu on PA-ALD Pd

Electroless deposition (ELD) of Cu is a facile means to test the catalytic activity of the palladium ultra-thin films. RBS can show that there is palladium on the surface and its aerial density (atoms/cm^2) but cannot discriminate whether the palladium is catalytic. Electroless deposition has an appeal for an "end of the roadmap" technology since the barrier layer/seed layer can be very thin <50 Å. Also, the surface chemistry issues worked out for atomic layer-deposited metals will help advance electroless-deposited metals such as CoWP. It is preferred to use PA-ALD to deposit palladium on TaNX or Ta surfaces because of their robust native oxide [8]. In situ deposition of the barrier and seed layers would not necessitate this additional complexity.

When Pd PA-ALD is undertaken on Ir, W, and Ta, it is consistently observed that the thickest films are deposited on W and Ta that possess a native oxide [8]. This was interpreted that the remote hydrogen plasma can remove the native oxide with relative ease and what is left is a pristine elemental surface. The noble metals, for example, Ir, do not possess a native oxide but attract adventitious carbon. This carbon is strongly bonded to the noble metal surface. Apparently, atomic hydrogen is not very effective in removing this carbon. The pristine metallic surface can both chemisorb metallorganics better and cleave molecular hydrogen to atomic hydrogen.

Electroless Cu deposition was undertaken on patterned TaN_X surfaces using a glyoxylic acid-based solution [16].

$CuSO_4.5H_2O$: 7.62 g/l
EDTA: 10.23 g/l
Glyoxylic acid monohydrate: 7 g/l
2, 2′- Dipyridine: 4 mg
Polyethylene glycol: 5 ml
Re-610: 5 ml
pH : ~12.5 w/tetramethylammonium hydroxide
Temperature: 50°C

The thickness of the Pd PA-ALD films was 20–40 Å resulting from 150 cycles at a deposition temperature of 80°C. The Cu electroless solution is formaldehyde, potassium, and sodium free appropriate for environmental issues and semiconductor manufacturing. The trenches in Fig. 12.7 are filled in "conformal" mode, which is appropriate for the deposition of a Cu seed layer. The deposition was allowed to proceed to just show the evolution. The trench geometry and the wafer size, e.g., 200 or 300 mm, would dictate the appropriate Cu seed layer thickness. A Cu seed layer can be deposited with ease within a few minutes. This can be undertaken on a

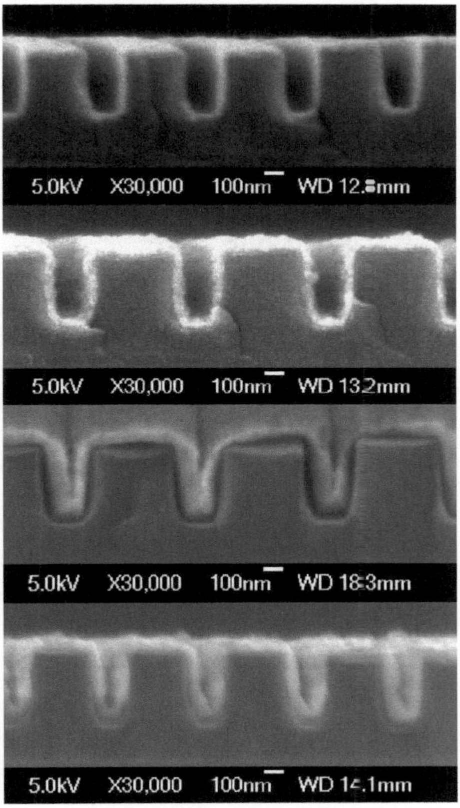

Fig. 12.7 FE-SEM images of electroless deposition of Cu on 130 nm 2.5:1 aspect ratio trenches with 20–40 Å Pd PA-ALD on 50 Å PVD TaNX. Starting at the *upper left* and proceeding clockwise: bare TaNX, 3 min electroless Cu, 5 min electroless Cu, 20 min electroless Cu

Pd layer as thin as 12 Å, which would minimally impact Cu in line resistivity due to diffusion of Pd into the Cu after thermal cycling.

The traditional method to deposit the Pd activation (catalytic) layer is via wet chemistry. This acid-bath chemistry is typically not stable and the Pd colloids resulting from this solution cannot easily achieve a 12–50 Å continuous Pd ultra-thin film over demanding aspect ratio trenches [17]. The Pd colloids can easily be 300–400 Å, which are not appropriate for the deposition of highly textured Cu films on a nanometer scale. Electroless copper can be deposited in "conformal" mode where Cu would be deposited as a seed layer for electrolytic copper. However, it may be desired to deposit copper such that copper fills the trench. It has been found that

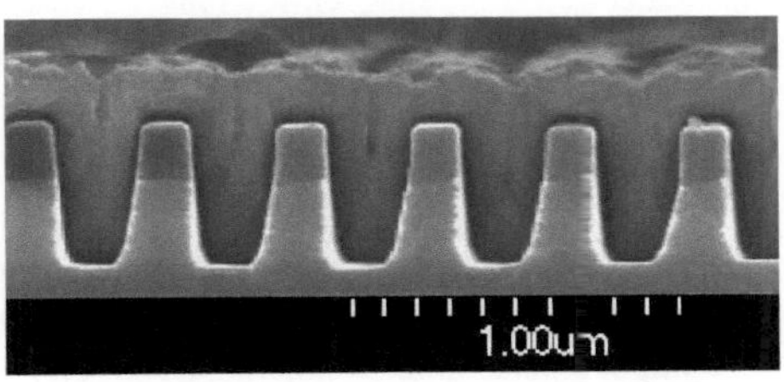

Fig. 12.8 FE-SEM images of electroless deposition of Cu film on TaNX at a solution temperature of 45°C for 30 min with 1 s of ultrasonic vibration at the beginning of the deposition

ultrasonic vibration, lowering the solution temperature, and the use of an appropriate surfactant help trench fill. There is still more work needed in this area to accelerate the deposition rate of the electroless Cu for trench fill; however, the work to date looks promising.

12.6 Conclusions

Metal ALD is very challenging due to limited precursor selection and the necessity of an organic reducing agent. Pd ALD is a significant improvement over the acid-bath solution route to Pd activation. Ultra-thin films of 12–50 Å that are highly conformal can be deposited via $Pd^{II}(hfac)_2$ and hydrogen via a remote hydrogen plasma or glyoxylic acid. Glyoxylic acid does not thermally cleave until 200°C. At 200°C $Pd^{II}(hfac)_2$ shows much desorption according to Langmuir kinetics and much chemical dissociation. The most thermally labile chemistry in the $Pd^{II}(hfac)_2$ molecule is the $-CF_3$ group that dissociates even at moderate temperatures of $\sim$130°C. Improvement is certainly possible with these ultra-thin metallic layers deposited via ALD. Electroless copper was deposited with ease on these Pd PA-ALD layers. Electroless copper can either be deposited in "conformal" mode or "trench fill" mode depending on the experimental process parameters employed. The conformal mode is appropriate for use of electroless Cu as a seed layer for electrolytic Cu. The trench fill mode is appropriate for demanding trenches as part of demanding three-dimensional or MEMS trenches or vias.

References

1. Senkevich, J. J.; Mitchell, C. J.; Yang, G.-R.; and Lu, T.-M.: Reduced sulfur-terminated silanes to promote the interaction of palladium(II) hexafluoroacetylacetonate with dielectric surfaces. Colloids Surf. A **221(1–3)**, 29 (2003)
2. Senkevich, J. J.; Tang, F.; Rogers, D.; Drotar, J. T.; Wang, G.-C.; Lu, T.-M.; Jezewski, C.; and Lanford, W. A.: Substrate independent palladium atomic layer deposition. Chem. Vap. Dep. **9(5)**, 258 (2003)
3. Adamson, A. W.: Physical Chemistry of Surfaces, 5th Ed., John Wiley & Sons, New York (1990)
4. Suntola, T.: Atomic Layer Epitaxy, Simpson, M. (ed.), Blackie, Chapman & Hall (1990)
5. Kim, G. Y.; Srivinastava, A.; Foote, D.; Londergan, A.; Karim, Z.; Ramanathan, S.; and Seidel, T.: AVS Science and Technology, ALD 2004, Abstract Book, Helsinki, (2004)
6. Senkevich, J. J.; Yang, G.-R.; Tang, F.; Wang, G.-C.; Lu, T.-M.; Cale, T. S.; Jezewski, C.; and Lanford, W. A.: Substrate-independent sulfur-activated dielectric and barrier-layer surfaces to promote the chemisorption of highly polarizable metallorganics. Appl. Phys. A **79(7)**, 1789 (2004)
7. Hagedorn, C. J.; Weiss, M. J.; and Weinberg, W. H.: Dissociative chemisorption of hydrogen on Ir(111): Evidence for terminal site adsorption. Phys. Rev. B **60(20)**, R14016–8 (1999)
8. Ten Eyck, G. A.; Senkevich, J. J.; Tang, F.; Tang, F.; Liu, D.; Pimanpang, S.; Karabacak, T.; Wang, G.-C.; Lu, T.-M.; Jezewski, C.; and Lanford, W. A.: Plasma-assisted atomic layer deposition of palladium. Chem. Vap. Dep. **11(1)**, 60 (2005)

9. Solanki, R. and Pathangey, B.: Atomic layer deposition of copper seed layers. Electrochem. Solid-State Lett. **3(10)**, 479 (2000)
10. Hsu, H.-H.; Lin, K.-H.; Lin, S.-J.; and Yeh, J.-W.: Electroless copper deposition for ultralarge-scale integration. J. Electrochem. Soc. **148(1)**, C47(2001)
11. Diamand, Y. S.: Electroless copper deposition using glyoxylic acid as reducing agent for ultralarge scale integration metallization. Electrochem. Solid-State Lett. **3(6)**, 279 (2000)
12. Jezewski, C.; Lanford, W. A.; Wiegand, C. J.; Singh, J. P.; Wang, P.-I.; Senkevich, J. J.; and Lu T.-M.: Inductively coupled hydrogen plasma-assisted Cu ALD on metallic and dielectric surfaces. J. Electrochem. Soc. **152(2)**, C60 (2005)
13. Toerndahl, T.; Ottosson, M.; and Carlsson, J.-O.: Growth of copper metal by atomic layer deposition using copper(I) chloride, water and hydrogen as precursors. Thin Solid Films **458(1–2)**, 129 (2004)
14. Zangwill, A. and Soven, P.: Density-functional approach to local-field effects in finite systems: Photoabsorption in the rare gases. Phys. Rev. A **21**, 1561 (1980)
15. Back, R. A. and Yamamoto, S.: Gas-phase photochemistry and thermal decomposition of glyoxylic. Acid. Can. J. Chem. **63**, 542 (1985)
16. Kim, Y.-S.; Kim, H.-I.; Cho, J.-H.; Seo, H.-K.; Dar, M. A.; Shin, H.-S.; Ten Eyck, G. A.; Lu, T.-M.; and Senkevich J. J.: Electroless copper on refractory and noble metal substrates with an ultra-thin plasma-assisted atomic layer deposited palladium layer. Electrochmica Acta. **51(12)**, 2400 (2006)
17. Hong, S. W.; Shin, C.-H.; and Park, J.-W.: Palladium activation on TaN_x barrier films for autocatalytic electroless copper deposition. J. Electrochem. Soc. **149(1)**, G85 (2002)

Part IV
Deposition Processes for ULSI Interconnects

This part presents the different deposition processes for ULSI interconnects. Chapter 13 describes the fabrication of copper interconnects via Damanscene process, which is an electrodeposition process combined with chemical mechanical polishing. Copper interconnection is able to support higher current density, and this makes it possible to accelerate further miniaturization of interconnects. Copper is expected to be a potential candidate for the interconnect material because of its lower resistivity (1.68 μΩ cm) compared with that of aluminum (2.65 μΩ cm). The other important advantage of copper interconnects is that copper offers much better electromigration resistance than aluminum. Copper plating chemistry and electrochemical process for seed layer formation are also presented in detail with respect to interconnect deposition. Chapter 14 presents the atomic layer deposition process for ULSI manufacturing. Atomic layer deposition (ALD) is a technique where precursors are introduced alternatively and a monolayer (or fraction thereof) is deposited on the surface at a time. The sequential introductions of all precursors, separated by purge steps, complete an ALD cycle. As the geometries continue to shrink for each successive technology node, scaling has both pushed target thickness of many films used in ULSI manufacturing to a few monolayers and created requirements for excellent step coverage in high aspect ratio features. The distinct properties of ALD such as precise thickness control and conformal deposition have attracted great interest from mainstream ULSI production and R&D groups. There are potential applications in many areas spanning from front-end-of-line (FEOL) to back-end-of-line (BEOL) processing. Some leading examples of ALD applications in ULSI are presented in the next few sections. Chapter 15 deals with the electroless deposition technique when the molecular level deposition is required. The use of electrodeposition (galvanic deposition) of copper in the fabrication of microchips has shown that sub-micrometer structures can be produced at the solid/liquid interface with great precision. Further improvements will make this method compatible with smaller and smaller scales; however, it will become more and more difficult to contact the structure to be metallized, especially when the molecular scale (< 10 nm) is approached. Here, electroless deposition (ELD) can offer an attractive alternative: The electrons that reduce solvated metal ions in galvanic deposition, provided by an external current source, are replaced by electrons from reductants that are added to the metal ion solution. If the reductants are oxidized by the growing metal, or their oxidation is catalyzed by the growing metal, the process is autocatalytic, and continuous plating ensues. I will provide examples for sub-10 nm metal plating by ELD, with and without templates, including biomolecules.

Chapter 13
Electrochemical Processes for ULSI Interconnects

Tetsuya Osaka, Madoka Hasegawa, Masahiro Yoshino, and Noriyuki Yamachika

13.1 Introduction

Copper is widely used as interconnecting material in ultra-large-scale integration (ULSI) circuits (Fig. 13.1). Fabrication of copper interconnection has been achieved by "Damascene process" [1], which is an electrodeposition process combined with chemical–mechanical polishing (CMP) [2]. Damascene process led to a remarkable change in the industry. Most manufacturers have now converted to this electrodeposited copper interconnect technology. Before the introduction of copper to interconnects, aluminum and aluminum–copper alloy were used as the interconnecting materials for many years. Aluminum interconnect layers are easily fabricated by subtractive etching process (Fig. 13.2a). In process, the interconnect layers are deposited by physical vapor deposition (PVD), followed by reactive ion etching (RIE). Aluminum is preferable as interconnects because this material does not diffuse into SiO_2 substrate and the layers adhere well to the substrate. However, resistivity of aluminum is relatively high (2.65 μΩ cm), and the layers suffer from the disadvantage of its poor electromigration resistance. With an increase in interconnects density and a decease in the dimensions of interconnects, the problem of an increase in latency, or *RC* delay, and electromigration became much more critical. Therefore, search for new interconnect materials for miniaturization of semiconductor devices was mainly focused on minimizing *RC* delay and electromigration [3–10]. Copper was expected to be a potential candidate for the interconnect material because of its lower resistivity (1.68 μΩ cm) compared with that of aluminum (2.65 μΩ cm). The other important advantage of copper interconnects is that copper offers much better electromigration resistance than that of aluminum. For similar dimension, the time-to-failure of copper interconnects was about a hundred times that of aluminum interconnects [11–13]. Therefore, copper interconnection is able to support higher

T. Osaka (✉)
Faculty of Science and Engineering, Waseda University, 3-4-1, Okubo, Shinjuku-ku, Tokyo 169-8555, Japan
e-mail: osakatets@waseda.jp

Y. Shacham-Diamand et al. (eds.), *Advanced Nanoscale ULSI Interconnects: Fundamentals and Applications*, DOI 10.1007/978-0-387-95868-2_13,

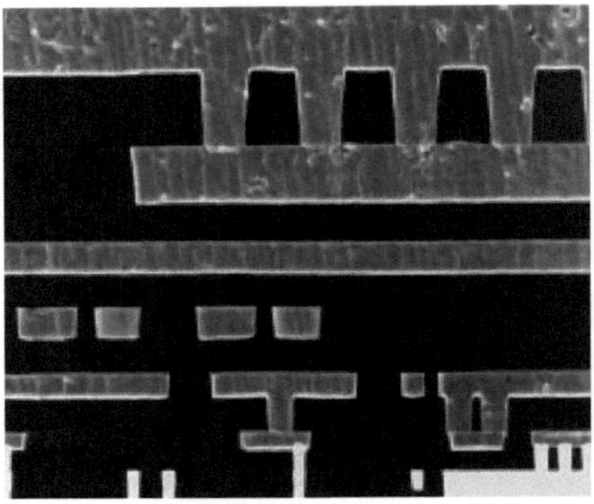

Fig. 13.1 Cross-sectional image of a 6-level copper interconnection structure fabricated by IBM (reproduced by permission of IBM J. Res. & Dev. [2])

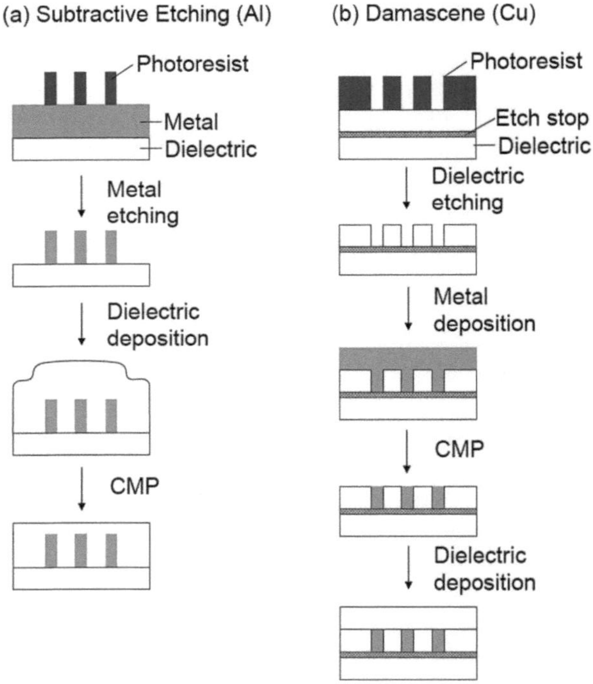

Fig. 13.2 Comparison of the subtractive etching process and the Damascene process

current density, and this makes it possible to accelerate further miniaturization of interconnects.

However, the application of copper to interconnection was not accomplished for many years because the traditional process for aluminum metallization was not applicable to copper. As described above, aluminum interconnects were fabricated by subtractive etching process, in which reactive ion etching is used for patterning of aluminum layer. Reactive ion etching is extremely difficult to be applied to copper interconnect fabrication because copper compound formed during the etching process is not volatile. Although technologies for direct subtractive etching of copper exist, these technologies require more than 210°C [14–17] which is higher than the application limit of most organic photoresist. These copper etch technologies are also more expensive. Therefore, the application of copper to interconnect required the new approach.

In 1997, IBM succeeded to apply copper to fully integrated interconnection by employing electrodeposition [1, 11]. This process named "Damascene process" after the ancient inlay technique developed in the city of Damascus as early as 12th century. In this ancient metal work, decorative metals such as goldsilver, or copper were hammered into grooves made in armaments or other metal objects after which the excess material was removed by polishing. For the interconnection fabrication (Fig. 13.2b), the process begins with deposition of SiO_2 or other dielectric material, followed by the etching of trench line or via-holes into the layer. Subsequently, a barrier layer, which is essential for copper interconnects to prevent copper from diffusing into substrate, is formed on the trenches and via-holes by dry process such as sputtering or CVD. TaN or TiN is usually employed as a barrier material. After the formation of a barrier layer, the trenches and vias are completely filled with copper. After the copper filling process, the excessive metal deposited outside of the trenches and the vias was removed by using CMP. The Damascene process, in which only trenches and vias are fabricated individual process, called "single Damascene" (Fig. 13.3a). On the other hand in

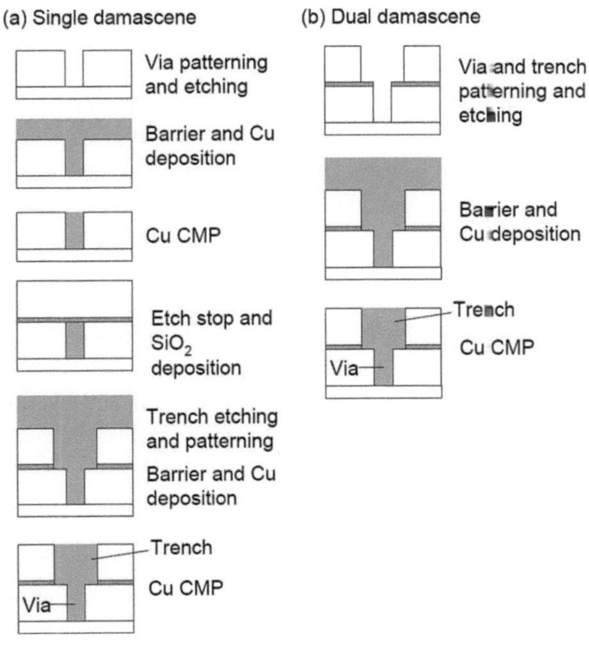

Fig. 13.3 Comparison of the single Damascene and the dual-Damascene processes

"dual-Damascene process" (Fig. 13.3b), two insulator levels are patterned filled and planarized. In this process, two levels of insulator and etch stop layers are patterned as trenches and vias. These trenches and vias are filled simultaneously by electrodeposition. Thus, this process reduces the number of process step, resulting in lower cost, less requirements for equipment and space, and potentially fewer defects.

One of the most important issues for the Damascene process was gap filling of trenches and vias without producing voids, which affect a capacitance per unit length and resistivity. Therefore, complete filling of trenches or vias with copper has been investigated extensively by employing several deposition techniques which included PVD, CVD, electrodeposition, and electroless deposition. Among these techniques, electrodeposition was employed for copper filling of Damascene structure because this process is capable of successful filling of trenches and vias. Furthermore, high-quality copper deposits were obtained by electrodeposition at room temperature, with high throughput.

In the Damascene copper plating process, first, the seed layer is uniformly deposited on the trench- and via-patterned insulator layer coated by a barrier layer. After formation of a seed layer, copper electrodeposition is carried out to fill the trenches and vias formed on the insulator layer. A copper layer covers entire surface after electrodeposition. Therefore, the excess deposits must be removed by planarization step such as CMP.

In this chapter, fundamental aspect of electrochemical deposition and recent progress in the electrochemical approaches for fabrication of ULSI copper interconnects are discussed.

13.2 Copper Plating Chemistry

13.2.1 Copper Electrodeposition for Trench Filling

Damascene copper electrodeposition and CMP were the key technologies which led to the implementation of copper interconnects. Most important requirement for the success of this process is its ability to fill trenches and vias completely without producing any defect such as voids and seams. Voids (or seams) are undesirable because they lead to an increase in resistivity, and enhance electromigration issues. Furthermore, trapped electrolyte in voids results in the corrosion of copper. Void-free filling was accomplished by using several bath additives in combination. In this section, the fundamental chemistry of electro copper deposition and the effect of additives in the Damascene copper electrodeposition are described.

13.2.1.1 Bath Composition for Void-Free Filling

Possible profile evolution in Damascene copper plating is classified into three types, namely, subconformal filling, conformal filling, and superfilling [1]. Figure 13.4

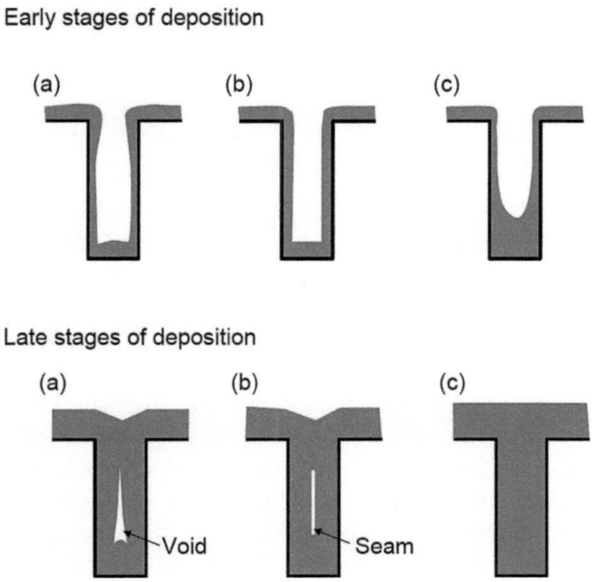

Fig. 13.4 Types of profile evolution during copper filling by electrodeposition: (**a**) subconformal deposition, (**b**) conformal deposition, and (**c**) superfilling

shows schematic illustrations of these three types of profile evolution with time of copper deposits. Subconformal filling results from preferential flow of current at trench openings, which is caused by geometric effect of electric field or substantial depletion of the cupric ions inside the feature. In subconformal deposition, copper is deposited faster at trench opening than at the bottom, which results in formation of voids. In conformal filling, copper is uniformly deposited at all point of a feature, resulting in formation of a seam. Superfilling is a void-free and seamless filling of trenches or vias. To achieve superfilling, deposition rate at the bottom is required to be higher than those at other portions of the feature. This mode of growth of copper is called "bottom-up" growth. Bottom-up growth of copper is accomplished by using several additives.

13.2.1.2 Copper Deposition Mechanism and Kinetics

Reduction of copper deposition on the cathode is considered to proceed by a two charge-transfer step reaction [18–22]:

$$Cu^{2+} + e^- \rightarrow Cu^+ \qquad E_0 = 0.153V, \tag{13.1}$$

$$Cu^+ + e^- \rightarrow Cu^0 \qquad E_0 = 0.52V. \tag{13.2}$$

The first reaction (the reduction of Cu^{2+} to Cu^+) is rate-determining step [18, 22], because the rate constant of the second step may be up to three orders of magnitude greater. In the conventional Damascene electrodeposition bath, the activity of cuprous ions is considered to be affected.

13.2.1.3 Chemistry of Copper Plating Bath

An acidic copper sulfate acidic bath is employed in the Damascene process because this bath is stable, inexpensive, non-toxic, and readily available. This bath contains copper sulfate as the source of copper ions and sulfuric acid as the electrolyte to provide conductivity to the solution. This bath also contains several additives in combination to achieve void-free filling, or "superfilling" of trenches and vias.

The typical copper sulfate concentration in commercial use for the interconnection fabrication is in the range from 0.2 to 0.6 mol/L. High concentrations of cupric ions are useful when a rapid deposition is required to achieve desired throughput. On the other hand low concentrations of copper ions are effective to improve uniformity of copper deposits across the wafer because in this concentration range, charge-transfer resistance is increased uniformly at all points on the wafer surface [23, 24]. For the interconnect applications, copper sulfate concentration is required to be sufficiently high enough to avoid the depletion of cupric ions within high-aspect-ratio features with an appropriate deposition rate.

Sulfuric acid is usually added to the plating bath to increase the conductivity of the solution [23]. More conductive solutions minimize variation of potential gradients within the plating bath, and thus, results in more uniform and geometry-independent interfacial kinetics [25]. Therefore, higher concentration of sulfuric acid is preferred for the process of interconnection fabrication. However, the solubility of copper sulfate decreased with an increase in sulfuric acid concentration. For example, the maximal solubility of copper sulfate is about 0.75 mol/L in the presence of 2 mol/L of sulfuric acid while the value decreases to about 0.5 mol/L in the solution containing sulfuric acid at concentration of 4 mol/L. Therefore, the concentration of sulfuric acid is optimized to increase conductivity of the solution while maintaining the sufficient concentration of cupric ions.

The most important issue for Damascene copper filling is to achieve the void-free filling of copper in trenches. In ordinary deposition, voids or seams are formed particularly in narrow and small trenches and vias because the deposition rate at the openings tends to be higher than other portions of trenches in the absence of additives. This is caused by the current distribution occurred on trenches (Fig. 13.5), in which current density is concentrated at the opening of trenches for the geometrical reason [26].

Additives are known to adsorb on the cathode surface and affect the rate of copper deposition as well as the microstructure and morphology of copper deposits. In the Damascene copper electrodeposition, superfilling was achieved by employing several organic bath additives in combination. Based on their function, these additives are categorized into three types: a deposition accelerator, a deposition inhibitor, and a leveler [23, 27, 28]. The accelerator is essentially an organic sulfur-containing compound while the suppressor is a glycol, which decreases the rate of copper

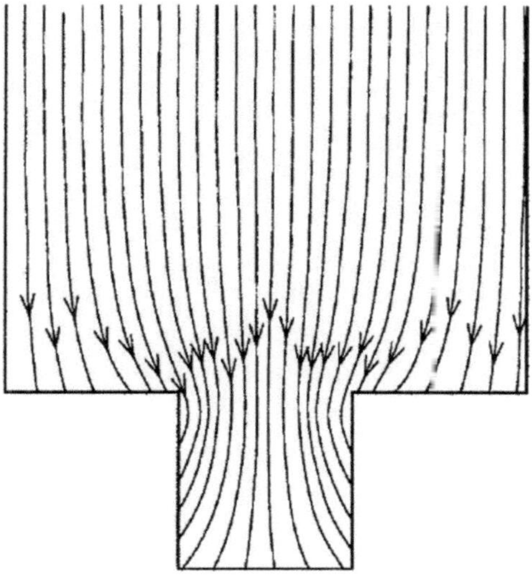

Fig. 13.5 Current density profiles above and inside a feature (reproduced by permission of The Electrochemical Society) [26]

deposition especially in the presence of chloride ions. The leveler is generally a kind of an azo dye compound. This additive acts as an inhibitor as well as leveler.

Table 13.1 lists an example of copper electroplating bath used in the trench-filling process. Conventional additives used to accomplish "superfilling" are chloride ions (Cl^-), polyethylene glycol (PEG), bis(3-sulfopropyl)disulfide (SPS) or 3-mercapto-1-propanesulfonate (MPS), and Janus Green B (JGB) as given in Fig. 13.6.

13.2.1.4 Reaction Mechanism: Effect of Additives

The addition of chloride ions, which catalyzing the rate-determining Cu^{2+}/Cu^+ reaction, is known to accelerate copper deposition [21, 30–33]. However, it is also known to interact with PEG [21, 34] and SPS when they are present together (Table 13.2). PEG is a kind of surfactant and is usually added to Damascene plating bath as a deposition inhibitor. This additive is known to inhibit copper deposition significantly in the presence of chloride ions, while PEG itself is a weak inhibitor. This compound adsorbs on metal surface and forms a polymeric PEG–Cl^- adlayer in the presence of chloride ions [21, 32, 34]. This PEG–Cl^- adlayer is considered to inhibit copper deposition strongly (Fig. 13.7). Recent studies suggested that this adlayer also contains cuprous ions. Yokoi et al. [21] suggested that this blocking

Table 13.1 Bath composition [29]

H_2SO_4	2.0 mol/L
$CuSO_4$ $5H_2O$	0.26 mol/L
Cl^-	50 ppm
PEG (MW 2000)	100 ppm
SPS	2 ppm
JGB	2 ppm

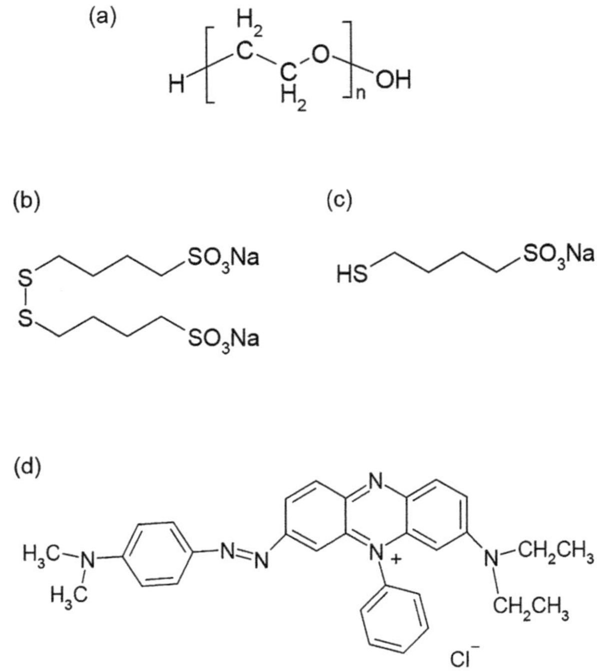

Fig. 13.6 Conventional additives used in the process of trench filling by electrodeposition: (**a**) polyethylene glycol (PEG), (**b**) bis(3-sulfopropyl)disulfide (SPS) (**c**) 3-mercapto-1-propanesulfonate (MPS), and (**d**) Janus Green B (JGB)

layer might consist of a complex of cuprous ion and PEG formed on specifically adsorbed chloride ions. This layer is considered to act as blocking layer and inhibits copper deposition.

Figure 13.8 shows cathodic polarization curves for the Cl–PEG–SPS baths. The acceleration effect of SPS was clearly observed in the polarization curves. Compared with the knowledge of the effect of suppressor, the mechanism of the accelerator such as SPS or its monomeric compoundMPS, is unclear. Both of them are reported to act as a strong accelerator in the presence of Cl^- [35].

For trench filling in the Damascene process, SPS and MPS exhibit similar effect on bottom-up growth of copper in trenches [36]. Thus, it is assumed that disulfide linkage in SPS is cleaved in the conditions, where copper electrodeposition takes place. It also indicated that cuprous ions and SPS form an adsorbed species and the complex formed by this reaction stabilizes the cuprous ions [37]. Because the reduction of Cu^{2+} to Cu^+ is the rate-limiting step of copper electrodeposition, the stabilization of cuprous ions leads to an increase in cuprous ions near the surface of copper, which results in the increase in the rate of copper deposition. Another model for the catalytic effect of the accelerator is that SPS displaces the PEG–Cl^- layer, which inhibits copper electrodeposition, and this competitive adsorption between an accelerator and an inhibitor results in the increase in the rate of copper deposition [38]. However, the detailed reactions of these additives still remain to be unclear because of the complexity of the interaction of additives. The possible reactions which are assumed to take place in the plating bath are shown in Table 13.2 [37].

Table 13.2 Reactions at the copper/electrolyte interface in copper plating baths containing Cl, PEG, and SPS or MPS (reprinted with permission of IBM Technical Journals) [37]

Copper comproportionation reactions	
{1}	$Cu + Cu^{2+} \rightleftharpoons 2Cu^{+}$
{2}	$Cu + Cu^{2+} + 2Cl^{-} \rightleftharpoons 2CuCl_{ad}$
{3}	$Cu + Cu^{2+} + 2MPS \rightleftharpoons 2Cu(thiolate)_{ad} + 2H^{+}$
Redox reactions involving SPS	
{4}	$2Cu^{+} + SPS + 2H^{+} \rightleftharpoons 2Cu^{2+} + 2MPS$
{5}	$4Cu^{+} + SPS \rightleftharpoons 2Cu^{2+} + 2Cu(I)(thiolate)_{ad}$
{6}	$2Cu^{2+} + 4MPS \rightleftharpoons 2Cu(I)(thiolate)_{ad} + SPS + 4H^{+}$
Surface adsorption reactions	
{7}	$2Cu_{s} + SPS \rightleftharpoons 2Cu(I)(thiolate)_{ad}$
{8}	$nCuCl_{ad} + HO((CH_2)_xO)_yH \rightleftharpoons \lfloor HO((CH_2)_xOCuCl)_n\,((CH_2)_xO)_{y-n}H\rfloor_{ad}$
Exchange reactions	
{9}	$CuCl_{ad} + MPS \rightleftharpoons Cu(I)(thiolate)_{ad} + Cl^{-} + H^{+}$
{10}	$\lfloor HO((CH_2)_xOCuCl)_n\,((CH_2)_xO)_{y-n}H\rfloor_{ad} + nMPS \rightleftharpoons HO((CH_2)_xO)_yH + nCl^{-} + nCu(I)(thiolate)_{ad}$
Complexation reactions	
{11}	$Cu^{+} + Cl^{-} \rightleftharpoons CuCl_{ad}$
{12}	$CuCl_{ad} + Cl^{-} \rightleftharpoons CuCl_2^{-}$
{13}	$CuCl_{ad} + MPS \rightleftharpoons CuCl(thiolate)^{-} + H^{+}$
{14}	$Cu^{+} + MPS \rightleftharpoons Cu(I)(thiolate)_{ad} + H^{+}$
{15}	$Cu^{+} + SPS \rightleftharpoons Cu(I)(SPS) + H^{+}$
{16}	$Cu(I)(thiolate)_{ad} + MPS \rightleftharpoons Cu(I)(thiolate)_2^{-} + H^{+}$
{17}	$Cu(I)(thiolate)_{ad} + Cl^{-} \rightleftharpoons Cu(thiolate)Cl^{-}$
{18}	$4Cu(I)(thiolate)_{ad} + Cu^{2+} \rightleftharpoons Cu[Cu(I)(thiolate)]_4^{2+}$
{19}	$Cu(I)(thiolate)_{ad} + Cu^{+} \rightleftharpoons Cu(I)(thiolate)Cu(I) + H^{+}$

JGB is regarded as a leveling agent [27, 29, 39, 40]. This additive increases the overpotential and inhibits the copper deposition (Fig. 13.9). The additive is also known to be consumed easily at copper surface during the electrodeposition.

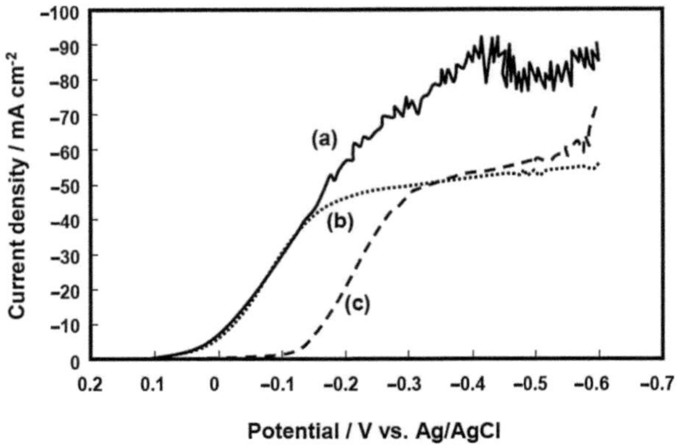

Fig. 13.7 Cathodic polarization curves for copper electrodeposition from (a) additive-free, (b) PEG, and (c) Cl–PEG baths (reproduced by permission of The Electrochemical Society) [29]

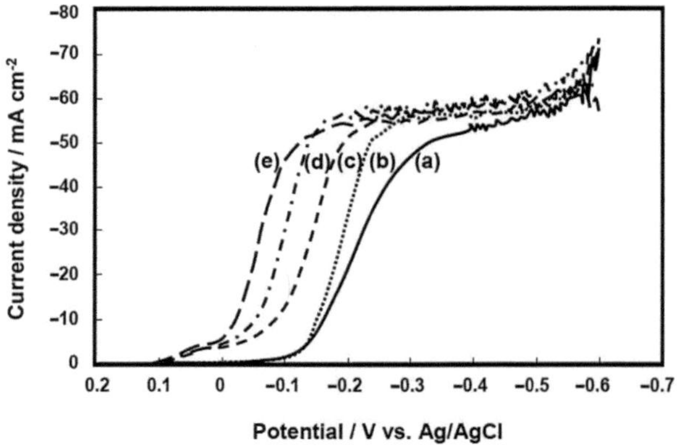

Fig. 13.8 Effect of SPS concentration on polarization characteristics for copper electrodeposition from Cl–PEG–SPS baths containing (a) 0, (b) 1, (c) 10, (d) 100, and (e) 1000 ppm of SPS (reproduced by permission of The Electrochemical Society) [29]

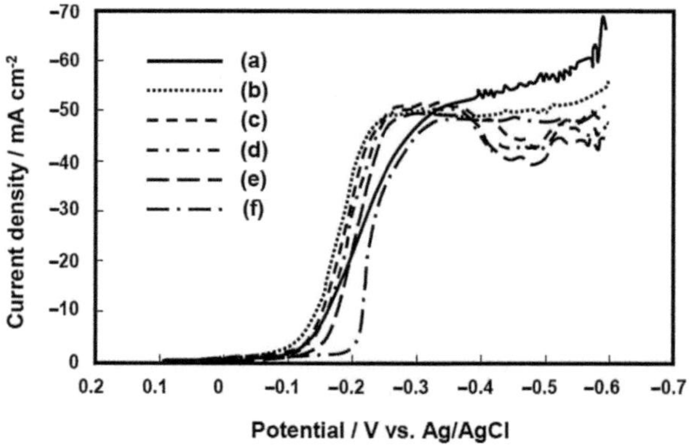

Fig. 13.9 Effect of JGB concentration on polarization characteristics for copper electrodeposition: (a) Cl–PEG, (b) Cl–PEG–SPS, and (c–f) Cl–PEG–SPS–JGB baths containing (c) 1, (d) 5, (e) 10, and (f) 50 ppm of JGB. SPS concentration was 5 ppm (reproduced by permission of The Electrochemical Society) [29]

13.2.1.5 Mechanism of Void-Free Filling Effect of Additives

For the application to submicrometer interconnects, it has been reported [27, 38] that the "bottom-up" growth, in which the deposition from the bottom of trenches was accelerated relative to that in the areas of openings of the trenches, has been shown to occur in the bath containing Cl^-, PEG, and SPS or MPS, in combination. Figure 13.10 shows the sequential images of copper deposited in trenches obtained from a bath without additives (additive-free bath, Fig. 13.10a) a bath containing Cl^- and PEG (Cl–PEG bath, Fig. 13.10b), and a bath containing Cl^-, PEG, and SPS (Cl–PEG–SPS bath, Fig. 13.10c). Copper deposition at the bottom is significantly accelerated by the addition of SPS to the Cl–PEG bath (Fig. 13.10c).

On the other hand the subconformal deposition and the conformal deposition were observed in the images of the specimens obtained from the additive-free bath (Fig. 13.10a) and the Cl–PEG bath (Fig. 13.10b), respectively. It has been suggested that the competitive adsorption between the accelerator and the inhibitor results in the bottom-up effect [38]. The bottom-up effect achieves the superfilling of copper in trenches, while it also brings about the "overfill" phenomenon, in which copper bumps are formed above the copper-filled trenches [38, 40].

Early electrodeposition models for bottom-up growth [1, 41–43] assumed that the deposition reaction is controlled by mass transfer of inhibiting additives

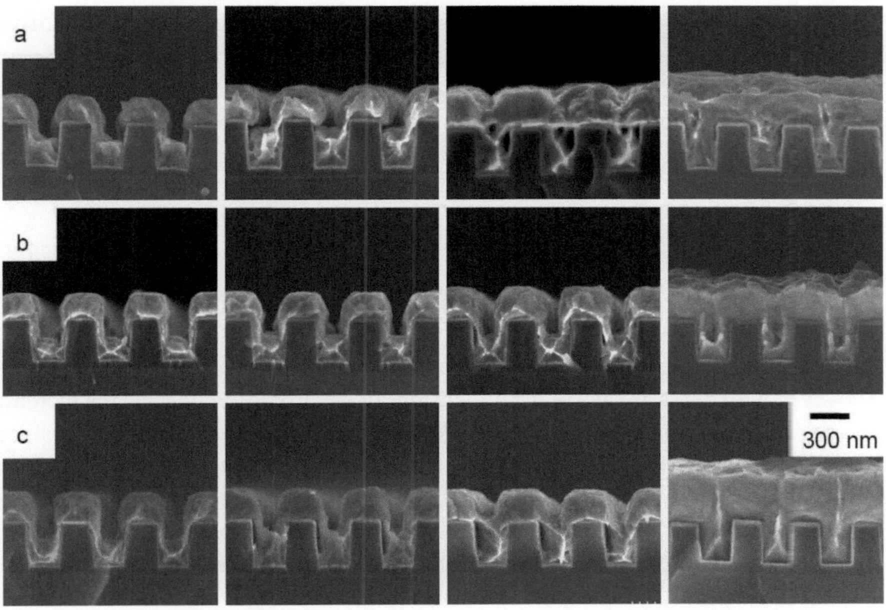

Fig. 13.10 Cross-sectional SEM images of copper electrodeposited in trenches for filling sequences from (**a**) the additive-free, (**b**) Cl–PEG, and (**c**) Cl–PEG–SPS baths. Deposition time, 15 s, 30 s, 60 s, and 90 s (*left to right*)

(transport-limited leveling model) [44–46]. This model is based on the model for surface leveling of the electrodeposits. The model considered based on this traditional transport-limiting leveling model considered the effect of inhibitor and its mass transfer, ignoring the effect of the accelerator. Because the concentration of additive is very low (10^{-3} to 10^{-6} mol/L), the adsorption of the additive is controlled by its diffusion from the bulk of the solution into trenches. These models predict the decrease in the deposition rate at the opening of trenches by considering the diffusion and consumption of inhibitor during electrodeposition. However, these models did not consistent with the experimental observations such as overfill phenomenon and shape evolution of the copper during filling [38, 47, 48]. This indicated that traditional transport-limited leveling model is not suitable for explanation of the mechanism of superfilling. Moffat et al. [49, 50] and West et al. [51] have independently proposed models in which the coverage of accelerator at the bottom of trenches is assumed to increase during the filling of trenches. In their models, the simple competitive adsorption between the inhibitor and the accelerator is considered. The model proposed by Moffat et al. [49, 50] assumes that variation of the accelerator surface coverage depends on local surface curvature, while that proposed by West et al. [51] considered that surface site density changes during the deposition. Their models predict not only the superfilling but also the "overfill" phenomenon, explaining successfully the fundamental aspects of the bottom-up growth, for instance, the presence of an incubation period before the occurrence of the bottom-up growth (Fig. 13.11) [33, 36, 47–50, 52, 53].

The overfill phenomenon itself is disadvantageous for the planarization step (CMP) which follows the copper deposition in the Damascene process. JGB is regarded to serve as a leveling agent, flattening these bumps of copper deposits on the surface, or to influence the filling properties [39, 40, 54]. Unlike the other additives, JGB affects copper deposition only at the outside of trenches because of its consumption during the deposition resulting in a lower concentration at bottom of trenches [29].

13.2.2 Copper Electroless Deposition for Trench Filling

The electrodeposition process is highly useful to achieve superfilling of copper. However, with the shrinkage of the dimensions of electronic circuits and the increase in the size of a wafer, some critical issues, such as the need for the deposition of a uniform seed layer of sputtered copper and the problem of uneven current distribution on the wafer, are becoming much more critical problems for this process. The filling of trenches and via-holes by electroless copper deposition is expected to be an effective alternative for dealing with these problems, because electroless deposition is, in principle, capable of forming a uniform thin film on large substrates with superior step coverage. Furthermore, electroless deposition is advantageous because it does not require a copper seed layer.

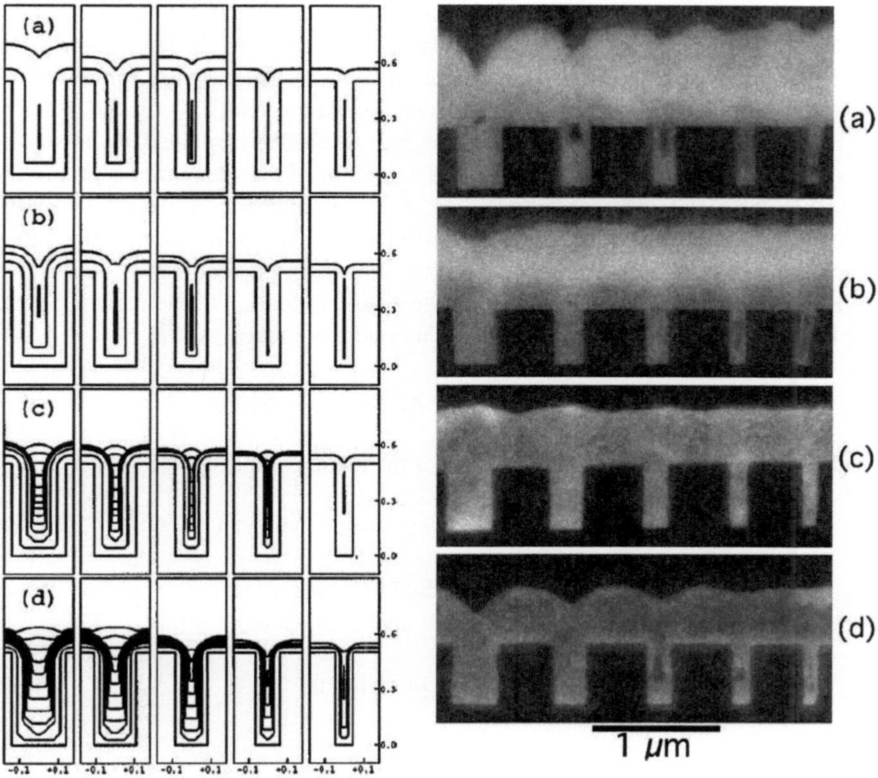

Fig. 13.11 Simulations (*left*) and SEM images (*right*) of copper deposition in trenches with MPS 0, 0.0005, 0.005, and 0.04 mol/m^3; overpotentials: –0.097, –0.301, –0.282, and –0.150 V [*top* (a) to *bottom* (d)]; and aspect ratios: 1.8, 2.3, 3.0, 4.0, 5.6 (*left to right*) (reproduced by permission of The Electrochemical Society) [2]

13.2.2.1 Mechanism of Electroless Copper Deposition

Electroless deposition solution contains metal ions, a reducing agent, a complexing agent, and some additives such as stabilizers and inhibitors. Formaldehyde (HCHO) and ethylenediaminetetraacetic acid (EDTA) are widely used as the reducing agent and the complexing agent, respectively, for electroless copper deposition. The overall reaction of electroless copper deposition in this solution is [55]

$$[\mathrm{Cu\,(II)\text{-}EDTA}]^{2-} + 2\mathrm{HCHO} + 4\mathrm{OH}^- \rightarrow \mathrm{Cu}^0 + 2\mathrm{HCOO}^- + 2\mathrm{H_2O} + \mathrm{H_2} + \mathrm{EDTA}^{4-}. \tag{13.3}$$

The deposition rate is described as follows:

$$\mathrm{r} = \mathrm{k}[C_{\mathrm{Cu}^{2+}}]^{\mathrm{a}}[C_{\mathrm{HCHO}}]^{\mathrm{b}}[C_{\mathrm{OH}^-}]^{\mathrm{c}}[C_{\mathrm{EDTA}^{4-}}]^{\mathrm{d}}\exp\left(-\frac{E}{T}\right), \tag{13.4}$$

where k is the rate constant at a given temperature; a, b, c and d are the reaction orders for the reactant; E is the activation energy; and T is the temperature. The deposition rate is affected by concentration of copper ions ($[C_{Cu^{2+}}]$), formaldehyde ($[C_{HCHO}]$), hydroxide ions ($[C_{OH-}]$), and complexing agent ($[C_{EDTA^{4-}}]$).

Electroless deposition reactions in general have been explained based on the mixed potential theory [56] in which the oxidation of the reducing agent and the reduction of metal ions are assumed to occur simultaneously on the substrate. According to this theory, superposition of polarization curves for the two partial reactions should yield the curve for the complete electroless bath.

Based on the mixed potential theory, the anodic oxidation of the reducing agent for Cu/formaldehyde electroless deposition system is written as [57]

$$2CH_2(OH)_2 + 2OH^- \leftrightarrow 2CH_2COOH + H_2 + 2H_2O + 2e^-. \tag{13.5}$$

The cathodic reaction is the reduction of the metal complex [57]:

$$[Cu\,(II)\text{-}EDTA]^{2-} + 2e^- \rightarrow Cu^0 + EDTA^{4-} \tag{13.6}$$

In alkaline aqueous solution, formaldehyde is hydrated as the methylene glycol. This compound is dissociated and exists as a glycolate anion [58].

$$HCHO + H_2O \leftrightarrow CH_2(OH)_2\,, \tag{13.7}$$

$$CH_2(OH)_2 + OH^- \leftrightarrow CH_2(OH)O^- + H_2O. \tag{13.8}$$

The glycolate anion is reported to accelerate the cathodic reaction in copper electroless deposition system [57, 59]. At low formaldehyde concentrations, diffusion of the methylene glycol anion from the bulk of the solution into copper surface is the rate-determining step.

Furthermore, in the electroless deposition bath, formaldehyde is decomposed by following side reaction (Cannizzaro reaction):

$$2HCHO + OH^- \rightarrow HCOO^- + CH_3OH. \tag{13.9}$$

The rate of this reaction increases with increases in temperature and pH. When temperature is more than 70°C, the rate of this reaction is three to four times that of the anodic oxidation of electroless deposition.

13.2.2.2 Superfilling by Electroless Copper Deposition

Superfilling has already been demonstrated to be achievable in an electroless copper plating bath containing small amounts of those additives which are used in the electrodeposition bath. Shingubara et al. [60–62] reported that superfilling was achieved by using bis(3-sulfopropyl)disulfide (SPS) as an inhibiting additive. Other researchers [63–65] have also achieved the superfilling by using SPS-containing

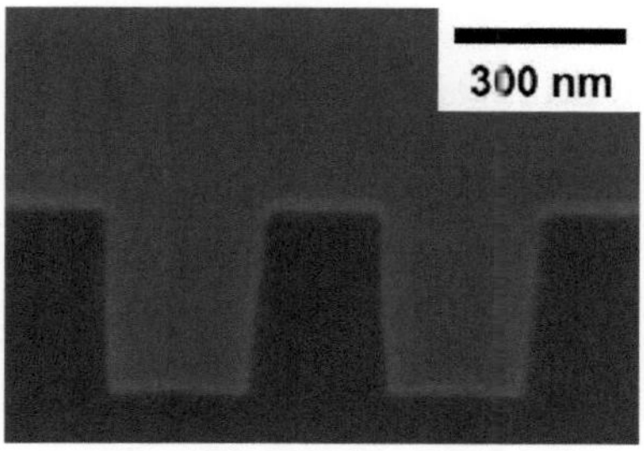

Fig. 13.12 Cross-sectional SEM image of copper deposited in trenches from an electroless copper plating bath containing HIQSA and PEG [66] (reproduced by permission of The Electrochemical Society) [66]

baths, although they reported that SPS accelerated copper deposition at very low concentrations.

Superfilling of trenches by electroless copper deposition was also achievable by using the combination of 8-hydroxy-7-iodo-5-quinoline sulfonic acid (HIQSA) [66] as an accelerating additive and polyethylene glycol (PEG) as an inhibiting additive (Fig. 13.12).

Figure 13.13 shows the deposition rates of copper measured on an unpatterned substrate in the baths containing various concentrations of HIQSA alone (Fig. 13.13a), PEG alone (Fig. 13.13b, open circles), and PEG with 3 ppm of HIQSA (Fig. 13.13b, filled squares).

For the HIQSA bath, the deposition rate was higher than that of the additive-free bath at all HIQSA concentrations. The deposition rate decreased significantly upon addition of PEG to the HIQSA bath, although it still tended to be higher than that of the PEG bath. It is important to note that copper deposition was accelerated by HIQSA only at very low concentrations (lower than 1 ppm) of PEG. The deposition rate of the bath containing 3 ppm of HIQSA without PEG was 10.1 μm h^{-1}, which was higher than that of the additive-free bath (6.48 μm h^{-1}). However, the deposition rate abruptly decreased when a few parts per million of PEG was added to the HIQSA bath. The result shows that the acceleration of copper deposition by HIQSA

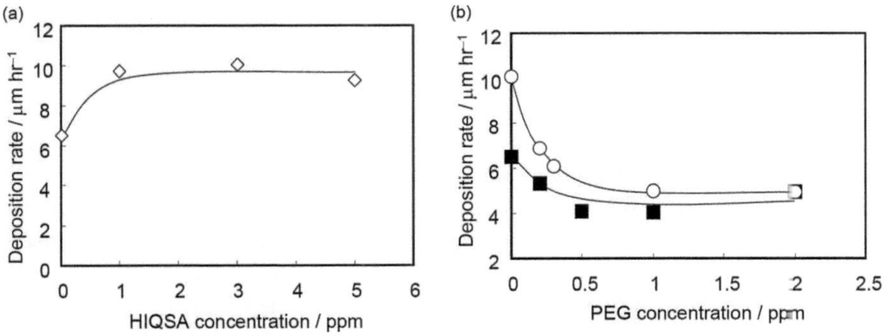

Fig. 13.13 Effect of HIQSA and PEG on the rate of electroless copper deposition in Bath A: (**a**) without PEG, (**b**) without HIQSA (), and with 3 ppm of HIQSA◯) (reproduced by permission of The Electrochemical Society) [66]

is significant only at PEG concentrations lower than 1 ppm, and it becomes insignificant in the presence of more than 1 ppm of PEG. On the other hand for the bath without HIQSA, the extent of the decrease in deposition rate with an increase in PEG concentration was relatively small. These results suggested that the improvement of the filling property observed in the simultaneous presence of HIQSA and PEG is relevant to the observed acceleration of copper deposition brought about by HIQSA at very low concentrations of PEG. The diffusion rate of PEG, which is a large polymer molecule, is most likely to be much smaller than that of HIQSA. Because the diffusion rate at trench bottom must be lower than that at the opening, a decrease in the concentration of PEG in the interior of trenches is expected to occur during the copper deposition. As mentioned earlier, the acceleration effect of HIQSA is significant only at very low concentrations of PEG. Therefore, the improvement of the filling property observed with the addition of both HIQSA and PEG is considered to be due to both the acceleration effect of HIQSA at the bottom of trenches, where the concentration of PEG is low, and the inhibition effect of PEG at trench openings.

13.3 Electrochemical Process for Seed Layer Formation

13.3.1 Electroless Deposition for Formation of Seed Layers

A conductive seed layer is essential for electrodeposition process. In the conventional process, a copper seed layer is deposited prior to superfilling by electrodeposition. The seed layer is required to be uniform in high-aspect-ratio and narrow trenches to maintain conductivity of the surface for electrodeposition. However, the copper film deposited by PVD on high-aspect-ratio trenches tends to be discontinuous or agglomerated. Usually, the film at the openings of narrow trenches is thicker than that at the bottom and the side walls. The increase in thickness at the trench openings leads to the nonuniform deposition inside the trenches, resulting in the formation of voids [23, 67]. To achieve conformal deposition even in high-aspect-ratio trenches, alternative deposition techniques such as electroless deposition, atomic layer deposition (ALD), and CVD have been investigated (Fig. 13.14).

Among these techniques, electroless deposition is promising process for forming seed layer because it achieves the excellent step coverage on high-aspect-ratio and narrow structures fabricated even on a large substrate. As noted in the previous section, electroless deposition is useful for fabrication of copper interconnects. In parallel with this, electroless deposition for seed layer fabrication has also been investigated intensively [68–74]. It has been demonstrated previously that a very thin seed layer with excellent step coverage on trenches was achieved by using electroless deposition technique. To achieve a thin and continuous copper seed layer by electroless deposition on a barrier layer, the development of the catalysis process as well as the optimization of bath composition is a key issue.

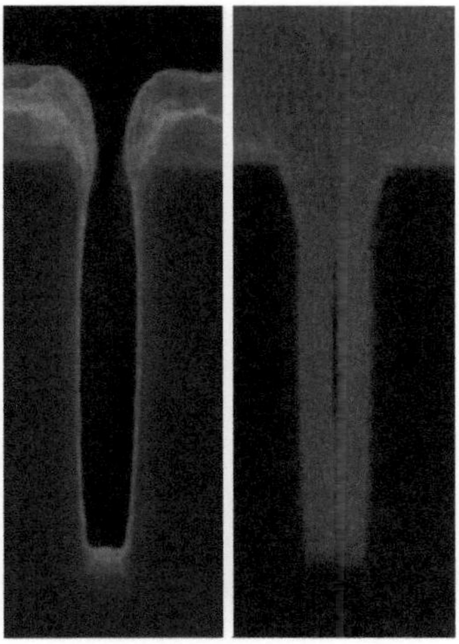

Fig. 13.14 Seed profile of a trench feature showing pinch off near the opening following PVD seed deposition and fill the result following plating [23]

Most researchers [69–74] focused their attention on the activation process to achieve direct deposition of a uniform and thin copper seed layer on the barrier materials such as TiN, TaN, and WN. In recent years, displacement deposition has been investigated for the activation of these materials [75, 76]. This activation process employs a $PdCl_2$/HF solution or a $PdCl_2$/HF/HNO_3 solution. In these solutions, two galvanic half cell reactions, which include the deposition of Pd particle and the removal of the oxide layer on the surface of substrate, take place simultaneously. Other techniques investigated for the catalysis of these barrier metal substrates are vacuum technologies such as ionized cluster beam (ICB) [72, 74, 77] and ALD [78, 79].

Copper seed layer is also required to possess low resistivity and the plating bath should be stable from the practical standpoint. Therefore, some stabilizers, such as 2,2'-dipyridine, CN^-, Neocuproine, and Rodamine, and the surfactants, such as polyethylene glycol, RE-610, and Triton-X-100, have been employed to the electroless plating bath [68].

13.3.2 Seedless Copper Electrodeposition on Barrier Materials

One alternative is elimination of copper seed layer to deal with the failures of copper seed layers which is responsible for many integration difficulties. However, copper electrodeposition on conventional barrier layers, such as TaN, TiN and WN, is very difficult because of their high resistivity and their highly stable oxide formed on the surface of the layer. Ru and other Pt-group material such as Ir, Os, Rh are expected to be a candidate as alternative barrier materials [27, 80–84]. Among them, Ru is particularly promising because bulk phase Ru and Cu are immiscible. Furthermore, the thermal and electrical conductivity of Ru are twice as much as those of Ta. This is preferable because barrier materials occupy larger fraction of the cross-sectional area of the conductors with a decrease in the dimensions of interconnects. Seedless superfilling of copper was demonstrated on sub-100 nm trenches with Ru barrier layers, which is deposited by PVD and ALD [85].

13.3.3 Electroless Deposition for Barrier Layer Formation and Seedless Copper Filling

Electroless deposition was also applied to the formation of Cu diffusion barrier layer [86–89]. For this process, Pd-catalyzed self-assembled monolayer (SAM) is applied as the adhesion/catalysis layer, which was the key for achieving uniform thin layers with sufficient adhesion to both SiO_2 and low-k substrates (Fig. 13.15a). NiB alloy was employed as a barrier layer. A conformal electroless NiB film was successfully formed on trench-patterned substrate with excellent step coverage by employing

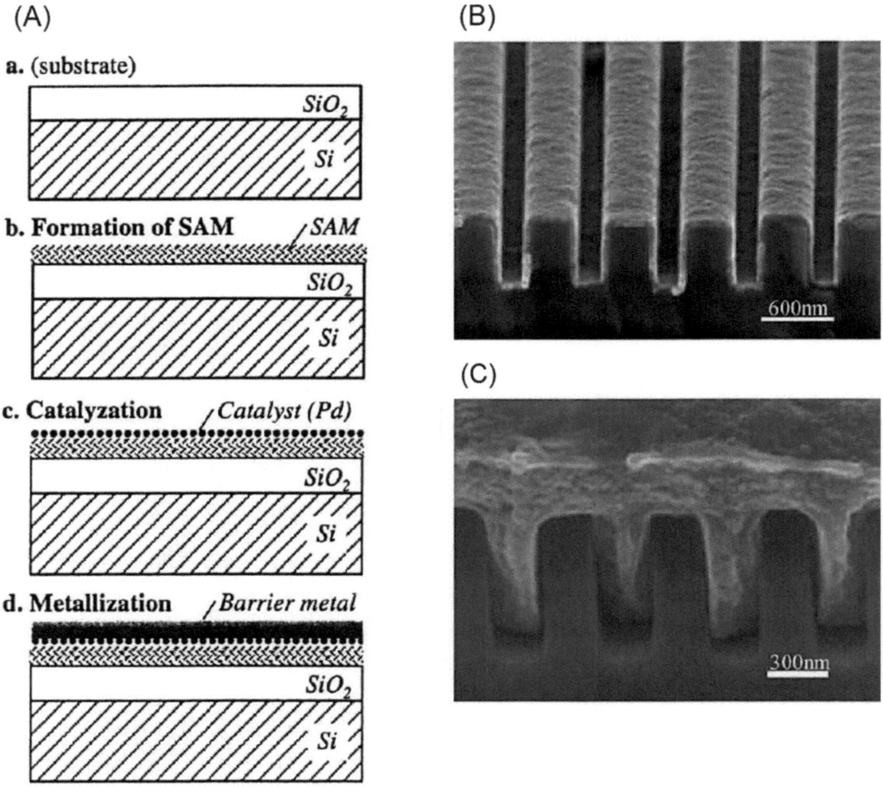

Fig. 13.15 (**a**) Schematic illustration of process for fabrication a electroless barrier layer, (**b**) SEM image of electroless NiB film deposited on trench-patterned substrate, and (**c**) SEM image of Cu electrodeposits on the NiB barrier layer [86, 88]

this process (Fig. 13.15b) [86]. This NiB film also exhibited sufficient barrier properties against copper diffusion into the substrate and acceptable thermal stability. The resistivity of the film was sufficiently low for copper electrodeposition. Consequently, seedless copper superfilling was achieved on this electroless NiB barrier layer (Fig. 13.15c).

References

1. Andricacos, P. C.; Uzoh, C.; Dukovic, J. O.; Horkans, J.; and Deligianni, H.: Damascene copper electroplating for chip interconnections. IBM J. Res. Dev. **42**, 567 (1998)
2. Andricacos, P. C.: Copper on-chip interconnections. The Electrochem. Soc. Interface **8**, 32 (1999)
3. Ismail, Y. I.; Friedman, E. G.; and Neves, J. L.: Exploiting the on-chip inductance in high-speed clock distribution networks. IEEE Trans Very Large Scale Integr (VLSI) Syst. **9**, 963 (2001)
4. Deutsch, A.; Kopcsay, G. V.; Coteus, P. W.; Surovic, C. W.; Dahlen, P. E.; Heckmann, D. L.; and Duan, D. W.: Frequency-dependent losses on high-performance interconnections. IEEE Trans. Electromagn. Compat. **43**, 446 (2001).
5. Ismail, Y. I.; Friedman, E. G.; and Neves, J. L.: Repeater insertion in tree structured inductive interconnect. IEEE Trans. Circuits Syst. II-Analog Digital Signal Process **48**, 471 (2001)
6. Deutsch, A.; Coteus, P. W.; Kopcsay, G. V.; Smith, H. H.; Surovic, C. W.; Krauter, B. L.; Edelstein, D. C.; and Restle, P. J.: On-chip wiring design challenges for gigahertz operation. Proc. IEEE. **89**, 529 (2001)
7. Ismail, Y. I.; and Friedman, E. G.: Effects of inductance on the propagation delay and repeater insertion in VLSI circuits. IEEE Trans Very Large Scale Integr (VLSI) Syst. **8**, 195 (2000)
8. Ismail, Y. I.; Friedman, E. G.; and Neves, J. L.: Equivalent Elmore delay for RLC trees. IEEE Trans. Comput-Aided Des. Integr. Circuits Syst. **19**, 83 (2000)
9. Sakurai, T.: Approximation of wiring delay in MOSFET LSI. IEEE J. Solid-State Circuit. **18**, 418 (1983)
10. Edelstein, D. C.; Saihalasz, G. A.; and Mii, Y. J.: VLSI on-chip interconnection performance simulations and measurements. IBM J. Res. Dev. **39**, 383 (1995)
11. Edelstein, D.; Heidenreich, J.; Goldblatt, R.; Cote, W.; Uzoh, C.; Lustig, N.; Roper, P.; McDevitt, T.; Motsiff, W.; Simon, A.; Dukovic, J.; Wachnik, R.; Rathore, H.; Schulz, R.; Su, L.; Luce, S.; and Slattery, J.: Full copper wiring in a sub-0.25 μm CMOS ULSI technology. In Technical Digest, IEEE International Electron Devices Meeting 773 (1997)
12. Venkatesan, S.; Gelatos, A. V.; Hisra, S.; Smith, B.; Islam, R.; Cope, J.; Wilson, B.; Tuttle, D.; Cardwell, R.; Anderson, S.; Angyal, M.; Bajaj, R.; Capasso, C.; Crabtree, P. Das, S.; Farkas, J.; Filipiak, S.; Fiordalice, B.; Freeman, M.; Gilbert, P. V.; Herrick, M. Jain, A.; Kawasaki, H.; King, C.; Klein, J.; Lii, T.; ReidK.; Saaranen, T.; Simpson, C.; Sparks, T.; Tsui, P.; Venkatraman, R.; Watts, D.; Weitzman, E. J.; Woodruff, R.; Yang, I.; Bhat, N.; Hamilton, G.; and Yu, Y.: A high performance 1.8 V, 0.20 μm CMOS technology with copper metallization. In Technological Digest, IEEE International Electron Devices Meeting 769 (1997)
13. Rosenberg, R.; Edelstein, D. C.; Hu, C. K.; and Rodbell, K. P.: Copper metallization for high performance silicone technology. Annu. Rev. Mat. Sci. **30**, 229 (2000)
14. Lee, S. K.; Chun, S. S.; Hwang, C. Y.; and Lee, W. J.: Reactive ion etching mechanism of copper film in chlorine-based electron cyclotron resonance plasma. Jpn. J. Appl. Phys. PART 1. **36**, 50 (1997)
15. HowardB. J. and Steinbruchel, C.: Reactive ion etching of copper in $SiCl_4$-based plasmas. Appl. Phys. Lett. **59**, 914 (1991)
16. Ohno, K.; Sato, M.; and Arita, Y.: Reactive ion etching of copper films in $SiCl_4$ and N_2 mixture. Jpn. J. Appl. Phys. PART 2. **28**, L, 1070 (1989)

17. Schwartz, G. C. and Schaible, P. M.: Reactive ion etching of copper films. J. Electrochem. Soc. **130**, 1777 (1983)
18. Krzewska, S.; Impedance investigation of the mechanism of copper electrodeposition from acidic perchlorate electrolyte. Electrochim. Acta. **42**, 3531 (1997)
19. Yoon, S.; Schwartz, M.; and Nobe, K.: Rotating ring-disk electrode studies of copper electrodeposition: effect of chloride ions and organic additives. Plat. Surf. Finish. **81**, 65 (1994)
20. Jardy, A.; Lasallemolin, A. L.; Keddam, M.; and Takenouti, H.: Copper dissolution in acidic sulphate media studied by QCM and rrde under ac signal. Electrochim. Acta. **37**, 2195 (1992)
21. Yokoi, M.; Konishi, S.; and Hayashi, T.: Adsorption behavior of polyoxyethyleneglycole on the copper surface in an acid copper sulfate bath. Denki Kagaku. **52**, 218 (1984)
22. Yokoi, M.; Konishi, S.; and Hayashi, T.: Mechanism of the electrodeposition and dissolution of copper in an acid copper sulfate bath I. The behavior of intermediate Cu^{+}. Denki Kagaku. **51**, 310 (1984)
23. ReidJ.; Copper electrodeposition: principles and recent progress. Jpn. J. Appl. Phys. PART 1. **40**, 2650 (2001)
24. Takahashi, K. M.: Electroplating copper onto resistive barrier films. J. Electrochem. Soc. **147**, 1414 (2000)
25. Broadbent, E. K.; McInerney, E. J.; Gochberg, L. A.; and Jackson, R. L.: Experimental and analytical study of seed layer resistance for copper Damascene electroplating. J. Vac. Sci. Tech. B. **17**, 2584 (1999)
26. Soukane, S; Sen, S.; and Cale, T. S.: Feature superfilling in copper electrochemical deposition. J. Electrochem. Soc. 149:C, 74 (2002)
27. Kelly, J. J. and West, A. C.: Leveling of 200 nm features by organic additives. Electrochem. Solid State Lett. **2**, 561 (1999)
28. Mirkova, L.; Rashkov, S.; and Nanev, C.: The leveling mechanism during bright acid copper plating. Surf. Tech. **15**, 181 (1982)
29. Hasegawa, M.; Negishi, Y.; Nakanishi, T.; and Osaka, T.: Effects of additives on copper electrodeposition in submicrometer trenches. J. Electrochem. Soc. **152**, C221 (2005)
30. Kelly, J. J.; and West, A. C.: Copper Deposition in the Presence of Polyethylene Glycol I. Quartz Crystal Microbalance Study. J Electrochem Soc. **145**, 3472 (1998)
31. Kelly, J. J. and West, A. C.: Copper deposition in the presence of polyethylene glycol II. Electrochemical impedance spectroscopy. J. Electrochem. Soc. **145**, 3477 (1998)
32. Stoychev, D. and Tsvetanov, C.: Behaviour of poly (ethylene glycol) during electrodeposition of bright copper coatings in sulphuric acid electrolytes. J. Appl. Electrochem. **26**, 741 (1996)
33. Moffat, T. P.; Wheeler, D.; and Josell, D.: Electrodeposition of copper in the SPS-PEG-Cl additive system I. Kinetic measurements: influence of SPS. J. Electrochem. Soc. **151**, C262 (2004)
34. Healy, J. P.; Pletcher, D.; and Goodenough, M.: The chemistry of the additives in an acid copper electroplating bath Part I. Polyethylene glycol and chloride ion. J. Electroanal. Chem. **338**, 155 (1992)
35. Kang, M.; Gross, M. E.; and Gewirth, A. A.: Atomic force microscopy examination of Cu electrodeposition in trenches. J. Electrochem. Soc. **150**, C292 (2003)
36. Moffat, T. P.; Wheeler, D.; Edelstein, M. D.; and Josell, D.: Superconformal film growth: mechanism and quantification. IBM J. Res. Dev. **49**, 19 (2005)
37. Vereecken, P. M.; BinsteadR. A.; Deligianni, H.; and Andricacos, P. C.: The chemistry of additives in Damascene copper plating. IBM J. Res. Dev. **49**, 3 (2005)
38. Moffat, T. P.; Bonevich, J. E.; Huber, W. H.; Stanishevsky, A.; Kelly, D. R.; StaffordG. R.; and Josell, D.: Superconformal electrodeposition of copper in 500–90 nm features. J. Electrochem. Soc. **147**, 4524 (2000)
39. Miura, S.; Oyamada, F.; Takada, Y.; and Honma, H.: ULSI wiring formation by copper electroplating in the presence of additives. Electrochemistry **69**, 773 (2001)
40. Taephaisitphongse, P.; Cao, Y.; and West, A. C.: Electrochemical and fill studies of a multicomponent additive package for copper deposition. J. Electrochem. Soc. **148**, C492 (2001)

41. Dukovic, J. O.: Feature-scale simulation of resist-patterned electrodeposition. IBM J. Res. Dev. **37**, 125 (1993)
42. West, A. C.: Theory of filling of high-aspect ratio trenches and vias in presence of additives. J. Electrochem. Soc. **147**, 227 (2000)
43. Takahashi, K. M. and Gross, M. E.: Transport phenomena that control electroplated copper filling of submicron vias and trenches. J. Electrochem. Soc. **146**, 4499 (1999)
44. Madore, C.; Agarwal, P.; and Landolt, D.: Blocking inhibitors in cathodic leveling III. Electrochemical impedance spectroscopy study. J. Electrochem. Soc. **145**, 1561 (1998)
45. Madore, C.; Matlosz, M.; and Landolt, D.: Blocking inhibitors in cathodic leveling I. Theoretical analysis. J. Electrochem. Soc. **143**, 3927 (1996)
46. Madore, C. and Landolt, D.: Blocking inhibitors in cathodic leveling II. Experimental investigation. J. Electrochem. Soc. **143**, 3936 (1996)
47. Josell, D.; Baker, B.; Witt, C.; Wheeler, D.; and Moffat, T. P.: Via filling by electrodeposition superconformal silver and copper and conformal nickel. J. Electrochem. Soc. **149**, C637 (2002)
48. Moffat, T. P.; Wheeler, D.; Witt, C.; and Josell, D.: Superconformal electrodeposition using derivitized substrates. Electrochem. Solid State Lett. **5**, C110 (2002)
49. Josell, D.; Wheeler, D.; Huber, W. H.; and Moffat, T. P.: Superconformal electrodeposition in submicron features. Phys. Rev. Lett. **87**, 016102 (2001)
50. Moffat, T. P.; Wheeler, D.; Huber, W. H.; and Josell, D.: Superconformal electrodeposition of copper. Electrochem. Solid State Lett. **4**, C26 (2001)
51. West, A. C.; Mayer, S.; and ReidJ.: A superfilling model that predicts bump formation. Electrochem. Solid State Lett. **4**, C50 (2001)
52. Wheeler, D.; Josell, D.; and Moffat, T. P.: Modeling superconformal electrodeposition using the level set method. J. Electrochem. Soc. **150**, C302 (2003)
53. Josell, D.; Moffat, T. P.; and Wheeler, D.: An exact algebraic solution for the incubation period of superfill. J. Electrochem. Soc. **151**, C19 (2004)
54. Haba, T.; Itabashi, T.; Akahoshi, H.; Sano, A.; Kobayashi, K.; Miyazaki, H.: Electrochemical and simulative studies of trench filling mechanisms in the copper Damascene electroplating process. Mater. Trans. **43**, 1593 (2002)
55. Lukes, R. M.: The chemistry of the autocatalytic reduction of copper by alkaline formaldehyde. Plating **51**, 1066 (1964)
56. Wagner, C. and Traud, W.: Über die deutung von korrosionsvorgängen durch überlagerung von elektrochemischen teilvorgängen und über die potenialbildung an mischelektroden. Z Elektrochem. **44**, 391 (1938)
57. Feldman, B. J. and Melroy, O. R.: The mechanism of electroless Cu deposition: extraction of the oxidative and reductive electrochemical half-cell currents from a complete bath. J. Electrochem. Soc. **136**, 640 (1989)
58. Schumacher, R.; Pesek, J. J.; and Melroy, O. R.: Kinetic analysis of electroless deposition of copper. J. Phys. Chem. **89**, 4338 (1985)
59. Okinaka, Y. and Osaka, T.: Advances in electrochemical science and engineering. In: Gerischer, H. and Tobias, C. W.: (Eds.) VCH Publishers Inc, Weinheim, 387 (1994)
60. Wang, Z. L.; Yaegashi, O.; Sakaue, H.; Takahagi, T.; and Shingubara, S.: Bottom-up fill for submicrometer copper via holes of ULSIs by electroless plating. J. Electrochem. Soc. **151**, C781 (2004)
61. Wang, Z. L.; Yaegashi, O.; Sakaue, H.; Takahagi, T.; and Shingubara, S.: Effect of additives on hole filling characteristics of electroless copper plating. Jpn. J. Appl. Phys. PART 1. **43**, 7000 (2004)
62. Shingubara, S.;Wang, Z. L.; Yaegashi, O.; Obata, R.; Sakaue, H.; and Takahagi, T.: Bottom-up fill of copper in deep submicrometer holes by electroless plating. Electrochem. Solid State Lett. **7**, C78 (2004)
63. Kim, J. J.; Kim, S. K.; and Kim, Y. S.: Catalytic behavior of 3-mercapto-1-propane sulfonic acid on Cu electrodeposition and its effect on Cu film properties for CMOS device metallization. J. Electroanal. Chem. **542**, 61 (2003)

64. Lee, C. H.; Lee, S. C.; and Kim, J. J.: Improvement of electrolessly gap-filled Cu using 2,2'-Dipyridyl and Bis-(3-sulfopropyl)-disulfide (SPS). Electrochem. Solid State Lett. **8**, C110 (2005)
65. Lee, C. H.; Lee, S. C.; and Kim, J. J.: Bottom-up filling in Cu electroless deposition using bis-(3-sulfopropyl)-disulfide (SPS). Electrochim. Acta. **50**, 3563 (2005)
66. Hasegawa, M.; Okinaka, Y.; Shacham-DiamandY.; and Osaka, T.: Void-free Trench-filling by electroless copper deposition using the combination of accelerating and inhibiting additives. Electrochem. Solid-State Lett. **9**, C138 (2006)
67. ReidJ.; Mayer, S.; Broadbent, E.; Klawuhn, E.; and Ashtiani, K.: Factors influencing Damascene feature fill using copper PVD and electroplating. Solid State Technol. **43**, 86 (2000)
68. Dubin, V. M.; Shacham-DiamandY.; Zhao, B.; Vasudev, P. K.; and Ting, C. H.: Selective and blanket electroless copper deposition for ultralarge scale integration. J. Electrochem. Soc. **144**, 898 (1997)
69. Lee, C. H.; Hwang, S.; Kim, S. C.; and Kim, J. J.: Cu electroless deposition onto Ta substrates. Electrochem. Solid State Lett. **9**, C157 (2006)
70. Kim, S. K.; Cho, S. K.; Kim, J. J.; and Lee, Y. S.: Superconformal Cu electrodeposition on various substrates. Electrochem. Solid State Lett. **8**, C19 (2005)
71. Kim, Y. S.; Bae, D. L.; Yang, H. C.; Shin, H. S.; Wang, G. W.; Senkevich, J. J.; and Lu, T. M.: Direct copper electroless deposition on a tungsten barrier layer for ultralarge scale integration. J. Electrochem. Soc. **152**, C89 (2005)
72. Wang, Z. L.; Sakaue, H.; Shingubara, S.; and Takahagi, T.: Influence of surface oxide of sputtered TaN on displacement plating of Cu. Jpn. J. Appl. Phys. PART 1 **42**, 1843 (2003)
73. Kim, J. J.; Kim, S. K.; Lee, C. H.; and Kim, Y. S.: Investigation of various copper seed layers for copper electrodeposition applicable to ultralarge-scale integration interconnection. J. Vac. Sci. Tech. B **21**, 33 (2003)
74. Wang, Z.; Ida, T.; Sakaue, H.; Shingubara, S.; and Takahagi, T.: Electroless plating of copper on metal-nitride diffusion barriers initiated by displacement plating. Electrochem. Solid State Lett. **6**, C38 (2003)
75. Hsu, H. H.; Hsieh, C. C.; Chen, M. H.; Lin, S. J.; and Yeh, J. W.: Displacement activation of tantalum diffusion barrier layer for electroless copper deposition. J. Electrochem. Soc. **148**, C590 (2001)
76. Patterson, J. C.; Nidheasuna, C.; Barrett, J.; Spalding, T. R.; Oreilly, M.; Jiang, X.; and Crean, G. M.: Electroless copper metallisation of titanium nitride. Appl. Surf. Sci. **91**, 124 (1995)
77. Wang, Z. L.; Yaegashi, O.; Sakaue, H.; Takahagi, T.; and Shingubara, S.: Highly adhesive electroless Cu layer formation using an ultra thin Ionized Cluster Beam (ICB)-Pd catalytic layer for sub-100 nm Cu interconnections. Jpn. J. Appl. Phys. PART 2. **42**, L, 1223 (2003)
78. Kim, Y. S.; Shin, J.; Cho, J. H.; Ten Eyck, G. A.; Liu, D. L.; Pimanpang, S.; Lu, T. M.; Senkevich, J. J., and Shin, H. S.: Surface characterization of copper electroless deposition on atomic layer deposited palladium on iridium and tungsten. Surf. Coat. Tech. **200**, 5760 (2006)
79. Kim, Y.; Ten Eyck, G. A.; Ye, D. X.; Jezewski, C.; Karabacak, T.; Shin, H. S.; Senkevich, J. J.; and Lu, T. M.: Atomic layer deposition of Pd on TaN for Cu electroless plating. J. Electrochem. Soc. **152**, C376 (2005)
80. Josell, D.; Witt, C.; and Moffat, T. P.: Osmium barriers for direct copper electrodeposition in Damascene processing. Electrochem. Solid State Lett. **9**, C41 (2006)
81. Josell, D.; Bonevich, J. E.; Moffat, T. P.; Aaltonen, T.; Ritala, M.; and Leskela, M.: Iridium barriers for direct copper electrodeposition in Damascene processing. Electrochem. Solid State Lett. **9**, C48 (2006)
82. Moffat, T. P.; Walker, M.; Chen, P. J.; Bonevich, J. E.; Egelhoff, W. F.; Richter, L.; Witt, C.; Aaltonen, T.; Ritala, M.; Leskela, M.; and Josell, D.: Electrodeposition of Cu on Ru barrier layers for Damascene processing. J. Electrochem. Soc. **153**, C37 (2006)
83. Zheng, M.; Willey, M.; and West, A. C.: Electrochemical nucleation of copper on ruthenium effect of Cl^-, PEG, and SPS. Electrochem. Solid State Lett. **8**, C151 (2005)

84. Radisic, A.; Cao, Y.; Taephaisitphongse, P.; West, A. C.; and Searson, P. C.: Direct copper electrodeposition on TaN barrier layers. J. Electrochem. Soc. **150**, C362 (2003)
85. Josell, D.; Wheeler, D.; Witt, C.; and Moffat, T. P.: Seedless superfill: copper electrodeposition in trenches with ruthenium barriers. Electrochem. Solid State Lett. **6**, C143 (2003)
86. Yoshino, M.; Nonaka, Y.; Sasano, J.; Matsuda, I.; Shacham-DiamandY.; and Osaka, T.: All-wet fabrication process for ULSI interconnect technologies. Electrochim. Acta. **51**, 916 (2005)
87. Osaka, T.; Takano, N.; and Yokoshima, T.: Microfabrication of electro- and electroless-deposition and its application in the electronic field. Surf. Coat. Tech. **169**, 1 (2003)
88. Osaka, T.; Takano, N.; Kurokawa, T.; Kaneko, T.; and Ueno, K.: Electroless nickel ternary alloy deposition on SiO_2 for application to diffusion barrier layer in copper interconnect technology. J. Electrochem. Soc. **149**, C573 (2002)
89. Osaka, T.; Takano, N.; Kurokawa, T.; and Ueno, K.: Fabrication of electroless NiReP barrier layer on SiO_2 without sputtered seed layer. Electrochem. Solid State Lett. **5**, C7 (2002)

Chapter 14
Atomic Layer Deposition (ALD) Processes for ULSI Manufacturing

Schubert S. Chu

14.1 Introduction

Atomic layer deposition (ALD) is a technique where precursors are introduced alternatively, and a monolayer (or fraction thereof) is deposited on the surface at a time [1–4]. The sequential introduction of all precursors, separated by purge steps, completes an ALD cycle. Figure 14.1 illustrates the steps that comprise an ALD cycle.

In step one of the ALD cycle, the first precursor (Precursor 1) is introduced into the reaction space and adsorbed onto the substrate surface (Fig. 14.1a). One key characteristic of ALD is that the adsorption of precursor is self-limiting; once the surface is saturated with adsorbed precursor, or if all available adsorptions sites are occupied, no additional precursor molecules are adsorbed onto the substrate. In order for a reaction to exhibit ALD characteristics, it is essential that the operating temperature is below that of the decomposition threshold for the precursors used in the reaction to maintain the self-limiting behavior. In the second step of the ALD cycle, excess precursor is removed from the reactor, leaving behind a Precursor 1 layer adsorbed onto the substrate (Fig. 14.1b). As the design of ALD reactors vary, the precursor removal can be achieved by either purging with an inert gas or simply evacuating the reactor space down to base pressure. A second reactant (Precursor 2) is introduced into the third step of the cycle to react with the Precursor 1 layer previously adsorbed onto the substrate surface (Fig. 14.1c). In the final step of the ALD cycle, volatile reaction byproducts, as well as excess precursor are removed from the reactor (Fig. 14.1d). The result is the desired molecules on the substrate surface. These four steps complete the ALD cycle which is repeated until the desired film thickness is reached.

Typically, films deposited by ALD have a deposition rate of less than one atomic layer per cycle, with rates ranging from 0.01 to 0.4 nm/cycle (calculated by dividing the total film thickness by the total number of cycles). The primary reason for this is

S.S. Chu (✉)
Global Product Manager at Applied Materials, Applied Materials, Inc., Santa Clara, CA 95054-3299, USA
e-mail: schubert_chu@amat.com

Y. Shacham-Diamand et al. (eds.), *Advanced Nanoscale ULSI Interconnects: Fundamentals and Applications*, DOI 10.1007/978-0-387-95868-2_14,

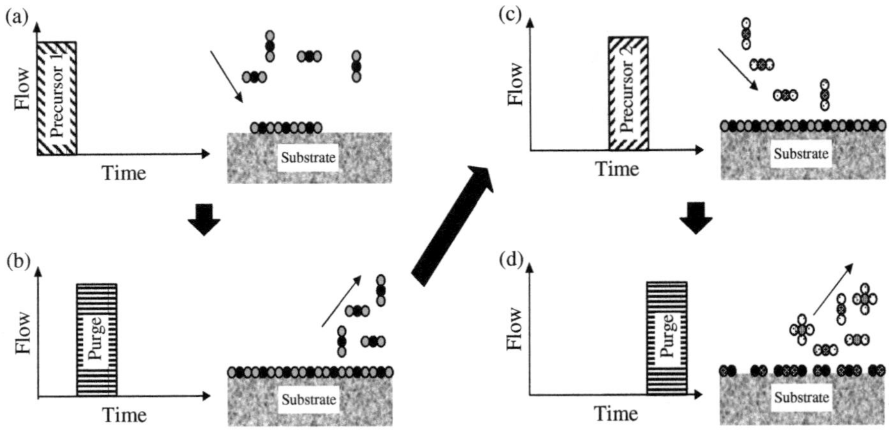

Fig. 14.1 Steps of an atomic layer deposition (ALD) cycle 1. Precursor 1 is introduced into the reactor and adsorbed onto the substrate. 2. Adsorption of Precursor 1 is saturated. Excess precursor is pumped or purged out of the reactor. 3. Precursor 2 is introduced into the reactor and reacts with Precursor 1. 4. Resultant molecule forms on the substrate. Reaction byproduct and excess precursor are pumped or purged out of the reactor.

steric hindrance of the precursor molecule due to the bulkiness of its ligands [5]. A second reason is related to the number of available adsorption and reaction sites [6].

Atomic layer deposition exhibits several highly desirable characteristics including uniform deposition, conformal step coverage, and insensitivity to precursor dosage. This is due to the dependence of ALD processes on self-limiting surface adsorption. Figure 14.2a illustrates the self-limiting behavior of atomic layer deposition. As precursor dose is increased (usually done by increasing precursor exposure time within the reactor), the deposition rate increases until saturation is reached. Further precursor exposure or higher concentration will not cause an increase in the deposition rate as additional precursor is unable to adsorb on the substrate. Commercial ALD reactors take advantage of this characteristic and supply excess precursors to ensure full surface coverage while maintaining uniform film properties. The self-limiting behavior depicted below is generally observed for both precursors used in an ALD reaction. However, a nominal increase in deposition rate with additional precursor exposure beyond the saturation point is sometimes observed in practice.

Atomic layer deposition processes are known to have wide temperature windows. Deposition rate dependency on process temperature as shown in Fig. 14.2b can be obtained for an ideal ALD process. In Zone I, the substrate temperature is insufficient to complete the reaction. The deposition rate will increase with increasing temperature as the reaction is closer to being complete. Zone II is often referred to as the "ALD process window" where the deposition rate is insensitive to process temperature. The implication of such temperature window is that small variations in the substrate temperature may be immaterial to the film uniformity, and therefore the requirements for substrate design in an ALD reactor can be simplified from that of a CVD reactor. If the substrate temperature is beyond the decomposition temperature

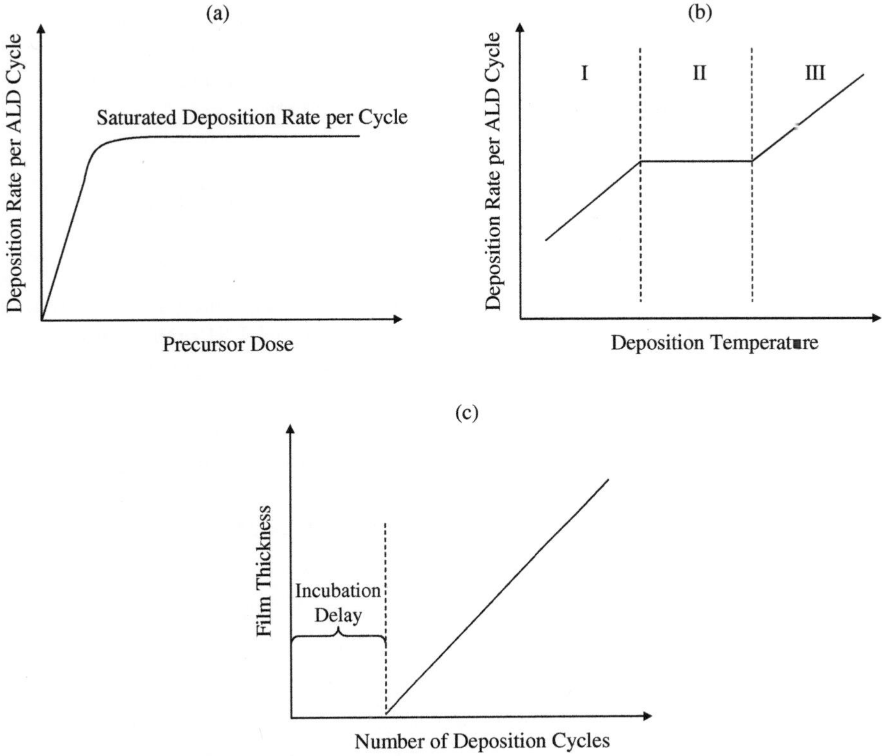

Fig. 14.2 Characteristics of atomic layer deposition. **a**. Self-limiting deposition rate as a function of precursor dose. **b**. Deposition rate versus deposition temperature – Zone I: incomplete reaction, Zone II: self-limiting ALD reaction, Zone III: increased deposition due to precursor decomposition. **c**. Deposition thickness linearly proportional to the number of deposition cycles

of the precursor, an increase in deposition rate can be expected with increased temperature and is depicted as Zone III. Films deposited in this temperature regime will exhibit diminished ALD characteristics as the reaction is increasingly more CVD-like. Expected consequences of deposition in Zone III include degradation in conformality as well as increase in film non-uniformity corresponding to precursor flow patterns across the surface. It is important to note that many ALD precursors actually do not exhibit a measurable temperature window even though other characteristics of ALD still exist.

As the deposition rate for each ALD cycle is self-limiting, the growth of film thickness can be "digitally" controlled by manipulating the number of deposition cycles. As shown in Fig. 14.2c, the deposition thickness is linearly proportional to the number of ALD cycles. Substrate surface condition is usually the primary factor in determining the extent of the delay in initiating deposition, typically called an incubation delay. In some cases, incubation delays can be minimized or eliminated by extended exposure of one of the precursors. In other systems, hydroxyl groups promote the chemisorption of precursor molecules [7], and substrates with hydroxyl surface termination generally have little to no incubation delay. In contrast, substrates with hydrophobic surfaces are difficult for ALD nucleation.

14.2 Applications of Atomic Layer Deposition in ULSI Manufacturing

As the geometries continue to shrink for each successive technology node, scaling has both pushed target thickness of many films used in ULSI manufacturing to a few monolayers and created requirements for excellent step coverage in high aspect ratio features. The distinct properties of ALD such as precise thickness control and conformal deposition have attracted great interest from mainstream ULSI production and R&D groups. There are potential applications in many areas spanning from front-end-of-line (FEOL) to back-end-of-line (BEOL) processing (Fig. 14.3). Some leading examples of ALD applications in ULSI are presented in the next few sections.

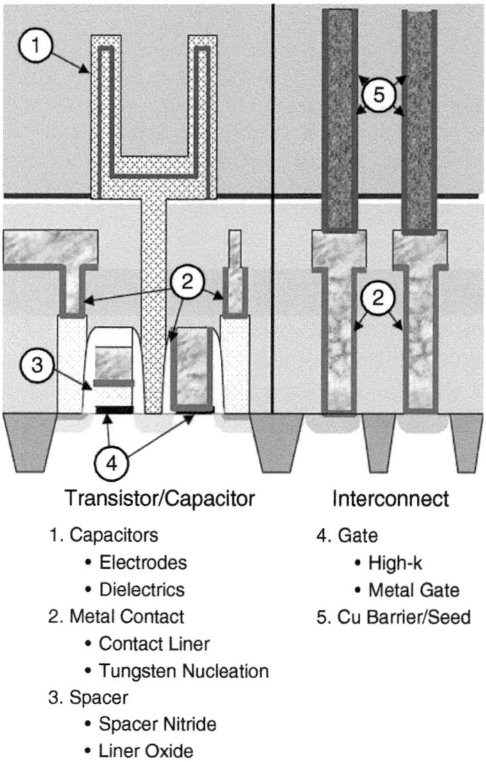

Fig. 14.3 Applications of atomic layer deposition in various areas of ULSI processing

14.2.1 Front-End-of-Line (FEOL) Applications

Some of the biggest technical challenges for continuing Moore's law exist in the front-end-of-line area. Many of the films used in FEOL are approaching thicknesses of just a few monolayers, and film properties and interfaces are becoming increasingly more critical. Films deposited by ALD are well suited to address these needs, and there are active research efforts to implement ALD in the areas of gate dielectric, gate electrode, and capacitors for DRAM.

As transistors continue to scale beyond 100 nm with an equivalent oxide thickness (EOT) below 1 nm, the requirements for gate dielectric leakage become more stringent. This requirement is more severe for low-power devices and for mobile applications as shown in Table 14.1 [8]. At the EOT required, conventionally grown silicon oxides or oxynitrides can no longer meet the gate dielectric leakage requirements. Dielectric films with much higher dielectric constants (κ) are needed to meet both the EOT and the gate dielectric leakage requirements [9, 10].

Table 14.1 Selected FEOL International Technology Roadmap for Semiconductors (ITRS) specifications for 45, 32, and 22 nm technology nodes [8]

Year	Node	Application	Physical gate length (nm)	Equivalent oxide thickness (nm)	Gate dielectric leakage at 100°C (nA/um)
2010	45 nm	High performance	18	0.7	330
		Low power	22	0.9	2.33
2013	32 nm	High performance	13	0.6	1000
		Low power	16	0.8	3.33
2016	22 nm	High performance	9	0.5	1670
		Low power	11	0.7	10

Several high-*k* materials such as aluminum oxide and zirconium oxide have been studied as potential candidates for gate dielectric material. Recently, hafnium oxide (HfO_2) has been shown to be a promising material for its high dielectric constant, wide band gap, and band offset from silicon. ALD HfO_2 can be deposited from a number of sources. Deposition using inorganic precursors such as $HfCl_4$ has been reported with H_2O [11, 12]. However, residual Cl is a potential cause for concern at the Si interface. Alternatively, metal organic precursors such as hafnium amido compounds [e.g., tetrakis(dimethylamino)hafnium (TDMAH), tetrakis(ethylmethylamino) hafnium (TEMAH), tetrakis(diethylamido)hafnium (TDEAH)] are gaining attention. Deposition of HfO_2 by TDEAH has been demonstrated with H_2O [13] and O_3 as oxidizers.

While high-*k* gate dielectric may be effective in reducing gate leakage, metal gate electrode eliminates the poly-depletion issues experienced with polysilicon gates. Further, metal electrodes also eliminate Fermi level pinning associated with polysilicon gates. Many materials [10] have been evaluated for NMOS and PMOS gates with recent focus on TiN [14] and Ru [15]. Materials selection for metal gate has been done primarily via physical vapor deposition (PVD) due to the relative simplicity of acquiring targets for sputtering pure materials. Key problems with thin

PVD films for this application are plasma damage to the gate dielectric, high film non-uniformity, and lack of deposition rate control. At present, ALD is thought to be an enabling technique for metal gate electrodes as it overcomes these limitations. However, identification of appropriate precursors and reactants may be required to enable desired materials systems.

DRAM makers have made the migration from polysilicon electrodes to metal–insulator–silicon (MIS) capacitors in recent years. The next generation of capacitors employs metal–insulator–metal (MIM) capacitors. TiN, deposited via a CVD reaction, has been the most common material for metal electrodes. However, as aspect ratio (AR) of the MIM capacitors and film conformality requirements increase and thermal budget becomes a consideration, ALD becomes an attractive technique for depositing both the metal electrode and the insulator film. In this application, TiN is typically deposited from $TiCl_4$ with NH_3 [16] at a temperature as low as 450°C. An alternative is to utilize tetrakis(dimethylamido)titanium (TDMAT) and NH_3 as precursors. TiN films can be deposited at a substantially lower temperature (180°C) with these precursors [17]. For the purposes of depositing the insulator, the most popular choice is Al_2O_3 deposited by using trimethyl aluminum (TMA) precursor [18]. While Al_2O_3 offers great leakage benefits, its dielectric constant (8–12) is lower than that of HfO_2 (20–30). HfO_2 is therefore attractive for the purposes of increasing capacitance. Some DRAM makers have recently chosen to use a laminated ALD Al_2O_3 and ALD HfO_2 approach in order to realize the benefits of low leakage of Al_2O_3 and high capacitance of HfO_2.

14.2.2 Middle-of-Line (MOL) Applications

Despite the migration of logic devices from aluminum to copper interconnects in the backend of line at 130 nm, tungsten contacts have extended through the 65 nm technology node and are anticipated to meet 45 nm requirements. Typically, contact formation starts with argon sputter clean followed by deposition of titanium and titanium nitride as liner, barrier, and glue layer. Tungsten then fills the remainder of the contact. While chemical vapor deposition (CVD) has been used for TiN and tungsten formation, ALD is receiving increased attention as the contact aspect ratio increased and resistance requirements became more stringent.

In logic circuits, TiN is typically by MOCVD TiN with tetrakis(dimethylamido) titanium (TDMAT) followed by a plasma treatment with N_2 and H_2. There have been reports of ALD TiN by alternating exposure of TDMAT and NH_3 [19]. However, this process has not gained wide acceptance. For DRAM applications, CVD TiN with $TiCl_4$ and NH_3 has been widely used. Within the DRAM segment, work on ALD TiN with $TiCl_4$ and NH_3 is quite active for its conformal deposition in high aspect ratio structures.

Tungsten filling of contracts traditionally has been performed by first depositing a CVD nucleation layer using WF_6 and SiH_4, followed by bulk CVD fill with WF_6 and H_2. Fundamentals of chemical vapor deposition of tungsten by using WF_6 along with H_2, SiH_4, Si_2H_6, and B_2H_6 have been thoroughly studied and reported [20]. While CVD tungsten has been adequate for legacy technology nodes, it is increasingly more difficult to obtain void-free contact fills for 65 nm node and below. The primary cause for incomplete contact fills is the poor step coverage of the nucleation layer, which can be as low as 50%. It is for this reason that atomic layer deposition of tungsten, especially as a nucleation layer to subsequent CVD bulk tungsten fill, has gained momentum in recent years. ALD tungsten with WF_6 has been reported with Si_2H_6 [6, 21] and B_2H_6 or SiH_4 [22] as reactants. Proposed reaction mechanism with B_2H_6 is illustrated in Fig. 14.4. The reaction between WF_6 and B_2H_6 exhibits

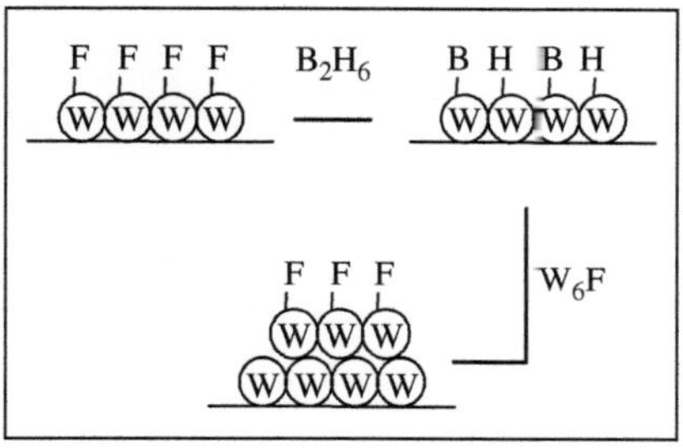

Fig. 14.4 Proposed ALD W reaction mechanism with WF_6 and B_2H_6 as precursors

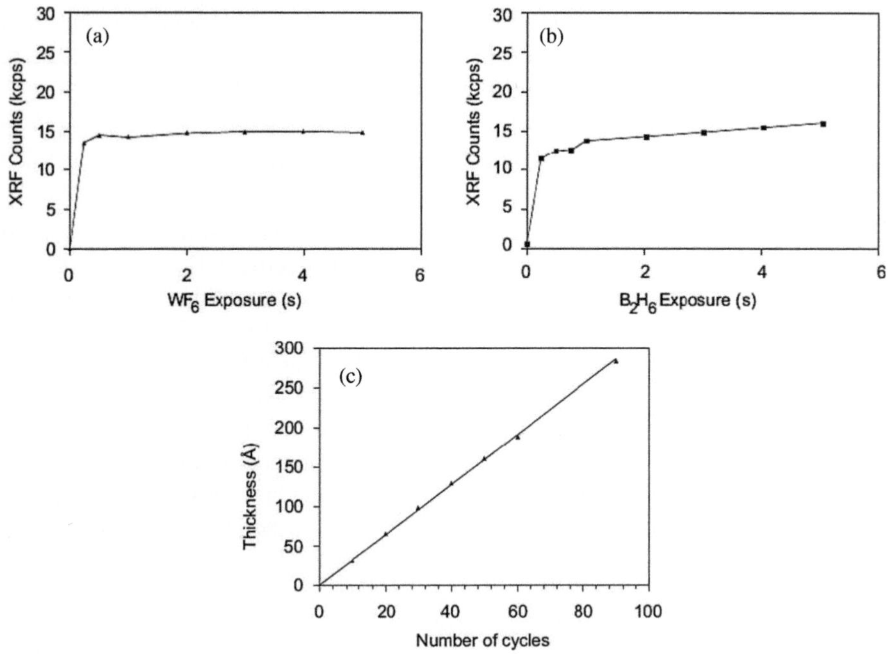

Fig. 14.5 Characteristic ALD behaviors as exhibited by W deposition with WF_6 and B_2H_6. (**a**) Self-limiting growth with WF_6 precursor exposure. (**b**) Self-limiting growth with B_2H_6 precursor exposure. (**c**) Linear growth proportional to the number of deposition cycles

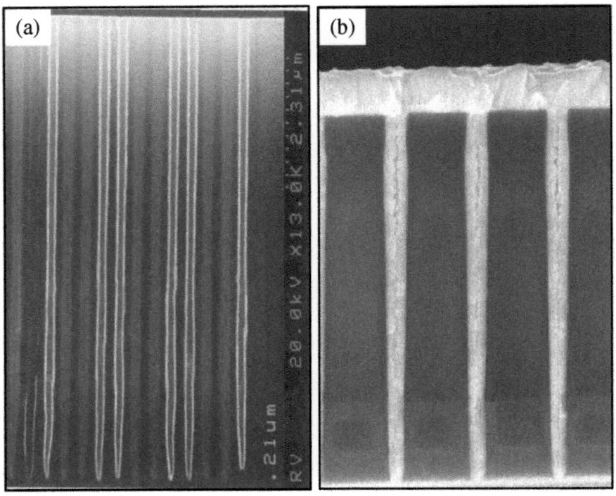

Fig. 14.6 Application of ALD W in high aspect ratio structures. (**a**) Conformal step coverage of ALD W in 50:1 aspect ratio features. (**b**) Nearly void-free fill with ALD W nucleation layer followed by CVD W bulk fill in 20:1 aspect ratio, 0.18 μm diameter contacts

classic ALD behaviors such as self-limiting growth as well as film thickness linearly proportional to the number of deposition cycles (Fig. 14.5). Conformal deposition of ALD W nucleation layers can be obtained in high aspect ratio structures as shown in Fig. 14.6a. These conformal nucleation layers are essential enablers for subsequent CVD bulk fill of W to be void-free as shown in Fig. 14.6b.

14.2.3 Back-End-of-Line (BEOL) Applications

RC delays in the BEOL interconnects necessitated the transition from aluminum to copper wiring (coupled with the migration to low-*k* dielectrics). However, as BEOL dimensions continue to shrink with each technology node, traditional physical vapor deposition (PVD) of barrier and seed for Cu plating is at risk of no longer being able to keep up with the demand for lower *RC* delay. The effective resistivity of Cu increases dramatically as a function of decreasing line width [23]. Modeling and experimental results confirm that the primary cause of the increase is dimensional; i.e., the fraction of line volume occupied by Cu as compared to the higher resistivity metal barrier must be maximized. The industry standard for Cu barrier is PVD of TaN and Ta. However, films deposited by PVD have non-conformal and asymmetrical coverage. A thick PVD film is required to obtain sufficient film thickness on every structure on a substrate. Compared to PVDbarriers deposited by ALD are conformal and occupy a smaller fraction of the line volume. Atomic layer-deposited TaN can be produced by alternating exposures of pentakis-(dimethylamido)tantalum (PDMAT) and NH_3. The growth of the TaN film on SiO_2 is shown to have no incubation delay and is linear with the number of deposition cycles (Fig. 14.7). Step coverage of ALD TaN is conformal (Fig. 14.8). The effective resistivity of the Cu

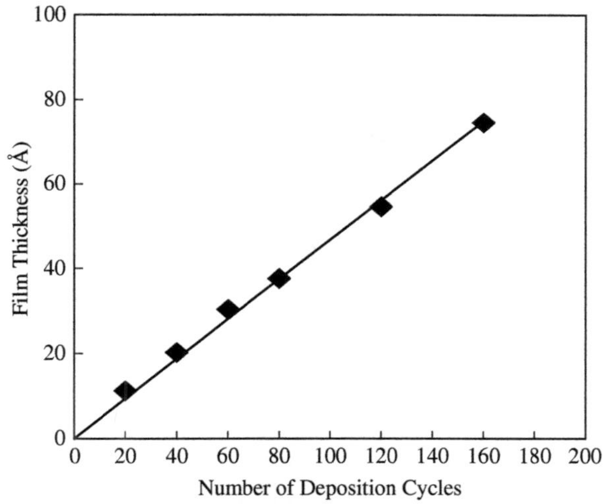

Fig. 14.7 ALD TaN grown on SiO_2. Deposition thickness is linear with the number of deposition cycles with no incubation delay

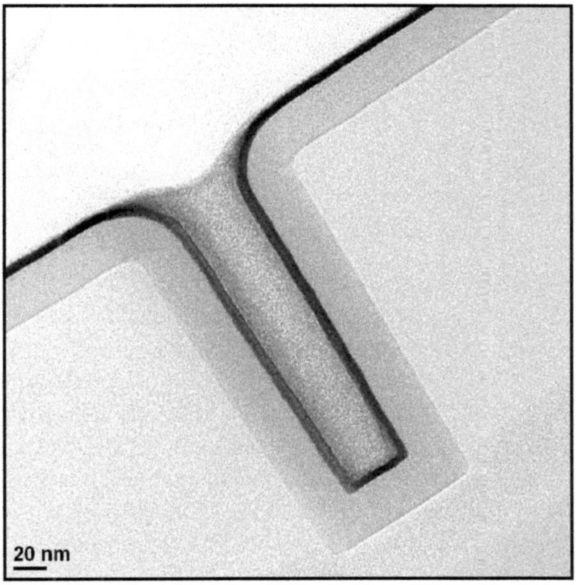

Fig. 14.8. Step coverage of ALD TaN is conformal on 5:1 aspect-ratio, 50 nm diameter structure

line that utilizes ALD TaN as the diffusion barrier is lower than that which utilizes PVD TaN/Ta as the diffusion barrier for Cu (Fig. 14.9). Similar line resistance advantages of utilizing an ALD TaN barrier have been reported elsewhere [25–27].

Besides the issue of maintaining suitable line resistance, an additional challenge for dual-Damascene metallization as dimensions continue to shrink is void-free fill by electroplating of Cu. One main inhibitor to obtaining void-free Cu fill is the overhang created by PVD of Cu seed layer. There have been a number of reports of ALD or CVD of Cu [28–31] issues with adhesion to the substrate. Furthermore,

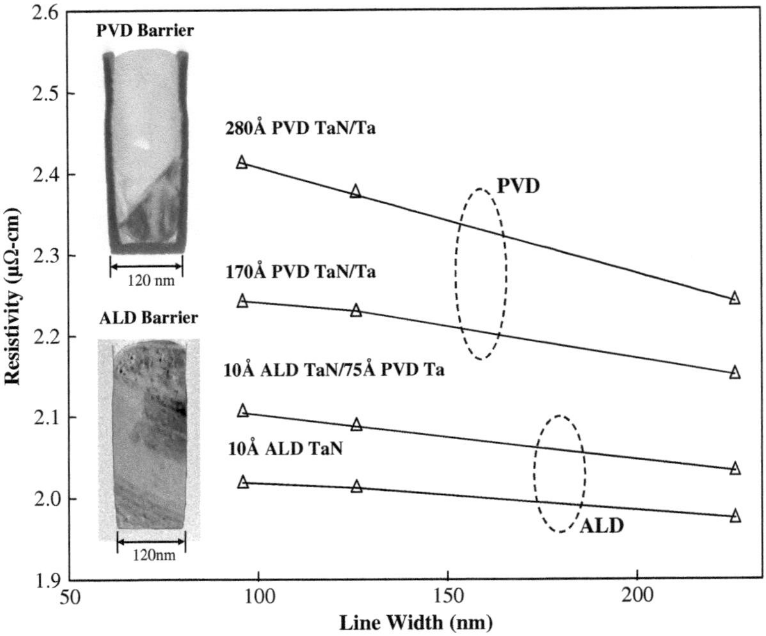

Fig. 14.9 Effective resistivity of Cu lines of various widths. Lines with ALD TaN barrier have lower resistivity than those with PVD TaN/Ta barrier

deposition temperature required for ALD or CVD of Cu drives the poorly adhered Cu to de-wet from substrate surface and agglomerate into large islands. The result is a rough film that is often discontinuous below the thickness of 100 nm. Since many BEOL features below the 65 nm technology node are smaller than 100 nm, the current ALD/CVD Cu technologies are not suitable as seed layers (Table 14.2). One potential solution to the issue of ALD/CVD Cu agglomeration on substrates is the deposition of an ALD ruthenium adhesion layer [32].

An alternative approach to eliminating Cu seed overhang is to eliminate the Cu seed layer altogether. This approach has gained interest with the rise of both direct and electroless plating of Cu. While it holds much promise,

Table 14.2 Selected BEOL International Technology Roadmap for Semiconductors (ITRS) specifications for 45, 32, and 22 nm technology nodes [24]

Year	Technology node	Metal 1 wiring pitch (nm)	Metal 1 aspect ratio	Conductor effective resistivity for Metal 1 (μΩ-cm)	Barrier thickness for Metal 1 (nm)
2010	45 nm	108	1.8	3.62	4
2013	32 nm	76	1.9	4.14	2.8
2016	22 nm	54	2	4.88	2

direct plating or electroless plating of Cu on the traditional Ta-based barrier metal has proven to be difficult. This led to renewed interest in atomic layer deposition of noble metals, of which ruthenium has been seen as the enabling material for direct and electroless plating of Cu [33]. Popular precursors used for depositing ruthenium include bis(cyclopentadienyl)ruthenium [$Ru(Cp)_2$] and its derivatives, tris(acetylacetoniate)ruthenium $Ru(acac)_3$, and tris(2,2,6,6,-tetramethyl-3,5-heptanedionato) ruthenium [$Ru(thd)_3$] [34–36]. Oxygen is the most frequently used reactant for depositing ruthenium. In a deposition scheme employing O_2 and $Ru(Cp)_2$, O_2 and $Ru(Cp)_2$ are alternatively introduced. The deposition rate is approximately 0.05 nm per cycle at a temperature of 250–400°C. The resultant film is typically very pure and suitable as an activation layer for electroless or direct plating of Cu. However, ALD Ru processes in the presence of oxygen have potential compatibility issues with BEOL interconnect materials such as Cu and Ta. Plasma-assisted atomic layer deposition (PAALD) of ruthenium has been reported and may be an alternative. Further discussion on plasma-assisted ALD Ru can be found in the next section.

14.3 Plasma-Assisted Atomic Layer Deposition (PAALD)

Plasma-assisted atomic layer deposition (PAALD), also known as plasma-enhanced ALD (PEALD), utilizes ionized reactant in the reaction. PAALD is usually used where the thermal reaction pathway does not exist or is not practical (e.g., unacceptable reactant toxicity). In many cases, the precursors simply have negligible reactivity with each other at deposition temperature without some other form of energy source. PAALD of Ru with bis(ethylcyclopentadienyl)ruthenium [$Ru(EtCp)_2$] and NH_3 plasma is one example. NH_3 has essentially no reactivity with $Ru(EtCp)_2$, as well as with other derivatives of $Ru(Cp)_2$, at temperatures compatible with BEOL interconnects ($\leq$400°C). However, deposition of Ru is enabled when $Ru(EtCp)_2$ and NH_3 plasma are alternatively exposed to the substrate [37]. Although good quality film can be deposited with these reactants, step coverage in high aspect ratio structures can be problematic for PAALD as it becomes increasingly more difficult to ensure sufficient amount of the ionized reactant species reach bottom of the high aspect ratio structures. Furthermore, it is difficult to ensure that the film properties on the structure sidewalls are consistent with the film properties on the field and bottom of the structures.

14.4 Challenges for Adapting Atomic Layer Deposition in ULSI Manufacturing

While there are numerous advantages to implementing atomic layer deposition in ULSI manufacturing, there are also significant hurdles. Throughput has historically been the foremost obstacle to adapting ALD in manufacturing settings. Recent advances in single-wafer reactors have significantly reduced the amount of time required to complete an ALD cycle. These advances come during the developments of components such as programmable logic controllers (PLC), reliable and fast-switching valves, well-designed reactor space, and efficient precursor delivery systems. Even with these improvements, there are still practical limits to the maximum film thickness suitable for ALD. In most cases, 5 nm is the threshold for films to be deposited by single chamber ALD processes. Throughput requirements also often necessitate limiting the exposure and/or purge times so that the reaction is substantially but not completely saturated. Therefore, although many commercial ALD reactors operate under the ALD principles and realize ALD benefits, they also exhibit some CVD-like behaviors.

The self-limiting nature of ALD has re-opened the opportunity of utilizing low vapor pressure liquid or even solid precursors that were unsuitable for CVD reactions. However, reliable and repeatable delivery of such precursors is another challenging aspect of implementing ALD in volume manufacturing. Many of the precursors discussed in this chapter are in solid form and/or have extremely low vapor pressure. Care must be taken in the precursor delivery design so as to provide sufficient precursor concentration into the reactor without the potential consequences of condensation in the delivery system. Further, plasma-assisted ALD made it possible to make use of precursors that were thought to have too little reactivity. However, plasma delivery hardware is complex. Furthermore, plasma-assisted reactions are far less sensitive to temperature. As a consequence, deposition on reactor walls is observed even in cold-wall reactors. Over time, the deposition on reactor walls can become sources for particle defects.

Finally, the cost of applying ALD in manufacturing is yet another hindrance. Although the cost of a finely engineered ALD reactor is substantial, it is not beyond the limits of semiconductor fabrication. However, the development and production costs of the precursors can be barriers in bringing ALD into the mainstream. Care must be taken in selecting precursors so that the benefits of ALD are not overshadowed by the high costs of the chemicals.

References

1. Suntola, T.: Atomic Layer Epitaxy. 16th International Conference on Solid State Devices and Materials, 647 (1984)
2. Suntola, T. and Antson, J.: US patent No. 4058430.
3. Ritala M. and Leskelä, M.: In: Handbook of Thin Film Materials, Volume 1, Deposition and Processing of Thin Films, Nalwa, M. S. (Eds.), Academic Press, San Diego, 103 (2001)
4. Suntola, T.: Atomic layer epitaxy. Thin Solid Films, **216**, 84 (1992)
5. Leskelä M.; and Nünistö, L.: In Atomic Layer Epitaxy. Suntola, T.; and Simpson, M. (Eds.), Blackie, Glasgow (1990)
6. George, S. M.; Ott, A. W.; and Klaus, J. W.: Surface chemistry for atomic layer growth, J. Phys. Chem., **100(31)**, 13121 (1996).
7. Matero, R.; Rahtu, A.; Ritala, M.; Leskelä, M.; and Sajavaara, T.: Effect of water dose on the atomic layer deposition rate of oxide thin films. Thin Solid Films **368(1)**, 1 (2000)
8. International Technology Roadmap for Semiconductors 2004 Update, Front End Process

9. Nouri, F.; Kher, S.; Narwankar, P.; Sharangpani, R.; Muthukrishnan, S.; Kraus, P.; Ahmed K.; Olsen, C.; Chua, T. C.; Cruse, J.; Hung, S.; Bae, S. H.; Kang, A.; Higashi, G.; and Miner, G.: Trends in gate stack engineering, Proceedings of IEEE International Conference on Integrated Circuit Design and Technology, 275 (2004)
10. Wilk, G. D.; Wallance, R. M.; and Anthony, J. M.: High-k gate dielectrics: current status and materials properties considerations. J. App. Phys. **89(10)**, 5243 (2001)
11. Delabie, A.; Puurunen, R. L.; Brijs, B.; Caymax, M.; Conard,T.; Onsia, B.; RicharcO.; Vandervorst,; Zhao, W. C.; Heyns, M. M.; and Meuris, M.: Atomic layer deposition of hafnium oxide on germanium substrates. J. App. Phys. **97(6)**, 064104 , 1 (2005)
12. Kim, H.; McIntyre, P. C.; and Saraswat, K. C.: Effects of crystallization on the electrical properties of ultrathin HfO2 dielectrics grown by atomic layer deposition. App. Phys. Lett. **82(1)**, 106 (2003)
13. Deshpande, A.; Inman, R.; Jursich, G.; and Takoudis, C.: Atomic layer deposition and characterization of hafnium oxide grown on silicon from tetrakis(diethylamino)hafnium and wafer vapor. J. Vac. Sci. technol. A. **22(5)**, 2035 (2004)
14. Fillot, F.; Morel, T.; Minoret, S.; Matko, I.; Maitrejean, S.; Guillaumot, B.; Chenevier, B.; and Billon, T.: Investigations of titanium nitride as metal gate material, elaborated by metal organic atomic layer deposition using TDMAT and NH3. Microelectron. Eng. **82**, 248 (2005)
15. Zhong, H.; Hong, S. N.; Suh, Y. S.; Lazar, H.; Heuss, G.; and Misra, V.: Properties of Ru-Ta alloys as gate electrodes for NMOS and PMOS silicon devices. International Electron Devices Meeting, 467 (2001)
16. Ritala, M.; Leskelä, M.; Rauhala, E.; and Haussalo, P.: Atomic layer epitaxy growth of TiN thin films. J. Electrochem. Soc. **142(8)**, 2731 (1995)
17. Min, J. S.; Park, H. S.; Koh, W.; and Kang, S. W.: Chemical vapor deposition of Ti-Si-N films with alternating source supply. Mater. Res. Soc. Symp. Proc. **564**, 207 (1999)
18. Ott, A. W.; Klaus, J. W.; Johnson, J. M.; and George, S. M.: Al2O3 thin film growth on Si(100) using binary reaction sequence chemistry. Thin Solid Films **292(1)**, 135 (1997)
19. Kim, H. K.; Kim, J. Y.; Park, J. Y.; Kim, Y.; Kim, Y. D.; Jeon, H.; Kim, and W. M. Metalorganic atomic layer deposition of TiN thin films using TDMAT and NH3. J. Korean Physical Soc. **41(5)**, 739 (2002)
20. Ohba, T.: Chemical-vapor-deposited tungsten for vertical wiring. MRS Bull. **20(11)**, 46 (1995)
21. Klaus, J. W.; Ferro, S. J.; and George, S. M.: Atomic layer deposition of tungsten using sequential surface chemistry with a sacrificial stripping reaction. Thin Solid Films **360(1)**, 145 (2000)
22. Yang, M.; Chung, H.; Yoon, A.; Fang, H.; Zhang, A.; Knepfler, C.; Jackson, M.; Byun, J. S.; Mak, A.; Eizenberg, M.; Xi, M.; Kori, M.; and Sinha, A.: Atomic layer deposition of tungsten film from WF_6/B_2H_6: Nucleation layer for advanced semiconductor devices, Advanced Metallization Conference, 655 (2001)
23. Rossnagel, S. M.; and Kuan, T. S.: Alteration of Cu conductivity in the size effect regime. J. Vac. Sci. Tech. B **22(1)**, 240 (2004)
24. International Technology Roadmap for Semiconductors 2004 Update, Interconnect.
25. Svedberg, L.; Prindle, C.; Brennan, B.; Lee, J.J.; Guenther, T.; Ryan, T.; Junker, K.; Grove, N.; Jiang, J.; Denning, D.; Shahvandi, R.; and Yu, K.: Electrical and physical characterization of atomic layer deposited thin films for copper barrier applications, Advanced Metallization Conference, 701 (2002)
26. Chung, H.; Chang, M.; Chu, S.; Kumar, N.; Goto, K.; Maity, N.; Sankaranarayanan, S.; Okamura, H.; Ohtsuka, N.; and Ogawa, S.: An ultra-thin ALD TaN barrier for high performance Cu interconnects. IEEE Intl. Symp. on Semi. Manuf. , 454 (2003)
27. Michaelson, L. M.; Thrasher, S. R.; Besling, W. F. A.; Rasco, M. ; Acosta, E.; Jiang, Z. Xi.; Kim, K.; Kirksey, S. H.; Rose, S. H.; and Vanypre, T.: ALD TaN reliability improvement in dual-Damascene structures. Advanced Metallization Conference, 699 (2004).

28. Martensson, P.; and Carlsson, J. O.: Atomic layer epitaxy of copper. Growth and selectivity in the Cu(II)-2,2,6,6-tetramethyl-3,5-heptanedionate/H2 process. J. Electrochem. Soc. **145(8)**, 2926 (1998)
29. Martensson, P. and Carlsson, J. O.: Atomic layer epitaxy of copper on tantalum. Chem. Vap. Dep. **3(1)**, 45 (1997)
30. Huo, J.; Solanki, R.; and McAndrew, J.: Characteristics of copper films produced via atomic layer deposition. J. Mat. Res. **17(9)**, 2394 (2002)
31. Marcadal, C.; RichardE.; Torres, J.; Palleau, J.; and Madar, R.: CVD process for copper interconnection. Microelec. Eng. **37**, 97 (1997)
32. Kwon, O. K.; Kim, J. H.; Park, H. S.; and Kang, S. W.: Atomic layer deposition of ruthenium thin films for copper glue layer. J. Electrochem. Soc. **151(2)**, 109 (2004)
33. Sun, Z. W.; He, R.; and Dukovic, J. O.: Direct plating of Cu on Ru: Nucleation kinetics and gapfill chemistry. Advanced Metallization Conference, 531 (2004)
34. Aaltonen, T.; Alen, P.; Ritala, M.; and Leskela, M.: Ruthenium thin films grown by atomic layer deposition. Adv. Mater. **15(1)**, 45–49 (2003)
35. Aaltonen, T.; Ritala, M.; Arstila, K.; Keinonen, J.; and Leskela, M.: Atomic layer deposition of ruthenium thin films from Ru(thd)3 and oxygen. Chem. Vap. Dep. **10(4)**, 215 (2004)
36. Kwon, O. K.; Kim, J. H; Park, H. S.; and Kang, S. W.: Atomic layer deposition of ruthenium thin films for copper glue layer. J. Electrochem. Soc. **151(2)**, 109 (2004)
37. Kwon, O. K.; Kwon, S. H.; Park, H. S.; and Kang, S. W.: Plasma-enhanced atomic layer deposition of ruthenium thin films. Electrochem. Solid State Lett. **7(4)**, 46 (2004)

Chapter 15
Electroless Deposition Approaching the Molecular Scale

A.M. Bittner

15.1 Introduction

Electroless deposition (ELD) encompasses quite a range of chemical deposition processes at the solid/liquid interface. Here I focus exclusively on autocatalytic ELD of metals, i.e., the plated metal catalyzes its own deposition, and hence the process is continuous as long as sufficient amounts of reactants are provided [1–5]. Such a reaction requires a catalytic site to start; usually this is a noble metal nanoparticle, while for technical applications mixed metal particles are employed. Obviously, the size of the nanoparticle must be smaller than the desired metal structure. The theoretical limit for common electroless Cu plating baths is in the atomic range, and indeed clusters with two or four atoms may act as nuclei; the smallest electroless (EL) metal structures are in the range of 2 nm (corresponding to several hundred atoms) [6–7].

Apart from scientists working on ULSI, there are at least five other distinct scientific communities interested in ELD: first, plating with or without current source is one of the best methods for surface coating, especially for nonconductive surfaces; second, structural biologists employ nanoparticles for visualization in electron microscopy, enlarging the particle diameter (enhancement) is of great importance; third, the photographic process uses the same enhancement for nanoparticles produced by photoreduction; fourth, the research on metal nanoparticles shares this interest, and even the synthesis of nanoparticles is mostly effected by a type of ELD; fifth, large (macro) molecules and biomolecules are increasingly used as templates for nanostructures, and ELD offers many advantages for a metallization since it does not require electrical contacting.

ELD is known as a technical process to fabricate metallic macro- and microstructures; the best known example is probably the Tollens reaction that was technically used to produce silver mirrors from Ag^{+} and dissolved reductants. Note that it is also

A.M. Bittner (✉)
Group leader "Self-Assembly" Asociacion CIC nanoGUNE Tolosa Hiribidea,
76, 20018 Donostia – San Sebastian, Spain
e-mail: a.bittner@nanogune.eu

Y. Shacham-Diamand et al. (eds.), *Advanced Nanoscale ULSI Interconnects: Fundamentals and Applications*, DOI 10.1007/978-0-387-95868-2_15,

possible to bind the reductants to a surface [8]. Typically, the object to be metallized is nonconductive and does not catalyze reactions such as metal ion reduction. To start ELD, the object must be coated with catalytically active centers such as noble metal particles. Since this is possible with very small clusters which can be produced by simple wet chemical recipes metallization on the sub-10 nm scale can be attempted. This first step, the so-called "sensitization," is also known as first step of galvanic deposition on nonconductive surfaces, such as copper electrodeposition on silicon wafers [9], which requires a seed layer. Autocatalytic deposition on sensitized objects principally continues without limitation, as long as the reactants (metal ions and reductant) are transported to the metal surface. The contrast to galvanic plating can be sketched as follows:

Galvanic electrodeposition: $M^{z+} + z\,e^- \rightarrow M$
Autocatalytic electroless deposition (ELD): $M^{z+} + z\,RH \rightarrow M + z\,R + z\,H^+$
Sensitization of nonconductive surfaces: ${M_{noble}}^{z+}{}_{ads} + z\,RH \rightarrow M_{noble,ads} + z\,R + z\,H^+$

Note that a hydrogen-containing reductant RH is assumed while baths can also be prepared with hydrogen-free reductants such as Co(II) (oxidized to Co(III)) [10]. Of great importance is the exact bath formulation: To a salt of the metal ion M^{z+} and the reductant RH one adds practically always complexants for M^{z+}, very often a pH buffer, and occasionally small amounts of additives (e.g., stabilizers, accelerators, inhibitors).

It is worth noting that other chemical deposition reactions are sometimes called ELD, too, e.g., displacement plating (deposition with simultaneous dissolution of the surface, useful, e.g., for some plating reactions on semiconductors) or contact plating (deposition on a substrate with simultaneous dissolution of a second piece of metal that contacts the substrate). Some mineralization processes (precipitation of a metal cation with an anion) can also be directed by oxidation reactions at noble metal particles, although they do not involve metal ion reduction (e.g., the deposition of ZnO from Zn(II) and reductants).

Why should electroless deposition on the nanoscale be interesting? Traditionally ELD was used to plate thick (microscale) layers, but already in the 1970s copper ELD was suggested for the formation of conductive pathways on microchips, and special bath formulations were developed [11]. A rather different driving force is the parallel bottom-up fabrication of nanoscale objects (Fig. 15.1). Quite often it is desirable to create nanoscale metal objects, or to enlarge already existing objects; in fact the enhancement of the latent image in the photographic process is an old example for "nanotechnology." One of the main driving forces to use ELD rather than alternative methods (mainly gas phase based) are simplicity and biocompatibility: Wet chemistry is employed, and more and more formulations turn out to be compatible with even functional biomolecules. A very exciting plan is to use ELD to contact nano-objects: Since it is extremely difficult to place nanoscale objects such as molecular switches with nanometer precision, a possible fabrication scheme

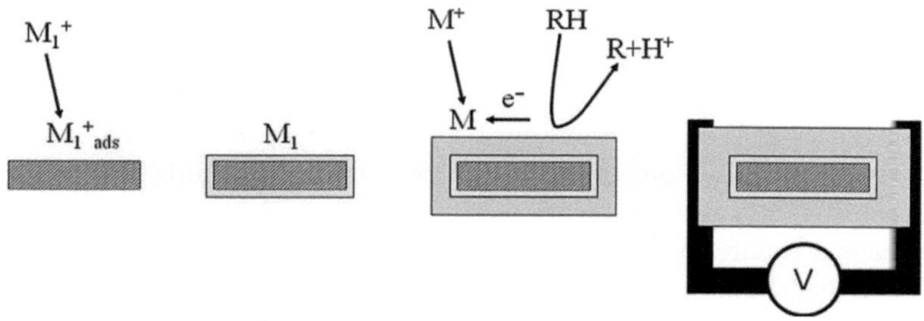

Fig. 15.1 Schematics of ELD on a nanoscale object (*dark*) and of "wiring up" the object by ELD. Sensitization by metal M1 (*gray*), deposition of metal M (*dark gray*) by reductant RH

involves placing the object in the vicinity (say up to 30 nm) of an electrical contact and thereafter use ELD to wire the object (Fig. 15.1).

In the following, sensitization and ELD on nanoscale objects are considered, in other words, the ELD production of 2D (layer), 1D (wire), or 0D (dot/particle) nanopatterns. "Nano" is here synonymous with "molecular scale," so sub-30 nm structures and especially the sub-10 nm range qualify. A short overview over employed ELD processes, over the nano-objects that can be used, especially biomolecules, and over various ELD bath compositions is presented. Relatively few theoretical descriptions are available on the molecular scale; altogether one has to state that only for few (macroscopic) ELD processes the mechanisms and the reactant transport to the surface are well understood.

15.2 Fundamentals

ELD is usually considered as combination of two electrochemical reactions, the reduction of the metal ion to the zero-valent state and the oxidation of the reductant. The electrons required by the first reaction must be completely supplied by the second, resulting in the overall reaction equation [1–4, 12]. The underlying idea is that for very small deposition rates each equation can be expressed as a Nernst equilibrium, thus defining two electrochemical potentials at the metal/liquid interface:

$$M^{z+} + z\,e^- \rightarrow M$$

$$z\,RH \rightarrow z\,R + z\,H^+ + z\,e^-$$

(Note that some rarely used reductants such as Co(II) do not contain hydrogen) Combining both reactions means that a "mixed potential" is developing. Although purely thermodynamic considerations cannot be valid for catalytic reactions, this simple approach is useful for finding possible ELD processes: The mixed potential must be positive (the free enthalpy of the overall process must be negative), so the metal/metal ion redox potential must be more positive than that of the RH oxidation potential. This means that the solution, the bath, is thermodynamically unstable; in fact suppression of homogeneous reactions in the bath by activation barriers for the catalytic processes is essential. The most obvious barrier is to add complexants such that the metal ion reduction potential is shifted in the negative direction. Simple, but effective methods to avoid homogeneous reduction in the bath are mixing metal ions and reductant freshly for each experiment and to avoid nucleation centers such as dust. Note that similar processes are employed to synthesize nanoparticles; in this case also some nonaqueous methods are known.

The mixed potential theory turns out to be at best a crude approximation: The two reactions often turn out to be coupled, so the reduction influences the surface such that the reduction is affected [13]. Obviously this coupling must be even stronger on the nanoscale where the two reactions must happen in immediate vicinity. In more detail, one has to include a kinetical analysis, which is accomplished by setting up a Butler–Volmer equation for each reaction (this includes a linear current–voltage relation close to each Nernst potential and an exponential one for larger deviations). The negative metal ion reduction current must now be balanced by the positive oxidation current, such that the overall zero current defines the mixed potential. In this way the action of additives can be separated into processes that effect the metal ion reduction and those that influence the oxidation of RH.

15.2.1 Steering Macroscopic Electroless Deposition

From the Nernst equations the concentration dependence of the kinetics can be deduced. The general trend is that these relations are quite well obeyed, such that increasing the reductant or the metal ion concentration will increase the rate. For some baths, empirical rate laws are available [4]. Certainly it is as yet impossible to connect (assumed or deduced) mechanisms with (measured) rate laws.

Another important issue is that side reactions are practically not avoidable [12]. Clearly the presence of protons or water at relatively negative potentials results in hydrogen evolution, which is also possible directly from a hydrogen-containing reductant. In fact the observation of hydrogen bubbles can be used as an indicator for deposition. The evolution and also the oxidation reaction imply that the pH changes during deposition. Quite often this is suppressed by working in buffers, ranging from slightly acidic to strongly basic conditions. Another important side reaction is the reduction of oxygen (if present). This process is under intense scrutiny for fuel cells, where kinetic barriers have to be kept as low as possible; obviously this can be achieved in ELD by the sensitization (and also by the plated metal). The reduction of water (or protons) to hydrogen and the reduction of oxygen consume some of the electrons provided by the oxidation, and thus suppress the metal deposition – employing terms known from galvanic deposition, the current efficiency is below 100%. Hence the reductant has to be present in higher concentration than the theoretical balance.

When a bath is contacted with a sensitized surface, the ELD reaction may take some time (seconds to minutes) to start. While this so-called "induction period" is of greatest importance for applying a bath, little is known about its physicochemical background. Clearly the deposits have a structure that differs from the structure during the later stages of plating [14]. More important, plating practice shows that organic or inorganic additives, present in much smaller amounts than the reactants, improve the process and the quality of the deposit [4]. When one adds a second metal ion species, one can achieve codeposition, even anomalous codeposition (metal ions are reduced above their Nernst potential). This feature makes nanoscale ELD especially attractive since many properties (mechanical, magnetic, electrical) can be finely tuned by depositing an alloy. For example, alloys with various advantageous features (corrosion resistance, conductivity) can be plated inside vias [5].

15.2.2 What Makes Nanoscale Deposition Special?

Many features of ELD, especially those mentioned above, can be directly translated into the nanoscale; however, some are not well known on small scales. First and most important, the sensitization requires colloids or particles well below the final feature size. This usually entails diameters around 1 nm, well established as tools in structural biology [15, 16], which means that the electronic structure of the respective bulk metal is strongly distorted. Luckily, the catalytic activity appears to be influenced much more by the ligand shell. Moreover, even small clusters consisting of less than five atoms show catalytic behavior. Only recently the first steps of sensitization (metal–metal bond formation) were modeled leading to a better understanding of the first stages of metal–metal bond formation [17].

The diffusion on the nanoscale deserves special attention [18]. Usually diffusion is thought to obey Fick's laws. On the other hand, there must be a limit for the minimal distance. This limit is roughly the molecular diameter of hydrated metal ions or

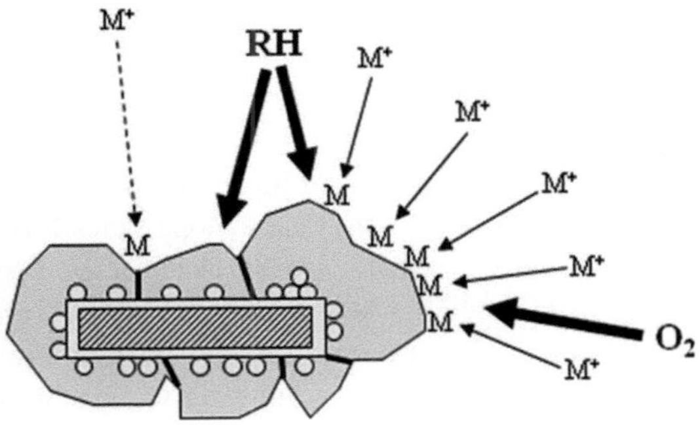

Fig. 15.2 Schematics of reactant diffusion and partial reactions on a nanoscale surface, sensitized by nanoparticles (*circles*). Reduction and oxidation likely have to take place at different surface sites. Oxygen competes with M^+ for reduction. The growth is highly anisotropic; coalescence of growing nuclei forms grain boundaries (*black*)

of reductants. Here a macroscopic theory cannot work any more (think of Brownian motion), see Fig. 15.2. Some observations suggest that the transport mechanism does indeed change, and small molecules such as oxygen are favored, resulting in stopping the deposition (oxygen is reduced instead of the metal ions) [19]. Another issue is that on the scale below about 2 nm electrochemical potentials cannot be defined sharply, since the electronic structure changes from a band to energy levels. This problem is equivalent to unification of macro- and microscopic descriptions of electronic and also chemical properties. For example, binding and reducing a metal ion on a triatomic metal cluster would best be described by a molecular theory based on orbitals; on a 10 nm metal particle one would apply colloid and electrochemical theories [20].

15.3 Scaling Down Macroscopic Electroless Deposition

Many researchers try to apply known ELD recipes to obtain nanoscale deposits. However, as often found in so-called "nanotechnology," processes and products are well developed, but known under different headings, often under different names in various sub-disciplines of chemistry and physics. For nanoscale ELD one should keep in mind that at least for typical distances above 10 nm one can assume that bulk behavior prevails, so scaling down standard recipes is logical. In fact this means that the same parameters as in macroscopic ELD can be varied and thus used for better control. Steering the process can thus be achieved with the same parameters, namely careful control of the induction period or changing concentrations, temperatures, pH, and so on. In addition one has to note that stopping ELD after short times (experimentally very nicely accessible, seconds to minutes) will result in nanoscale films (more or less 2D). Since additives are not well investigated for nanoscale ELD, they will not be discussed.

15.3.1 Sensitization, Pd and Pd/Sn Colloids

The simplest technical sensitization can be obtained by sequential dipping into Pd(II) and Sn(II) solutions. More complex formulations apply a colloidal or nanoparticle suspension with preformed Pd/Sn particles. It is generally accepted that Sn(II) reduces Pd(II) to Pd metal, and that nanoparticles form during this process. These particles are coated by Sn(II) and Sn(IV) species. For nanoscale ELD, a nanoscale sensitization is crucial; whenever agglomerates form, their size will determine the minimal size of the deposit. The ideal scenario involves pure noble metal particles of only some tens of atoms; but even with such simplified systems (compared to technical baths), not much insight has been gained. The Dressick group has published a range of papers, showing various ways to produce thin and uniform layers of small nanoparticles for ELD [21, 22]. Recently the metallization of biomolecules has renewed the interest in sub-2 nm noble metal particles for sensitization [20, 23]. One should be aware that the particles always carry adsorbed molecules, in some cases even very strongly adsorbed, so that metallization starts with removal of adsorbates. This is slightly different for ELD processes that are "self-sensitizing" – here the ligand has to be removed. Note that here Ag–hydroquinone baths have become very popular (Tables 15.1 and 15.2).

Table 15.1 Typical bath constituents used for molecular scale ELD. Usually a pH buffer is added

$AuCl_4^-$	NH_2OH
$PdCl_4^{2-}$, $PtCl_4^{2-}$	$H_2PO_2^-$
Ag^+	Hydroquinone $HO–C_6H_4–OH$
Ni^{2+}	$H_2PO_2^-$
Co^{2+}, Ni^{2+}, Cu^{2+}	$(CH_3)_2NHBH_3$

Table 15.2 Biological and related templates, sensitization methods, ELD bath components, structure type, and width (applied around 25°C; buffering around neutral pH)

[44] M 13 phage	5 nm Au colloids	0.3 mM $HAuCl_4$ 4 mM NH_2OH	~40 nm wire
[48] CPMV crystal	Glutaraldehyde; buffer + 20 mg/mL K_2PdCl_4	Buffer 10 mg/mL K_2PtCl_4 20 mg/mL NaH_2PO_2	10 nm wires
[37] λ-DNA/RecA (protein)	Glutaraldehyde; $AgNO_3$ in NH_3	60 mg/mL KSCN 23 mg/mL $KAuCl_4$; 5.5 mg/mL hydroquinone in buffer	50–100 nm wire
[49] DNA/telomerase (protein)	OH-succinimid-nanogold	Buffer 0.5 mM $HAuCl_4$ 0.5 mM NH_2OH	Chain of 10–40 nm particles
[42] G-actin	OH-succinimid-nanogold	0.15 mM $HAuCl_4$ 0.15 mM NH_2OH	80–150 nm high wire
[6] S layer protein	No sensitization	~1.9 mg/mL protein 0.17 mM K_2PtCl_4 (aged) 0.02%NaN_3	2 nm particles
[40] Peptide on bolaamphiphile tube	No sensitization	5 mM bolaamphiphile Peptide 12.5 mM $CuCl_2$ 0.25 M $NaBH_4$ pH 6	10 nm particles
[50] Bolaamphiphile tube	No sensitization	0.08 M $NiCl_2$ 0.1 M NaH_2PO_2 pH 6, citrate buffer	1 μm (diameter) tube

Table 15.2 (continued)

[51] Peptide on bolaamphiphile tube	No sensitization	20 μmol bolaamphiphile 4 μmol peptide 5 mg $AuClP(CH_3)_3$ 50 μmol N_2H_4 pH 11.5	6 nm particles
[39] Phospho-lipid/polyelectrolyte microtubule	$PdCl_4^{2-}$ with Cl^- at pH 5	Commercial Cu bath	~400 nm spiral (coil width)
[38] Three-helix bundle DNA	Glutaraldehyde; $AgNO_3$ in NH_3	Commercial Ag bath (Ag(I) and hydroquinone)	20–50 nm wires
[41] Microtubule	Glutaraldehyde; $AgNO_3$; $NaBH_4$	3 mM $AgNO_3$ 20 mM hydroquinone	40 nm wire
[7, 46] Tobacco mosaic virus	$PdCl_4^{2-}$ or $PtCl_4^{2-}$ with Cl^- at pH 5.5	0.18 M $Ni(CH_3COO)_2$ 0.28 M lactic acid 0.034 M $(CH_3)_2NHBH_3$, or 0.16 M $CoSO_4$ 0.15 M Na succinate 0.007 M $(CH_3)_2NHBH_3$, or 0.032 M $CuSO_4$ 0.04 M Na_2EDTA 0.05 M triethanolamine 0.067 M $(CH_3)_2NHBH_3$, All at pH 6–8	3–4 nm wire; for HPO_4^{2-} pretreatment 100–300 nm wires

15.3.2 Confinement Plating

Plating a hollow structure inside, at best completely filling it, is a very good way for shape control, and it can work on the nanometer scale [24]. However, the surface chemistry of the walls has to be very well defined and/or modified. Some examples for quite large structures can be found; of greatest commercial impact could be vias and trenches in ULSI [5, 25]. Efforts to plate carbon nanotubes inside have not been successful, the reason being most likely that the strongly hydrophobic graphene sheets hinder the entry of solvent and dissolved species. Indeed, hydrophobic molecules such as C_{60} or neutral metal complexes are known to enter the orifice. Obviously, this constraint is relaxed for very wide tubes (tens of nanometers inner diameter), as shown for nickel ELD of 50 nm wide wires [26]. Of course plating on the outer sheath is possible after chemical modification that allows for sensitization [27]. Due to reasons that will be discussed in relation with biological templates, it is very hard to obtain uniform coatings (smooth tubes) on the nanoscale.

Concerning the confinement templates, nanoscale holes can be produced by various means. The most popular substrates are track etch membranes [28] and porous alumina [29]. In both cases, the hole walls require careful chemical modification, followed by sensitization. The Martin group has shown that pores can be narrowed down to about 1 nm diameter in a defined way, such that the gold tubes allow the passage of molecules only in single file [30].

Finally, ELD can also be used to create, rather than exploit confinement. Microscale contacts, which can easily be produced lithographically, can be narrowed by ELD to sub-50 nm gaps. Recently, Ah et al. showed that 50 nm gaps can be narrowed down to about 1 nm [31] in analogy to the galvanic method developed by the Tao group [32]. This allows for straightforward measurement of tunneling currents and can be the base for molecular electronics at the solid/liquid interface.

15.3.3 Biomolecules as Templates

Building on "enhancement," various schemes for biosensors have been developed. They all have in common that gold-tagged biomolecules, e.g., antibodies or DNA, are available, and that enhancement up to coalescence of the gold particles changes optical (absorption) or electrical (conductivity) properties [33]. Technical applications include staining of protein gels used in microbiology. Inspired from this, gold or silver nanoparticles used for immunostaining or as tags for DNA and RNA can be enhanced [15, 16]. The methods are closely related to "enhancement" of small metal particles as in the photographic process: The very small metal particles that are created by illumination of Ag^+ in organic matrices have to be enlarged for visualization. The standard reductant is hydroquinone: Selectively attacking only those spots where Ag nanoparticles are present, further Ag^+ is reduced until a layer of large particles is formed which appears black.

Further examples (see Table 15.2) are immunoblots that can be used to detect antigens, based on enhancement of small Au nanoparticles bound to relevant biomolecules (antigens or antibodies) [34]. DNA pairing can be detected when one of the strands is labeled with gold [35]. A quite recent approach is based on better control by so-called "enzyme metallography": An enzyme is linked to the DNA to be sensed, and ELD progresses very specifically at the enzyme [36]. Due to its chemical and physical resilience, and due to its wire-like structure, and of course due to its biological relevance, DNA is a favorite template for 1D structures [35, 37, 38]. The sensitization of such a small molecule is not straightforward; any low-density coverage or any predominance of one very active nucleus will induce inhomogeneous growth (e.g., 30 nm large grains). Indeed 10 nm-scale structures appear to be impossible, despite considerable effort. The reason is likely that the sensitization is not sufficiently homogeneous. This could be improved by direct binding of defined metal clusters to DNA bases, but even then one cannot expect perfect growth (same speed for each nucleus). However, the template can be improved: Recently DNA tiles (made up from three DNA double strands) were used as templates for very uniform wires [38]. Relatively little is known about nanoscale plating on other 2D systems. Mertig et al. used bacterial S layers, known for their chemical and physical resilience to create square-shaped lattices of metal nanoparticles [6].

Phospholipids can assemble into relatively large tube structures, again platable by ELD. The resulting spiral structures are unique [39]. Peptide structures have the big advantage of self-assembly from subunits (either peptides or amino acids), which opens up completely new fabrication methods. Moreover, the highly complex chemistry allows for chemical fine-tuning: The Matsui group employs His-rich peptides for strong bonds to the Au ions in their ELD bath. Related non-biological systems are tubular assemblies of organic molecules, mainly amphiphiles [40]. Protein fibers are quite comparable although usually more sensitive. Behrens et al. employed self-assembling microtubules as wire templates [41] and the Willner group used actin [42].

The next level of complexity is based on self-assembling protein aggregates, the best known of which are amyloid fibers and viruses. Scheibel et al. used ELD to create relatively thick wires on amyloid fibers [43]. The Belcher group has much

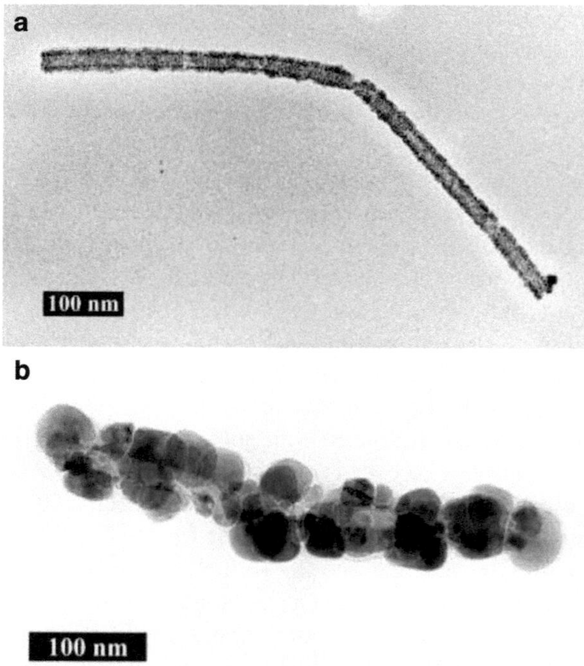

Fig. 15.3 (*top*) Palladium nanoparticles on the surface of two tobacco mosaic virions (virus particles) and (*bottom*) metal wire (> 50 nm diameter) plated on tobacco mosaic virions (reprinted with permission from Wiley-VCH) [7]

experience with the M13 phage, a rod-like bacterial virus. Genetic engineering and a combinatorial method (phage display) are employed to create Au-binding peptide surface groups. Binding gold nanoparticles followed by enhancement on this quite resilient system resulted in gold wires [44]. The Bittner group used the tobacco mosaic virus (TMV), a 300 nm long and 18 nm wide protein tube with a buried RNA. First the virions were sensitized with Pd(II) or Pt(II) in presence of phosphate traces. At this stage one can visualize the sensitization by reducing the metal ions; nanoscale particles form (Fig. 15.3). As soon as the sensitized virions are suspended in ELD baths, the outer surface is plated with metal, e.g., Ni or Co. The grains are quite large, such that the resulting diameter is above 50 nm [7].

15.3.4 Confinement Plating in Biomolecules

As yet, very few examples for ELD in a nanoscale cavity inside biomolecular templates exist. Reches and Gazit proved the growth of 30 nm Ag wires inside diphenylalanine tubes [45]. The fascinating aspect is that these tubes self-assemble from the monomer, a relatively small molecule. The Bittner group has shown that the tobacco mosaic virus (TMV) (see above) can be sensitized with Pd(II). When phosphate is removed from the suspension, electroless plating of Co, Ni, and Cu resulted in wires up to several 100 nm lengths (Fig. 15.4). The diameter of the wires is limited by the 4 nm diameter of the inner channel of TMV [46, 47]. Clearly such structures cannot be produced by adsorption of metal ions and subsequent reduction, rather a continuous autocatalytic growth is required. The very high specificity for the channel is likely due to detachment of large particles that grow on the outer sheath. Note that the analysis of such structures can be quite difficult – the chemical identity was proven with energy-filtering TEM close to the resolution limit of the method [46]. However, careful selection of the proper sensitization and ELD baths is the most crucial part. For example, standard Cu baths are based on formaldehyde and work at pH>11, which would destroy any biomolecules, including TMV. Balci et al. used a formulation based on dimethylamine borane, $(CH_3)_2NHBH_3$, that works at neutral pH, but necessitates nitrogen-containing ligands. Apart from specially synthesized macrocycles, EDTA qualifies [11, 46].

For the 20 nm thick silver wires inside peptide tubes [45], enzymatic removal of proteins/peptides is simple. Such a digestion/removal of the biomolecular template is one more example for the usefulness of biochemical reactions for nanofabrication. Note that Reches' silver wires grow inside many micrometers long self-assembling peptide tubes, hence their aspect ratio is very high; contacting was possible and proved Ohmic behavior [45]. Generally self-assembly of molecular structures with cavities, followed by ELD, has potential to create very well-defined sub-10 nm metal structures, while the template can bind to that part of the substrate surface where the structure is desired. On the other hand, when a regular superstructure of such nanostructures is searched for, one can employ self-assembly of the templates themselves (rather than self-assembly of subunits). Falkner et al. showed that Pd(II)-sensitized

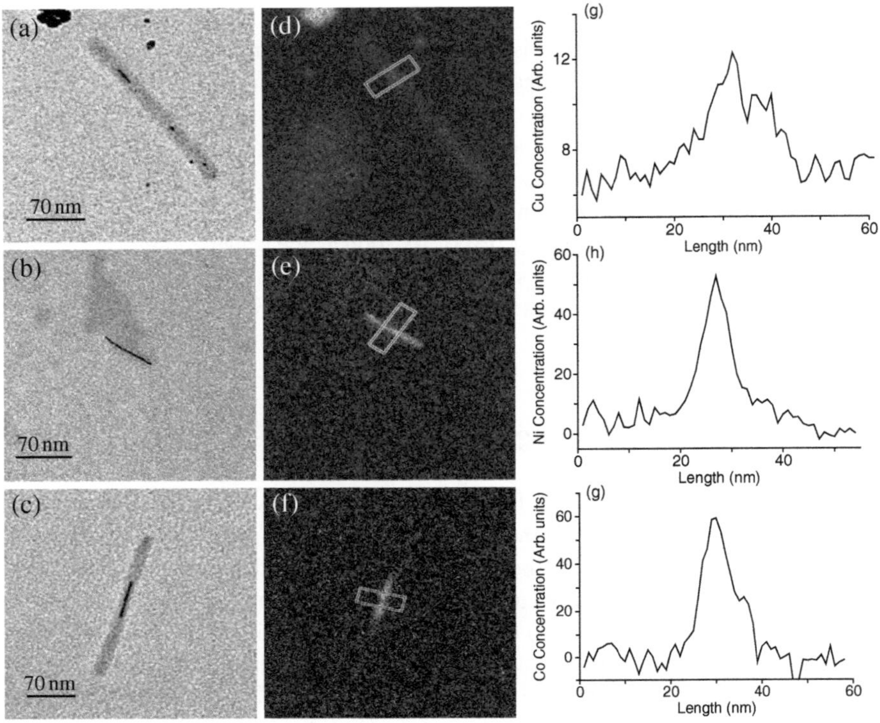

Fig. 15.4 Cobalt, nickel, and copper wires of 3 nm diameter plated inside tobacco mosaic virions, analyzed by energy-filtering transmission electron microscopy. *Left*: micrographs, *middle*: elemental maps, *right*: line scans averaged from the *white boxes* in the elemental maps (reprinted with permission from Elsevier) [46]

cowpea mosaic virions can be filled by ELD with Pt nanoparticles of 2.7 nm diameter and that biochemical crosslinking before the ELD process results in a well-ordered 3D arrangement, corresponding to the structure of the virus crystal [48]. A related process is the plating of Au nanotubes inside pores in polymer membranes, as shown by Wirtz et al. [30].

References

1. Paunovic, M. and Schlesinger, M.: Fundamentals of electrochemical deposition. Wiley Interscience: New York (1998)
2. Mallory, G. O. and Hajdu, J. B.: Electroless Plating: Fundamentals and Applications. American Electroplaters and Surface Finishers Society: Orlando (1990)
3. O′Sullivan, E. J.: Fundamental and Practical Aspects of the Electroless Deposition Reaction. In Advances in Electrochemical Science and Engineering, Alkire, R. C.; Kolb, D. M., Eds. Wiley-VCH: Weinheim **7**, 225 (2002)
4. Djokic, S. S.: Electroless Deposition of Metals and Alloys. In Modern Aspects of Electrochemistry, Conway, B. E.; White, R. E., Eds. Kluwer: New York, **35**, 51 (2002)

5. Shacham-Diamand, Y.; Inberg, A.; Sverdlov, Y.; Bogush, V.; Croitoru, N.; Moscovich, H.; and Freeman, A.: Electroless processes for micro- and nanoelectronics. Electrochimica Acta **48(20–22)**, 2987 (2003)
6. Mertig, M.; Kirsch, R.; Pompe, W.; and Engelhardt, H.: nanocluster array on S layer. Eur. Phys. J. D **9**, 45 (1999)
7. Knez, M.; Sumser, M.; Bittner, A. M.; Wege, C.; Jeske, H.; Martin, T. P.; and Kern, K.: Spatially Selective Nucleation of Metal Clusters on the Tobacco Mosaic Virus. Adv. Funct. Mater. **14**, 116 (2004)
8. Grummt, U.-W.; Geissler, M.; and Schmitz-Huebsch, T.: Chemical deposition of silver nanoclusters on self-assembled organic monolayers. A strategy to contact individual molecules. Chem. Phys. Lett. **263**, 581 (1996)
9. Andricacos, P. C.: Copper on-chip interconnections. The Electrochemical Society – Interface, Spring, 32 (1999)
10. Vaskelis, A.; Stalnionis, G.; and Jusys, Z.: Cyclic voltammetry and quartz crystal microgravimetry study of autocatalytic copper(II) reduction by cobalt(II) in ethylenediamine solutions. J. Electroanal. Chem. **465**, 142 (1999)
11. Jagannathan, R. and Krishnan, M.: Electroless plating of copper at a low pH level. IBM. J. Res. Develop. **37**, 117 (1993)
12. Van den Meerakker, J. E. A. M.: On the mechanism of electroless plating. II. One mechanism for different reductants. J. Appl. Electrochem. **11**, 395 (1981)
13. Wiese, H. and Weil, K. G.: Separation of partial processes at mixed electrodes. J. Electroanal. Chem. **228**, 347 (1987)
14. Inberg, A.; Zhu, L.; Hirschberg, G.; Gladkikh, A.; Croitoru, N.; Shacham-Diamand, Y.; and Gileadi, E.: Characterization of the initial growth stages of electroless Ag(W) films deposited on Si(100). J. Electrochem. Soc. **148**, C784 (2001)
15. Pohl, K. and Stierhof, Y.-D.: Action of gold chloride ("Gold Toning") on silver-enhanced 1 nm gold markers. Microscopy Res. Technique **42**, 59 (1998)
16. Baschong, W. and Stierhof, Y.-D.: Preparation, use, and enlargement of ultrasmall gold particles in immunoelectron microscopy. Microscopy Res. Technique **42**, 66 (1998)
17. Ciacchi, L.; Pompe, W.; and De Vita, A.: Initial nucleation of platinum clusters after reduction of K2PtCl4 in aqueous solution: A first principles study. J. Am. Chem. Soc. **123**, 7371 (2001)
18. Van der Putten, A. M. T. and de Bakker, J. W. G.: Geometrical effects in the electroless metallization of fine metal patterns. J. Electrochem. Soc. **140**, 2221 (1993)
19. Kind, H.; Bittner, A. M.; Cavalleri, O.; Kern, K.; and Greber, T.: Electroless deposition of metal nanoislands on aminothiolate-functionalized Au(111) electrodes. J. Phys. Chem. B **102(39)**, 7582 (1998)
20. Bittner, A. M.: Clusters on soft matter surfaces. Sur. Sci. Rep. **61**, 383–428 (2006)
21. Brandow, S. L.; Chen, M.-S.; Wang, T.; Dulcey, C. S.; Calvert, J. M.; Bohland, J. F.; Calabrese, G. S.; and Dressick, W. J.: Size-controlled colloidal Pd(II) catalysts for electroless Ni deposition in nanolithography applications. J. Electrochem. Soc. **144**, 3425 (1997)
22. Dressick, W.; Kondracki, L.; Chen, M.; Brandow, S.; Matijevic, E.; and Calvert, J.: Characterization of a colloidal Pd(II)-based catalyst dispersion for electroless metal deposition. Coll. Surf. A **108**, 101 (1996)
23. Bittner, A. M.: Biomolecular rods and tubes in nanotechnology. Naturwissenschaften **92**, 51 (2005)
24. Huczko, A.: Template-based synthesis of nanomaterials. Appl. Phys. A-Mater. Sci. Process. **70(4)**, 365 (2000)
25. Chen, M. S.; Brandow, S. L.; and Dressick, W. J.: Additive channel-constrained metallization of high-resolution features. Thin Solid Films **379(1–2)**, 203 (2000)
26. Li, J.; Moskovits, M.; and Haslett, T. L.: Nanoscale electroless metal deposition in aligned carbon nanotubes. Chem. Mater. **10(7)**, 1963 (1998)
27. Ang, L. M.; Hor, T. S. A.; Xu, G. Q.; Tung, C. H.; Zhao, S. P.; and Wang, J. L. S.: Decoration of activated carbon nanotubes with copper and nickel. Carbon **38**, 363–372 (2000)

28. Fink, D.; Petrov, A. V.; Rao, V.; Wilhelm, M.; Demyanov, S.; Szimkowiak, P.; Behar, M.; Alegaonkar, P. S.; and Chadderton, L. T.: Production parameters for the formation of metallic nanotubules in etched tracks. Radiat. Meas. **15**, 751 (2003)
29. Kordas, K.; Toth, G.; Levoska, J.; Huuhtanen, M.; Keiski, R.; Härkönen, M.; George, T. F.; and Vähäkangas, J.: Room temperature chemical deposition of palladium nanoparticles in anodic aluminium oxide templates. Nanotechnology **17**, 1459 (2006)
30. Wirtz, M.; Yu, S. F.; and Martin, C. R.: Template synthesized gold nanotube membranes for chemical separations and sensing. Analyst **127(7)**, 871 (2002)
31. Ah, C. S.; Yun, Y. J.; Lee, J. S.; Park, H. J.; Ha, D. H.; and Yun, W. S.: Fabrication of integrated nanogap electrodes by surface-catalyzed chemical deposition. Appl. Phys. Lett. **88**, 1331161 (2006)
32. He, H.; BoussaadS.; Xu, B.; Li, C.; and Tao, N.: Electrochemical fabrication of atomically thin metallic wires and electrodes separated with molecular-scale gaps. J. Electroanal. Chem. **522**, 167 (2002)
33. Niemeyer, C. M.: Nanoparticles, proteins, and nucleic acids: biotechnology meets materials science. Angew. Chem. Int. Ed. **40**, 4128 (2001)
34. Powell, R. D.; Halsey, C. M. R.; Liu, W.; Joshi, V. N.; and HainfeldJ. F.: Giant platinum clusters: 2-nm covalent metal cluster labels. J. Struct. Biol. **127**, 177 (1999)
35. Taton, T. A.; Mirkin, C. A.; and Letsinger, R. L.: Scanometric DNA array detection with nanoparticle probes. Science **289**, 1757 (2000)
36. Möller, R.; Powell, R. D.; HainfeldJ. F.; and Fritzsche, W.: Enzymatic control of metal deposition as key step for a low-background electrical detection for DNA Chips. Nano Lett. **5**, 1475 (2005)
37. Keren, K.; Krueger, M.; GiladR.; Ben-Yoseph, G.; Sivan, U.; and Braun, E.: Sequence-specific molecular lithography on single DNA molecules. Science **297**, 72 (2002)
38. Park, S. H.; Barish, R.; Li, H.; Reif, J. H.; Finkelstein, G.; Yan, H.; and LaBean, T. H.: Three-helix bundle DNA tiles self-assemble into 2D lattice or 1D templates for silver nanowires. Nano Lett. **5**, 693 (2005)
39. Price, R. R.; Dressick, W. J.; and Singh, A.: Fabrication of nanoscale metallic spirals using phospholipid microtubule organizational templates. J. Am. Chem. Soc. **125**, 11259 (2003)
40. Banerjee, I. A.; Yu, L.; and Matsui, H.: Cu nanocrystal growth on peptide nanotubes by biomineralization: Size control of Cu nanocrystals by tuning peptide conformation. Proc. Natl. Acad. Sci. USA **100**, 14678 (2003)
41. Behrens, S.; Wu, J.; Habicht, W.; and Unger, E.: Silver Nanoparticle and Nanowire Formation by Microtubule Templates. Chem. Mater. **16**, 3085 (2004)
42. Patolsky, F.; Weizmann, Y.; and Willner, I.: Actin-based metallic nanowires as bio-nanotransporters. Nat. Mater. **15**, 692 (2004)
43. Scheibel, T.; Parthasarathy, R.; Sawicki, G.; Lin, X. M.; Jaeger, H.; and Lindquist, S. L.: Conducting nanowires built by controlled self-assembly of amyloid fibers and selective metal deposition. Proc. Natl. Acad. Sci. **100(8)**, 4527 (2003)
44. Huang, Y.; Chiang, C.-Y.; Lee, S. K.; Gao, Y.; Hu, E. L.; DeYoreo, J.; and Belcher, A. M.: Programmable assembly. Nano Lett. **5**, 1429 (2005)
45. Reches, M. and Gazit, E.: Casting metal nanowires within discrete self-assembled peptide nanotubes. Science **300**, 625 (2003)
46. Balci, S.; Bittner, A. M.; Hahn, K.; Scheu, C.; Knez, M.; Kadri, A.; Wege, C.; Jeske, H.; and Kern, K.: Copper nanowires within the central channel of tobacco mosaic virus particles. Electrochim. Acta **51/28**, 6251 (2006)
47. Knez, M.; Bittner, A. M.; Boes, F.; Wege, C.; Jeske, H.; Maiß, E.; and Kern, K.: Biotemplate Synthesis of 3-nm Nickel and Cobalt Nanowires. Nano Lett. **3**, 1079 (2003)
48. Falkner, J. C.; Turner, M. E.; Bosworth, J. K.; Trentler, T. J.; Johnson, J. E.; Lin, T.; and Colvin, V. L.: Virus crystals as nanocomposite scaffolds. J. Am. Chem. Soc. **127**, 5274 (2005)
49. Weizmann, Y.; Patolsky, F.; Popov, I.; and Willner, I.: Telomerase-generated templates for the growing of metal nanowires. Nano Lett. **4**, 787 (2004)

50. Matsui, H.; Pan, S.; Gologan, B.; and Jonas, S. H.: Bolaamphiphile nanotube-templated metallized wires. J. Phys. Chem. B **104**, 9576 (2000)
51. Djalali, R.; Chen, Y.-F.; and Matsui, H.: Au nanocrystal growth on nanotubes controlled by conformations and charges of sequenced peptide templates. J. Am. Chem. Soc. **125**, 5873 (2003)

Part V
Modeling

Part V presents the modeling of superconformal electrodeposition for the application in the ULSI interconnect technology. Superconformal electrodeposition enables the void-free filling of high aspect ratio features such as trenches or vias in the Damascene metallization process. Superconformal electrodeposition, also known as superfill, occurs when particular combinations of chemical additives are included in an electrolyte. The additives enable preferential metal deposition at the bottom surface of the feature, which leads to "bottom-up" filling before the side walls close off, avoiding the formation of voids or seams in the feature. In order to predict and characterize this behavior, governing equations must be derived and solved. In general, this requires the numerical solution of a coupled set of partial differential equations (PDEs) for a full description. FiPy is an open source, PDE solver that has been used to numerically solve the superfill problem. It is written in Python and based on a standard finite volume approach. FiPy includes numerical algorithms that track moving and deforming interfaces, such as an electrode–electrolyte interface, that commonly occur in materials science problems. The chapter will begin with a broad introduction to gap-fill modeling with particular emphasis on the superfill process. Next, the governing equations will be introduced followed by the numerical aspects of interface tracking for the superfill problem. Finally, a description of how to use FiPy for solving the superfill problem will be given with an example.

Chapter 16
Modeling Superconformal Electrodeposition Using an Open Source PDE Solver

D. Wheeler and J.E. Guyer

16.1 Introduction

A critical aspect in the manufacturing of semiconductor devices is the filling of trenches and vias using metal electrodeposition. This filling process has been characterized extensively with semi-empirical mathematical and computational models, although much work is still required in order to employ less empirical models. However, these models have not been used in a predictive capacity in industry due to the time frame required for code development and experimental design and lack of available modeling software. To overcome these issues, an open source tool called FiPy [1] was developed for solving PDEs that commonly occur in materials science problems. It seeks to address the issues of coding practice and open source development by employing modern coding techniques and providing a flexible coding framework to rapidly pose, prototype, and share models of superfill and general deposition processes.

In the past, the understanding of feature fill has been based on the diffusion-driven theory of leveling. Leveling theory depends on a spatial drop in the concentration of leveling agent within the trench cavity, which provides a differential in the adsorption rate of the agent. The adsorption rate influences the deposition rate, for which a number of mechanisms have been proposed [2–4, 29]. As a general rule, as feature sizes are reduced the concentration gradient within the trench is reduced and the effect of the leveling agent is curtailed. This general rule, of course, depends on many factors, but in particular on the additive chemistry [5] and ratio of trench depth to boundary layer depth [6]. Although the description of superfill modeling presented in this chapter uses a different mechanism to describe deposition, generic gap-fill modeling described by leveling theory is historically relevant as it shares many of the same underlying numerical algorithms required to model phenomena such as deforming interfaces, adsorption dynamics, and material transport. We will

D. Wheeler (✉)
Materials Science and Engineering Laboratory, National Institute of Standards and Technology, Gaithersburg, MD 20899, USA
e-mail: daniel.wheeler@nist.gov

Y. Shacham-Diamand et al. (eds.), *Advanced Nanoscale ULSI Interconnects: Fundamentals and Applications*, DOI 10.1007/978-0-387-95868-2_16,

review some of the early papers in the development of gap-fill modeling, most of which use leveling theory. This is by no means an extensive literature review, but rather reflects the authors' view of the most relevant publications. One of the earliest papers for predicting a discontinuous electrode geometry in combination with the solution of the Laplace equation was Alkire et al. [7]. The finite element method (FEM) was used to predict the potential in the electrolyte. The potential was then used in the Butler–Volmer equation to update the electrode position.

Hume et al. [8] predicted shape change on the micro-profile level, only modeling the diffusion within the convective boundary layer. The Tafel equation was solved to predict the current density using the steady-state concentration profile of the reacting species.

The boundary element method (BEM) was used to solve Laplace's equation in the bulk electrolyte for the steady-state reacting species. The BEM has the benefit of reducing the computational load when compared with the FEM. A more complete model for the gap-fill process than those that had been previously published was provided by Dukovic et al. [2]. They were the first authors to implement a detailed description for the advancement of the electrode profile at the feature level. Their method solved the equations for both the potential and leveling species in order to predict the deposition rate in the trench. Madore et al. [3] extended this work by assuming an area-blocking mechanism for the leveling agent between the electrolyte and the electrode surface, which led to a better understanding of the correlation between additive coverage and deposition rate. The blocking coverage was dependent on the ratio of the leveling agent flux to the metal ion flux. Andricacos et al. [4] used the same methods as Dukovic and Madore to model superfill, but with an empirical modification of leveling theory to create a new form for the blocking mechanism independent of the metal ion flux, using this to draw a distinction between superfill and traditional leveling. West [6] used a pseudo-one-dimensional leveling theory model to investigate the parameter space of feature fill using an additive blocking mechanism based directly on the concentration of leveling agent at the interface. Leveling theory suffers from three main deficiencies when used to explain superconformal filling:

1. As noted by West [6], as feature sizes are reduced the ability of the leveling agent to induce bottom-to-top fill becomes diminished in contrast to experimental evidence.
2. The simple area blockage mechanism, which underpins leveling theory, requires empirical alteration in order to reconcile simulation and experimental profiles, in particular, the distinctive regions of coverage on the electrode surface during filling, manifesting as discrete flat panels, cannot be explained.
3. The over-bump that forms above features, caused by what has become known as "momentum plating," cannot be predicted by the action of a leveling agent.

Due to the deficiencies of leveling theory, when used to explain superconformal filling, a new theory was first proposed by Moffat et al. [9] and later, independently, by West et al. [10]. This theory postulates, as in leveling theory, that a species in the electrolyte can adsorb onto the electrode/electrolyte interface and influence the deposition rate.

However, the new theory diverges from leveling theory by presupposing that once the additive is adsorbed onto the interface (i) competition occurs between rate accelerating and rate suppressing additives with the rate acclerating additive being more tightly bound to the surface and (ii) the additive floats freely on the interface with the characteristics of a traditional surfactant. On a fluid–fluid interface, a surfactant floats between two materials and the surfactant coverage changes as a result of the deformation of the interface. Applying this concept to a filling feature, it is manifest that an accelerating species, acting as a surfactant, will develop a high coverage at the bottom of the fill profile and thus, induce bottom-to-top fill. The rate of change of the surfactant coverage is proportional to the local curvature, and hence, the important function behind this theory has become known as the curvature-enhanced accelerator coverage (CEAC) mechanism. The theory has gained wide acceptance due to its predictive power. In particular, the CEAC mechanism predicts the filling of small features. The theory also predicts the discrete flat panels that occur in the feature profile during filling and the momentum plating that causes the over-bump above the filled feature.

The model used by Moffat et al. [9] was based on a two-dimensional Lagrangian approach that tracked the interface. It was assumed that a fractional coverage of accelerator surfactant was adsorbed onto the interface during the initial period of conformal filling. Their Lagrangian approach used nodal points to explicitly track the position of the interface and accelerator concentration but ignored the bulk transport in the electrolyte, focusing only on the CEAC mechanism.

Although Lagrangian interface tracking methods are extremely accurate, they suffer from a lack of robustness often requiring complex remeshing techniques that have to account for special cases of extreme interface distortion or altering interface topology. Eulerian techniques, where a field variable governs the position of the interface on a fixed mesh, handle distortion and topology changes implicitly and link well with the finite volume method (FVM) and FEM, which are generally solved on a fixed mesh. However, Eulerian techniques tend to be less accurate and require more memory than Lagrangian techniques. Nevertheless, the Eulerian level set method (LSM) has been shown to be a successful tool for modeling deposition processes in a variety of mechanisms in a trio of papers by Adalsteinsson et al. [11–13]. Due to the LSM's robustness and its easy coupling to the FVM, Josell et al. [14, 15] employed the LSM to model superfill. They implemented a model that involved the full bulk diffusion for each species, the adsorption dynamics for each species, and the tracking of the position of the interface along with the coverage of the adsorbed accelerator surfactant. They accurately accounted for the concentration of surfactant material as the interface evolved a difficult proposition within an Eulerian framework. Both the Lagrangian and Eulerian approaches described above have been used in conjunction with the CEAC mechanism to simulate a variety of processes and mechanisms including copper [15], silver [16], and gold [17] electrodeposition; chemical vapor deposition [18]; catalyst deactivating leveler [19]; accelerator surface diffusion [20]; and the brightening mechanism [21].

The numerical solution of some of the superfill models described above requires discretization of the governing equations and encoding (writing computer code) on a computer in order to obtain a discrete solution. Until recently, this process has been done in a somewhat ad hoc manner without employing modern software engineering tools and techniques leading to a host of difficulties in model dissemination, code reuse, code efficiency, and new model development among other issues. The process of encoding the model is fraught with difficulties that are of no interest to materials scientists interested in the superfill mechanism. In order to address some of these concerns the development of FiPy was undertaken in 2004.

The FiPy framework provides a high-level language for posing coupled sets of PDEs in conjunction with a moving interface across which the coefficients of the PDEs change rapidly. FiPy examples have been customized to solve a number of such problems including the superfill problem. The remainder of the chapter will be concerned with a description of the governing equations and discretization required for a numerical solution of the superfill problem along with a FiPy tutorial example.

It should be emphasized that the FiPy example presented in this chapter requires the use of parameters obtained from experimental studies with specific electrolyte chemistries. In order to make useful progress in modeling superfill for any given electrolyte chemistry, a parallel experimental effort will be required to obtain the necessary parameters for the model. There is considerable literature on determining parameters for superfill modeling [22, 23], and it will not be alluded to further here.

16.2 Governing Equations

The objective of the superfill model presented here is to predict the dynamic position of the electrode/electrolyte interface in order to optimize the filling process and avoid the formation of seams or voids within the feature. In principle, to achieve this goal, only the interface velocity and those electrolyte species that influence the velocity require inclusion in the model. These species, together with the interface velocity, form the dependent variables for the system. The domain of interest for solving the dependent variables consists of the electrolyte between the electrode interface and the bulk electrolyte outside the boundary layer. This domain shall be designated as Γ and its boundary as, $\partial\Gamma$. The boundary can be subdivided into the electrolyte/electrode interface, $\partial\Gamma_e$; the interface between the electrolyte bulk and the boundary layer, $\partial\Gamma_b$; and the symmetry boundaries, $\partial\Gamma_s$; such that (see Fig. 16.1)

$$\partial\Gamma = \partial\Gamma_e + \partial\Gamma_b + \partial\Gamma_s \tag{16.1}$$

In general, models such as these have species concentration fields governed by transport equations in the electrolyte, Γ, with given boundary conditions on $\partial\Gamma$, forming a traditional boundary value problem (BVP). Indeed, this model can be characterized in such a way but with two important subtleties. Firstly, our interest

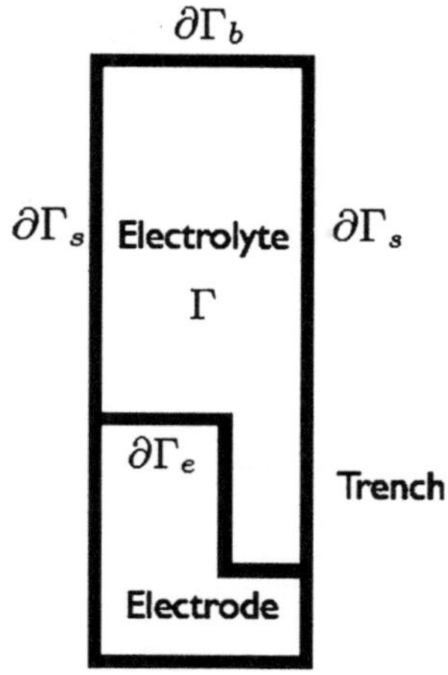

Fig. 16.1 The solution domain, Γ, and its boundary, Γ

is in the evolution of the electrolyte/electrode interface location, $\partial\Gamma_e$, which is not stationary, but continuously changing, forming a moving BVP. Secondly, the interface velocity is dependent on the coverages of additive surfactants at the interface. The evolution of the surfactants requires the solution of transport equations on $\partial\Gamma_e$ to account for the influence of adsorption and interface deformation dynamics.

The species in the electrolyte that are relevant to this model consist of both the metal ions and the dilute additives that affect interface motion. Previous work has generally focused on modeling only one additive in the bulk. In this analysis a more general set of equations for the CEAC mechanism in the case of multiple additives will be presented. The concentration of a species in the electrolyte is denoted by c_i, where $i = a$ for an electrolyte additive species or $i = m$ for metal ions. The symbol a can represent accelerator or leveling additive. An important aspect of the model is the additive adsorption and subsequent segregation on the evolving surface to form surfactants. When an additive is adsorbed onto the interface, the proportional coverage, normalized to the number of available sites, for the given additive is denoted by θa and the proportion of suppressed interface is denoted by θv. The combined local coverage of the adsorbed additives must satisfy

$$\left.\begin{array}{l}\theta_v + \sum_a \theta_a = 1\\ \theta_v \geq 0\\ \theta_a \geq 0\end{array}\right\} \quad \text{on } \partial\Gamma_e \tag{16.2}$$

Equation (16.2) is used to determine θ_v. In practice, θ_v is taken to be 1 initially as suppressor additive is adsorbed onto the surface extremely rapidly. The normal interface velocity can be written in terms of these variables, such that

$$V_n = V_n(c_m, \theta_a, \theta_v) \qquad \text{on } \partial\Gamma_e \tag{16.3}$$

The choice of expression for the velocity generally has a physical basis, but the parameters for the expression are empirically determined. The common form for the velocity expression is based on a linear interpolation between states of saturated coverage for each additive and the supressed state, such that

$$V_n = \frac{c_m}{c_m^\infty}(v_v\theta_v + \sum_a v_a\theta_a) \qquad \text{on } \partial\Gamma_e \tag{16.4}$$

where $c^\infty{}_m$ is the concentration in the bulk. The parameters v_a and v_v represent the ideal interface velocities in the cases of complete coverage of additive i and suppressed interface, respectively, with negligible boundary layer ($c_m = c^\infty{}_m$).

As mentioned above, the adsorbed species must be accounted for with the interface conservation equation; see Stone et al. [24] for a first principles derivation. The interface conservation equation for a surfactant is given by

$$\frac{d\theta_a}{dt} = \kappa V_n\theta_a + \nabla_s \cdot D_a^s \nabla_s\theta_a + F_a^+ - F_a^- \qquad \text{on } \partial\Gamma_e \tag{16.5}$$

where κ denotes the curvature, D_a^s denotes the surface diffusion coefficient, F_a^+denotes the adsorption flux, and F_a^- denotes the consumption flux. The term involving curvature can be thought of as the mathematical description of the CEAC mechanism.

Since, the curvature term has no implicit mechanism for ensuring that Eq. (16.2) holds, an artificial mechanism allowing surfactants of greater surface affinity to displace those with less surface affinity must be introduced. The displacement is in fact part of the CEAC mechanism [19]. To account for this, Eq. (16.5) is modified under some conditions, such that

$$\frac{d\theta_a}{dt} = -\sum_{a^+}\frac{d\theta_{a^+}}{dt} \text{ when } \sum_{a^+}\theta_{a^+} = 1 \qquad \text{and } \kappa > 0 \text{ on } \partial\Gamma_e \tag{16.6}$$

where a^+ denotes additives with more surface affinity than additive a. In this chapter the surface diffusion will be ignored; see Josell et al. [20] for an examination of the CEAC mechanism in conjunction with surface diffusion. The consumption source is generally determined by an empirical formula based on the local coverage and interface velocity such that

$$F_a^- = F_a^-(\theta_a, V_n) \qquad \text{on } \partial\Gamma_e \tag{16.7}$$

The adsorption flux is based on Langmuir adsorption, where the adsorption rate is proportional to the number of available sites, such that

$$F_a^+ = c_a k_{av}\theta_v + \sum_i (c_a k_{ai}\theta_i) - (c_i k_{ai}\theta_a) \qquad \text{on } \partial\Gamma_e \tag{16.8}$$

where i denotes all additives other than additive a and k_{ai} and k_{av} denote the adsorption coefficients for species a onto species i and supressed interface, respectively. The first term on the right-hand side is the adsorption flux of species a onto a supressed interface, the second term is the adsorption flux of species a onto species i, and the third term represents the desorption flux of species a due to the adsorption of species i.

The variables c_a and c_m, required by Eqs. (16.3), (16.4), and (16.8), need to be determined by solving transport equations in Γ. For the case when both convection and electromigration are negligible in the electrolyte boundary layer, the transport equation for the metal ions and additive species is given by

$$\frac{\partial c_j}{\partial t} = \nabla \cdot D_j \nabla c_j \qquad \text{for } j = a \text{ and } j = m \text{ in } \Gamma \tag{16.9}$$

where D_j represents the diffusion coefficient for each additive. The boundary condition at the edge of the boundary layer is

$$c_j = c_j^{\infty} \qquad \text{for } j = a \text{ and } j = m \text{ on } \partial\Gamma_b \tag{16.10}$$

and on the symmetry planes

$$\vec{n} \cdot \nabla_{C_j} = 0 \qquad \text{For } j = a \text{ and } j = m \text{ on } \partial\,\Gamma_S \tag{16.11}$$

where $\vec{n}$ points into the domain Γ. At the electrode interface the boundary condition is given by

$$D_j \vec{n} \cdot \nabla_{C_j} = \begin{cases} \frac{V_n}{\Omega_m} & \text{if } j = m \\ \gamma_a C_a \left(k_{av}\,\theta_v + \sum_j k_{ai}\theta_i \right) & \text{if } j = a \end{cases} \quad \text{on } \partial\Gamma_e \tag{16.12}$$

where i denotes all additives other than additive a, Ω_m denotes the metal ion molar volume. The site density γ_a is the surface coverage required to saturate the surface with adsorbate a.

The last term in Eq. (16.8) can account for either desorption or deactivation depending on the species.

16.3 Level Set Equations

This section will present an overview of the LSM. The LSM is an algorithmic approach for handling moving interfaces in an Eulerian framework. The method works by creating an artificial interface depth that is maintained throughout the simulation preventing numerical smearing. This is in contrast to sharp interface methods such as the volume of fluid method where the interface integrity is not implicitly maintained.

Other methods that use a diffuse interface include the phase field method (PFM) where the equation governing the interface motion is physically derived from first principles.

The PFM has been used successfully to model the electrochemical interface at both small length scales (see Guyer et al. [25, 26]) and at large length scales.

In the LSM, a scalar variable ϕ is defined over the entire region. The set of locations $\phi = 0$ (i.e., the zero level set) defines the position of the interface. The variable ϕ is continuous, smooth, and monotonic in the direction normal to the interface and is governed by the advection equation, given by

$$\frac{\partial \phi}{\partial t} + V_n |\nabla \phi| = 0 \qquad \text{in } \Gamma \tag{16.13}$$

During the evolution of the interface, ϕ is maintained as a distance function, such that

$$|\nabla \phi| = 1 \qquad \text{in } \Gamma \tag{16.14}$$

with $\phi = 0$ on $\partial \Gamma^e$. Preserving ϕ as a distance function allows the distance between any point in the solution domain and the interface to be known. This maintains the "diffuse" interface at a steady depth, eliminating numerical smearing. The distance function is initialized once at the start of the simulation using the fast marching method. Thereafter the use of extension velocities and the correct discretization of the advection equation maintain the distance function. The extension velocities are calculated from

$$\nabla \phi \cdot \nabla v_{ext} = 0 \qquad \text{in } \Gamma \tag{16.15}$$

with $v_{ext} = V^n$ on $\partial \Gamma_e$. The diffusion coefficient for Eq. (16.9) can be modified to vary smoothly across the interface, such that

$$D_i^* (\phi) = \begin{cases} D_i & \text{if } -\phi < \alpha, \\ \frac{D_i}{2} \left(1 + \sin \frac{\phi \pi}{2\alpha}\right) & \text{if } -\alpha \leq \phi \leq \alpha, \\ 0 & \text{if } \phi < -\alpha, \end{cases} \tag{16.16}$$

where 2α is the chosen interface depth and $\phi > 0$ in the electrolyte. The numerical details of the LSM are considerable, but well documented and will not be described further [15, 27].

16.4 Numerical Discretization

In this section, the numerical discretization for a simplified version of Eq (16.5) will be given. The numerical procedures for evaluating the other equations in this chapter have been described elsewhere. The equations in Section 16.2, other than Eq. (16.5), are solved using a standard finite volume approach [1, 21]. Likewise, the numerical solution for the level set equations (Eqs. (16.13) and (16.14)) is given in Sethian [27].

The simplified version of Eq. (16.5) that will be discretized is given by

$$\frac{d\theta}{dt} = \kappa V_n \theta \qquad \text{on } \partial\Gamma_e \tag{16.17}$$

Equation (16.17) represents Eq. (16.5) without the adsorption and consumption source terms. The numerical discretization for these terms uses a standard approach and they are omitted for clarity. For the numerical scheme proposed here, both the governing equation, Eq. (16.17), and the dependent variable θ will be transformed from surface to volume quantities.

The interface coverage θ can be transformed from a surface coverage to a volume concentration using

$$\int_S \theta d\theta = \int_V \Theta dV \tag{16.18}$$

where S is the area of interface $\partial\Gamma_e$ within the control volume defined by V. Equation (16.18) can be discretized but first an arbitrary decision must be made to choose which cells the interface occupies. Here, the cells adjacent to and lying in front of the interface will be used. A member of this set of cells will be labelled P and must satisfy $\phi_P > 0$ and $\phi_A \phi_P \leq 0$, where A represents a neighboring cell to cell P. The discretized version of Eq. (16.18) is given by

$$\theta_P S_P = \Theta_P V_P \tag{16.19}$$

where V_P is the volume of cell P, S_P is the area of interface $\partial\Gamma_e$ associated with cell P, Θ_P is the volume concentration of the surfactant, and θ_P is the coverage of the surfactant. The area S^P is calculated by

$$S_P = -\sum_F A_F \delta_F \vec{n}_{IF} \cdot \vec{n}_F \tag{16.20}$$

where F are the faces surrounding cell P, A_F is the area of the face, and $\vec{n}_F$ is the outward pointing normal from the face. The interface delta function for a face is defined by

$$\delta_P = \begin{cases} 1 & \text{if } \phi_A \phi_P < 0 \\ 0 & \text{otherwise} \end{cases} \tag{16.21}$$

and the interface normal at the face is defined by

$$\vec{n}_{IF} = \frac{(\nabla\phi)_P + (\nabla\phi)_A}{|(\nabla\phi)_P + (\nabla\phi)_A|} \tag{16.22}$$

assuming regular, equispaced cells. Since Eq. (16.17) is being solved in an Eulerian framework, it needs to be transformed into an equation on Γ. The surfactant conservation equation can be recast as a volume integral [21], such that

$$\int_V \left(\frac{\partial \theta}{\partial t} + V_n \nabla \cdot (\vec{n\theta}) \right) \delta(\phi) |\nabla \phi| dV = 0 \tag{16.23}$$

and then discretized such that

$$\theta_P = \theta_P^O - \frac{V_n (\delta t)^P}{V_P} \sum_F [\vec{n_{IF}} \cdot \vec{n}_F \theta_F A_F] \tag{16.24}$$

where $\delta(\phi)$ has been chosen to be non-zero for cells on the interface. Equation (16.24) is not solved at every time step; it is only solved when cells are removed from the interface region as a result of the interface advancing. Thus, the time step $(\delta t)_P$ is not the standard time step, but a local time step. The local time step is chosen such that

$$(\delta\tau)_P = \frac{V_P}{V_n S_P} \tag{16.25}$$

The quantity V_P/S_P can be thought of as the distance moved by the interface to cross cell P. In this scheme Θ_P is used instead of θ_P. Thus, θ_P and $(\delta t)_P$ can be substituted out of Eq. (16.24) using Eqs. (16.19) and (16.25), such that

$$\theta_P = \theta_P^0 - \frac{1}{V_P} \sum_F \left[\vec{n}_{IF} \cdot \vec{n}_F \frac{\Theta_A V_A}{S_A} A_F \right] \tag{16.26}$$

$$-\sum_{F_A} \left[\frac{\vec{n}_{IF_A} \cdot \vec{n}_{F_A} A_{F_A}}{S_{F_A}} \right] = 1 \tag{16.27}$$

using an upwind scheme to choose $\underline{\theta}F = \underline{\theta}A$. The above equation is solved after every time step when an adjacent cell changes sign and leaves P. Since the scheme presented here is entirely conservative, F_A represents the adjacent (A) cell faces. The scheme demonstrates first-order accuracy for a number of simple numerical test cases. For example, one test case involves a circle expanding from an initial diameter of 1/2 and an initial coverage of 1 to become a circle with a final diameter of 3/4 and a final coverage of 2/3. As the grid is refined this test case demonstrates first-order convergence for L_1, L_2, and L_∞ norms. The test case has a final maximum error of less than 1% when a cell size of 1/256 is used. It should be noted that an analytical analysis to ascertain the formal accuracy of the scheme has not been undertaken.

16.5 FiPy Example

This section will present a FiPy example for interface motion due to a pre-adsorbed surfactant. It acts as a simple tutorial for the use of FiPy from which more complex examples can be developed. There are many more detailed examples available with the FiPy distribution [1]. In the following text, ">>>" denotes the Python command line and "..." denotes the Python prompt for a command continuing from the previous line.

Any command that follows these symbols can be typed into the Python interpreter or placed in a batch file and executed as a batch process. If you have no experience with the Python programming language, please refer to the online tutorial [28].

The code in this example is an explicit explanation of the input commands required to set up and run a simplified superfill problem. A trench of aspect ratio 2 evolves with an initial coverage of 0.1 and the interface velocity is a linear function of coverage. The first line imports the FiPy module into Python.

```
>>> from fipy import *
```

The following code builds the mesh with nx * ny elements.

```
>>> cellSize = 0.02

>>> cellsBelowTrench = 10

>>> spaceAboveTrench = 1.0

>>> trenchDepth = 1.0

>>> ny = cellsBelowTrench + int((trenchDepth
                               + spaceAboveTrench) / cellSize)

>>> trenchSpacing = 1.0

>>> nx = int(trenchSpacing / cellSize / 2.0)

>>> mesh = Grid2D(dx=cellSize, dy=cellSize, nx=nx, ny=ny)
```

The Grid2D object has been used to create a two-dimensional equally spaced grid. A DistanceVariable object is required to store the position of the interface. It calculates its value to be a distance function (i.e., holds the distance at any point in the mesh from the electrolyte/metal interface at $\phi = 0$) using Eq. (16.14). Thus, ϕ is created and is initially set to –1 everywhere:

```
>>> numberOfCellsInNarrowBand = 10
>>> narrowBandWidth = numberOfCellsInNarrowBand * cellSize
>>> distanceVar = DistanceVariable(mesh=mesh,
...                                value=-1,
...                                narrowBandWidth=narrowBandWidth,
...                                hasOld=True)
```

The following code then fixes the position of the trench by setting $\emptyset = 1$ in the electrolyte region and $\phi = -1$ in the metal region:

```
>>> bottomHeight = cellsBelowTrench * cellSize
>>> trenchHeight = bottomHeight + trenchDepth
>>> aspectRatio = 2.0
>>> trenchWidth = trenchDepth / aspectRatio
>>> sideWidth = (trenchSpacing - trenchWidth) / 2
>>> x, y = mesh.getCellCenters(),
>>> distanceVar.setValue(1, where=(y > trenchHeight)
...                         |((y > bottomHeight)
...                         & (x < (nx * cellSize - sideWidth))))
```

Since the electrode and electrolyte regions have been initializedthe distance function can be calculated, $|\nabla\phi| = 1$.

```
>>> distanceVar.calcDistanceFunction(narrowBandWidth=1e10)
```

The DistanceVariable has now been created to mark the interface. Another variable needs to be created that governs the coverage of an additive species. The following code demonstrates how to set up the surfactant as represented by θ_a in Eq. (16.5).

```
>>> catalystVar=SurfactantVariable(value=0.1,name='coverage',
...                         distanceVar=distanceVar)
```

Equation (16.4) represents a general form for a deposition rate V_n based on a number of additives. Here, we will simply use the realtionship that $V_n = 0.01 + \theta$.

```
>>> depositionRateVariable=0.01+catalystVar.getInterfaceVar()
```

The extension velocity^{vext} uses Eq. (16.15) to spread the velocity at the interface to the rest of the domain.

```
>>> extensionVelocityVariable = CellVariable(mesh=mesh,
... value=depositionRateVariable)
```

The surfactantEquation is created to govern the 'catalystVar' coverage.

```
>>> surfactantEquation = SurfactantEquation
                          (distanceVar=distanceVar)
```

The variable ϕ is advected by Eq. (16.13), which is labeled by 'advectionEquation' in this example.

```
>>> advectionEquation = buildHigherOrderAdvectionEquation(
... advectionCoeff=extensionVelocityVariable)
```

We can now create a viewer for the 'catalystVar'.

```
>>> viewer = MayaviSurfactantViewer(distanceVar,
...                                 catalystVar.getInterfaceVar(),
...                                 limits={ 'datamax' : 1.0,
...                                 'datamin' : 0.0 },
...                                 title='catalyst_coverage',
...                                 smooth=1)
```

The 'levelSetUpdateFrequency' defines how often to call the 'distanceEquation' to reinitialize the 'distanceVariable' to satisfy Eq. (16.14). The following loop does a number of time steps determined by 'numberOfSteps'. The time step is calculated using the Courant–Freidricks–Levy number, denoted by 'cflNumber', and the maximum extension velocity. The output from this simulation can be seen in Fig. 16.2.

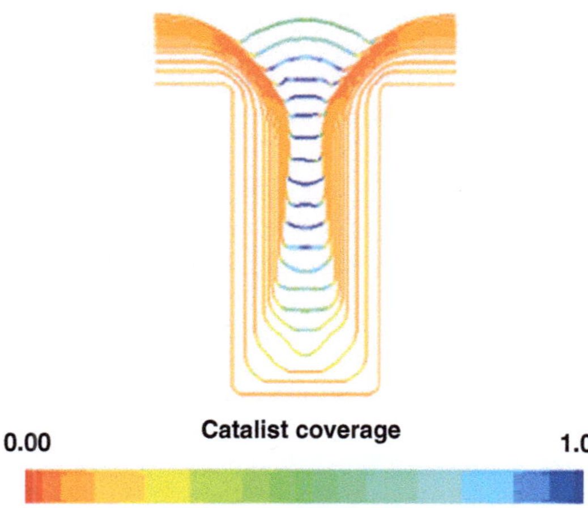

Fig. 16.2 The plot shows the FiPy output when running the code example. The trench profiles show the coverage of catalyst plotted as isocontours every 20 time steps during the simulation (*See* Color Insert)

```
>>> cflNumber = 0.2
>>> levelSetUpdateFrequency = int(0.7 * narrowBandWidth \
...                              /(cellSize * cflNumber * 2))
>>> numberOfSteps = 400
>>> for step in range(numberOfSteps):
...
...   if step % 20 == 0:
...             viewer.plot()
...   if step % levelSetUpdateFrequency == 0:
```

```
...    distanceVar.calcDistanceFunction(deleteIslands = True)
...    extensionVelocityVariable.setValue(depositionRateVariable())
...    distanceVar.updateOld()
...    catalystVar.updateOld()
...    distanceVar.extendVariable(extensionVelocityVariable)
...    dt = cflNumber * cellSize / max(extensionVelocityVariable)
...    advectionEquation.solve(distanceVar, dt=dt)
...    surfactantEquation.solve(catalystVar, dt=dt)
```

16.6 Summary

This chapter started with a broad introduction to gap-fill modeling. It went on to give the general form of the governing equations for superfill with an arbitrary number of additive species. These equations can be solved with a standard finite volume methodology and the moving interface can be tracked with the ubiquitous level set treatment.

The difficulty lies in the discretization of the surfactant equation. Hence, Section 16.4 focused on a detailed description of the discretization for Eq. (16.17). The discretization is first-order accurate, but more importantly for the superfill problem, it is completely conservative. The final section gave a simple FiPy example for modeling a trench geometry.

The example has many of the details removed in order to demonstrate the essential FiPy commands required to model a moving interface with a simple relationship between surfactant coverage and interface velocity (more complex examples are available online [1]). This demonstration shows the power and flexibility of FiPy as a tool for modeling deposition processes. It should be noted that because FiPy uses the Python scripting language, many freely available scientific modules previously written in Python can be used in conjunction with FiPy, creating an environment where experimental and simulation analysis can be undertaken in tandem.

References

1. Gyyer, J. E.; Wheeler, D.; and Warren. J. A.: FiPy: Partial differential Equations with Python, Comput. Sci. Eny. **11(3)**, 6–15 (2004)
2. Dukovic, J. O.; and Tobias, C. W.: Simulation of Leveling in Electrodeposition. J. Electrochem. Soc. **137(12)**, 3748 (1990)
3. Madore, C.; Matlosz, M.; and Landolt, D.: Blocking inhibitors in cathodic leveling 1. Theoretical analysis. J. Electrochem. Soc. **143(12)**, 3927 (1996)
4. Andricacos, P. C.; Uzoh, C.; Dukovic, J. O.; Horkans, J.; and Deligianni, H.: Damascene Copper Electroplating for Chip Interconnections. IBM J. Res. Dev. **42(5)**, 567 (1998)
5. Vereecken, P. M.; BinsteadR. A.; Deligianni, H.; and Andricacos, P. C.: The chemistry of additives in damascene copper plating. IBM J. Res. Dev. **49(1)**, 3 (2005)
6. West, A. C.: Theory of filling of high-aspect ratio trenches and vias in presence of additives. J. Electrochem. Soc. **147(1)**, 227 (2000)
7. Alkire, R.; Bergh, T.; and Sani, R. L.: Predicting electrode shape change with use of finite-element methods. J. Electrochem. Soc. **125(12)**, 1981 (1978)

8. Hume, E. C.; Deen, W. M.; and Brown, R. A.: Mass-transfer analysis of electrodeposition through polymeric masks. J. Electrochem. Soc. **131(6)**, 1251 (1984)
9. Moffat, T. P.; Wheeler, D.; Huber, W. H.; and Josell, D.: Superconformal electrodeposition of copper. Electrochem. Solid State Lett. **4(4)**, C26 (2001)
10. West, A. C.; Mayer, S.; and ReidJ.: A superfilling model that predicts bump formation. Electrochem. Solid State Lett. **4(7)**,C50 (2001)
11. Adalsteinsson, D. and Sethian, J. A.: A level set approach to a unified model for etching, deposition, and lithography I: Algorithms and two-dimensional simulations. J. Comput. Phys. **120**, 128 (1995)
12. Adalsteinsson, D. and Sethian, J. A.: A level set approach to a unified model for etching, deposition, and lithography II: Three-dimensional simulations. J. Comput. Phys. **122**, 348 (1995)
13. Adalsteinsson, D. and Sethian, J. A.: A level set approach to a unified model for etching, deposition, and lithography III: Redeposition, reemission, surface diffusion and complex simulations. J. Comput. Phys. **138**, 193 (1997)
14. Josell, D.; Wheeler, D.; Huber, W. H.; and Moffat, T. P.: Superconformal electrodeposition in submicron features. Phys. Rev. Lett. **87(1)**, 016102 (2001)
15. Wheeler, D.; Josell, D.; and Moffat, T. P.: Modeling superconformal electrodeposition using the level set method. J. Electrochem. Soc. **150(5)**, C302 (2003)
16. Moffat, T. P.; Baker, B.; Wheeler, D.; Bonevich, J. E.; Edelstein, M.; Kelly, D. R.; Gan, L.; Stafford G. R.; Chen, P. J.; Egelhoff, W. F.; and Josell, D.: Superconformal electrodeposition of silver in submicrometer features. J. Electrochem. Soc. **149(8)**, C423 (2002)
17. Josell, D.; Wheeler, D.; and Moffat, T. P.: Gold superfill in submicrometer trenches: Experiment and prediction. J. Electrochem. Soc. **153(1)**, C11 (2006)
18. Josell, D.; Wheeler, D.; and Moffat, T. P.; Superconformal deposition by surfactant-catalyzed chemical vapor deposition. Electrochem. Solid State Lett. **5(3)**, C44 (2002)
19. Moffat, T. P.; Wheeler, D.; Kim, S. K.; and Josell, D.; Curvature Enhanced Adsorbate Coverage Model for Electrodeposition. J. Electrochem. Soc. **153(2)**, C127 (2006)
20. Moffat, T. P.; Josell, D.; and Wheeler, D.: Superfilling when adsorbed accelerators are mobile. J. Electrochem. Soc. **154(4)**, D208 (2007)
21. Wheeler, D.; Moffat, T. P.; McFadden, G. B.; Coriell, S.; and Josell, D.: Influence of a catalytic surfactant on roughness evolution during film growth. J. Electrochem. Soc. **151(8)**, C538 (2004)
22. Moffat, T. P.; Wheeler, D.; Edelstein, M. D.; and Josell, D.: Superconformal film growth: Mechanism and quantification. IBM J. Res. Dev. **49(1)**, 19 (2005)
23. Moffat, T. P.; Wheeler, D.; and Josell, D.: Quantifying Competitive Adsorption Dynamics in Superfilling Electrolytes. In Deligianni, H.; Mayer, S. T.; Moffat, T. P.; and Stafford, G. R. (eds): Electrodeposition in ULSI and MEMS Fabrication, The Electrochemical Society, Inc., 23 (2005)
24. Stone, H. A.: A simple derivation of the time-dependent convective-diffusion equation for surfactant transport along a deforming interface. Phys Fluids A-Fluid Dyn. **2(1)**, 111 (1990)
25. Guyer, J. E.; Boettinger, W. J.; Warren, J. A.; and McFadden, G. B.: Phase field modeling of electrochemistry I: Equilibrium. Phys. Rev. E **69**, 021603 (2004)
26. Guyer, J. E.; Boettinger, W. J.; Warren, J. A.; and McFadden, G. B.: Phase field modeling of electrochemistry II: Kinetics. Phys. Rev. E **69**, 0216016 (2004)
27. Sethian, J. A.: Level Set Methods and Fast Marching Methods. Cambridge University Press (1996)
28. van Rossum, G.: Python Tutorial. URL http://docs.python.org/tut /.
29. Jordan, K. G. and Tobias, C. W.: The effect of inhibitor transport on leveling in electrodeposition. J. Electrochem. Soc. **138(5)**, 1251 (1991)

Part VI
Electrochemical Process Integration

Part VI focuses on advanced multilevel interconnects, contributed by distinguished authors in the following sections: "Introduction to Electrochemical Process Integration for Cu Interconnects" (Chapter 17), "Damascene Concept and Process Steps" (Chapter 18), "Advanced BEOL Technology Overview" (Chapter 19), "Lithography for Cu Damascene Fabrication" (Chapter 20), "Physical Vapor Deposition Barriers for Cu Metallization PVD Barriers" (Chapter 21), "Low-*K* dielectrics" (Chapter 22), "CMP for Cu Processing" (Chapter 23), "Electrochemical View of Copper Chemical Mechanical Polishing" (CMP) (Chapter 24), "Copper Post-CMP Cleaning" (Chapter 25). Chapter 17 gives brief introduction to electrochemical process integration for Cu interconnects. Chapter 18 describes about the damascene process for Cu interconnect formation. There are two damascene processes: single and dual. Using Cu as an alternative for Al was a drastic change for reducing the interconnect resistance. Technological challenges are being faced in the development of low dielectric-constant (*k*) dielectrics for reducing the parasitic capacitance of Cu interconnects given in a separate section. This chapter also describes the need of damage-free damascene process. Chapter 19 presents the overview for the BEOL technology for 90 nm and 65 nm/45 nm node and presents the low-*k* material for each technology node and metallization issues. This part also presents the process development tools for establishing the robust process and the characterization development for patterned low-*k* films. Both items will be necessary for future BEOL process development. Chapter 20 deals with the lithography for copper damascene fabrication. For the DD interconnect fabrication, interconnect trench and via holes are patterned in the ILD film sequentially. The multi-hard-mask (MHM) processes are innovated to avoid the oxygen plasma damage observed in the conventional photo-resist (PR) mask process, however, the line top spreading (LTS) and the line edge roughness (LER) are the main issues. The lithography sequence such as the via-first or the trench-first processes affects the misalignment margin between the lines and vias. We should design the lithography process sequence carefully by taking the material combination and the misalignment margin into considerations. Chapter 21 describes the necessity of diffusion barrier layer and gives the thermophysical characteristics which play significant role in the selection of metallurgy for barrier material selection. It also includes the barrier formation technique and its kinetics. Chapter 22 gives the details about low-*k* dielectrics such as basic properties and innovations of low-*k* material and process. Introduction of low-*k* dielectric film, especially porous film, is a key factor to reduce the parasitic capacitance (C_{int}) among multi-layer interconnects in leading-edge ULSI devices. The mechanical properties as well as the electrical reliability depend on the physical and chemical structure in the porous film. New innovations have been implemented such as EB/UV post-curing process and molecule structural design to stabilize the porous films. Reduction in the *k*-values of barrier dielectrics or capping films becomes important to decrease C_{int} effectively especially for 45/32 nm nodes. Chapter 23 presents the CMP review for Cu processing and includes general principals for Cu CMP and planarization. CMP (chemical mechanical polisher) is now regarded as an essential process among semiconductor device manufacturing processes. This CMP was first introduced by IBM in early 1980s for planarization of logic devices. Chemical mechanical polishing or planarization (CMP) aims at the removal of overburden copper after its electrochemical deposition. CMP primer objective is to achieve a global planarization of patterned surface. CMP appears to be the most promising pattern delineation technique. Chapter 24 gives electrochemical view of copper chemical mechanical polishing (CMP) in detail using different solutions such as ammonium hydroxide, nitric acid, peroxide and carbonate and sorbate-based solutions along with respective polarization curves. Chapter 25 describes copper post-CMP cleaning process as it is an important step in damascene technology utilizing copper on-chip interconnects. This process can significantly improve wafer surface quality by removing different defects and contaminations from copper layers and ILD. Etching of wafer surface plays an important role in abrasive particles' removal from the copper surface layer and interlayer dielectric. Copper layers are mostly contaminated with abrasive particles which determine strategy and procedure of post-CMP cleaning. HNO_3-based solutions are highly effective in copper layers etching. Simultaneous etching of silicon oxide and copper layers can be efficiently conducted in diluted HF solutions. However, copper etching in aqueous solutions is usually accompanied with a deposition of corrosion products (Cu(1) compounds). Formation of such deposition is being determined electrochemically subsequent to an etching process (at OCP) in some of the recommended post-CMP cleaning solutions. These results imply that one should be very cautious in choosing the most appropriate etching solution for Cu post-CMP cleaning.

Chapter 17
Introduction to Electrochemical Process Integration for Cu Interconnects

Takayuki Ohba

17.1 Introduction

This chapter focuses on advanced multilevel interconnects, contributed by distinguished authors in the following sections: Damascene Concept and Process Steps (Nobuyoshi Kobayashi), Advanced BEOL Technology Overview (Takashi Yoda and Hideshi Miyajima), Lithography for Cu Damascene fabrication (Yoshihiro Hayashi), Physical Vapor Deposition Barriers for Cu metallization PVD Barriers (Junichi Koike), Low-*k* dielectrics (Yoshihiro Hayashi), CMP for Cu Processing (Manabu Tsujimura), Electrochemical View of Copper Chemical Mechanical Polishing (CMP) (D. Starosvetsky and Y. Ein-Eli), and Copper Post-CMP Cleaning (D. Starosvetsky and Y. Ein-Eli).

Electrochemistry is the principal science involved in various stages of semiconductor manufacturing. In fact, applications involving electrochemistry have expanded from wafer processing to packaging, in terms of the processing equipment itself and in terms of the materials used. This is due to the growth of the semiconductor market to more than $300B as a result of the ubiquitous adoption of mobile devices, game sets, and audiovisual equipment in the multimedia marketplace [1]. In case of wafer processing, more than 1000 process steps are used and each step is precisely controlled by high resolution at an atomic level and manufacturability. According to this, the electrochemical processes must be controlled also by lithography, low-*k* dielectrics, etching, barrier metal, and CMP; taking integration suitability into account.

From a consumer perspective, cost and performance are the major concerns. These concerns are likely to be addressed by high-performance ultra-low-power LSI with lower impact on Greenhouse issues. There are two main competing technologies for implementing system LSIs: the System on a Chip (SoC) and System in a Package (SiP) design concepts. The SiP may include the three-dimensional

T. Ohba (✉)
The University of Tokyo, 7-3-1, Hongo, Bunkyo-ku, Tokyo 113-0033, Japan
e-mail: ohba@ducr.u-tokyo.ac.jp

Y. Shacham-Diamand et al. (eds.), *Advanced Nanoscale ULSI Interconnects: Fundamentals and Applications*, DOI 10.1007/978-0-387-95868-2_17,

integration (3DI) in near future. It provides similar functions, but the SiP technique is advantageous in terms of turnaround time and cost of manufacturing [2, 3]. Both SoC and SiP are, however, limited by the interconnects between transistors and between chips, due to the requirements of faster signal propagation and fabrication flexibility.

In the case of SoC, the processing speed of micro-processing units (MPUs) in such LSIs has been increased by using a combination of copper (Cu) interconnects and low permittivity (low-k) dielectrics [4–6]. The Cu interconnects are fabricated by electrochemical processes, i.e., electrochemical plating (ECP). For example, more than 12 levels of Cu interconnects are provided in 45 nm node MPUs. Cu/low-k interconnects have also been investigated for high-speed data transfer between chips in SiP and will be widely applied to 3DI as well as circuit board technology.

Unlike transistor speedwhich is limited only by the gate architecture, the operating speed of a system is determined by the delay time of interconnects between devices, global wiring, and chip sets. The total delay time (τ_{Total}) can be expressed in terms of gate delay (τ_{Gate}) and interconnects delay (RC) as $\tau_{Total} = \tau_{Gate} + RC$. At currently employed chip scales, such as 45 nm nodes and beyondthe delay time due to the interconnects is significant and is determined by the interconnect resistance (R) and wiring capacitance (C) following RC product based on the equation $V_{out}(t) = V_{dd}\{1 - \exp(-t/RC)\}$, where output voltage signal (V_{out}) saturates exponentially with RC. The V_{out} becomes $0.632V_{dd}$ at $t = RC$, where t is the time constant. In case of signal propagation, if there is a long distance between transistor blocks in a large chip operating at a high clock frequency for example, the signal transmission length becomes only 1.5 cm at the 100 ps of 10 GHz operation frequency, assuming specific dielectric constant $(\varepsilon) = 4$ and specific magnetic permeability $(\mu) = 1$ [7].

Accordingly, the major challenge in interconnects fabrication is how to achieve high-speed signal integrity and lowering RC product. Since the Joule's heating (Q) taking place with current flow, followed by $Q = RI^2$, accelerates both electro and stress migration phenomena, the RC product is anticipated to be low to ensure the reliability of interconnects. Thus, the wiring resistance is reduced by using Cu instead of aluminum (Al) because of its low electrical resistance (1.67 μΩ cm compared with 2.62 μΩ cm for Al). Also, the higher melting point of Cu (1083.4°C vs. 660°C) makes it more resistant to migration failures than Al [8]. Cu interconnects provided by the Damascene technique is now widely used [9].

Alternatively, for reducing the dielectric constant, porous materials such as fluorine-doped SiO_2 ($k = 4$–3.1) and silicon-oxy-carbide (SiOC: 3–2.6), (2.5) have been developed instead of SiO_2 (4.2) [10, 11, 12]. Their dielectric constants are varied by the dielectric polarization of atomic bond and neighborhood atoms, which are roughly correlated to trends observed for the electronegativity difference such as A–B bond ($\Delta\chi_P^{A-B}$), i.e., $\Delta\chi_P^{Si-O} = 1.7$, $\Delta\chi_P^{Si-C} = 0.7$, and $\Delta\chi_P^{O-F} = 0.5$. In addition, an effective dielectric constant decreases with increasing porosity of dielectrics, so-called ultra low-k, resulting in $k = 2$ at the porosity larger than 30%. However, their mechanical strength and diffusion resistance are low compared to pristine dielectrics, due to low film density. Thus, conventional CMP and etching processes cannot be used without optimization. An open pore may act as diffusion path, which causes moisture uptake and Cu diffusion. Therefore, in case of porous low-*k materials* application for Cu Damascene process, total optimization must be developed not only for the electrical characteristics but also for the materials suitability for patterning and metallization.

Usually, a Damascene interconnect consists of a barrier layer, seed/electrode Cu, and embedded electrochemically plated Cu (ECP-Cu) [13]. A thin layer of Ta or TaN (< 10 nm) is used as a barrier layer to prevent Cu diffusion. A leakage current between Cu lines occurs when there is excessive Cu diffusion into the dielectric layer, including the interface. Damascene interconnects have several features: lower level (local) interconnects have narrow line widths and spaces, such as 80 nm pitch (40 nm line width/40 nm space) at 45 nm node; and global interconnects connecting over long distances between transistor blocks, such as 1600 nm pitch, are 1–10 mm long. The leakage current along the line tends to increase due to residual defects such as particles and contamination, resulting in electrical short failures. The leakage current must be kept low, typically less than 10^{-10} A, even in ultra narrow lines.

A layer of seed Cu ranging from 20 to 100 nm in thickness is formed by sputtering and is used as an electrode for the ECP-Cu. The barrier layer must be controlled to be very thin, otherwise the relative volume of the Cu line decreases, and thus, the electrical resistivity becomes higher than the expected value according to the design pattern. Thus, there is a trade-off between barrier characteristics and electrical performance. A further challenge is to form the film structure of the barrier layer continuously, even at very small thicknesses on the order of several nanometers. In general, the film structure depends on the grain size and film thickness. An island-like structure is formed with decreasing film thickness, reaching the size of a grain. This island-like layer, however, has poor barrier properties, because the Cu atoms diffuse easily through the grain boundaries.

In the early stage of Cu wiring development, Cu dry etching, as well as Al patterning, was studied. However, it is very difficult to form fine and narrow Cu patterns without degrading the pattern profile and electrical resistivity. The main reason for this is the low vapor pressure of the Cu-chloride byproducts and the low passivity of the Cu surface which significantly differ from Al. Low-vapor-pressure compounds (byproducts) produced in the dry etching process tend to remain at the Cu surface. Those residual Cu byproducts impede the etching reaction, thus saturating the etching rate. When excessive Cu byproducts are generated, they form particles sticking on the wafer and/or the wall of the process chamber. Poor passivity causes thick Cu oxide formation in ambient oxygen and moisture, especially in the ashing process (resist removal by oxygen plasma).

The embedded process, i.e., Damascene technique, was therefore proposed to overcome these difficulties. In this process, Cu is embedded electrochemically into a trench and via pattern formed on a dielectric layer. The overfilled Cu is polished-off with chemical mechanical polishing (CMP). In the case of Cu filling using sputtering, void formation is unavoidable due to physical limitations such as the shadowing effect occurring at the top corners of the Damascene pattern. The shadowing is significant at the small feature size and high aspect ratio at around 180 nm node. Sputtering is however used in the formation of thin film layers such as barrier layers (Ta and TaN) and seed Cu for ECP-Cu with improving the ion flux and re-sputtering technique.

One challenge in the Damascene process is how to optimize the multilayer photomask structure and etching process sequence. The processes used and the number of steps involved in the Damascene method vary depending on the type of dielectric used, i.e., organic or inorganic materials. In particular, in dry etching processes, the photolithography resist masks and etch-stop layers for organic and inorganic materials are very different from each other. Because the etchant gases used in the dry etching process are different, for example, N_2/H_2 is used as an organic low-k etchant and C_xF_y is used as an inorganic etchant, the etching selectivity of the resist mask for the stopping layer is different. The trade-offs between critical dimension (CD) control of the trench patterning, sputtering damage, and etching residue are common concerns in the etching process. Since there is both a physical and a chemical trade-off, it takes time to optimize the etching profile and material combinations. Recently, an additional issue has been arisen regarding line-edge-roughness (LER) along the line pattern. An excess LER causes an electrical short between wires and increasing effective line resistance. For instance, in case of 40 nm narrow line, LER must be kept below 2 nm.

The interconnect processes described above involve many process steps, sometimes more than 600 (2/3 of the entire wafer process), for 10–12 multilevel interconnects used in manufacturing typical 300 mm wafers. Furthermore, all process steps, including deposition and patterning, are strongly interrelated; for example, the electrochemical Cu process is affected by the uniformity of the seed and/or barrier layer underneath. The uniformity of barrier layer is affected by chemical property of dielectrics surface. The chemical property of dielectrics surface is varied with hydrocarbon residues formed by etching process. The electrochemically deposited Cu in turn affects the reliability of the CMP process, as well as the electrical characteristics of the interconnects. Therefore, to improve current and future semiconductor manufacturing, it is necessary to study the process details and physical/chemical behaviors of not only electrochemical processes, but also other interactive processes. This is the reason for the huge level of financial investment and research carried out worldwide on semiconductors, both in industry and in academia.

References

1. Ohba, T.: Multilevel Interconnect Technologies in SoC and SiP for 100-nm Node and Beyond. Proc. of IEEE 6th International Conf. on Solid-State Integrated Circuit Technol. (ICSICT), Bing-Zong Li et al. Eds. 46 (2001)
2. Sakurai, T.: Interconnection from Design Prospective. Proceedings of Advanced Metallization Conference 2000, Edelstein. D.; Dixit, G.; Yasuda, Y.; and Ohba. T., Eds. Mat. Res. Soc. PA, 53 (2001)
3. Suga, T. and Otsuka, K.: A new era of system integration and packaging. J. Electron. Packaging **13(7)**, 621 (2000)

4. Bohr, M. T.: Interconnect scaling – The real limiter to high performance ULSI. IEEE IEDM Tech. Dig. 241 (1995)
5. Venkatesan, S.; Gelatos, A. V.; Misra, V.; Smith, B.; Islam, R.; Cope, J.; Wilson, B.; Tuttle, D.; Cardwell, R.; Anderson, S.; Angyal, M.; Bajaj, R.; Cappasso, C.; Crabtree, P.; Das, S.; Farkas, J.; Fiordalice, B.; Freeman, M.; Gilbert, P.V.; Herrick, M.; Jain, A.; Kawasaki, H.; King, C.; Klein, J.; Lii, T.; ReidK.; Saaranen, T.; Simpson, C.; Sparks, T.; Tsui, P.; Venkatraman, R.; Watts, D.; Weitzman, E. J.; Woodruff, R.; Yang, I.; Bhat, N.; Hamilton, G.; and Yu, Y.: A High Performance 1.8 V 0.20 μm CMOS Technology with Copper Metallization. IEEE IEDM Tech. Dig. 769 (1997)
6. Edelstein, D.; Heidenreich, J.; Goldblatt, R.; Cote, W.; Uzoh, C.; Lustig, N.; Roper, P.; McDevitt, T.; Motsiff, W.; Simon, A.; Dukovuc, J.; Wachnik, E.; Rathore, H.; Schultz, R.; Su, L.; Luce, S.; and Slattery, J.: Full Copper Wiring in a sub-0.25 μm CMOS ULSI Technology. IEEE IEDM Tech. Dig. 773 (1997)
7. Masu, K.: GHz Interconnect in ULSI, Technical Report of IEICE. The Inst. of Electronics. Inf. Commun. Eng. **101(1)**, 87 (2001)
8. Misawa, N.; Ohba, T.; and Yagi, H.: Planarized copper multilevel interconnections for ULSI applications. MRS Bull. XIX **8**, 63 (1994)
9. Kaanta, C. W.; Bombardier, S. G.; Cote, W. J.; Hill, W. R.; Kerszykowski, G.; Landis, H. S.; Poindexter, D. J.; Pollard, C. W.; Ross, G. H.; Ryan, J. G.; Wolff, S.; and Cronin, J E.: Dual Damascene: A ULSI Wiring Technology. Proc. of 8th Int. IEEE VLSI Multilevel Interconnection Conf. 144 (1991)
10. Kudo, H.; Yoshie, K.; Yamaguchi, S.; Watanabe, K.; Ikeda, M.; Kakamu, K.; Hosoda, T.; Ohhira, K.; Santoh, N.; Misawa, N.; Matsuno, K.; Wakasugi, Y.; Hasegawa, A.; Nagase, K.; and Suzuki, T.: Copper Dual Damascene Interconnects with Very Low-k Dielectrics Targeting for 130 nm Node. Proc. of IEEE Int. Interconnects Conf. (IITC) 270 (2000)
11. Ohba, T.: A Study of Current Multilevel Interconnect Technologies for 90 nm Nodes and Beyond, Fujitsu Sci. Tech. J. 13 (2002)
12. Nakai, S.; Kojima, M.; Misawa, N.; Miyajima, M.; Asai, S.; Inagaki, S.; Iba, Y.; Ohba, T.; Kase, M.; Kitada, H.; Satoh, S.; Shimizu, N.; Sugiura, I.; Sugimoto, F.; Setta, Y.; Tanaka, T.; Tamura, N.; Nakaishi, M.; Nakata, Y.; Nakahira, J.; Nishikawa, N.; Hasegawa, A.; Fukuyama, S.; Fujita, K.; Hosaka, K.; Horiguchi, N.; Matsuyama, H.; Minami, T.; Minamizawa, M.; Morioka, H.; Yano, E.; Yamaguchi, A.; Watanabe, K.; Nakamura, T.; and Sugii, T.: A 65 nm CMOS Technology with a High-Performance and Low-Leakage Transistor (A 0.55-μm^2 6T-SRAM Cell and Robust Hybrid-ULK/Cu interconnects for Mobile Multimedia Applications). IEEE IEDM Tech. Dig. 285 (2003)
13. Dubin, V. M.; Shacham-DiamandY.; Zhao, B.; Vasudev, P. K.; and Ting, C. H.: Selective and blanket electroless copper deposition for ultralarge scale integration. J. Electrochem. Soc. **144**, 898 (1997)

Chapter 18
Damascene Concept and Process Steps

Nobuyoshi Kobayashi

18.1 Damascene Process

A Cu interconnect was first introduced to manufacturing in 1997 [1]. The Damascene process has been used for Cu interconnect formation because of the difficulty in Cu dry etching. There are two Damascene processes, as shown in Fig. 18.1: single and dual. In the single Damascene process, trenches and via contacts (Vias) are formed one step at a time. In the dual Damascene process, they are formed simultaneously. Fewer steps make the dual process favorable for manufacturing, so it has been extensively used. There are several fabrication methods, such as via first and trench first, in the dual Damascene process. These methods depend on the lithography mask, the materials of interlevel dielectrics, and other factors. The typical steps of the via-first method are

1. deposition of interlevel and diffusion barrier dielectric films;
2. via formation in the interlevel dielectrics using lithography and dry etching;
3. trench formation in the interlevel dielectrics with resist-filled vias;
4. deposition of barrier metal film (Ta/TaN films, etc.) and the Cu seed layer by physical vapor deposition (PVD), ionized sputtering, etc.;
5. Cu filling of trenches and vias using electroplating;
6. removal of unnecessary Cu films around the trench region using chemical mechanical polish (CMP); and
7. deposition of interlevel dielectrics and diffusion barrier dielectric films.

Using Cu as an alternative for Al was a drastic change for reducing the interconnect resistance. Technological challenges are now being faced in the development of low dielectric constant (k) dielectrics for reducing the parasitic capacitance of Cu interconnects.

N. Kobayashi (✉)
Process Integration Technology, R&D, ASM, Japan
e-mail: n.kobayashi@asm.com

Y. Shacham-Diamand et al. (eds.), *Advanced Nanoscale ULSI Interconnects: Fundamentals and Applications*, DOI 10.1007/978-0-387-95868-2_18,

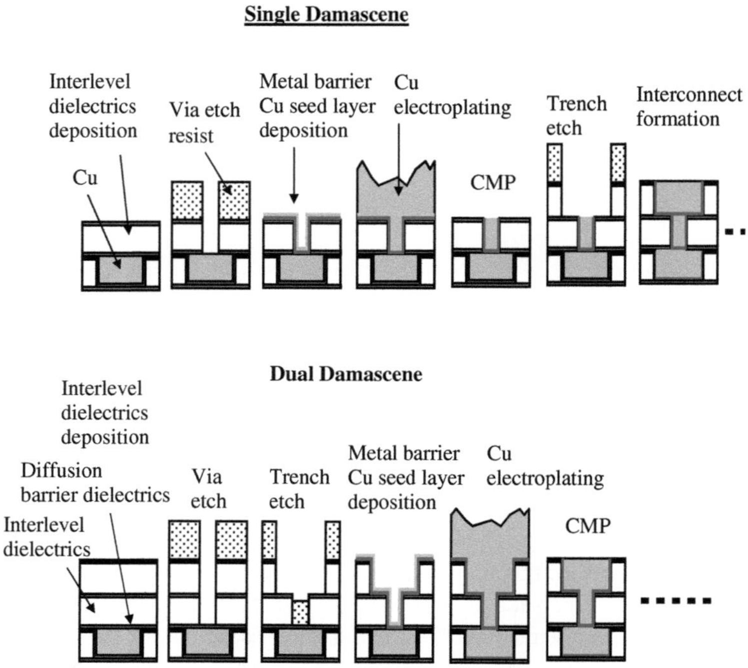

Fig. 18.1 Damascene process flow for the single (*upper sketch*) and dual Damascene methods (*lower sketch*)

Note that the dielectric films (insulators) consist of interlevel and diffusion barrier dielectrics, which prevent Cu diffusion in a transistor region.

Historically, the dielectric constant of interlevel and diffusion barrier dielectrics has been reduced by introducing new low-*k* materials. In the diffusion barrier layers of 65 nm node and beyond, SiCN or SiC film with a *k*-value of 4–6 was introduced as an alternative for an SiN film with *k*-value of 7–8. As an alternative for interlevel dielectrics of SiO_2, new low-*k* materials have been investigated for 90 nm node technology and beyond, which will be discussed in the next section. Generally speaking, low-*k* materials are of low mechanical and chemical resistance against interconnect processes, so measures for overcoming these problems have been developed.

In 45 nm node technology, an increase in the electrical resistivity of a Cu interconnect is remarkable (>10%) since interconnect width (~70 nm) approaches to the electron mean free path of Cu (~40 nm) [2]. Thin barrier metal (i.e., a few nanometers thick) effectively reduces Cu interconnect resistivity, because the electrical resistivity of Ta/TaN film is about 100 times higher than that of Cu film. Instead of PVD, chemical vapor deposition (CVD) and atomic layer deposition (ALD) have been intensively investigated for thin metal barrier formation. Since interconnect reliability strongly depends on the Cu grain size and Cu/Ta(TaN) interface properties, Cu electroplating with new additives and pre-treatment before metal and dielectrics deposition has been developed. CMP slurry has been developed for

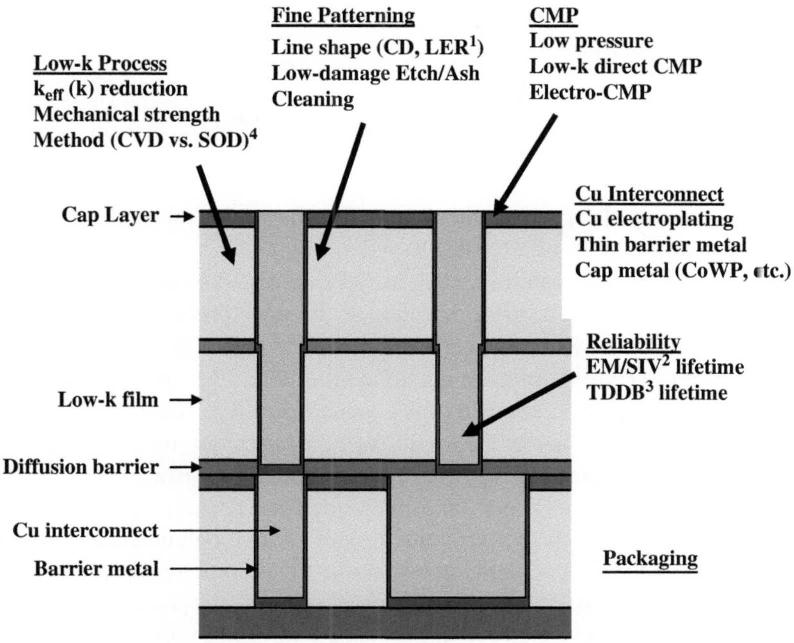

Fig. 18.2 Technological challenges for Cu interconnects in the 65 nm node technology and beyond

low-pressure CMP to suppress the mechanical friction on a low-*k* film without reducing productivity.

The technological challenges for 65 nm node Cu interconnects and beyond are summarized in Fig. 18.2.Developing the robust low-*k* materials ($k < 2.5$) is critical for reliable low-*k*/Cu integration. Low-*k* diffusion barrier dielectrics ($k < 4.0$) should be developed to obtain a target value of effective dielectric constant (k_{eff}) less than 2.7. Moreover, low-pressure CMP can be compatible with the mechanical and chemical properties of low-*k* material. Electro-CMP has been proposed to ultimately reduce the CMP pressure [3]. Since the interconnect pitch is reduced to less than 140 nm using ArF immersion lithography, conventional resist-mask and hard-mask patterning methods have been developed to reduce the induced damages to low-*k* material, critical dimension (CD) control of interconnect, and process cost. Needless to say, the interconnect reliability parameters such as electromigration (EM), stress-induced void (SIV), time-dependent dielectric breakdown (TDDB) are the most critical issues in the low-*k*/Cu integration. In addition, compatibility of process with conventional packaging is important because the mechanical strength of low-*k* material is reduced.

1. CD: critical dimension, LER: line edge roughness
2. EM: electromigration, SIV: stress-induced void

3. TDDB: time-dependent dielectric breakdown
4. SOD: spin on dielectrics

18.2 Low-*k* Integration Challenges

Many low-*k* materials, such as those that are shown in Table 18.1, have been investigated for use in Cu interconnects.

However, only a few materials, such as fluorine-doped oxide (SiOF) have been widely used in the manufacturing because of the poor process compatibility and interconnect reliability. Recently, carbon-doped oxide (SiOC) has been introduced for the most advanced 90 nm node manufacturing. For a 65 nm node and beyond, there are two approaches to lowering the *k*-value below 2.5. One uses organic polymer, such as aromatic ether or CF*x*, and the other uses porous material that contains nanometer-level pores in the film. Typical problems of using organic polymer are mechanical weaknesses, poor adhesion to dielectrics, and large thermal expansion coefficients. Hybrid dielectrics consisting of organic polymers and SiOC films were investigated for their potential in overcoming these problems [4]. Porous SiOC has been widely investigated due to its greater process compatibility than organic polymers. However, there are several problems with using porous SiOC in manufacturing. The deposition method of porous SiOC films is plasma-enhanced CVD (PECVD) and spin on dielectrics (SOD). A typical porous SiOC film formed by SOD method is porous methyl silsesquioxane (p-MSQ) film. Generally speaking, a porous low-*k* film has weaker mechanical and chemical properties than a non-porous one. Figure 18.3 shows the correlation between the dielectric constant and Young's modulus of p-MSQ films with film porosity [5].

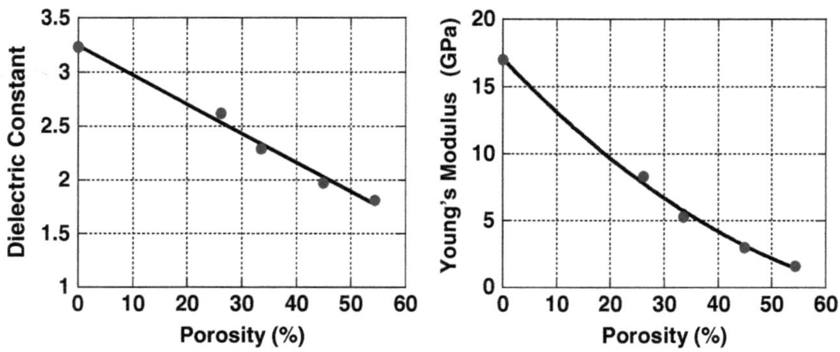

Fig. 18.3 The correlation between dielectric constant (*k*) and Young's modulus with porosity of p-MSQ film. Porosity was changed by the change of porogen content. Film of 500 nm thick was deposited by SOD

Nanometer-level pores were formed in the film by removing porogen after curing at 450°C. With an increase in porosity, the dielectric constant decreases linearly, however, Young's modulus decreases more rapidly. The nanometer-level pores in

Table 18.1 Low-*k* materials which have been investigated for Cu interconnects

	Interlevel dielectrics		Deposition method	*k*	Chemical structure	Thermal stability (°C)	Problems
Inorganic dielectrics	SiO_2		PECVD	4.0	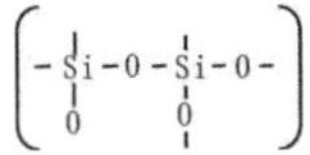	>100	–
	SiOF		PECVD	3.4–3.5		>750	F instability
	HSQ (hydrogen silsesquioxane)		SOD	2.8–3.0		400	Thermal stability O_2 plasma
	Porous silica		SOD	1.8–2.4		–	Water absorption
	SiOC		PECVD	2.8–2.9	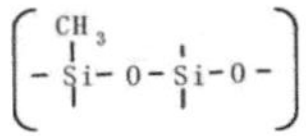	700	O_2 plasma
	Porous SiOC		PECVD	2.3–2.5		–	Mechanical strength
	MSQ (methyl silsesquioxane)		SOD	2.7–2.9	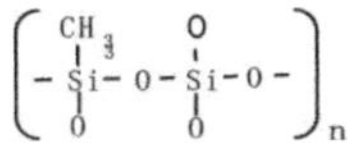	700	O_2 plasma
	Porous MSQ		SOD	1.8–2.5			Mechanical strength
Organic dielec-tric	Organic polymers	Polyimide	SOD	3.0–3.5	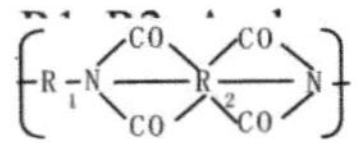	450	Mechanical strength O_2 plasma
		Aromatic ether, etc.	SOD	2.6–2.9	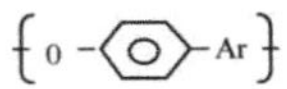	450	Adhesion Mechanical strength
		Parylane, etc.	SOD, plasma polymer-ization	2.2–3.0	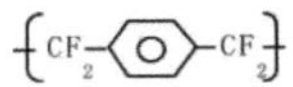	400	O_2-free Adhesion
		CFx, etc.	PECVD	2.0–2.4	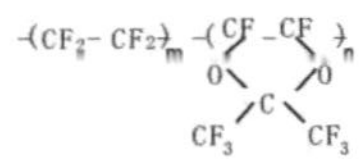	300	O_2 free Adhesion Thermal stability
	Borazine compound		PECVD	2.0–2.3	BN (C)	400	Fine patterning

these p-MSQ films were directly observed by three-dimensional tomography using TEM [6], and their average size was less than 2 nm. In this case, connected pores (open channels) were gradually increased when porosity was increasedas shown in Fig. 18.4.

In contrast to the porogen methodan MSQ film developed using the nano-clustering method is robust and has smaller pores [7]. On the other handp-SiOC PECVD film has pores smaller than the nanometer level (nano-pores), which cannot be directly observed by TEM. Thus, we have investigated the molecular modeling

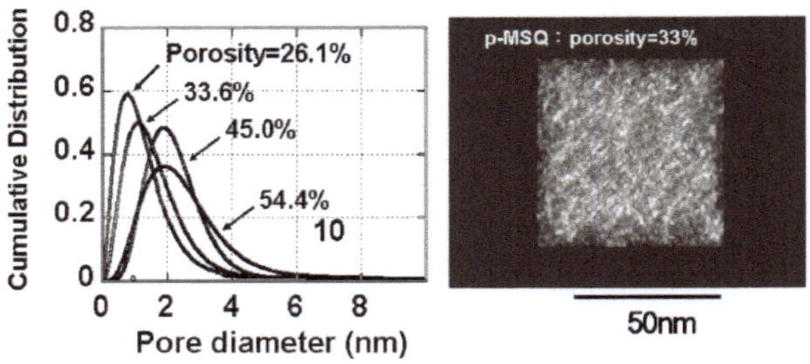

Fig. 18.4 Distribution of pore diameters measured by small angle X-ray scattering (SAXS) as a function of the p-MSQ film porosity (*left*). Pores of p-MSQ films observed by three-dimensional tomography of TEM (*right*) (the pores correspond to the *white area*)

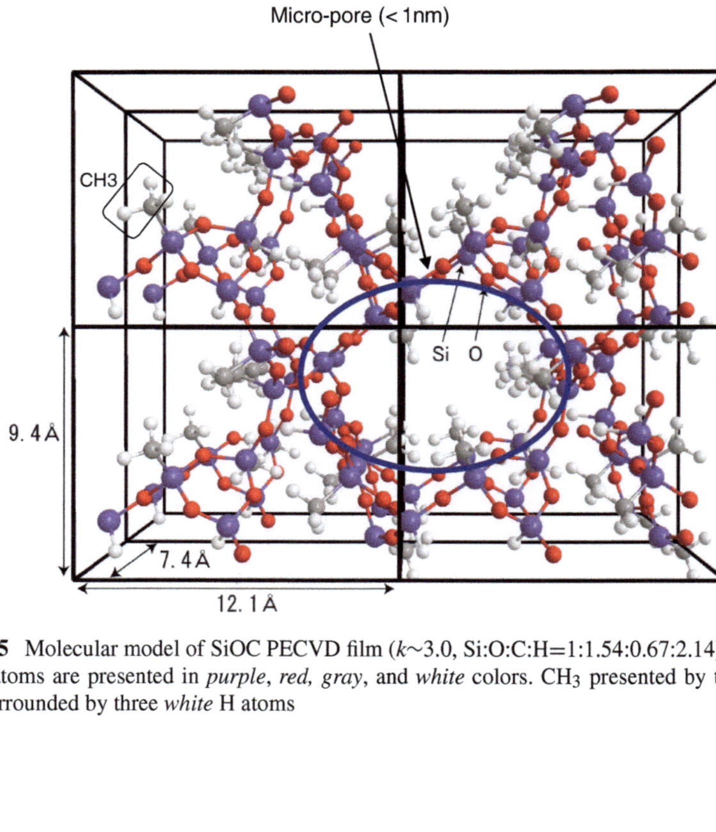

Fig. 18.5 Molecular model of SiOC PECVD film (k~3.0, Si:O:C:H=1:1.54:0.67:2.14). Si, O, C, and H atoms are presented in *purple*, *red*, *gray*, and *white* colors. CH_3 presented by the *gray* C atom surrounded by three *white* H atoms

of a p-SiOC PECVD film based on the film of chemical composition, Si:O:C:H. The method uses three-dimensional periodic molecular structures of tens of atoms in a unit cell to represent the polymer networks of p-SiOC film. Figure 18.5 shows an example of the p-SiOC PECVD model (k=3.0) that can explain the experimental dielectric constant, Young's modulus, and FT-IR absorption spectra [8]. In this case, there are small pores which consist of Si–O ring structures decorated with Si–CH_3 and Si–H chemical bonds. The ultra violet (UV) light and electron beam (EB) irradiation in proper conditions cause restructuring of the Si–O rings, which results in an improvement in the mechanical strength of the p-SiOC and p-MSQ films [9, 10].

In addition, the electrical properties of porous low-k materials are strongly influenced by the damages incurred during the fabrication process, such as CMP, metal deposition, and dry etching. For example, fluorine gas is incorporated into a p-SiOC film during dry etching using fluorine-based chemistry, so it can react with water in Cu electroplating, resulting in a void formation [11]. To fabricate Cu interconnects, the process must be compatible with using conventional wet chemicals.

18.3 Damage-Free Process

In order to use porous low-k dielectrics in manufacturing, we need to overcome the problems described in the previous section. In this section, we discuss new process for integrating porous SiOC to suppress mechanical, deposition, and plasma damages that occur when fabricating devices.

During CMP of a Cu film, strong frictional force between the polishing pad (slurry) and the Cu film often results in delamination from low-k dielectrics.

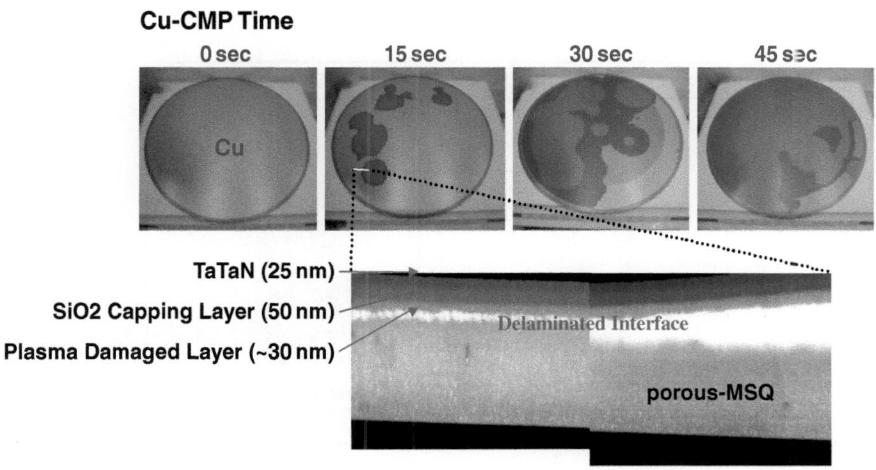

Fig. 18.6 Delaminated area of Cu film during CMP observed by cross-sectional TEM

Figure 18.6 shows an example of a delaminated area of a Cu film observed in a TEM cross section.

A delaminated interface exists between plasma-damaged layer and p-MSQ film.

The sample shown was a 300-mm blanket wafer with multi-layer films of electroplated Cu/Ta/TaN/SiO_2/p-MSQ (k=2.3). The SiO_2 surface was plasma treated before Ta/TaN PVD to improve adhesion. The delaminated area increased with the Cu-CMP time. It should be noted that the delaminated interface exists at the boundary region between the plasma-modified layer and the p-MSQ film, not at the Ta/TaN/SiO_2 interface. This indicates that cohesive delamination occurs in the p-MSQ film; therefore, the mechanical properties of a p-MSQ film, such as Young's modulus, correlate with CMP delamination [12]. There was no delamination when the modulus of a p-MSQ film was larger than 8 GPa at a conventional CMP pressure of 25 kPa.

ALD of TaN is promising for the thin metal barrier formation method to reduce interconnect resistance. However, the adhesion of ALD-TaN film with low-k and Cu films is worse than that of PVD-Ta/TaN film; therefore, an adhesion promoter (e.g., thin PVD-Ta) is needed to ensure SIV reliability [13]. Moreover, metal penetration into pores is a great concern when using porous low-k. Metal penetration is caused by diffusion of gaseous material into pores, which easily occurs when the porosity is increased to form open pores. To suppress metal penetration, the porosity, pore size, and distribution randomness should be reduced. However, this is a trade-off with reducing the dielectric constant, so pore-sealing technology is important to reliably integrate materials with high porosity (ultra low-k material, $k < 2.3$). Figure 18.7 shows an example of pore-sealing technology that suppresses metal penetration by using dense SiC films deposited on the sidewalls of p-MSQ films.

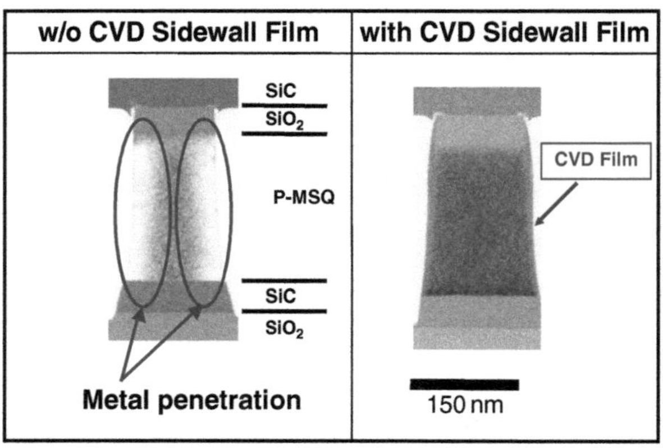

Fig. 18.7 Suppression of metal penetration during TaN ALD in p-MSQ film by using CVD sidewall film. *White* region corresponds to metal penetration (*left picture*)

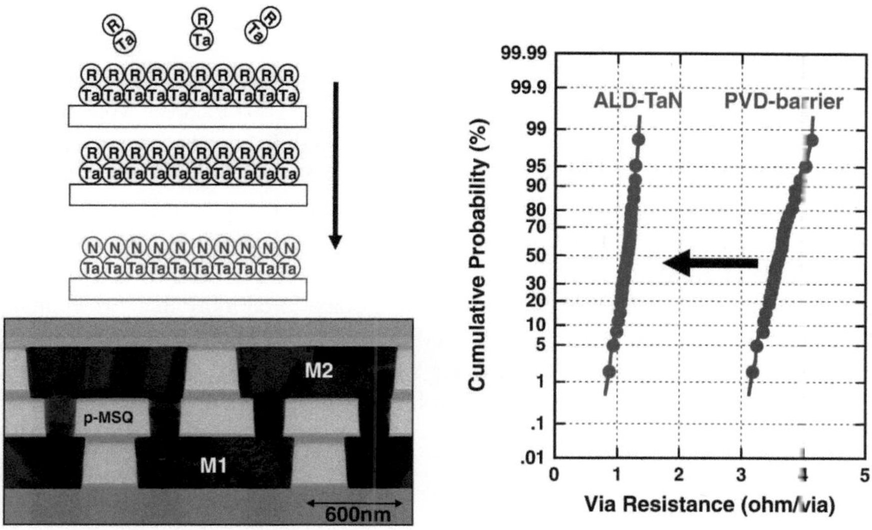

Fig. 18.8 Comparison of via resistance between ALD-TaN and PVD-Ta/TaN. Thin (~1 nm) ALD barrier metal can reduce via resistance from 3 to 1 ohm. *Left figures* correspond to the schematic concept of ALD-TaN and cross-sectional TEM picture of Cu interconnect using ALD-TaN barrier

Using a 1.5 nm thick ALD-TaN barrier metal with a pore-sealed p-MSQ film, the 200-nm via resistance was reduced approximately by 30% compared to a conventional PVD-Ta (TaN) barrier metal, as shown in Fig. 18.8.

The CH_xF_x chemistry for SiO_2 dry etching has been continuously used for porous SiOC film. Optimizing gas mixtures using CH_xF_y and a carrier gas (N_2, Ar, etc.) is necessary for precisely balancing CD control and reducing damage. Resist ash has been critical in fabricating devices using plasma. Ion-enhanced ash using various kinds of gases (O_2, NH_3, etc.) at low temperatures (~20°C) has been developed for this purpose. However, this kind of ash increases k-values when it is applied to porous low-k materials ($k < 2.5$). A new ash that uses a gas mixture of H_2/He at high temperatures (>250°C) has been developed to extend the conventional process of resist mask to a 65 nm node and beyond [14]. The rate of this new ash process had activation-type temperature dependence.

For temperatures higher than 250°C, the temperature dependence was approximately more than one order of magnitude higher than that of the conventional low-temperature process. High-temperature ash reduced the ash time; this resulted in no increase in k-value of a p-SiOC film. This is because extracting the Si–CH_3 chemical bonds from a p-SiOC film was reduced during plasma exposure, and radicals in the H_2/He plasma had little effect on the k-value. Another promising approach for suppressing plasma damage is dry etching using a hard mask. Since hard mask suffers from trench facet formation during dry etching, hard-mask materials need to have high selectivity in dry etching with low-k materials. Metal (or metal oxide) hard mask has been investigated for etching of ultra low-k materials [15].

18.4 Biography

Dr. Nobuyoshi Kobayashi was the program manager of BEOL technology at Selete (Semiconductor Leading Edge Technologies, Inc.) from 2001 to 2006, where porous low-*k* and Cu metal processes and their integrated module have been developed for logic 65 to 45 nm node applications. Before joining Selete, from 1980 to 2006, he worked with FEOL and BEOL process technologies for memory and logic applications at Hitachi Ltd and Renesas Technology Corp. He is now the executive director of R&D at ASM Japan K.K. and managed to develop plasma-enhanced CVD and ALD processes and tools. He authored and coauthored more than 80 technical papers and owned more than 30 patents in this field. He was the general chairman of 1999 Advanced Metallization Conference and 2000 IEEE-IITC. He graduated and received the degree of master and PhD in Physics from the University of Tokyo in 1977 and 1980, respectively.

References

1. Edelstein, E.; Heidenreich, J.; Goldblatt, R.; Cote, W.; Uzoh, C.; Lustig, N.; Roper, P.; McDevitt, T.; Motsiff, W.; Simon, A.; Dukovic, J.; Wachnik, R.; Rathore, H.; Schulz, R.; Su, L.; Luce, S.; and Slattery, J.: Full copper wiring in a sub-0.25 um CMOS ULSI technology. IEDM Tech. Dig. 773 (1997)
2. Steinhogel, W.; Schindler, G.; Steinlesberger, G.; and Engelhardt, M.: Size-dependent resistivity of metallic wires in the mesoscopic range. Phys. Rev. B, Condens. Matter **66**, 0754141 (2002)
3. Economikos, L.; Wang, X.; Sakamoto, X.; Ong, P.; Naujok, M.; Knarr, R.; Chen, L.; Moon, Y.; Neo, S.; Salfelder, J.; Duboust, A.; Manens, A.; Lu, W.; Shrauti, S.; Liu, F.; Tsai, S.; and Swart, W.: Integrated electro-chemical mechanical planarization (Ecmp) for future generation device technology. Proc. IEEE IITC, 233 (2004)
4. Kajita, A.; Usui, T.; Yamada, M.; Ogawa, E.; Katata, T.; Sakata, A.; Miyajima, H.; Kojima, A.; Kanamura, R.; Ohoka, Y.; Kawashima, H.; Tabuchi, K.; Nagahata, K.; Kato, Y.; Hayashi, T.; Kadomura, S.; and Shibata, H.: Highly Reliable Cu/low-k Dual-Damascene Interconnect Technology with Hybrid (PAE/SiOC) Dielectrics for 65 nm-node High performance eDRAM. Proc. IEEE IITC, 9 (2003)
5. Misawa, K.; Sone, S.; Shin, H. J.; Inukai, K.; Sudo, Y.; Kondo, S.; Yoon, B. U.; Tokitoh, S.; Yoneda, K.; Yoshie, T.; Ohashi, N.; and Kobayashi, N.: High-Modulus Porous MSQ Films for CU/Low-k Integration (keff<2.7). Extended Abstracts of International Conference on Solid State Devices and Materials (SSDM). 256 (2003)
6. Ogawa, S.; Shimanuki, J.; Shimada, M.; Nasuno, T.; Inoue, Y.; and Mori, H.: 3-Dimensional TEM Stereo Observation Technology for Characterization of Pores in Low-k Film. Proc. IEEE IITC, 100 (2003)
7. Nakamura, T.; and Nakashima, A.: Robust Multilevel Interconnects with a Nano-clustering Porous Low-k (k<2.3). Proc. IEEE IITC, 175 (2004)
8. Tajima, N.; Hamada, T.; Ohno, T.; Yoneda, K.; Kobayashi, N.; Hasaka, T.; and Fnoue, M.: First-Principle Molecular Model of PECVD SiOCH Film for the Mechanical and Dielectric Property Investigation. Proc. IEEE IITC, 66 (2005)
9. Yoneda, K.; K.; Kato, M.; Kondo, S.; Kobayashi, N.; Matsuki, N.; Matsushita, K.; Ohara, N.; Fukazawa, A.; and Kimura, T.: Impacts of UV Cure for Reliable Porous PECVD SiOC Integration. Proc. IEEE IITC, 220 (2005)

10. Yoda, T.; Nakasaki, Y.; Hashimoto, H.; Fujita, K.; Miyajima, H.; Shimada, M.; Nakata, R.; Kaji, N.; and Hayasaka, N.: Structural Studies of High-Performance Low-k Dielectric Materials Improved by Electron-Beam Curing. Jpn. J. Appl. Phys. **44**, 75 (2005)
11. Jacobs, T.; Brennan, K.; Carpio, R.; Mosig, K.; Jing-Cheng Lin; Cox, H.; Mlynko, W.; Fourcher, J.; Bennett, J.; Wolf, J.; Augur, R.; and Gillespie, P.: Voiding in Ultra Porous Low-k Materials Proposed Mechanism, Detection and Possible Solutions. Proc. IEEE IITC, 236 (2002)
12. Kondo, S.; Nasuno, T.; Ogawa1, S.; Tokitou, S.; Yoon, B. U.; Namiki, A.; Sone, Y.; Misawa, K.; Yoshie, T.; Yoneda, K.; Shimada, M.; Sone, S.; Shin, H.J.; Ohashi, N.; Matsumoto, I.; and Kobayashi, N.: The Delamination Mechanism of Porous Low-k Film during the Cu CMP process. Extended Abstract SSDM, 250 (2003)
13. Higashi, K.; Yamaguchi, H.; Omoto, S.; Sakata, A.; Katata, T.; Matsunaga, N.; and Shibata, H.: Highly reliable PVD/ALD/PVD stacked metal structure for 45 nm-node copper dual damascene interconnects. Proc. IEEE IITC, 6 (2004)
14. Matsushita, A.; Ohashi, N.; Inukai, K.; Shin, H.J.; Sone, S.; Sudou, K.; Misawa, K.; Matsumoto, I.; and Kobayashi, N.: Low Damage Ashing using H_2/He Plasma for Porous Ultra Low-k. Proc. IEEE IITC, 147 (2003)
15. Fox, R.; Hinsinger, O.; RichardE.; Sabouret, E.; Berger, T.; Goldberg, C.; Humbert, A.; Imbert, G.; Brun, P.; Ollier, E.; Maurice, C.; Guillermet, M.; Monget, C.; Plantier, V.; Bono, H.; Zaleski, M.; Mellier, M.; Jacquemin, J.-P.; Flake, J.; Sharma, B.G.; Broussous, L.; Farcy, A.; Arnal, V.; Gonella, R.; Maubert, S.; Girault, V.; Vannier, P.; Reber, D.; Schussler, A.; Mueller, J.; and Besling, W.: High performance k=2.5 ULK Backend Solution Using Improved TFHM Architecture, Extendible to the 45 nm Technology Node. IEDM Tech, Dig., 81 (2005)

Chapter 19
Advanced BEOL Technology Overview

T. Yoda and H. Miyajima

19.1 Introduction

Performance interconnects degrade with shrinking dimensions of the ULSI device. The scaling rules below depict these dimensions. To develop high-performance ULSI devices, minimizing the *RC* delay of interconnects is crucial. Decreasing power dissipation is also critical for high-performance system-on-chip (SOC) devices.

Trends of effective dielectric constant (k_{eff}) in ITRS2003 are shown in Fig. 19.1 [1]. The dotted line in this figure represents the targeted k_{eff} value in ITRS2002. The target value in each generation increased since ITRS2003. This is because we cannot obtain the low-*k* dielectric material with a *k*-value, which satisfies the k_{eff}-value of less than 3.0. Additionally, we can only obtain materials of low mechanical strength. Based on this reason, the k_{eff}-value region of less than 3.0 is called the "Red Brick Wall," which we realize as having no actual solution. In this case, the ideal resolution would be a device maker who could mass produce a robust interconnects technology with ultra low-*k* materials.

Next, we discuss integration issues for Cu/low-*k* processes. Figure 19.2 depicts the schematic of Cu/low-*k* (especially porous low-*k*) integration issues. The primary issues in this process are the improvement of low-*k* materials for trench and via dielectric. Other issues associated with this process are crack resistance, thermal stability, and Young's modulus. The metal CMP process also affects the low-*k* integration (i.e. large shear stress during the CMP process causes delamination at the interface and higher friction force induces defects such as scratches and particle contamination). Additionally, as the number of wiring levels increase, the increase of mechanical strength of low-*k* materials and the adhesion strength of each interface

T. Yoda (✉)
Advanced ULSI Process Engineering Department,
Process & Manufacturing Engineering Center,
Toshiba Corporation Semiconductor Company,
8 Shinsugita-cho, Isogo-ku, Yokohama 235-8522, Japan
e-mail: takahashi.yoda@toshiba.co.jp

Y. Shacham-Diamand et al. (eds.), *Advanced Nanoscale ULSI Interconnects: Fundamentals and Applications*, DOI 10.1007/978-0-387-95868-2_19,

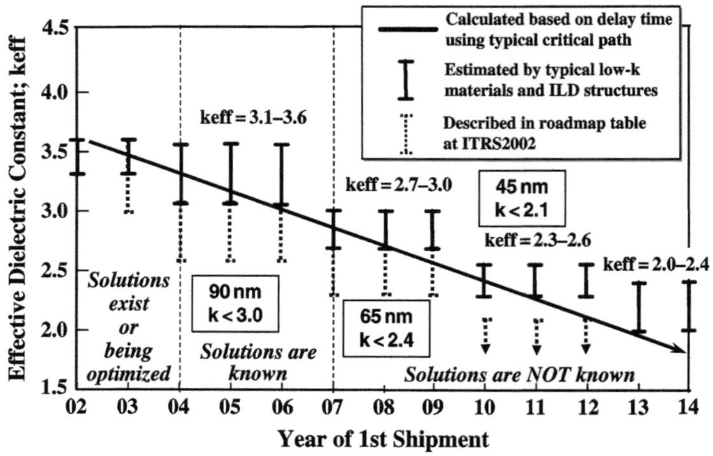

Fig. 19.1 Trends in effective dielectric constant (k_{eff}) in the ITRS2003 [1]

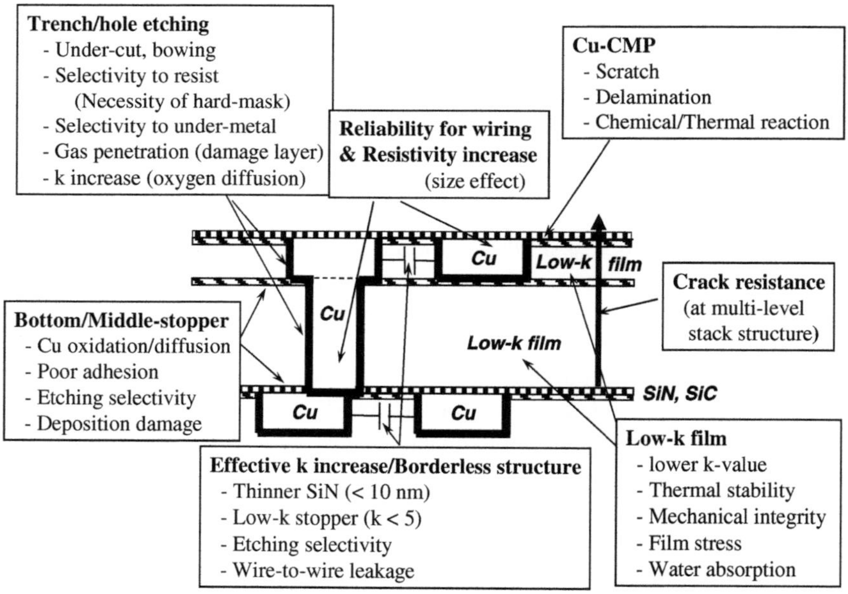

Fig. 19.2 Copper Damascene module issues for porous materials

should be significant. The number of wirings for 45 nm and 32 nm technology nodes will be at most 10 or more.

Integration issues related to the reactive ion etching (RIE) process are also important. The hard-mask process using SiN and SiO_2 is necessary in regard to the patterning process for 130 nm node and beyond. This is because conventional resist mask is insufficiently strong as RIE mask. The RIE processes with these hard masks can cause the trench bottom and sidewall damage of low-*k* materials. The damaged layers in porous low-*k* materials are extremely crucial (i.e., water adsorbs at the damaged layer and then penetrates). The penetrated water or the Si–OH bond in low-*k* materials increases the *k*-value. The process gases used in metallization processes easily penetrate into porous low-*k* materials. In order to improve step-coverage in deposition of barrier layer for Cu diffusion in dielectric materials, conventional physical vapor deposition (PVD) is replaced by chemical vapor deposition (CVD) or atomic layer deposition (ALD). In this case, gaseous precursors of CVD and ALD depositions penetrate inside the pores of low-*k* materials, increase the *k*-value, and degrade the reliability of wiring metals.

19.2 Low-*k* Materials for Each Technology Node

As shown in Fig. 19.3 , there are two kinds of solutions to decrease a dielectric constant (k) of materials.

First, decreasing the density of the films is effective. To decrease the *k*-value as 2.8–3.1, a hybridized material approach is commonly used. Substituting oxygen atoms for substitution group such as $-CH_3$ makes a steric hindrance in the materials. Typical materials are hydrogen silsesquioxane (HSQ) and methyl silsesquioxane (MSQ). To realize the *k*-value less than 2.5, porous materials are necessary. By introducing several nanometer size pores into the low-*k* films, *k*-value can dramatically decrease.

The second solution is to decrease the polarizability of the film. The fluorine-doped silicon oxide (SiOF) films were proposed as intermetal dielectric films with low dielectric constant. Typical *k*-value using SiOF is 3.4–3.8. Substituting oxygen atoms for fluorine atoms lowers the dielectric constant due to the decrease of the concentration of highly polarizable Si–OH bonds [2] and the change of the bonding structure of the film to a less polarizable geometry [3].

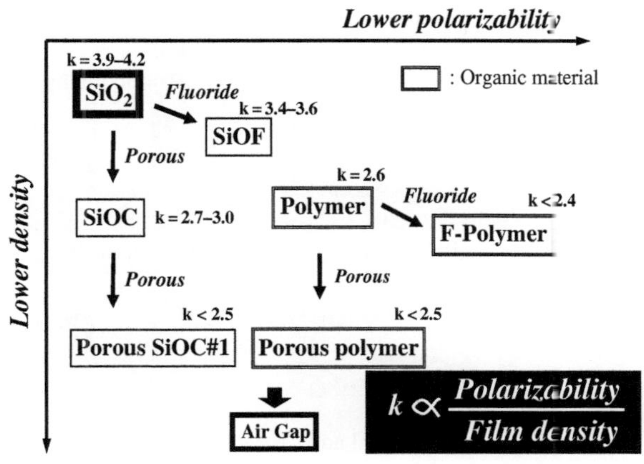

Fig. 19.3 The variety of low-*k* materials. *k* increases with polarizability and density

Organic film (k = 2.6–3.0) is another approach whereas the combination of organic film and fluorination produces the film with k-value of 2.4–2.9. Typical organic material is poly-arylene-ether (PAE), polyimides and benzocyclobutene (BCB). Fluorinated organic materials commonly used are fluorinated PI, Parylene-F, and CFx. In this chapter, we consider SiOF low-k film as an example of fluorinated materials and MSQ and SiOC low-k-hybridized films. PAE films for the application in Damascene structure are also reviewed.

19.3 Metallization Issues

Besides the above mentioned decrease of RC delay by low-k integration, other issues exist. The resistance increase of Cu wiring is an issue in regard to the wiring shrinkage. As shown in Fig. 19.4, the resistivity of Cu wires with a line width less than 100 nm increases remarkably.

This phenomenon is called "Size Effect" [4–6]. As the wiring dimension shrinks, an inelastic scattering of electron at the sidewall of the Damascene structure becomes dominant. In particular, for the intermediate lines in which the effect of resistivity of wiring (R) and RC delay is dominant, decreasing the wiring resistivity is difficult. The mean free path (MFP) of an electron in a material (λ) is related to this phenomenon. The size-effect-related resistivity is determined by the ratio between λ_{MFP} and the line width (W_{Line}). λ_{MFP} value of the conduction electrons in

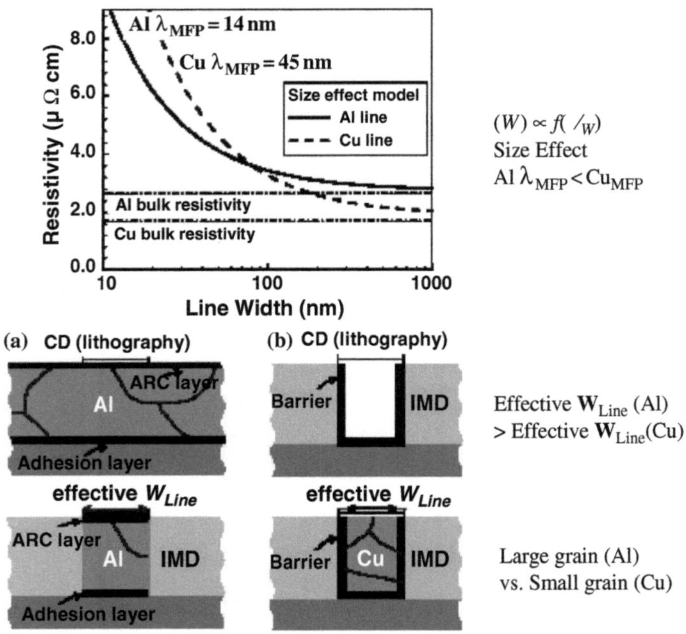

Fig. 19.4 Comparison of Al and Cu wiring [4]

Al is much smaller than that of Cu [6]. When the wiring width becomes three orders of the electron MFP, surface scattering and grain boundary scattering become dominant factors and the interface roughness between barrier metal and Cu and the texture of Cu strongly affect the increase of wiring resistivity. For the above-mentioned reason, the resistivity curves for Cu Damascene wiring and for Al RIE wiring cross at less than 100 nm dimension as shown in Fig. 19.4 [4]. Furthermore, the Al wiring process by RIE leads to a larger effective cross-sectional area of highly conductive Al material as shown in Fig. 19.4a. While in Cu Damascene wiring scheme, an effective line width W_{Line} becomes smaller than Al wiring, because Cu wiring is surrounded by the barrier metal. Additionally, after heat treatment and recrystallization, grain size in Al wiring is larger than within a limited trench of Cu Damascene structure.

Stress-induced voiding (SiV) is also a serious concern for wiring reliability [7–9]. It has been reported that two sites exist for void formation around via hole of Cu wiring. The first one is under the via connected to wide lines (Fig. 19 5). It is attributed to the diffusion of supersaturated vacancies in wide lines made by electrochemical deposition (ECD) of Cu. Wiring design improvement approach such as multiple via structure is one of the solution. The second one is the pull up of Cu materials inside the via hole connected to the wide upper lines. Improved adhesion strengths of Cu on barrier metals can improve the pull up mode degradation.

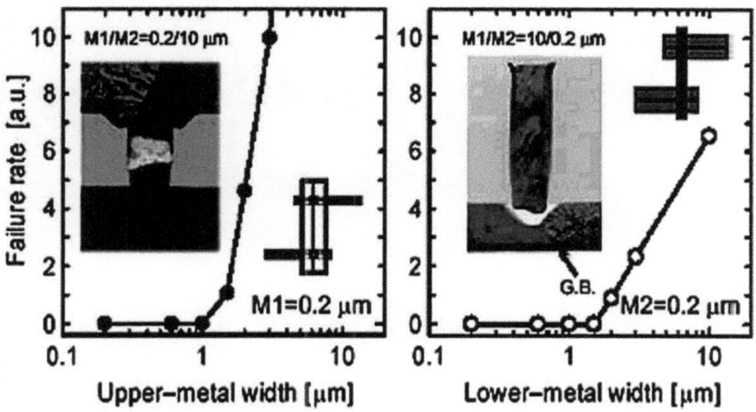

Fig. 19.5 TEM analysis of pattern dependence of SiV failure mode [8]

19.4 BEOL Technology for 90 nm Technology Node

19.4.1 Introduction

For the previous 130 nm technology node, most LSI manufacturers used PECVD–SiOF film ($k = 3.4$–3.6) as the first low-k material [10, 11]. The SiOF film is an inorganic material and it was easy to replace the original PECVD–SiO_2 film with it. However, free F liberated from SiOF film caused serious issues concerning film delamination, making it difficult to ramp up production. Alternative SiOF/SiN low-*k* materials, C-doped SiO (SiOC) ($k<3.0$) and SiC(N) ($k<5.0$), were developed for 90 nm technology node low-*k*/Cu interconnects [12, 13]. Although SiOC material resolved the issues associated with liberated F, it caused new problems because of the organic characteristics and lower film density. Successful realization of interconnects requires several newly developed technologies: low-*k* SiOC/low-*k* SiC(N) deposition technique, low-damage RIE/ashing process, precise control for CMP dishing/thinning, etc.

A typical interconnect structure for 90 nm technology node [14, 15] is shown in Fig. 19.6. Maximum metal layers consist of 11 levels or more. Focusing on the characteristics of low-*k* SiOC and SiC(N) materials, this chapter discusses the integration issues associated with using low-*k* SiOC.

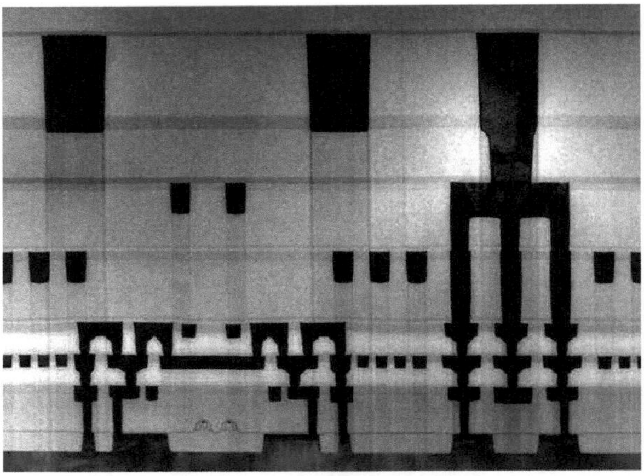

Fig. 19.6 Cross section SEM of 90 nm node 6 metal level interconnect scheme. The inter-level dielectrics are: SiOC ($k = 2.9$) for M2 and M3, SiOF ($k = 3.4$) for M4 and M5, and M6 is with SiO2 ($k = 4.0$) [14]

*19.4.2 Interconnects Structure and Low-*k *Film Characteristics*

The dielectric constant of the material is generally proportional to the film density and polarizability. F doping into SiO_2 decreases *k*-value of SiO_2 because of lower polarizability of F than O. On the other handC doping decreases film density of SiO_2 film. As C in the film exists mainly in the methyl group and creates bond with Si (CH_3-Si), SiOC material has both the conventional inorganic characteristics and the organic characteristics with lower film density. These characteristics pose new problems. The plasma process for the photomask removal damages the film and moisture absorption is observed. Water uptake increases k-value and degrades the reliability of Cu wiring. In addition, since density, mechanical strength, and adhesion strength of the film decrease, an advanced CMP and IC packaging techniques are required.

SiOC film is deposited by the PECVD method using organic silane precursor and oxidizer. Trimethylsilane (3MS, Me_3-Si-H), tetramethylsilane (4MS, Me_4-Si),

Table 19.1 Characteristics of SiOC and SiCN films

	SiOC	SiCN
k-value	2.9	4.9
RI	1.42	1.93
Hardness (GPa)	1.9	12.2
Modulus (GPa)	11.0	83.7
Stress (MPa)	42	–227

tetramethyl-cyclo-tetrasiloxane (TOMCATS^{TM} $(MeHSiO)_4$) or another organic silane is used as a SiOC precursor. PECVD N-doped SiC film was developed as an alternative for SiN ($k = 7.0$) barrier film. N-doped SiC (SiCN, $k = 4.9$) is widely used because of its excellent performance: good Cu diffusion barrier characteristics, high adhesion strength to Cu, high etching selectivity to SiOC, and low leakage current. As a CVD precursor of SiCN, 3MS or 4MS is generally used with NH_3. Table 19.1 shows film characteristics of typical SiOC and SiCN films.

SiCN deposition requires pretreatment using NH_3 plasma in order to remove CuO on Cu wiring. This pretreatment provides high adhesion strength between SiCN and Cu and robust reliability of Cu wiring, mainly with respect to TDDB (time-dependent dielectric breakdown) and EM (electro-migration) performances.

19.4.3 Dual Damascene Process

The dual Damascene scheme for 90 nm node low-*k*/Cu interconnects involves several major technical choices: via-first or trench-first scheme, and with or without SiO_2-cap structure. The via-first scheme has the advantages of very simple procedure and wide margin for misalignment of lithography. However, it requires several actions for plasma damage reduction during photoresist ashing process and for the suppression of shell-like residue around via hole. Although SiO_2 cap makes it easy to control the failure of CMP, it emerged as one of the causes of the resist poisoning. Next, newly developed techniques to resolve these issues are discussed. In addition, the advantage of SiCN barrier for EM performance is shown.

1. Via-first dual Damascene patterning

A via-first dual Damascene approach without middle etch stop for intermediate layer is explained in this chapter. For the via-first scheme, undesirable crown-shaped fence is formed when a conventional resist mask process with anti-reflective layer (ARL) is used (Fig. 19.7a). The ARL remaining in the hole acts as a mask in the trench-etching process and forms a shell-like residue. Due to the poor etching selectivity of ARL to photoresist mask, it is very difficult to etch back the ARL deeply in the hole.

Fig. 19.7 DD shape comparison by cross second SEM images [16]

The stacked mask process (S-MAP) for the trench-etching process can resolve the above issue [16]. After via hole etching was completedthe via hole was filled with bottom resist material followed by top resist/SOG coating. The bottom resist can be etched off deep in the hole, because of its good selectivity to the SOG mask. Excellent etch profile was obtained by S-MAP as shown in Fig. 19.7b .

In the Cu/low-*k* Damascene module scheme, care must be taken regarding the photoresist ashing process. SiOC film can be easily attacked by the O_2 plasma ashing process. The damaged layer causes serious problems, such as increasing of wire-to-wire capacitance value, leakage current increase due to moisture absorption, and peeling off of the stacked films by outgas. Therefore, the damaged layer must be etched off by chemicals, such as HF or NH_4F. Nevertheless the CD (critical dimension) value varies after the etch-off process. Minimizing the damaged layer thickness in the plasma ashing process is crucial. Notable improvement was observed by using the H_2O gas ashing process with better CD control. A cross section of the via hole pattern shows good etch profiles.

2. Capping SiO_2 process

A SiO_2/SiOC stacked film has been investigated for reliable and robust Cu/low-*k* process module. SiOC with capping SiO_2 layer exhibits good electrical performance of wire-to-wire leakage current. Direct SiOC CMP process causes micro-scratches and wire-to-wire leakage current.

However, DUV photoresist can be badly poisoned by using the stacked SiO_2/SiOC structure. N–H species through vias during trench lithography can cause the poisoning [17]. This issue is typically observed as undeveloped resist residue at the trench pattern over the vias facing the via-free zone (Fig. 19.8). Various solutions have been studied to eliminate issues associated with resist poisoning. These include optimizing pretreatments for the SiCN deposition, removing N-containing materials from the capping-SiO_2 deposition, and reducing the sensitivity of the resist to N–H species.

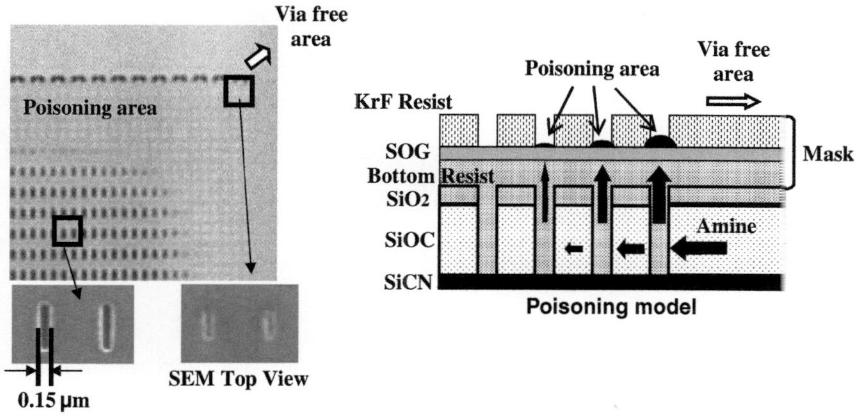

Fig. 19.8 Photoresist poisoning process. Amine in low-density SiOC film moves to via holes due to the presence of dense SiO_2 cap layer [17]

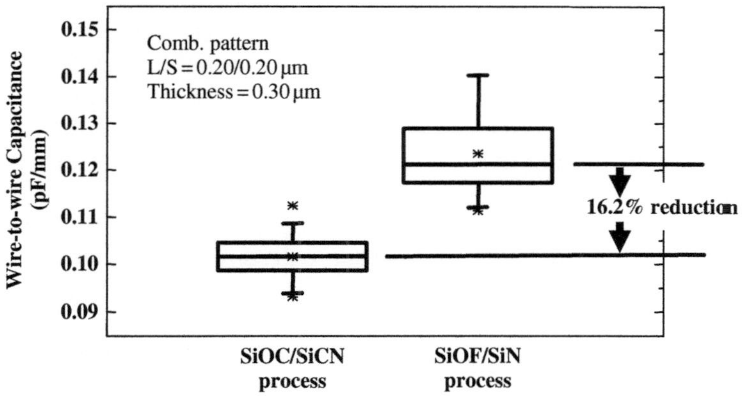

Fig. 19.9 Impact of the SiOC/SiCN introduction on wire-to-wire capacitance [17]

Figure 19.9 shows the impact of introducing SiOC/SiCN material for 90 nm node BEOL. Capacitance reduction of 16.2% was observed.

3. Barrier/etch-stop SiCN process

For the barrier/etch-stop layers, SiN ($k = 7.0$) film was replaced by PECVD SiCN film ($k = 4.9$) to reduce effective k-value. In addition, SiCN film can stabilize the Cu/SiCN interface and exhibits better EM performance. Interface between Cu and barrier/etch-stop layer could be the dominant diffusion path [18].

According to the cumulative EM failure distribution MTF (mean time to failure) for the sample with PECVD SiC was approximately two times as long as that for the sample with SiN. EM-induced void location was investigated and all of the voids were nucleated over via hole. Figure 19.10 shows cross-sectional TEM images of EM test structures containing p-SiC and p-SiN cap layers. No significant difference

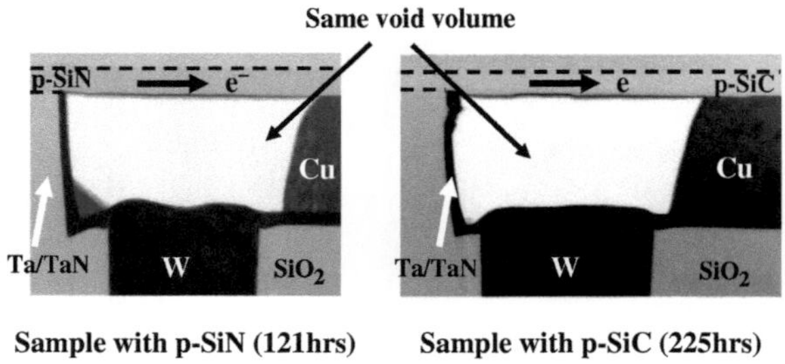

Fig. 19.10 Cross second TEM images of the test structure after EM testing. EM induces void nucleated at the cathode edge and grew toward the anode direction for the same void volume [19]

was observed in EM-induced void between two samples. The mechanisms of these failures are confirmed to be the same.

A modified edge liftoff test (m-ELT) was conducted to study the interfacial adhesion between Cu and the capped barrier/etch-stop layer. The critical stress intensity factor (K_{IC}) of Cu/SiCN is found to be better than that of Cu/SiN. For both SiCN and SiN samples, MTF and line width shows linear relationship. The dominant diffusion path is the interface between Cu and the barrier/etch-stop layer, not the interface between TaN/Ta and Cu. Activation energy values for the samples with p-SiC and p-SiN were almost the same. Therefore, the pre-exponential factor (D_0) of diffusivity (D) is closely related to the strength of adhesion. D_0 for SiCN is lower than that for SiN. Consequently, MTF of the sample with SiCN is confirmed to be longer than that with SiN [19].

It was found that MTF of Cu interconnects with p-SiC cap is approximately two times as long as that with p-SiN probably due to the difference of adhesion between Cu and the cap layer. In addition, it is also found that dominant diffusion path is the interface between Cu and p-SiN/p-SIC for Cu interconnects [19].

PECVD SiCN film shows excellent performance for k reduction, Cu diffusion barrier, high etching selectivity to SiOC, low leakage current, and realization of the robust reliability of Cu wiring because of high adhesion strength to Cu.

19.5 BEOL Technology for 65 nm/45 nm Technology Node

19.5.1 Introduction/Integration Issues

For 65 nm technology node and beyondin order to achieve an *RC* delay target and to minimize cross talk, high-performance system-on-chip (SOC) technologies require a Cu/ultra low-*k* dielectric with lower dielectric constants than those used in 90 nm generation. General target k-value is less than 2.6. Interlayer dielectrics satisfying this goal include spin-on dielectrics and C-doped SiO_2 [20–23] whose k-value may be reduced through the introduction of pore. Porous materials have low plasma resistance, high moisture uptake, high gas, and metal penetrations, degraded low mechanical strength and poor adhesion. These material issues lead to many module problems in LSI. To overcome material issues, three key technologies are important. The first one is material design engineering: small and tight pore size distribution, high plasma resistance, and high mechanical strength. The second one is process optimization and improvement: low damage RIE/ashing, low pressure CMP, advanced cure techniques with EB (electron beam) [24–26] or UV (ultra violet) [27–29], pores sealing, and plasma damage restoration techniques. The third one is interface engineering targeting to high adhesion strength and low leakage current. Low mechanical strength and low adhesion strength cause film delamination failure and cracking during CMP or high-temperature curing steps, and hence interconnect reliability problems. Spin-on-dielectric (SOD) low-*k* materials need high thermal budget (typically 60 min at 370–425°C for methyl-silsesquioxane (MSQ)) and they degrade the reliability of Cu wiring. Advanced post-cure technique improves mechanical strength, adhesion strength, and plasma resistance for porous low-*k* materials when plasma CVD and the spin-on technique are used. And it decreases thermal budget for SOD curing.

For 65 nm node onwardseveral kinds of module structure and process scheme are being developed to achieve reliable interconnects. Hybrid schemes were proposed for 65 and 45 nm node devices: porous SiOC (trench layer)/dense SiOC (via layer) stack and porous poly-arylene (PAr, for trench layer)/porous SiOC (via layer). Their main advantage is good controllability of DD shape by high etch selectivity [15, 30–34]. Figure 19.11 shows TEM image of 65 nm node BEOL using PAr/SiOC hybrid scheme [32].

Fig. 19.11 Cross section SEM of 65 nm node 6 metal level interconnect scheme. The inter-level dielectrics are: Par ($k = 2.65$)/SiOC ($k = 2.5$) for M2, M3, and M4, SiOC ($k = 2.9$) M5 and M6 is with SiO2 ($k = 4.0$) [14]

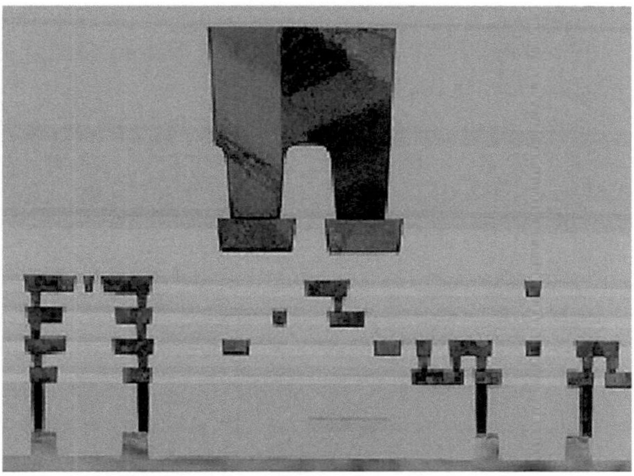

In this chapter, the impact of advanced post-cure method on the characteristics of porous material and low-*k*/Cu module integration are discussed.

19.5.2 Application of Advanced EB Curing Process

Table 19.2 shows the typical low-*k* materials for 65 and 45 nm node interlayer dielectrics (ILDs). Most of them contain pore structure and exhibit very low mechanical strength. Post-cure technology with EB or UV irradiation is examined

Table 19.2 Typical low-*k* materials for 65 nm node module and beyond

Low-*k* material	Deposition method	Component	k-Value	Modulus (GPa)	Average pore size (nm)
Porous MSQ	Spin-on	Si, O, CH	2.0–2.5	2.0–5.0	< 2.0
Porous SiOC	PECVD	Si, O, CH	2.0–2.5	2.0–5.0	< 2.0
Porous HSQ	Spin-on	Si, O, H	2.0	3.0	< 3.0
Porous Par	Spin-on	C, H	2.2–2.4	3.2	< 2.0
CF	PECVD	C, F	2.0–2.4	< 8.0	n/a

mainly to improve mechanical strength of Si–O network-based materials. Next, EB cure impact on porous MSQ film will be discussed.

1. Porous low-*k* MSQ evaluation

Recently, EB cure and UV cure were developed to improve the film characteristics of porous low-*k* materials deposited by PECVD or spin-on methods.

Figure 19.12 shows the impact of advanced curing methods on porous methyl-silsesquioxane (p-MSQ) material with $k = 2.2$. This figure shows dielectric constant and modulus of an EB-cured p-MSQ film as functions of EB total dosage. Modulus increases monotonously with the increase in dosage, and k-value was almost constant at the range from 0.2 to 0.5 mC/cm^2 dose. From this result, the EB cure method realizes improved p-MSQ with 1.5 times higher modulus (E = 9.0 GPa) and low dielectric constant ($k = 2.25$) at 0.5 mC/cm^2 dose. For the common SOD materials, high thermal budget is a serious issue. Figure 19.13 shows the relationship between *k*-value/Young's modulus and cure time in the case of using the total dosage of 500 μC/cm^2 constant, and dose rate is different for each cure time. As a

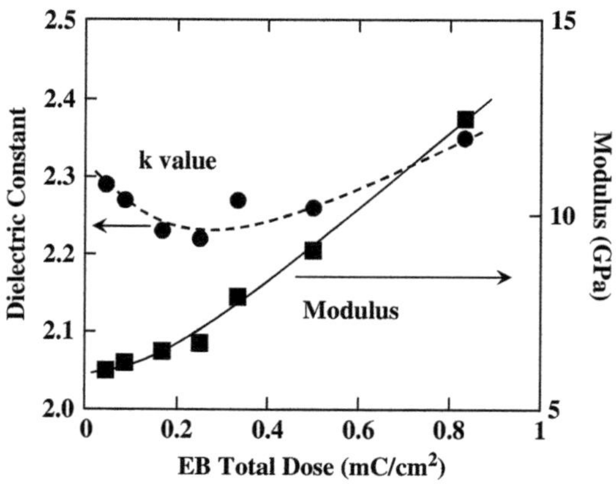

Fig. 19.12 Dielectric constant and elastic modulus as function of the EB total dose [22]

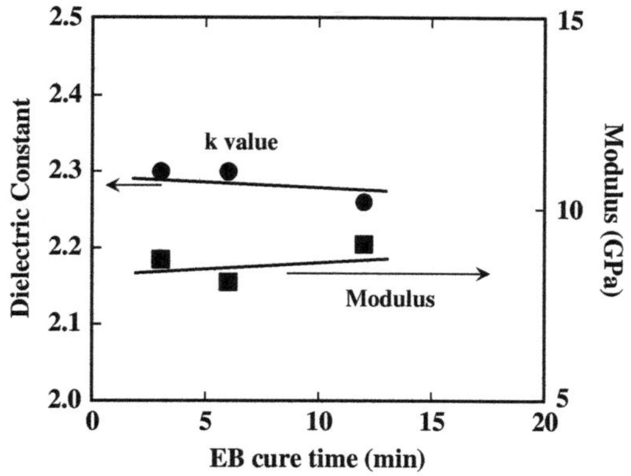

Fig. 19.13 Dielectric constant and elastic modulus as function of the EB curing time (total Electron beam dose is 0.5 mC/cm^2) [22]

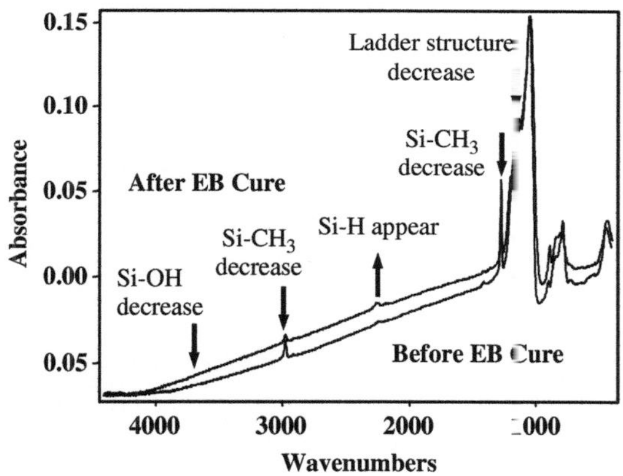

Fig. 19.14 FT-IR spectrum of p-MSQ before and after EB cure [22]

result, it was found that both the modulus and dielectric constant of the EB-cured p-MSQ film are unaffected by EB dose rate and EB cure drastically reduces p-MSQ curing time from 60 to 3 min.

Figure 19.14 shows the FT-IR absorption spectra before and after the EB curing of the p-MSQ films. The spectrum of the MSQ films is characterized by a relatively strong Si–O peak at 1049 cm^{-1}, two Si–CH_3 peaks at 778 and 1276 cm^{-1}, and two Si–H peaks at 888 and 2247 cm^{-1}. The shoulder of the 1139 cm^{-1} absorption peak is indicative of some caged Si–O bond structure.

The spectrum after the EB curing shows enhanced Si–H bond peaks at 888 and 2247 cm^{-1} and has a lower peak with a shoulder at 1139 cm^{-1}. A decrease in the height of the 1276 cm^{-1} peak of the Si–CH_3 bond was also observed. From these results, it is concluded that the cage Si–O bond structure and Si–CH_3 bonds are broken by EB curing, and a Si–O random network is constructedas shown in the previous report for dense MSQ films.

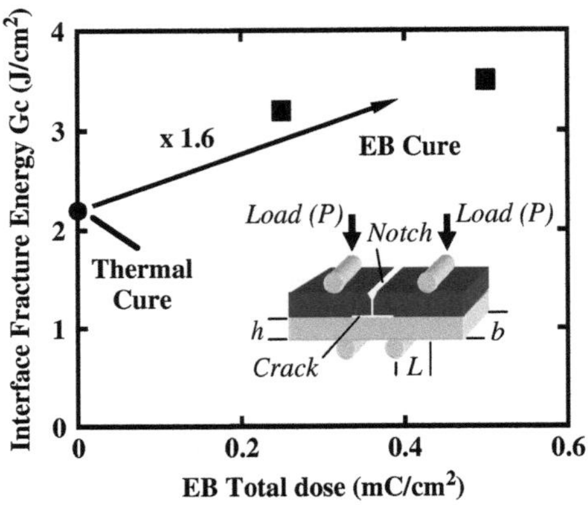

Fig. 19.15 Interface fracture energy (Gc) by four-point bending vs EB total dose [22]

The low adhesion strength between stacked dielectric structures is one of the serious problems concerning porous material [35, 36]. Figure 19.15 shows the effect of EB total dose on interface fracture energy, G_c (measure of adhesion strength), at the interface between the p-MSQ and PECVD SiCN films (G_c increases to 3.5 J/m^2, 1.6 times higher than that realized in the thermal curing. G_c in the four-point bending technique is defined below (Equation 19.1).

$$G_c = \frac{21(1-\nu^2)P^2L^2}{16E_{Si}b^2\,h^3}, \tag{19.1}$$

where P and L represent the load and the distance between the inner and outer loading lines, respectively, b is the beam width, h is the half thickness, and E and ν are the elastic modulus and Poisson's ratio of Si, respectively.

For EB cure technology, the most serious concern was transistor damage by electron irradiation and charging. The influence of EB curing on the performance of a transistor was evaluated using a p-channel MOSFET TEG and the cumulative distributions of gate leakage current and V_{th} were evaluated. No significant degradation in the performance of the MOSFET has been observed. Thus, it is concluded that the EB curing does not degrade the performance of a MOSFET due to optimum electron beam depth control.

2. Integration of PAr/SiOC hybrid scheme

For 65 and 45 nm node Cu/low-*k* module, hybrid structure with PAr/SiOC stack has been proposed as one of the promising process schemes.

With this scheme it is very easy to control DD profile by using high-etch selectivity between PAr and SiOC. On the other handto improve interfaces behavior, reliable interface engineering is required and an advanced cure technique is necessary for this scheme in order to improve adhesion strength of each interface.

Figure 19.16 shows the performance of EB cure for 65 nm node hybrid structure. EB cure is applied after PAr (k = 2.6) and MSQ (k = 2.6) stack fabrication and

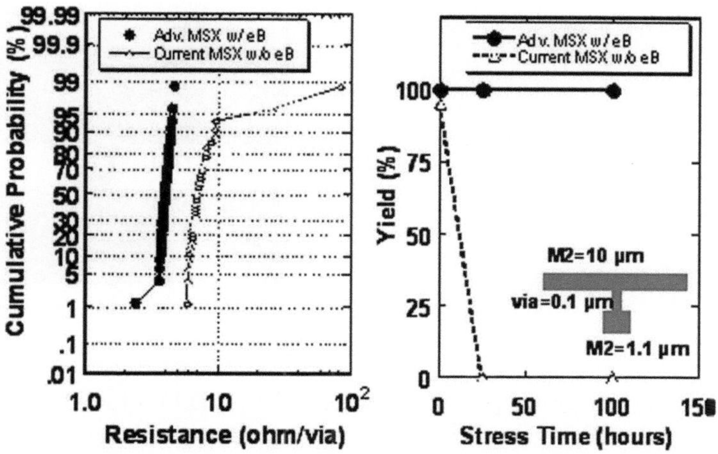

Fig. 19.16 Via resistance and via yield of SiV test results. *Left*: Via resistance of porous SOD, *right*: SiV performance. EB cure improves SiV performance of MSQ material [23]

essentially increases via reliability. EB cure accelerates PAr polymerization reaction and improves its adhesion strength as well as in the case of SiOC material. Figures 19.17 and 19.18 show the results for 45 nm low-k module with porous PAr ($k = 2.3$)/porous CVD–SiOC ($k = 2.3$) hybrid structure.

These results show that porous SiOC film has to be modified to be suitable for 32 nm node. EB cure provides excellent performance in low-k integration irrespective of film material, its formation methodand k-value. By using the combination of hybrid DD structure and EB cure, a reliable chip package was fabricated. Using 8 ML module with EB curing, flip chip package reliability was confirmed by TCT (temperature cycle test, –55°C to 125°C, 1000 cycles), PCT (pressure cooker test, 110°C/85% H, 500 h), THB (temperature humidity bias, 85°C/85% H/2.5 V,

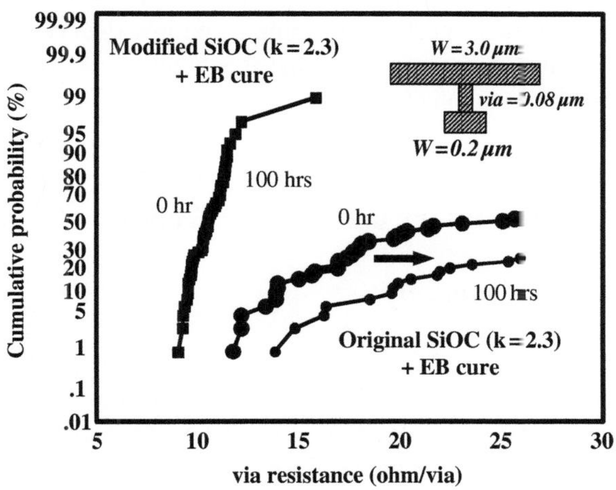

Fig. 19.17 SiV performance of 45 nm node hybrid DD with M2 wide pattern. It shows that SiOC has to be modified to be compatible with EB cure (via size 76 nm) [21]

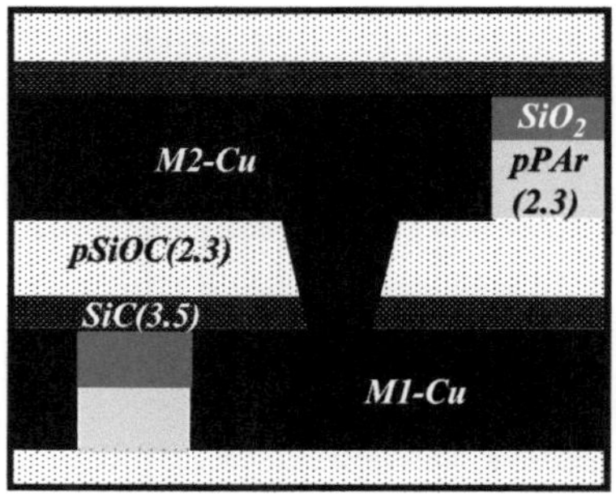

Fig. 19.18 Hybrid DD of porous Par ($k = 2.3$)/porous SiOC (k = 2.3) stack structure [21]

500 h), and HTS (high temperature storage, 150°C, 500 h). From these results, it is concluded that highly manufacturable interconnects are obtained using hybrid scheme and EB cure technology.

Recently, as well as EB cure, the UV cure method has also been widely examined with a view to its application as an advanced post-cure method. Although it is necessary to optimize UV cure process for the target low-k material, curing mechanism and impact on low-k film are very similar for different materials. It is thought that these post-cure technologies will occupy an important position in 65 nm node low-k/Cu interconnects and beyond. In addition, porous material easily absorbs moisture during Cu Damascene integration process because of its low-density structure. As moisture in low-k material degrades the reliability of Cu wiring, moisture control is the key technology to achieve robust Cu interconnects. Not only high plasma-resist material, but also effective moisture elimination using dummy pattern, oxidation resistance barrier metal technology, plasma damage restoration technique, and pore sealing technique are required to realize reliable Cu interconnects for 65 and 45 nm nodes [33, 34, 37].

While most of the porous low-k materials need the EB/UV post-cure to improve mechanical stress, newly developed SOG material was proposed to use for the mass production without EB/UV post-cure. Material design using small silica clusters realizes high mechanical strength and low k-value ($k = 2.3$) [38]. It is expected as the future technology from the standpoint of the balance of device performance and low-cost manufacturing.

19.6 Summary and Future Trends

19.6.1 Summary

In summary, in regard to high-performance logic devices beyond the 90 nm node, numerous challenges for Cu/low-k module developments have been produced. As a result of this investigation, it has become apparent that the following four technical developments are crucial. They are:

1. Low-k integration (low-k material design, process optimization, and interface engineering (advanced curing of EB and UV-curing)).
2. Cu wiring scaling (size effect, thin barrier metal (CVD and ALD depositions)).
3. Cu wiring reliability (dummy pattern design as a design for manufacturing (DFM), cap metal, Cu alloy, and new materials as carbon nanotube [39]).
4. Low-cost processing (low-cost low-k deposition, seedless Cu and self-consistent Cu alloy, etc.)

In addition, from the manufacturing point of view, to establish the robust process integration for pattern variations at the early stage of the process development which in turn can realize higher chip yield at the mass production will be crucial.

Furthermore, to assist the robust process development, characterization system for the local structure in patterned materials will be required.

In this section, first, we will discuss the process development tools for establishing the robust process. Next, the characterization development for patterned low-k films will be discussed. Both items will be necessary for future BEOL process development.

19.6.2 Robust Process Development

For the advanced VLSI technology, numerous of studies related to the pattern-dependent systematic failures in ICs were reported [40–43]. As an example, the interconnect line-width-dependent SiV (stress-induced voiding) failure [41] is one of the well-known Cu/low-k interconnect issues. Photoresist poisoning issues described in Section 19.4.3 are also the pattern-dependent systematic failure. As the interconnect feature size shrinks and the product pattern design becomes more complicatedits effect on failures becomes stronger causing many yield and reliability problems. The influence from the surrounding neighbor patterns is especially pronounced. Facing these issues, it is essential for the advanced Cu/low-k IC's development to examine IC's pattern design-dependent characteristics and take them into account for the product chip designing, and to setup the robust manufacturing process. For this purpose, several hundreds of different shape four-point probe Kelvin via/interconnect test structures (we define these Kelvin pattern as "Sea of Kelvin" pattern) are designed [44].

Cu/Low-k dual Damascene wafers of 65 nm node are fabricated with this multiple-pattern "Sea of Kelvin" mask set. Hybrid style dual Damascene architecture is employed for this generation [31].

There are some pattern-dependent systematic problems, which are not detectible with the conventional interconnect characterization methods. Several pattern parameter dependencies are identified for the via resistance (Rc) variation. They are, for example, M1 line width (wider lines lead to larger V2 Rc variation) and M1 line length (longer lines are worse). Evaluating the influence from the surrounding neighbor patterns is crucial.

Fig. 19.19 SiV stress test results – surrounding pattern

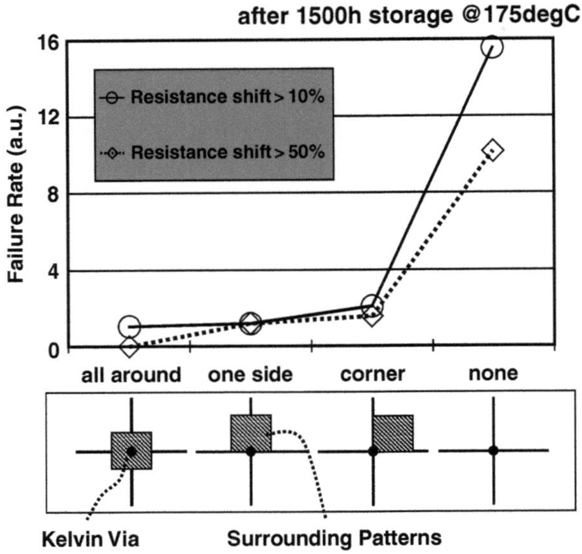

The effect of surrounding patterns was examined using SiV technique. More details are given below.

The wafers are stressed at 175°C temperature for 500 h. The surrounding dummy allocation spot dependency is examined as shown in Fig. 19.19. Failure rate increased with the degree of Kelvin via facing to open area. The worst results were obtained for isolated structure.

Based on these results, locating the trench dummy patterns in any open space area is very important to reduce the SiV failures. The trench dummy patterns are thought to be working as moisture ventilation windows for low-*k* dielectrics and protecting the PVD barrier film from being partly oxidizedwhich would induce the barrier/Cu layer adhesion degradation and causes the SiV failure (Fig. 19.20).

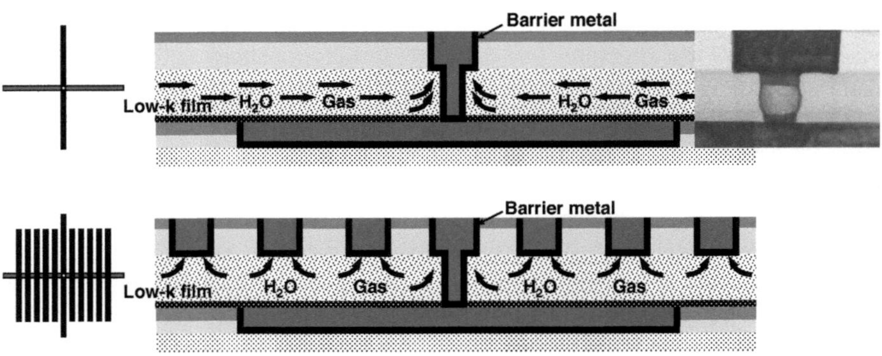

Fig. 19.20 Mechanism of the surrounding pattern effect

Newly developed interconnect characterization method of multiple-pattern four-point probe Kelvin structure is confirmed to be very effective to quantify the various pattern-dependent Cu/low-*k* process integration schemes. This is an important evidence that the dummy pattern placement in the product chip open area is very effective not only for CMP or RIE process uniformity improvement but also for the interconnect electrical stability enhancement.

*19.6.3 Patterned Low-*k* Films Characterization*

Numerous studies have been reported using Fourier transformed infrared spectroscopy (FT-IR) and Raman spectroscopy to characterize the blanket low-*k* structures. However, only few studies for characterization of patterned low-*k* structure were done. A structural analysis for nanometer-order local area helps the advanced process integration study.

Conventional electron energy loss spectroscopy (EELS) related to inner shell electron excitation to conduction bands has been used to characterize the composition of the elements. On the contrary, valence electron energy loss spectroscopy (V-EELS) is related to the valence electron excitation. V-EELS corresponds to optical constant and electron density of state (DOS) of the element. Additionally energy-filtered image of energy loss electrons of V-EELS can provide two-dimensional characterization of the local structure of patterned low-*k* materials.

Electrical and physical properties of low-*k* dielectric films as the interlayer dielectric are easily affected by changes in their structures, namely "damages," after several processes such as dry etch, ashing, pre-clean, and metal deposition. These damages can degrade the dielectric constant and influence time-dependent dielectric breakdown (TDDB) of the low-*k* dielectric. To understand the damage distribution and the structure change of patterned low-*k* film, a transmission electron microscopy (TEM) method might be useful. However, it is difficult to observe them by conventional TEM because of contrast insensitivity for the structural changes in amorphous materials such as the low-*k* films.

EELS combined with scanning TEM (STEM) was applied to two-dimensional characterization of the damages in trench-patterned low-*k* films. A narrow energy-filtering technique in a valence EELS region was developed to observe damages. A schematic diagram of TEM/EELS measurements for low-*k* trench structure is depicted in Fig. 19.21 [45].

The V-EELS from a whole region of the MSQ type low-*k* films which was dry etched followed by ashing using NH_3 and He/H_2 gases showed significant difference at the two regions of A (4–5 eV) and B (9 eV). In the NH_3-ashing specimen, weak diffused intensity was seen at region A, while intensity at region B was entirely suppressed.

On the other handalmost the same spectra were seen in the He/H_2-ashing sample and the reference (Fig. 19.22) [46]. These results correspond to the damages induced by plasma processes which decompose Si–$CH_3$and C–H bonds in the MSQ

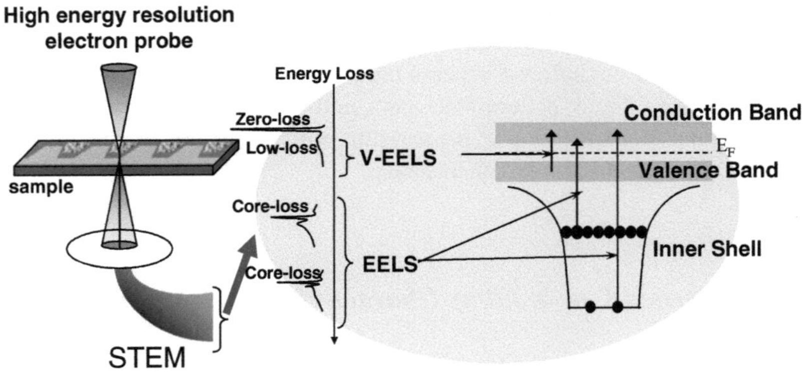

Fig. 19.21 Schematic diagram of TEM EELS [45]

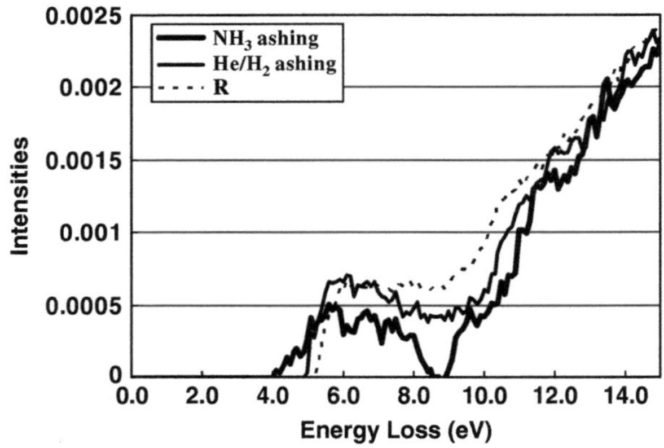

Fig. 19.22 Valence ELS from whole area of porous MSQ film [46]

type low-*k* films. Energy-filtered (EF) images at each energy region can provide two-dimensional damage characterization.

An extensive approach for estimating two-dimensional distribution of dielectric constants by using Kramers–Kronig analysis (KKA) has been reported [45, 47]. The advanced characterization method for local structure of patterned materials is confirmed to be effective for future BEOL process development.

19.7 Biography

Takashi Yoda is a senior manager for the interconnect technology (including metallization, dielectric, and CMP) and dry etching technology at Toshiba Corporation. He received an M.S. in Material Science and Engineering from Tohoku University and Ph.D. in Electron Devices from Tokyo Institute of Technology in Japan.

Hideshi Miyajima is a chief specialist in the ILD & CMP development group at Toshiba Corporation. His focus is on development of plasma CVD technology and low-*k* dielectric technology. He received M.S. in Energy Sciences from Tokyo Institute of Technology in Japan.

References

1. Shibata, H.: Chip-level Interconnect Scaling and Copper/Low-k Process Integrational Technology for 65 nm Node and beyond (invited). The Electrochemical Society of Japan, Proc. of the 67th Symposium on Semiconductors and Integrated Circuits Technology, 42–, 47 (2004)
2. Yoshimaru, M.; Koizumi, S.; and Shimokawa, K.: Structure of fluorine-doped silicon oxide films deposited by plasma-enhanced chemical vapor deposition. J. Vac. Sci. Technol. A **15** , 2908 (1997)
3. Yoshimaru, M.; Koizumi, S.; and Shimokawa, K.: Interaction between water and fluorine-doped silicon oxide films deposited by plasma-enhanced chemical vapor deposition. J. Vac. Sci. Technol. A **15** , 2915 (1997)
4. Steinlesberger, G.; Schindler, G.; Engelhardt, M.; Steinhogel W.; and Traving, M.: Aluminum Nano Interconnects. IEEE International Interconnect Technology Conference, 51–, 53 (2004)
5. Jiang, Q. T.; Tsai M. H.; and Havemann, R. H.: Line Width Dependence of Copper Reliability. IEEE International Interconnect Technology Conference, 227–, 229 (2001)
6. Schindler, G.; Steinhogel, W.; Steinlesberger, G.; Traving, M.; and Engelhardt, M.: Microstructure of cu damascene nano-interconnects. Proceedings of Advanced Metallization Conference (AMC) , 397 (2002)
7. Ogawa, E. T.; McPherson, J. W.; Rosal, J. A.; Dickerson, K. J.; Chiu, T. C.; Tsung, L. Y.; Jain, M. K.; BonifieldT. D.; Ondrusek J. C.; and McKee, W. R.: Stress-induced voiding under vias connected to wide Cu metal leads. International Reliability Physics Symposium 2002 (USA) 312 (2002)
8. Kawano, M.; Fukase, T.; Yamamoto, Y.; Ito, T.; Yokogawa, S.; Tsuda, H.; Kunimune, Y.; Saitoh, T.; Ueno, K.; and Sekine, M.: Stress relaxation in Dual-damascene Cu Interconnects to Suppress Stress-induced Voiding. IEEE International Interconnect Technology Conference 2003, 210 (2003)
9. Yoshida, K.; Fujimaki, T.; Miyamoto, K.; Honma, T.; Kaneko, H.; Nakazawa H.; and Morita, M.: Stress induced voiding phenomena for actual CMOS LSI interconnects, 2002 International Electron Devices Meeting. Technical Digest (USA) 753 (2002)
10. Usami, T.; Shimokawa, K.; and Yoshimaru, M.: Low dielectric constant interlayer using fluorine-doped silicon oxide. Jpn. J. Appl. Phys. **33(1B)**, 408(1994)
11. Miyajima, H.; Katsumata, R.; Nakasaki, Y.; Nishiyama, Y.; and Hayasaka, N.: Water absorption properties of fluorine-doped SiO_2 films using plasma-enhanced chemical vapor deposition. Jpn. J. Appl. Phys. **35(12A)**, 6217 (1996)
12. Loboda, M. J.: New solutions for intermetal dielectrics using trimethylsilane-based PECVD processes. Microelectron. Eng. **50(1–4)**, 15 (2000)
13. Sugiarto, D.; Lee, P.; Oka, N.; Iwasaki, N.; and Ishikawa, Y.: Integration of Carbon Doped Oxide-CVD low-k dielectric film for damascene Cu interconnection, Advanced Metallization Conference 2001 (AMC 2001). Proceedings of the Conference (USA) 337 (2001)
14. Inohara, M.; Tamura, I.; Yamaguchi, T.; Koike, H.; Enomoto, Y.; Arakawa, S.; Watanabe, T.; Ide, E.; Kadomura, S.; and Sunouchi, K.: High performance copper and low-k interconnect technology fully compatible to 90 nm-node SOC application (CMOS4). 2002 International Electron Devices Meeting. Technical Digest (USA) 77 (2002)
15. Miyajima, H.; Watanabe, K.; Fujita, K.; Ito, S.; Tabuchi, K.; Shimayama, T.; Akiyama, K.; Hachiya, T.; Higashi, K.; Nakamura, N.; Kajita, A.; Matsunaga, N.; Enomoto, Y.; Kanamura, R.; Inohara, R.; Honda, K.; Kamijo,H.; Nakata, R.; Yano, H.; Hayasaka, N.; Hasegawa, T.; Kadomura, S.; Shibata, H.; and Yoda, T.: Challenge of low-k materials for 130, 90, 65 nm node interconnect technology and beyond. 2004 International Electron Devices Meeting (USA) 329 (2005)

16. Abe, J.; Hayashi, H.; Kishigami, D.; Sato, Y.; Shiobara, E.; Shibata, T.; Onishi, Y.; and Ohiwa, T.: A new stacked mask process (S-MAP) utilizing spun-on carbon film for sub-130 nm etching. Proceeding of International Symposium on Dry Process (DPS2001) (Japan) , 187 (2001)
17. Higashi, K.; Nakamura, N.; Miyajima, H.; Satoh, S.; Kojima, A.; Abe, J.; Nagahata, K.; Tatsumi, T.; Tabuchi, K.; Hasegawa, T.; Kawashima, H.; Arakawa, S.; Matsunaga, N.; and Shibata, H.: Manufacturable copper/low-k SiOC/SiCN process technology for 90 nm-node high performance eDRAM. Proceedings of the IEEE 2002 International Interconnect Technology Conference (USA) 5 (2002)
18. Hu, C.-K.; Rosenberg, R.; Rathore, H. S.; Nguyen, D. B.; and Agarwala, B.: Scaling effect on electromigration in on-chip Cu wiring. Proceedings of the IEEE 1999 International Interconnect Technology Conference (USA) 267 (1999)
19. Hatano, M.; Usui, T.; Shimooka, Y.; and Kaneko, H.: EM lifetime improvement of Cu Damascene interconnects by p-SiC cap layer. Proceedings of the IEEE 2002 International Interconnect Technology Conference (USA) 212 (2002)
20. Grill, A. and Patel, V.: Ultra low-k dielectrics prepared by plasma-enhanced chemical vapor deposition. Appl. Phys. Lett. **79(6)**, 803 (2001)
21. Miyajima, H.; Masuda, H.; Idaka, T.; Shimayama, T.; Kagawa, Y.; Tabuchi, K.; Yano, H.; Hasegawa, T.; Kadomura, S.; and Yoda, T.: Material design of balance between high mechanical strength and high plasma resistance for porous PE-CVD SiOC film (k=2.3), Advanced Metallization Conference 2005 (AMC 2005). Proceedings of the Conference (USA) 297 (2005)
22. Fujita, K.; Miyajima, H.; Nakata, R.; and Miyashita, N.: Notable Improvement in Porous Low-k film Properties using Electron-Beam Cure Method. Proceedings of the IEEE 2003 International Interconnect Technology Conference (USA) 106 (2003)
23. Miyajima, H.; Fujita, K.; Nakata, R.; Yoda, T.; and Hayasaka, N.: The application of simultaneous ebeam cure methods for 65 nm node Cu/low-k technology with hybrid (PAE/MSX) structure. Proceedings of the IEEE 2004 International Interconnect Technology Conference (USA) 222 (2004)
24. Shimada, M.; Miyajima, H.; Nakata, R.; and Yoda, T.: High-performance low-k dielectric using advanced EB-cure process. Proceeding of 2001 international conference on solid state devices and materials (Japan) , 416 (2001)
25. Yoda, T.; Nakasaki, Y.; Hashimoto, H.; Fujita, K.; Miyajima, H.; Shimada, M.; Nakata, R.; Kaji, N.; and Hayasaka, N.: Structural Studies of High-Performance Low-*k* Dielectric Materials Improved by Electron-Beam Curing. Jpn. J. Appl. Phys. **44(1A)**, 75 (2005)
26. Ohnishi, T.; Nagaseki, K.; Shimada, M.; Miyajima, H.; Nakata, R.; Yamaguchi, M.; Murase, J.; and Hata, H.: Advanced EB-cure process and equipment for low-k dielectric. Proceeding of 2001 IEEE International Symposium on Semiconductor Manufacturing (USA) 325 (2001)
27. Yoneda, K.; Kato, M.; Kondo, S.; Kobayashi, N.; Matsuki, N.; Matsushita, K.; Ohara, N.; Fukazawa, A.; and Kimura, T.: Impacts of UV cure for reliable porous PECVD SiOC integration [IC interconnect applications]. Proceedings of the IEEE 2005 International Interconnect Technology Conference (USA) 220 (2005)
28. Furusawa, T.; Miura, N.; Matsumoto, M.; Goto, K.; Hashii, S.; Fujiwara, Y.; Yoshikawa, K.; Yonekura, K.; Asano, Y.; Ichiki, T.; Kawanabe, N.; Matsuzawa, T.; and Matsuura, M.: UV-hardened high-modulus CVD-ULK material for 45-nm node Cu/low-k interconnects with homogeneous dielectric structures. Proceedings of the IEEE 2005 International Interconnect Technology Conference (USA) 45 (2005)
29. Fujita, K.; Miyajima, H.; Nakao, S.; Sakanaka, T.; Nakata, R.; Yano, H.; and Yoda, T.: Comparison between UV and EB cure method for porous PAr/porous MSX hybrid structure. Extended Abstracts of the 2005 International Conference on Solid State Device and Materials (Japan) , 298 (2005)
30. Kanamura, R.; Ohoka, Y.; Fukasawa, M.; Tabuchi, K.; Nagahata, K.; Shibuki, S.; Muramatsu, M.; Miyajima, Usui, H. T.; Kajita, T.; Shibata, H.; and Kadomura, S.: Integration of Cu/low-k dual-Damascene interconnects with a porous PAE/SiOC hybrid structure for 65 nm-node high

performance eDRAMS. 2003 Symposium on VLSI Technology. Digest of Technical Papers (Japan) , 107 (2003)
31. Kajita, A.; Usui, T.; Yamada, M.; Ogawa, E.; Katata, T.; Sakata, A.; Miyajima, H.; Kojima, A.; Kanamura, R.; Ohoka, Y.; Kawashima, H.; Tabuchi, K.; Nagahata, K.; Kato, Y.; Hayashi, T.; Kadomura, S.; and Shibata, H.: Highly reliable Cu/low-k dual-damascene interconnect technology with hybrid (PAE/SiOC) dielectrics for 65 nm-node high performance eDRAM. Proceedings of the IEEE 2003 International Interconnect Technology Conference (USA) 9 (2003)
32. Honda, K.; Kanda, M.; Ishizuka, R.; Moriuchi, Y.; Matsubara, Y.; Habu, M.; Yoshida, T.; Matsuda, S.; Kittaka, H.; Miyajima, H.; Hachiya, T.; Kajita, A. Usui, T.; Nagashima, N.; Kanamura, R.; Okamoto, Y.; Yamada, S.; and Noguchi, T.: Integration of interconnect process highly manufacturable for 65 nm CMOS platform technology (CMOS5). 2004 Symposium on VLSI Technology. Digest of Technical Papers (USA) 62 (2004)
33. Matsunaga, N.; Nakamura, N.; Higashi, K.; Yamaguchi, H; Watanabe,T.; Akiyama, K.; Nakao, S.; Fujita, K.; Miyajima, H.; Omoto, S.; Sataka, A.; Katata, T.; kagawa, Y.; Kawashima, H.; Enomoto, Y.; Hasegawa, T.; and Shibata, H.: BEOL process integration technology for 45 nm node porous low-k/Copper interconnects. Proceedings of the IEEE 2005 International Interconnect Technology Conference (USA) 6 (2005)
34. Fujimaki, T.; Higashi, K.; Nakamura, N.; Matsunaga, N.; Yoshida, K.; Miyawaki, N.; Hatano, M.; Hasunuma, M.; Wada, J.; Nishioka, T.; Akiyama, K.; Kawashima, H.; Enomoto, Y.; Hasegawa, T.; Honda, K.; Iwai, M.; Yamada, S.; and Matsuoka, F.: Mechanism of moisture uptake induced via failure and its impact on 45 nm node interconnect design. 2005 International Electron Devices Meeting (USA) 191 (2005)
35. Dauskardt, R. H.; Lane, M.; Ma, Q.; Krishna, N.: Adhesion and debonding of multi-layer thin film structures. Eng. Fract. Mech. **61(1)**, 141 (1998)
36. Kitsutaka, H.; Suzuki, K.; Inoto, H.; Kawakami, M.; Honda, K.; Hasunuma, M.; Ito, S.; Miyajima, H.; Fujita, K.; Kaneko, H.; Yoda, T.; Oyamatsu, H.; Yamada, S.; Matsuoka, F.; and Noguchi, T.: Improvement of interfacial adhesion in chip-package interaction of 65 nm node SoC. Advanced Metallization Conference 2004 (AMC 2004) (USA) 265 (2005)
37. Nakamura, N.; Yoshizawa, T.; Watanabe, T.; Miyajima, H.; Nakao, S.; Yamada, N.; Fujita, K.; Matsunaga, N.; and Shibata, H.: A plasma damage resistant ultra low-k hybrid dielectric structure for 45 nm node copper dual-damascene interconnects. Proceedings of the IEEE 2004 International Interconnect Technology Conference (USA) 228 (2004)
38. Nakamura, T. and Nakashima, A.: Robust Multilevel Interconnects with a Nano-clustering Porous Low-k (k<2.3). Proceedings of the IEEE 2004 International Interconnect Technology Conference (USA) 175 (2004)
39. Nihei, M.; Horibe, M.; Kawabata, A.; and Awano, Y.: Carbon nanotube vias for future LSI interconnects. IEEE International Interconnect Technology Conference 2004 (USA) 251 (2004)
40. Ogawa, E. T.; McPherson, J. W.; Rosal, J. A.; Dickerson, K. J.; Chiu, T. C.; Tsung, L. Y.; Jain, M. K.; BonifieldT. D.; Ondrusek, J. C.; and McKee, W. R.: Stress-induced voiding under vias connected to wide Cu metal leads. 2002 International Reliability Physics Symposium(USA) 312 (2002)
41. Kawano, M.; Fukase, T.; Yamamoto, Y.; Ito, T.; Yokogawa, S.; Tsuda, H.; Kunimune, Y.; Saitoh, T.; Ueno, K.; and Sekine, M.: Stress relaxation in Dual-damascene Cu Interconnects to Suppress Stress-induced Voiding. IEEE International Interconnect Technology Conference 2003 (USA) 210 (2003)
42. Yoshida, K.; Fujimaki, T.; Miyamoto, K.; Honma, T.; Kaneko, H.; Nakazawa, H.; and Morita, M.: Stress induced voiding phenomena for actual CMOS LSI interconnects. 2002 International Electron Devices Meeting (USA) 753 (2002)
43. Huang, T.C.; Yao, C.H.; Wan, W.K.; Hsia, C. C.; and Liang, M.S.: Numerical modeling and characterization of the stress migration behavior upon various 90 nanometer Cu/Low-k interconnects. IEEE International Interconnect Technology Conference 2003 (USA) 207 (2003)

44. Okazaki, M.; Hatano, M.; Yoshida, K.; Shibasaki,S.; Kaneko, H.; Yoda T.; and Hayasaka, N.: Sea of Kelvin Multiple-pattern arrangement interconnect characterization for Low-k/Cu dual Damascene and its findings. IEEE International Interconnect Technology Conference 2004 (USA) 211 (2004)
45. Shimada, M.; Otsuka, Y.; Harada, T.; Tsutsumida, A.; Inukai, K.; Hashimoto, H.; and Ogawa, S.: 2-Dimentional distribution of Dielectric Constants in Patterned Low-k Structures by a nm-probe STEM/Valence EELS (V-EELS) Technique. IEEE International Interconnect Technology Conference 2005 (USA) 88 (2005)
46. Otsuka, Y.; Harada, T.; Ito, T.; Tsutsumida, A.; Shimada, M.; Inukai, K.; Hashimoto, H.; and Ogawa, S.: TEM-EELS 2-Dimensinal Characterization of Damage in Patterned Low-k Dielectric Films. Advanced Metallization Conference Asian Session 2004, Mat. Res. Soc. ULSI-XX(Japan) , 425 (2005)
47. Lo, S.-C.; Kai, J.-J.; Chen, F.-R.; Chang, L.; Chen, L.-C.; Chiang, C.-C.; Ding, P.; Chin, B.; Zhang. H.; and Chen, F.: Four-dimensional dielectric property image obtained from electron spectroscopic imaging series. J. Electron Microsc. **50(6)**, 497 (2001)

Chapter 20
Lithography for Cu Damascene Fabrication

Yoshihiro Hayashi

20.1 Introduction

As CMOS transistors are scaled down, interconnects to link them are also shrunk to reduce the line pitches [1, 2]. Figure 20.1a shows the technology trends of the CMOS gate length (Lg) and the minimum line pitches (P_{int}). In 130 nm node LSIs, the Lg and P_{int} have been approximately 100 and 400 nm, respectively. By device scaling, Lg and P_{int} in 45 nm node LSIs will be shrunk to 40–50 and 140 nm, respectively. These small features are patterned by photo-lithography process, in which photo-sensitive resist is coated on silicon wafer and is exposed by laser light. The resist is developed to make fine patterns of the resist. Figure 20.1b shows trends of the minimum pattern resolution (R) and the depth of focus (DOF). To reduce the resolution, the wavelength of exposure light should be shortened using the large diameter lens (NA number), however the DOF decreases drastically. This requires flat-surface interconnects such as Cu Damascene interconnects for multi-level stacks.

The Cu Damascene interconnects are fabricated by Cu deposition and chemical-mechanical polishing (CMP) on the interconnect dielectrics having opened patterns of line trenches and via-holes as shown in Fig. 20.2a,b. These open patterns in the dielectrics are transferred from the patterns of photoresist by dry-etching process. In this section, photo-lithography process including dry etching followed is explainedand the process issues and their solutions are described especially for scaled-down Cu dual Damascene interconnects in low-k dielectrics.

20.2 Lithography Process for Cu Damascene Interconnects

To fabricate the Cu Damascene interconnects, there are several ways of lithography sequences depending on the interconnect structure such as single Damascene (SD) or dual Damascene (DD), the interlayer dielectric (low-k) materials and the

Y. Hayashi (✉)
ULSI Fundamental Research Laboratory, Microelectronics Research Laboratories,
NEC Electronics Corporation, 1120, Shimokuzawa, Sagamihara, Kanagawa 229, Japan
e-mail: hayashi@mel.cl.nec.co.jp, y-hayashi@az.jp.nec.com

Y. Shacham-Diamand et al. (eds.), *Advanced Nanoscale ULSI Interconnects: Fundamentals and Applications*, DOI 10.1007/978-0-387-95868-2_20,

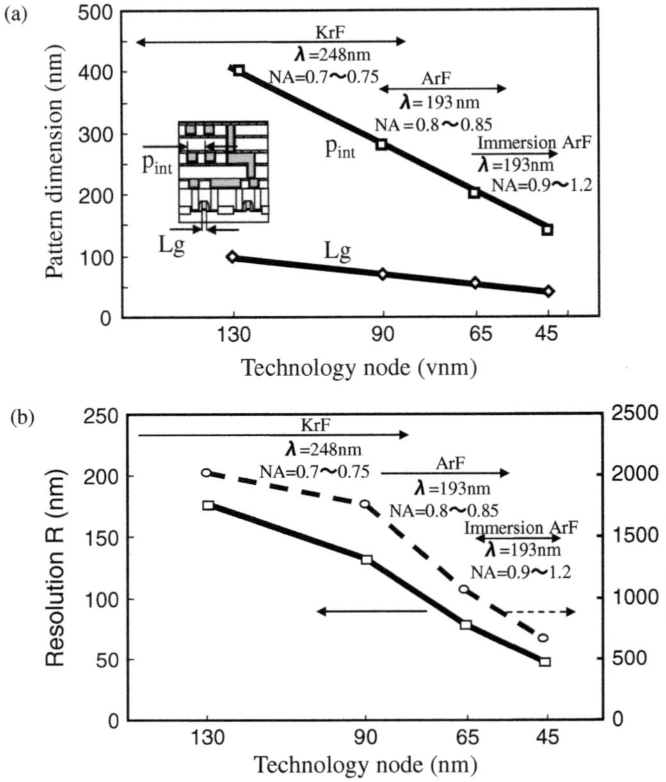

Fig. 20.1. Technical trends of (**a**) pattern dimensions of CMOS gate length and interconnects pitches and (**b**) the minimum pattern resolution and the depth of focus using KrF or ArF laser lights through optical lens with various diameter (NA)

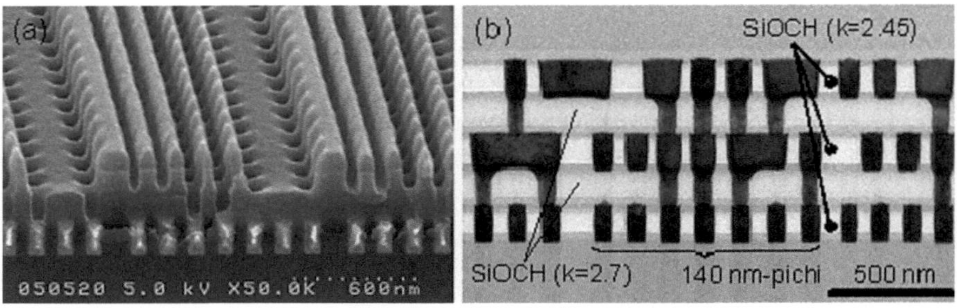

Fig. 20.2 SEM micrographs of (**a**) opened patterns of line trench and via-holes in low-*k* SiOCH films and (**b**) triple-layered Cu dual Damascene (DD) interconnects for 45 nm node LSIs with the minimum line pitch and the via-diameter of 140 nm and 70 nmϕ, respectively [3]

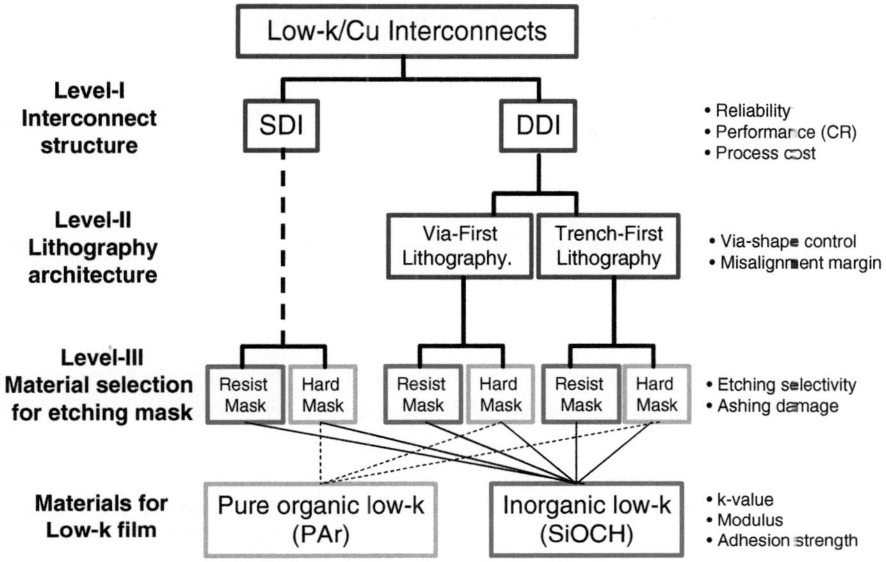

Fig. 20.3 Fabrication process for Cu Damascene interconnects in pure organic ILD film or inorganic ones

etching–mask materials such as photoresist or inorganic films, so called as "hard mask" as illustrated in Fig. 20.3. The interconnect structures are categorized into two types such as SD and DD. The Cu SD interconnects, in which the lines and vias are fabricated separately, are easy to control the depth profiles of the interconnects and vias, or essentially their electrical properties independently. In general, the via reliabilities for electro-migration (EM) and stress-induce voiding (SiV) are superior to those of DD due to the discontinuity of Cu metals between the lines and vias by existence of thin barrier metals such as TaN/Ta [4–8]. The Cu DD interconnects have advantages of lower via resistance and smaller process steps than SD. Most of the Cu interconnects in advanced LSIs have the DD structure to reduce the process cost of multi-level interconnects.

In the case of DD interconnects, the interconnect trench and via-holes are patterned in the ILD film sequentially. This means that we have two choices: "via-first followed by trench" or "trench-first followed by via." As described in latter, the lithography process step affects the DD profile in ILD film as well as the alignment margin between the via and the line. Etching-mask materials such as organic photoresist mask (RM) or inorganic mask (hard mask: HM) are also selected. The etching-mask selection depends strongly on the chemical composition of ILD films. When a pure organic ILD film is used, HM or multi-layer HM should be selected. HM process is superior to RM, one to avoid damages of the etched surface by oxygen plasma during the RM removal. The defect density in HM deposited by plasma-CVD process, however, is larger than that in MR, causing serious electrical defects such as the line shortage. We should select the best process and materials for the Damascene interconnect fabrication.

20.2.1 Process with Photoresist (PR) Mask

In the most case, via-first sequence for Cu DD fabrication is utilized. In Fig. 20.4 the via-first process is illustrated with RM for inorganic low-*k* ILD films such as plasma-CVD SiOCH film or porous SiOCH film. First, bottom anti-reflection coating (BARC) and photo resist are coated on the ILD stacks of capping film, via-ILD, etch-stop (ES), line-ILD, and hard mask (HM), and then via-hole patterns are exposed in the photo resist. Here, HM is just put on low-*k* line-ILD film that protects the surface from mechanical stress. Via-hole patterns through ILD stacks up to capping film are etched using the photoresist mask patterned. After the photo resist has been removed by oxygen plasma, BARC is coated again into the via-holes to flat the surface. The second photoresist is coated on BARC to pattern the line trenches in line-ILD. The photo resist and BACR are removed by oxygen plasma, and the capping film at the via bottom is removed by etch-back process. After barrier metal (Ta/TaN) and Cu seed layer are sputtered, Cu film is deposited by electro-chemical plating deposition (ECD). The Cu film and the barrier metal on the ILD surface were removed selectively by chemical-mechanical polishing (CMP). Finally, capping film is deposited on the Cu lines.

This process sequence with PR mask has a substantial issue of oxygen plasma damage to the low-*k* ILD films during resist removal as shown in Fig. 20.5 Here, the samples were dipped in diluted HF solution to etching SiO_2 layer, which had been converted from SiOCH by the oxygen plasma attack. It is clearly seen that the sidewalls of the Cu lines were etched by the oxygen plasma attacks. In other words, especially for the process sequence with PR mask, low-*k* ILD with high resistance

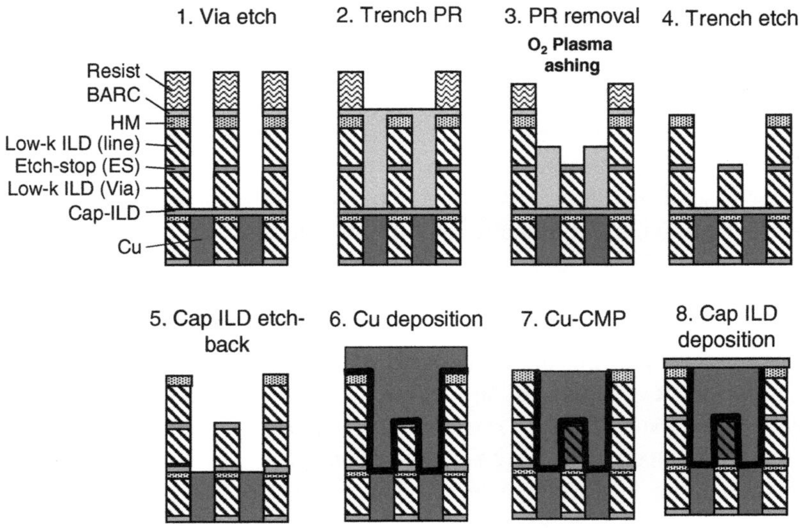

Fig. 20.4 Via-first process with RM for inorganic low-*k* ILD films such as plasma-CVD SiOCH films

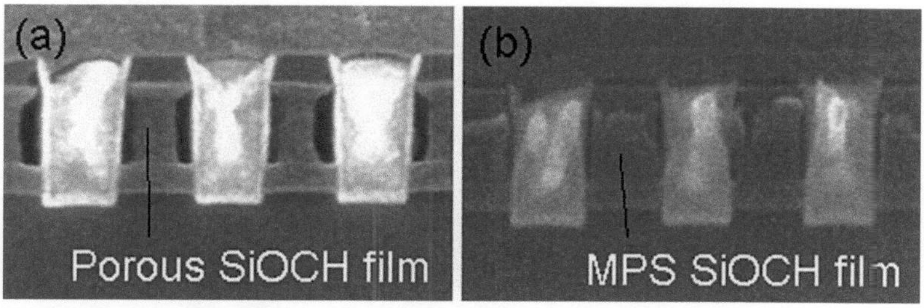

Fig. 20.5 Cross-sectional SEM micrographs of Cu Damascene interconnects in (**a**) conventional plasma-CVD-derived porous SiOCH film and (**b**) molecular pore-stacking (MPS) SiOCH film. The samples were dipped in diluted HF solution to dissolve the damaged SiOCH portion

to oxygen plasma attack is needed such as molecular pore-stacked (MPS) SiOCH films [3, 9–11]. The other issues on the PR mask process are sometimes observed such as "poisoned vias" and "fenced vias."

As shown in Fig. 20.6, the poisoned vias are caused in chemically amplified resist for DUV exposure, which contains a special chemical component releasing acid by DUV exposure [12, 13]. The acid partially breaks polymer network of the photo resist to enhance the developing rate. When the ILD stacks have nitrogen compounds such as SiN or SiCN, alkaline chemicals such as NHx are synthesized during plasma etching of via-holes to be absorbed in low-*k* ILD film. The

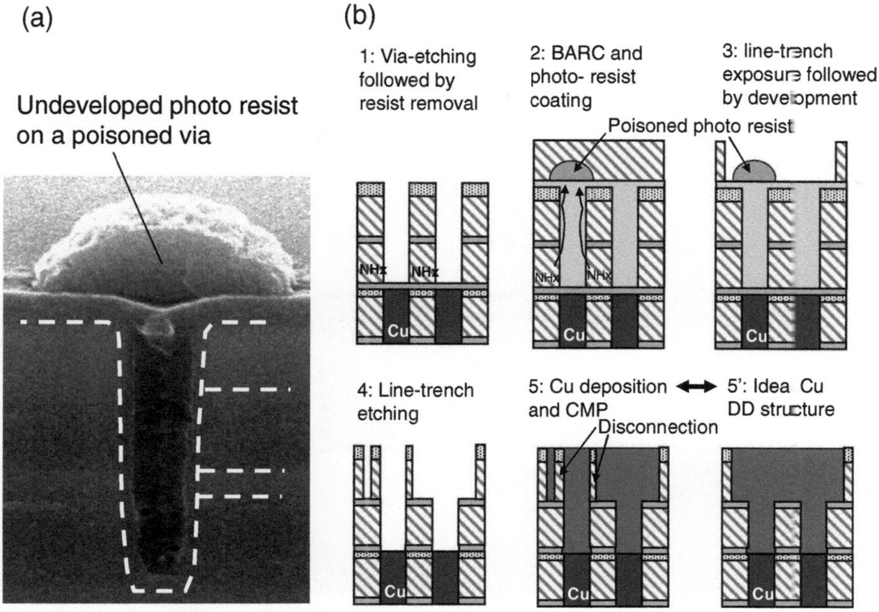

Fig. 20.6 Via poisoning occurred in via-first process with RM: (**a**) SEM micrograph and (**b**) mechanism for via poisoning failure in Cu DD interconnects

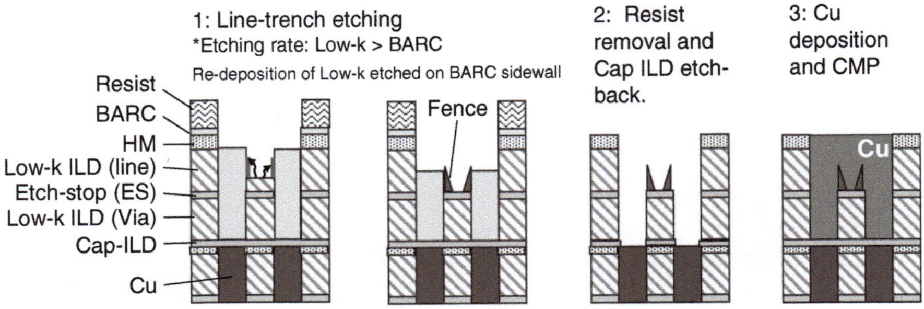

Fig. 20.7 Formation mechanisms of the fenced vias in the via-first process with RM

alkaline chemicals absorbed are released again to be diffused through the BARC to the photoresist. The alkaline chemicals kill the function of the acid promoters in the photoresist, remaining undeveloped areas on the via-holes during the line trench exposure. The undeveloped photoresist becomes undesirable etching masks during the line trench patterning, resulting in open failures in Cu DD interconnects. The photoresist poisoning is eliminated by controlling the etching gas and the ILD composition carefully.

The other issue is the fenced vias, which is created during the line trench etching as shown in Fig. 20.7. During the trench etching, the via-holes have been filled with BARC, of which the etching rate is slower than the low-*k* ILD. Therefore, some portions of the low-*k* ILD etched are deposited again on the BARC sidewall to make fences. The Cu DD interconnects with the fenced vias are unstable under thermal stress around the complicated via-profiles [14]. No stress-concentrated spot was detected in the tapered vias as shown in Fig. 20.8a, while large stress gradient was observed at the top of fenced via (Fig. 20.8b). Here, assume that the sample is cooled down to 25°C from 350°C, where the stress was released to be free. The

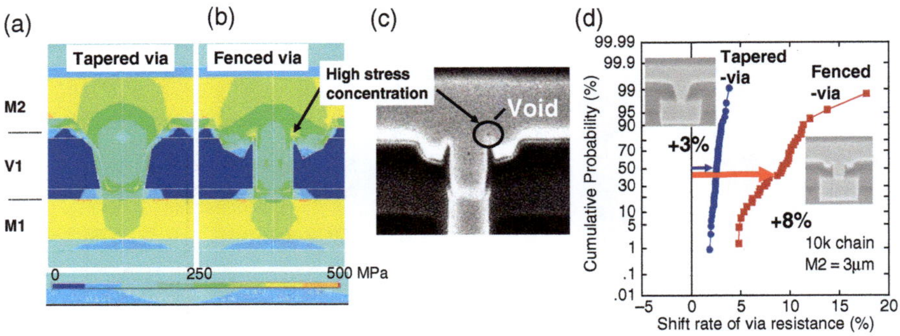

Fig. 20.8 Mises stress distributions and electrical properties of Cu DD interconnects with (**a**) ideal tapered via, (**b**) fenced via, (**c**) cross-sectional SEM micrograph of the fenced via after 20 times thermal cycle test (RT to 350°C), and (**d**) the distributions of the change in the 100 nmϕ via resistance after 20 cycles of HTC test from 25°C to 350°C [14]

cross-sectional SEM micrograph in Fig. 20.8c indicates a small void in Cu film at the top portion of fenced via. Figure 20.8d shows the distributions of the change in the 100 nmϕ via resistance after 20 cycles of HTC test from 25°C to 350°C. The average resistance of the fenced vias increased 8%, while the tapered vias suppressed the resistance increment within 4%. Fenced via degrades the thermal stability.

Namely, it is important to control the via-profile by modifications of the etching condition for the line trenches as well as the chemical composition of BACR, which reduces the etching rate difference from that of the low-*k* ILD.

20.3 Processes with Hard Masks (HM)

The MHM processes are innovated to avoid the oxygen plasma damage to the low-*k* sidewall etched. Here, multi-layer hard masks (MHM) with several inorganic films are stacked on low-*k* ILDs for the line isolation. In case of the via-first process as illustrated in Fig. 20.9a, MHM such as triple layers of SiO2/SiN/SiO2 are put on the ILD stacks followed by BARC and photoresist coating [15, 16]. Via-hole patterns are etched in the MHM by the photoresist (PR) exposed. Note that, at this stage, the bottom HM is not etched to cover line-ILD film. Therefore, no damage is introduced into the low-*k* ILD film, when the photo resist is removed by oxygen plasma. Then, the second BARC and photo resist are coated on the MHM to expose the line trench patterns, and the line trench patterns are etched into the MHM. The photo resist on MHM is removed eventually to engrave DD patterns of via-holes

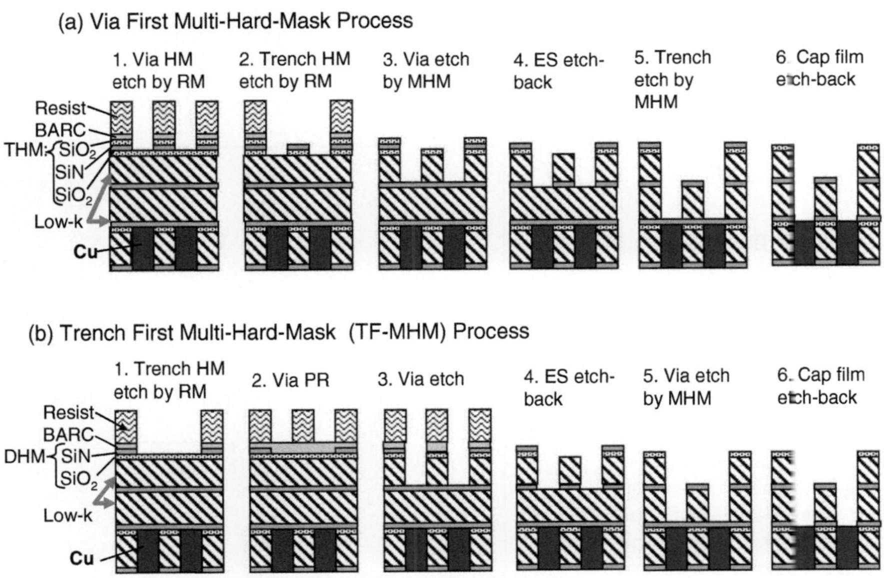

Fig. 20.9 Multi-hard mask (MHM) processes of (**a**) the via-first sequence with triple-layered HM (THM) and (**b**) the trench-first sequence with double-layered HM (DHM)

and line trenches into MHM. Next, the via-patterns are etched in the low-*k* ILD for the lines by MHM, and the ES at the via bottoms are removed by etch-back, simultaneously etching of HM in the line trench areas. By using line trench pattern of MHM and via-patterns of ES, their patterns are transferred to the line-ILD and via-ILD simultaneously. Finally, the capping film is removed by etch-back. The other MHM process has the trench-first sequence with dual hard mask (DHM) of SiN/SiO2 as illustrated in Fig. 20.9b. Here, the trench patterns are etched in top-layered hard mask in MHM by photoresist. After the resist removal, BARC and the second photo resist are coatedand the via-patterns are exposed. By using the PR

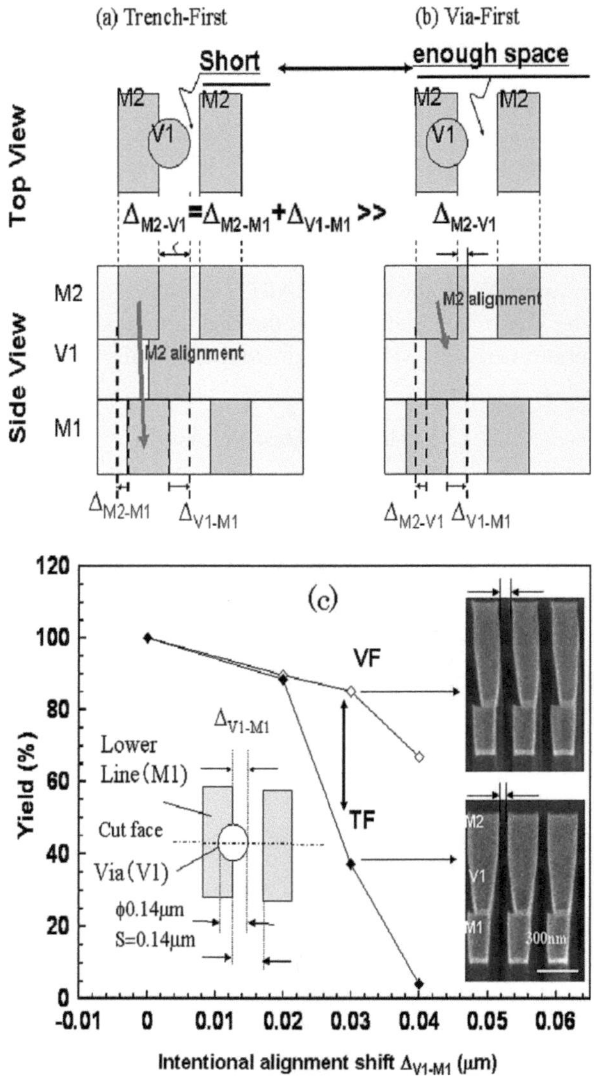

Fig. 20.10 Illustrate the alignment shift between *upper lines* (M2) and vias (V1), $\Delta_{M2\text{-}V1}$ (**a**) in the trench-first and (**b**) via-first (VF) processes, and (**c**) the yield of 0.28 µm-pitched line with 0.14 µm vias as a function of the intentional alignment shift to the via and M1. The VF-MHM process is superior in the misalignment margin to the TF-MHM one, even though the MHM structure is more complex than that in TF-MHM [15]

mask, via-patterns are etched in the line-ILD. Then, ES at the via bottoms as well as the HM in the line trench area are removed simultaneously. The line trench in the line-ILD and vias in the via-ILD are etchedfinally the capping film is removed by etch-back. The MHM process is applicable for organic low-k film.

The trench-first (TF) process is realized with much simpler double-layered HM (DHM) than the via-first (VF) process with triple-layered HM (THM), but the misalignment margin between via and line is much narrow. Figure 20.10 illustrates the alignment shift between upper lines (M2) and vias (V1), $\mathbf{\Delta_{M2\text{-}V1}}$, in the trench-first and via-first (VF) processes. In the trench-first process, the misalignment ($\mathbf{\Delta}_{M2\text{-}V1}$) between the upper lines (M2) and the vias (V1) is large principally, because both M2 and V1 are aligned to the lower lines (M1) independently. In the worst case such as the V1 misalignment opposite to the direction of M2 misalignment, the $\mathbf{\Delta}_{M2\text{-}V1}$ increases significantly. In the via-first process, on contrast, this V1 is aligned to M1 followed by the alignment of M2 to V1, minimizing $\mathbf{\Delta}_{M2\text{-}V1}$. Figure 20.10c shows the yield of 0.28 μm pitched line with 0.14 μm vias as a function of the intentional alignment shift. The yield in the VF-MHM process was much higher than that in the TF-MHM. Especially, the yield in the TF-MHM drastically decreases when the

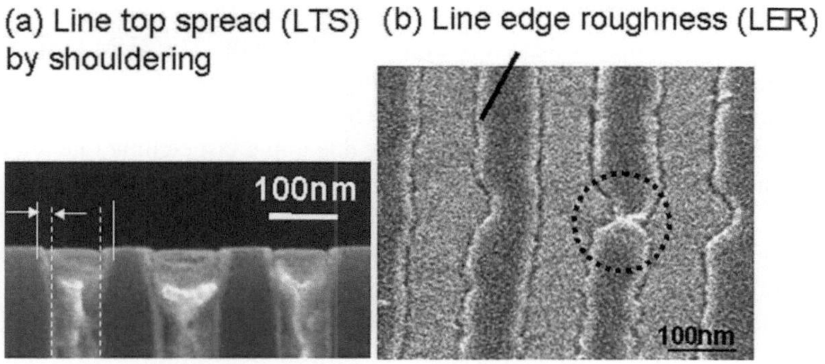

Fig. 20.11 SEM micrographs of Cu DD interconnects with (**a**) the line top spreading (LTS) and (**b**) the line edge roughness (LER). These LTS and LER cause the line-to-line shortage or leakage failure in the scaled-down narrow pitched lines

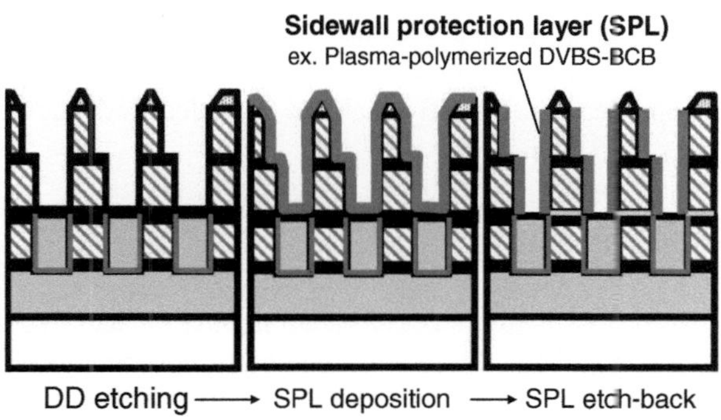

Fig. 20.12 Process sequence of SPL formation to reduce the LTS and LER

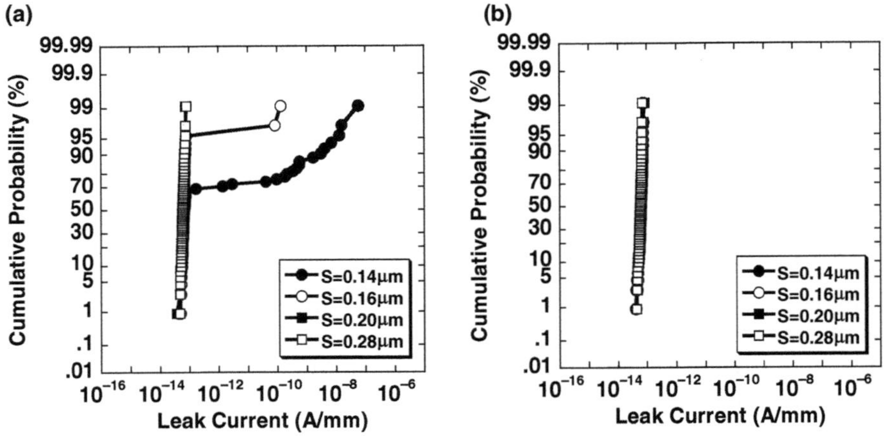

Fig. 20.13 The line-to-line leakage of the spaces from 0.14 to 0.28 μm (**a**) without and (**b**) with SPL of plasma-polymerized BCB [17]. The SPL eliminates the line-to-line leakages due to reductions in LTS and LER

intentional alignment shift was greater than 0.03 μm. These facts prove that the alignment margin in the VF-MHM is larger than that in the TF-MHM [15].

In addition to the misalignment, there are big issues on the MHM process such as line top spreading and line edge roughness. The line top spreading (LTS) is caused by shouldering of the thin MHM pattern edge due to the fast etching rate relative to the pattern center, narrowing the line space as shown in Fig. 20.11a. The line edge roughness is caused by a lack of endurance of PR to etching plasma, roughing the PR sidewall, or eventually the MHM sidewall. The line edge roughness of MHM is transferred to the sidewall of low-*k* film etched as shown in Fig. 20.11b. A solution to overcome these issues is to coat the sidewall of line trench by thin sidewall protection layer (SPL), such as plasma-polymerized BCB film [17]. As shown in Fig. 20.12, thin SPL film is deposited conformably on the low-*k* ILD patternedand the SPL is etch-backed to be remained just on the low-*k* sidewall. In case of the 5 nm thick BCB, the LTS was reduced from 30 to 20 nm and LER was decreased from 40 to 10 nm. Fig. 20.13 shows the leakage currents between the adjacent lines with several line spaces. The line-to-line leakage failure without SPL was observed with the line spaces smaller than 0.16 μm, while no leakage failure was detected with SPL. The SPL will be more important not only to avoid the line-to-line leakage but also to keep the TDDB reliability in scaled-down devices [10].

20.4 Summary

For the DD interconnect fabrication, interconnect trench and via-holes are patterned in the ILD film sequentially. The multi-hard mask (MHM) processes are innovated to avoid the oxygen plasma damage observed in the conventional photoresist (PR) mask process, but the line top spreading (LTS) and the line edge roughness (LER) come to the front issues. The lithography sequence such as the via-first or the trench-first processes affects the misalignment margin between the lines and vias. We should design the lithography process sequence carefully by taking the material combination and the misalignment margin into considerations.

Acknowledgments The author would like to acknowledge Drs. F. Ito, H. Ohtake (now in Tohoku University), J. Kawahara, N. Inoue, M. Tagami, M. Tada (now in NEC), M. Ueki, K. Hijioka, M. Abe, T. Takeuchi, T. Onodera, S. Saito, and N, Furutake all in LSI Fundamental Research Laboratory for research on advanced LSI BEOL technologies.

References

1. Hayashi, Y.: Impacts of Low-*k* Film on Sub-100 nm-node, ULSl Devices. IEEE Intl. Interconnect Tech. Conf., (San Francisco, USA), 145 (2002)
2. Maex, K.; Baklanov, M.R.; Shamiryan, D.; Lacopi, F.; Brongersma, S.H.; Yanovitskaya, Z.S.: Low dielectric constant materials for microelectronics. J. Appl. Phys, **93(11)**, 8793 (2003)
3. Abe, M.; Tada, M.; Ohtake, H.; Furutake, N.; Narihiro, M.; Arai, K.; Takeuchi, T.; Saito, S.; Taiji, T.; Motoyama, K.; Kasama, Y.; Arita, K.; Ito, F.; Yamamoto, H.; Tagami, M.; Tonegawa, T.; Tsuchiya, Y.; Fujii, K.; Oda, N.; Sekine M.; and Hayashi, Y.: A robust 45 nm-node, dual damascene interconnects with high quality cu/barrier interface by a novel oxygen absorption process. IEEE Intl. Electron Device Meeting, Tech. Washington DC, USA, Digest, 77 (2005)
4. Ogawa, E.T.; McPherson, J.W.; Rosal, J.A.; Dickerson, K.J.; Chiu, T.–C.; Tsung, L.Y.; Jain, M.K.; BonifieldT.D.; and Ondrusek, J. C.: Stress-induced voiding under vias connected to wide Cu metal leads. Proc. IEEE 40th Annual Intl. Reliability Physics Symp. San Jose, USA, 312 (2002)
5. Abe, M.; Furutake, N.; Saito, S.; Inoue N.; and Hayashi, Y.: Effects of the Metallurgical Properties of Upper Cu Film on Stress-Induced Voiding (SIV) in Cu Dual-Damascene Interconnects. Japanese J Appl. Phys. **44(4B)**, 2294 (2005)
6. Tonegawa, T.; Hiroi, M.; Motoyama, K.; Fujii, K.; and Miyamoto, H.: Suppression of bimodal stress-induced voiding using high-diffusive dopant from Cu-alloy seed layer. IEEE Intl. Interconnect Tech. Conf. San Francisco, USA, 216 (2003)
7. Furusawa, T.; Miura, N.; Matsumoto, M.; Goto, K.; Hashii, S.; Fujiwara, Y.; Yoshikawa, K.; Yonekura, K.; Asano, Y.; Ichiki, T.; Kawanabe, N.; Matsuzawa, T.; and Matsuura, M.: UV-hardened high-modulus CVD-ULK material for 45-nm node Cu/low-*k* interconnects with homogeneous dielectric structures. IEEE Intl. Interconnect Tech. Conf. San Francisco, USA, 45 (2005)
8. Miyajima, H.; Fujita, K.; Nakata, R.; Yoda, T.; and Hayasaka, N.: The application of simultaneous ebeam cure methods for 65 nm node Cu/low-*k* technology with hybrid (PAE/MSX) structure. IEEE Intl. Interconnect Tech. Conf. San Francisco, USA, 222 (2004)
9. Hayashi, Y.; Harada, Y.; Itoh, F.; Takeuchi, T.; Tada, M.; Tagami, M.; Ohtake, H.; Hijioka, K.; Saito, S.; Onodera, T.; Hara D.; and Tokudome, K.: Novel molecular-structure design for PECVD porous SiOCH films toward 45 nm node, ASICs with k = 2.3. IEEE Intl. Interconnect Tech. Conf. San Francisco, USA, 225 (2004)
10. Tada, M.; Ohtake, H.; Narihiro, M.; Itoh, F.; Taiji, T.; Tohara, M.; Motoyama, K.; Kasama, Y.; Tagami, M.; Abe, M.; Takeuchi, T.; Arai, K.; Saito, S.; Furutake, N.; Onodera, T.; Kawahara, J.; Kinoshita, K.; Hata, N.; Kikkawa, T.; Tsuchiya, Y.; Fujii, K.; Oda, N.; Sekine, M.; and Hayashi, Y.: Feasibility study of a novel molecular-pore-stacking (MPS), SiOCH film in fully scale-down, 45 nm-node Cu damascene interconnects. Symp. VLSI Technol. (Kyoto, JPN), Dig., 18 (2005)

11. Tagami, M.; Ohtake, H.; Tada, M.; Ueki, M.; Ito, F.; Taiji, T.; Kasama,Y.; Iwamoto, T.; Wakabayashi, H.; Fukai, T.; Arai, K.; Saito, S.; Yamamoto, H.; Abe, M.; Narihiro, M.; Furutake, N.; Onodera, T.; Takeuchi, T.; Tsuchiya, Y.; Oda, N.; Sekine, M.; Hane, M.; and Hayashi, Y.: High-performance Cu-interconnects with Novel Seamless Low-*k* SiOCH Stacks (SEALS) Featured by Compositional Modulation Process for 45 nm-node ULSI Devices. Symp.VLSI Tech., 134 (2006)
12. Li, H.Y.; Su, Y.J.; Tsang, C.F.; Sohan, S.M.; Bliznetsov, V.; and Zhang, L.: Process improvement of 0.13 μm Cu/Low K (Black Diamond) dual Damascene interconnection. Microelectron. Reliab. **45(7–8)**, 1134 (2005)
13. Mehta, S. S.; Qin, S. H.; Roy M. M.; Singh, N.; and Kumar, R.: Resist pattern peeling assessment in DUV chemically amplified resist. Microelectron. J. **35(5)**, 427 (2004)
14. Ohtake, H.; Saito, S.; Tagami, M.; Tada, M.; Abe, M.; Furutake, N.; and Hayashi,Y.: Via-profile controlledporous low-*k*/Cu DDIs with high thermal stability. In Ext. Abstr. 2005 Int. Conf. Solid State Dev. Mat., 300 (2005)
15. Ohtake, H.; Tagami, M.; Arita, K.; and Hayashi, Y.: Misalignment tolerated Cu dual Damascene interconnects with low-*k* SiOCH film by a novel via-first, multi-hard-mask process for sub-100-nm node ASICs. IEEE Intl. Electron Device Meeting, Tech. Washington DC, USA, Digest, 853 (2003)
16. Ueki, M.; Narihiro, M.; Ohtake, H.; Tagami, M.; Tada, M.; Ito, F.; Harada, Y.; Abe, M.; Inoue, N.; Arai, K.; Takeuchi, T.; Saito, S.; Onodera, T.; Furutake, N.; Hiroi, M.; Sekine, M.; and Hayashi, Y.: Highly reliable, 65 nm-node Cu dual Damascene interconnects with full porous-SiOCH (k = 2.5) films for low-power ASICs. Symp. on VLSI Technology Hawaii, USA, Digest, 60 (2004)
17. Kawahara, J.; Nakano, A.; Saito, S.; Kinoshita, K.; Onodera, T.; and Hayashi, Y.: High performance Cu interconnects with low-*k* BCB-polymers by plasma-enhanced monomer-vapor polymerization (PE-MVP) method. Symp. VLSI Technol. Kyoto, JPN, Dig., 45 (1999)

Chapter 21
Physical Vapor Deposition Barriers for Cu metallization – PVD Barriers

Junichi Koike

21.1 Necessity of Diffusion Barrier Layer

Cu is an interstitial impurity in Si and its diffusivity in Si is faster than other transition metals and of the order of 10^{-5} to 10^{-7} cm^2/s in the temperature range of 200–500°C [1]. Electronically, Cu is a deep-level dopant in Si and forms various donor and acceptor levels, inducing current leakage [2, 3]. In a multilayered device structure, Cu diffuses through a dielectric layer and reaches a Si substrate under electric bias field [4]. In order to prevent Cu diffusion, a barrier layer is necessary at an interface between Cu and the dielectric layers. By the use of high-resistivity barrier an metal, the effective resistivity of interconnect lines increases with the advancement of the technology node as shown in Fig. 21.1 [5, 6]. For a fixed barrier thickness of 10 nm, for example, effective resistivity increases rapidly from 2.35 μΩ cm for the 65 nm node to 2.85 μΩ cm for the 32 nm node. Meanwhile, the effective resistivity of 2.2 μΩ cm should be maintained in order to minimize *RC* delay [7]. This recommendation by the International Technology Roadmap for Semiconductors (ITRS) determines a required barrier thickness at a given technology node. In the 32 nm node, the barrier thickness should be 3.5 nm, approximately 10 atomic layers to prevent interdiffusion between Cu and the dielectric layer. In order to achieve this requirement, a proper barrier material should be deposited using proper deposition techniques and conditions. Wang et al. summarized the published data as of the year 1993 together with their investigation of TiW barrier [8]. Kaloyeros and Eisenbraun [9] published an excellent review of barrier materials as of 2000. In their review article, advantages and limitations of various barrier materials were described in detail based on numerous experimental works by others. Readers can find in this article how and why Ta/TaN barrier had come to use for the Cu interconnect. Since then, technology has rapidly advanced along the line of the ITRS roadmap. Once the technology node entered to a sub-micrometer range, barrier

---

J. Koike (✉)
Department of Materials Science, Tohoku University, Sendai 980-8579, Japan
e-mail: koikej@material.tohoku.ac.jp

Y. Shacham-Diamand et al. (eds.), *Advanced Nanoscale ULSI Interconnects: Fundamentals and Applications*, DOI 10.1007/978-0-387-95868-2_21,

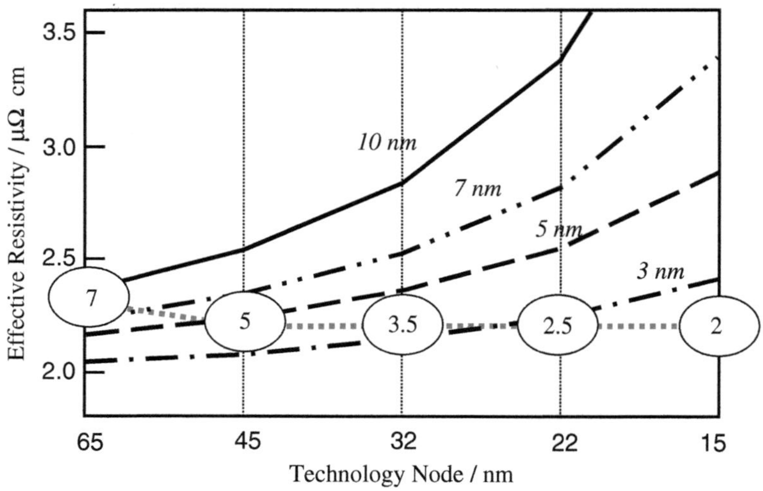

Fig. 21.1 Effective resistivity of the Cu/Ta/TaN interconnect for future technology node [6]. Numbers in italic are hypothetical barrier thickness of a fixed value. A *gray dotted line* is a required value of effective resistivity by ITRS. Numbers in circles are required barrier thickness for a given technology node

thickness becomes a critical issue to ensure the expected performance and reliability of advanced devices. Barrier materials and processes need to be revisited from fundamental viewpoint. In this chapter, the issues of physical vapor deposited (PVD) barrier will be discussed in terms of metallurgical and thermodynamic aspects.

21.2 Metallurgy for Barrier Material Selection

Conventional criteria for the selection of barrier metals are based on the following thermo-physical characteristics: (1) mutual solid solubility between barrier metals and Cu is negligibly small; (2) barrier metals do not form intermetallic compounds with Cu; (3) the electrical resistivity of barrier metals is smallest possible; and (4) barrier metals have high melting point, having a slow self-diffusivity. These parameters in bulk form are summarized in Table 21.1 including the information about the maximum solid solubility of barrier metals in Cu, the possibility of compound formation with Cu, the melting point, and the resistivity of barrier metals.

Elements in gray-colored cells have small values of the maximum solid solubility and do not form any intermetallic compounds with Cu. Although the maximum solubility values are finite, they all have negligibly small values below 500°C. One can expect no driving force for Cu to diffuse into the layers composed of these elements which can then act as a diffusion barrier layer. It is noted that vanadium is an exception. The solubility of V in Cu is negligible but the solubility of Cu in V has a relatively large value of 7.4 at.% at ambient temperature. The large solubility on the vanadium side leads to Cu diffusion through V [10]. Among the elements in the

Table 21.1 Related parameters for diffusion barrier

Element	Ta
Maximum solubility (at.%)	0
Compound formation (Yes or No)	N
Melting temperature (°C)	3015
Resistivity at 20°C (μΩ cm)	13.1

Li 22 N 181 9.32	Be 13.7 Y 1287 3.25											B 0.29 N 45 4512	C 0.04 N	N	O 0.03 Y		
Na 22 N 97.8 4.75	Mg 6.93 Y 649 4.30											Al 19.7 Y 660 2.74	Si 11.15 Y 1414	P 3.5 Y	S 0.02 Y		
K 63.2 7.19	Ca 0 Y 843 3.35	Sc 0.5 Y 1538 46.8	Ti 8 Y 1667 47	V 0.17 N 1902 19.9	Cr 0.89 N 1857 12.9	Mn 100 N 1244 136	Fe 4.6 N 1536 9.8	Co 8 N 1495 5.2	Ni 100 N 1453 7.04	Cu	Zn38. 3 Y 420 6.17	Ga 20.6 Y 30 14.85	Ge 12 Y 938	As 6.83 Y 817 29	Se 0 Y 221		
Rb 12.51	Sr 0 Y 769 21.5	Y 0 Y 1522	Zr 0.12 Y 1855 45	Nb 0.1 N 2467 14.5	Mo 0.06 N 2620 5.3	Tc 0 N 2155	Ru 0 N 2250 7.37	Rh 100 N 1963 4.78	Pd 100 Y 1555 10.55	Ag 14.1 N 962 1.61	Cd 2.07 Y 321 7.28	In 10.9 Y 157 8.75	Sn 9.1 Y 232 ~11	Sb 5.8 Y 631 41.3	Te 0 Y 450		
Cs 0 N 29.8 19.96	Ba 0 Y 725 39	La 0 Y 921 79	Hf 0.4 Y 2227 30.6	Ta 0 N 3015 13.1	W 0 N 3400 5.3	Re 0 N 3180 16.9	Os 0 N 3030 8.4	Ir 8 N 2410 5.07	Pt 100 Y 1772 10.42	Au 100 Y 1064 2.20	Hg 5 Y -39 95.9	Tl 0 N 304 16.4	Pb 0 N 327 21.0	Bi 0 N 271 116	Po 254 46		

Table 21.2 Resistivity and heat of formation of nitrides

	TiN	VN	ZrN	NbN	HfN	TaN	WN
ρ ($\mu\Omega$ cm)	22–130	86	12–14	200	32	135	–
ΔH (kJ/mol)	337	217, 264	365	234, 249	369	252, 271	15–22

gray-colored region, Ta has been most commonly used. Ta thin films generally have polycrystalline structure whose grain boundaries can act as fast diffusion paths of Cu. Even under the presence of grain boundaries, Cu diffusion in Ta is reported to be very slow with a high activation energy of approx. 2.3–2.5 eV [11, 12]. Holloway et al. reported that the Ta barrier is stable up to 550°C for 30 min. Above this temperature, Cu diffuses through Ta and forms Cu_3Si at the Ta/Si interface [12]. Bias thermal stressing (BTS) test was also reported on the Ta barrier and the TiN barrier. It was found that the Ta barrier can withstand the BTS test for much longer time than the TiN barrier, especially when they were annealed [13].

Even better barrier property was reported on Ta nitride films. This is considered to be due to their densely packed crystal structure. The barrier property improves with increasing nitrogen concentration to a stoichiometric 1:1 compound. Transition-metal nitrides are often called interstitial nitrides in which nitrogen atoms occupy the interstitial sites of the closely packed structure of transition metals. For this reason, other nitrides are possible choices for a diffusion barrier. Representative nitrides are listed in Table 21.2 with their bulk resistivity values and heat of formation at 25°C. HfN and ZrN may seem attractive because of their small resistivity values. However, both Hf and Zr have tendency to form intermetallic compounds with Cu and may not be good choices for barrier materials. The most common nitride barriers are TaN, TiN, and WN. The heat of formation of TaN and TiN are large negative, indicating their good thermal stability. On the other hand, WN is not expected to be as stable as the other nitrides and has a tendency to dissolve and form silicide. The barrier property of these nitrides depends on nitrogen concentration and microstructure. TiN_x tends to form a columnar grain structure with its grain boundaries running through the film thickness, leading to failure [14]. Tungsten-rich WN_x fails at about 450°C because of grain-boundary diffusion associated with crystallization to W and W_2N. In contrast, nitrogen-rich WN_x remains amorphous and stable up to 600°C [15]. TaN_x has various stable compounds and has better barrier property than pure Ta [16]. Barrier property improves with increasing N concentration [17]. Failure occurs by grain-boundary diffusion [18].

Transition-metal carbides are also interstitial carbides and are candidate materials for a diffusion barrier. Representative carbides of transition metals are listed in Table 21.3. Similar to the nitrides, WC has less thermodynamic stability than the other carbides. Some nitrides and carbides show low resistivity values. Nevertheless, both nitrides and carbides do not adhere well with Cu because of the absence of strong bonds between Cu–C or Cu–N.

Adhesion of Cu with nitrides and carbides can be understood in terms of wettability [19]. The work of adhesion of a liquid phase on a solid substrate can be given by the Young–Dupré equation:

Table 21.3 Resistivity and heat of formation of carbides

	TiC	VC	ZrC	NbC	HfC	TaC	WC
ρ (μΩ cm)	70–173	150–160	50–64	74–254	60	20–175	53
ΔH (kJ/mol)	184	101	202	138, 186	–	143, 203	26, 38

$$W_a^0 = \sigma_{LV} (\cos\theta + 1) \tag{21.1}$$

where σ_{LV} is the surface energy of the liquid phase and θ is contact angle between the liquid and the solid. The surface energy of the liquid phase scales linearly with the heat of evaporation and is loosely in proportion with melting temperature. The high contact angle is generally attributed to a weak van der Waals bonding. In contrast, the low contact angle is attributed to hybridization of electronic states or to metallic bonding.

Contact angle is influenced by the occurrence of interface reaction. For metal/metal systems, contact angle is below 50°. While for metal/oxide, contact angle is dependent on mutual reactivity. Non-reactive oxides form a contact angle of 100–140°, while reactive oxides form a contact angle of 0–90°. The former includes Cu/MgO [20], Cu/Al_2O_3 [21] and has small adhesion energy with Cu. The latter includes Cu/NiO and Cu/Fe_3O_4 [22]. These reactive oxides have large adhesion energy with Cu. However, this reaction involves with the reduction and dissolution of the oxides and may lead to the formation of freely migrating metallic atoms such as Ni and Fe. The reactivity effect is also expected on the adhesion of metal/fluoride, metal/nitride, and metal/carbide systems. The contact angle of Cu/CaF_2 [23] takes a large value of 140°. Non-reactive nitrides such as AlN [24] and Si_3N_4 [25] form a contact angle of 120–135°. Addition of Ti into Cu decreases contact angle with Si_3N_4 due to the interface reaction to form TiN at elevated temperatures [26]. In Cu/Ta, adhesion of Cu degrades with the formation of a non-reactive surface oxide of Ta substrate. On the other hand in Ta/Cu, adhesion of Ta improves with the formation of a reactive surface oxide of Cu substrate. Interestingly in Cu/WC, a very low contact angle of 10–30° can be observed. This is due to the metallic nature of electrons in WC [27]. Other conductive carbides show similar result. The effect of oxygen in TiN shows similar results with Al and Cu [28, 29]. Exposure of TiN to air before Al deposition led to the improvement in barrier reliability because of the Al reaction with oxygen. In contrast, the same air exposure of TiN does not show any beneficial effect for Cu because of the absence of the interface reaction.

These results suggest that the surface condition of the substrate is extremely important. Refractory metals, nitrides, and carbides are easily oxidized. Once their surface is oxidized, contact angle increases and adhesion energy decreases. For instance in Cu/W, contact angle was 10° when the W surface was purified with electron beam [30]. Whereas, contact angle increased to 50° when heat treatment was done at 900°C for 30 min under a vacuum of 1×10^{-3} Pa [31]. This fact suggests that conductive the nitrides and carbides can have good adhesion strength with

Cu as well the as good diffusion barrier property. Since adhesion strength is proportionally related to stress-migration and electro-migration resistance [32], a metallic overlayer is deposited as an adhesion promoter. A typical combination of the adhesion promoter and the nitride barrier layer is Ta/TaN. In this case, special care is needed to prevent the formation of a non-reactive surface oxide on Ta.

Although the Ta/TaN barrier has been most commonly used, the bilayer structure has a disadvantage in the further reduction of barrier thickness. In addition, conventional sputtering technique has its own limitation to form a thin and conformal barrier layer. Various new techniques have been investigated to overcome the difficulty in the conventional process.

21.3 Deposition Technique

Sputter deposition has been widely used as a physical vapor deposition (PVD) method. It has been used to deposit a barrier layer and a seed layer in Damascene trenches and vias [33]. Because sputtered atoms have a wide angular divergence, they tend to stick on a flat substrate surface and at the corner of via and trench openings. This leads to closure, or "pinching off", of the vias and the trenches and prevents atom supply into the narrow vias and trenches. Various efforts have been made to improve the directionality of sputtered atoms. The angular divergence can be reduced by the use of long throw sputtering (LTS) operated with a longer target-to-substrate (T–S) distance and at a lower sputtering pressure than conventional sputtering [34, 35]. The longer T–S distance prevents the arrival of sputtered atoms traveling at a large ejection angle from the target. The lower sputtering pressure reduces scattering events with Ar gas and encourages well-oriented sputtered atoms to arrive at the substrate. However, the directionality of the LTS becomes insufficient with the increase of the wafer size, leading to the asymmetric deposition of sputtered atoms within the vias and the trenches located away from the wafer center [36, 37]. In order to improve the asymmetry problems, a collimator can be placed between the substrate and the sputter target [38–40]. The collimator is made of a metal plate having an array of through-holes. This collimated sputtering is subject to a substantial atom deposition on the collimator and, in turn, a very slow deposition rate on the substrate, which increases the overall process cost. Thus far, the directional sputtering techniques have not been utilized successfully for barrier deposition.

An alternative technique is ionized sputter deposition with which sputtered atoms are ionized by an inductively coupled RF antenna placed over a target and are pulled into vias and trenches under a biased field [41, 42]. Using this method, a significant improvement can be obtained in side-wall and bottom-wall coverage [43]. When the ionized atoms are deposited on the bottom wall with a certain kinetic energy, already deposited atoms are resputtered and redistributed within the vias and the trenches. On the other hand, a small fraction of the non-ionized atoms are deposited on the upper side walls, leading to a good coverage of the entire side walls as well as of the bottom walls. However, this method cannot be applied to deposit a thinner barrier layer in a smaller feature size. The resputtering process leads to the formation of a void or a weak structure in the thin diffusion barrier.

Other methods have been proposed to improve the conformal formation of the barrier layer. Chemical vapor deposition (CVD) has been attempted using both inorganic and metal organic precursors [44–46]. However, deposited films tend to have a low density and contain F, C, or Cl impurities originated from the precursors, leading to a poor reliability. Conformal coverage can be further improved by the use of atom layer deposition (ALD) technique [47]. The details of this technique will be discussed in the next section. An alternative technique is self-assembled monolayer (SAM) deposition [48–50]. Polymer bilayers of polyethyleneimine (PEI) and polyacrylic acid (PAA) layers were adsorbed onto the SiO_2/Si substrate by sequential immersion in the polyelectrolyte solutions. Bias thermal annealing (BTA) test at 200°C and a 2 MV cm^{-1} electric field indicated that the Cu/polymer (1.5–2 nm thick)/SiO_2/Si samples showed a factor five increase in the failure time in comparison with Cu/SiO_2/Si samples. However, the SAM barrier has not been compared with conventional PVD barrier and its reliability is not known.

21.4 Self-Forming Technique

The concept of the self-formation of the barrier layer appeared as zero barrier thickness in the 1999 version of the ITRS road map. Basic idea is to add an alloying element to Cu, let it migrate to Cu/dielectric interface, react with the dielectric, and form a stable phase that can act as a diffusion barrier layer [51, 52]. In order to select proper alloying elements, thermodynamic driving force and kinetic factors are to be considered. In the following discussion, silicon oxide is considered for a representative dielectric material.

21.4.1 Driving Force

Two types of driving force should be considered: the driving force for the formation of an X oxide and the driving force for the expulsion of an alloying element from the Cu alloy. The first type corresponds to the standard free energy of oxide formation, $\Delta G°$, and can be found in the Ellingham diagram shown in Fig. 21.2 If the $\Delta G°$ values of some elements are not available in the diagram, heat of formation, ΔH, can be used instead and is readily available in various reference books [53]. In Fig. 21.2, Mg, Al, and Ti form more stable oxides than Si. If these metallic elements are in contact with SiO_2, the reduction reaction of SiO_2 may occur. In fact in Cu–Mg alloy on SiO_2, heat treatment at 600°C leads to the formation of MgO at the interface and accompanied the reduction of SiO_2, making Si atoms free to migrate

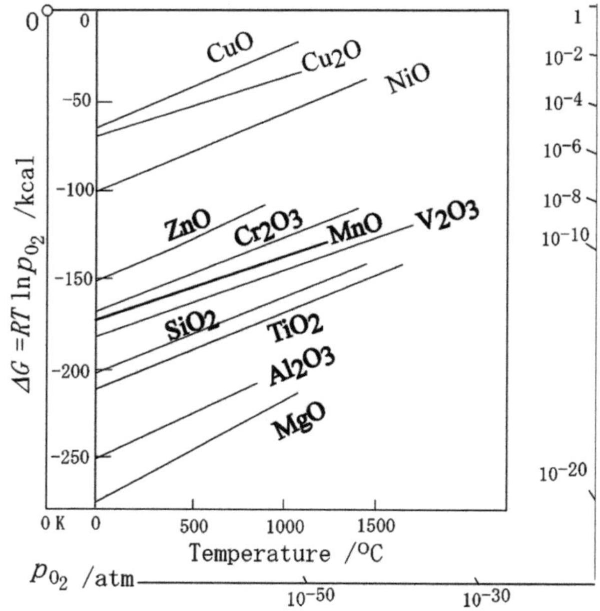

Fig. 21.2 Ellingham diagram showing the standard free energy of oxide formation at various temperatures and oxygen partial pressures

into the overlaying Cu film [54]. In contrast, such reaction may not occur for the metals of moderate ΔG° values, which may be preferable for the alloying element.

The second type of the driving force is related to the activity coefficients of alloying elements in Cu. The activity coefficient, γ, is a measure of deviation from a state of ideal solid solution and represents excess chemical potential. The chemical potential of an alloying element X in an ideal solid solution of Cu–X is given by

$$\mu_X = \mu_{Cu}^0 + RT \ln N_X \tag{21.2}$$

Here, μ_{Cu}^0 is the chemical potential of the pure Cu matrix and N_X is the concentration of the alloying element X. When there is deviation from the ideal state by the presence of chemical interaction, the chemical potential of X either increases or decreases with respect to that in Eq. (21.2). This is given by

$$\mu_X = \mu_{Cu}^0 + RT \ln a_X = \mu_{Cu}^0 + RT \ln N_X + RT \ln \gamma_X \tag{21.3}$$

Here, a_X is activity and is given by the product of concentration N_X and activity coefficient γ_X. The activity coefficient is 1 for ideal state, while larger or smaller than 1 for non-ideal state. When it is larger than 1, the chemical potential increases from that of the ideal state. This suggests that the alloying element would leave Cu if more stable configurations are available by reacting with a neighboring phase to the alloy. One such example is the migration of X to the film surface followed by

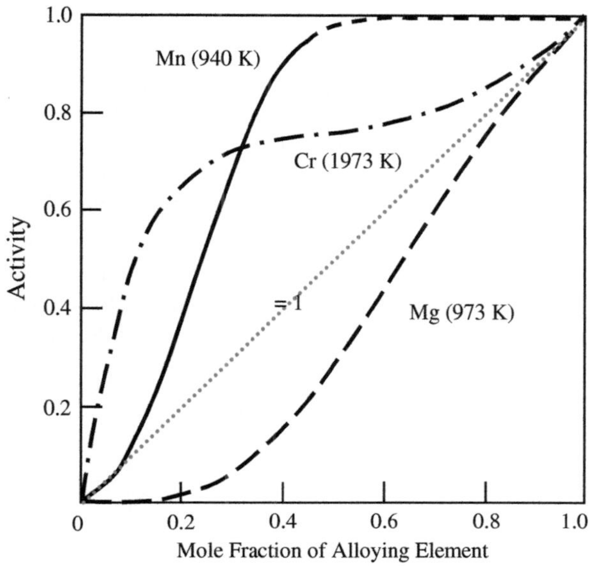

Fig. 21.3 Activity of Cr, Mg, and Mn in a Cu solid solution. Temperature in parentheses is measured temperature of each element [55–57]

the formation of X oxide under oxygen-containing atmosphere. Another example is migration to the dielectric interface followed by the reaction and formation of complex oxide including X. In this way, the total free energy of the system, O_2-including atmosphere and Cu–X/SiO_2 therein, can be minimized. On the contrary, when the activity coefficient is smaller than 1, chemical potential decreases from that of the ideal state. In this case, the alloying element tends to remain as a Cu solid solution unless the first type of driving force is very strong. In this regard, the third term in Eq. (21.3), $RT\,\mathbf{ln}\,\gamma_X$, is considered as a driving force to expel the alloying element X and to make the Cu–X alloy to pure Cu under the presence of, for example, oxygen atmosphere or an underlying oxide layer.

Figure 21.3 shows activity as a function of solute concentration for Cu–Mg [55], Cu–Cr [56], and Cu–Mn [57] alloys. Reported data of activity coefficients in Cu solid solution are summarized in Table 21.4 [55–61]. The slope of the curve corresponds to the activity coefficient. The activity coefficient of Mg is much smaller than 1, while that of Cr is much larger than 1. Therefore, Mg tends to remain in

Table 21.4 Metal classification by activity coefficient [55–61]

Activity coefficient	Elements in Cu solid solution
$\gamma < 1$	Al, As, Au, Ga, Ge, Mg, Sb, Se, Si, Sn, Te, Zn, Zr
$\gamma \sim 1$	Cd, In, Mn
$\gamma > 1$	Tl, Pb, Bi, Ag, Ni, Cr

Cu and has a wide solid-solution range as well as to form intermetallic compounds, leading to resistivity increase. Possible elements of this kind are, for instance, Al, Ti, Sr, Y, Zr, In, and Sn. In contrast, Cr tends to move out from Cu and has negligible solid-solution range without the formation of intermetallic compounds. Alloying elements of this kind having negligible solid solubility tend to segregate not only to surface and interface but also to grain boundaries [62]. If grain-boundary segregation occurs, film resistivity would increase and is not preferable for interconnect application. Possible elements of this kind are Cr, Nb, Mo, Tc, Ru, Cs, Ta, W, Re, Os, Tl, Pb, and Bi. In Contrast to the two types, Mn is soluble compared to Cu and takes an intermediate value of the activity coefficient in a low concentration range. In this case, Mn is expected to have a weak interaction with Cu and may have no strong reduction reaction with SiO_2.

21.4.2 Kinetics

Even if there is a driving force, kinetic factors play an important role. When a passivation oxide is formed, its growth is rate controlled either by reaction or by diffusion. In the case of the self-forming barrier, the diffusivity of the alloying elements should not be a rate-controlling factor. The alloying elements should diffuse into the reaction interface before noticeable interdiffusion takes place between Cu and SiO_2. In this regard, the diffusivity of the alloying element should be faster than the self-diffusivity of Cu. The diffusivity of alloying elements in Cu can be divided into three categories [63]. The first category is IV_A–VII_A transition metals. The diffusivity of Ti, V, Cr, and Mn was experimentally determined and is faster than Cu. The second category is VIII_A metals whose diffusivity is slower than Cu. The third category is I_B–VI_B elements whose diffusivity is faster than Cu.

As a choice of alloying elements, Mn satisfies all the conditions of oxide formation energy, activity coefficient, segregation tendency, and diffusivity. Koike et al. actually examined the possibility of self-forming barrier process with Cu–Mn alloy. They found that a barrier layer of 3–4 nm thick was formed after annealing at 450°C [64, 65]. Under a proper annealing atmosphere, residual Mn in the alloy film was found to migrate to film surface and formed Mn oxide, leading to a decrease in film resistivity. The self-forming barrier process was further investigated in a two-layered dual Damascene test structure of the 90 nm technology node. Usui et al. reported a substantial improvement in stress- and electromigration resistance and a better electrical property with smaller via resistance [66].

References

1. Graff, K.: Metal Impurities in Silicon-Device Fabrication, Springer-Verlag, Berlin, 29 (1999)
2. Collins, C. B. and Carlson, R. O.: Properties of silicon doped with iron or copper. Phys. Rev. **108**, 1405 (1957)
3. Toyama, N.: Copper impurity levels in silicon. Solid State Electron. **26(1)**, 37 (1983)

4. Shacham-Diamond, Y.; Dedhia, A.; Hoffstetter, D.; and Oldham, W. G.: Copper transport in thermal SiO_2. J. Electrochem. Soc. **140(8)**, 2427 (1993)
5. Kapur, P.; McVittie, J. P.; and Saraswat, K. C.: Technology and reliability constrained future copper interconnects-part I: Resistance modeling. IEEE Trans. Electron Devices **49(4)**, 590 (2002)
6. Shibata, H.: Practical roadmap and approach of multi-level interconnect technology for realizing over GHz system-on-chip. Proceedings of International Symposium on ULSI Process Integration of the 199th Electro-Chemical Society Meeting. 430 (2001)
7. International Technology Roadmap for Semiconductors (2003)
8. Wang, S-Q.; Suthat, S.; Hoeflich, K.; and Burrow, B. J.: Diffusion barrier properties of TiW between Si and Cu. J. Appl. Phys. **73(5)**, 2301 (1993)
9. Kaloyeros, A. E. and Eisenbraun, E.: Ultrathin diffusion barriers/liners for gigascale copper metallization. Annu. Rev. Mater. Sci. **30**, 363 (2000)
10. Iwamori, S.; Miyashita, T.; Fukuda, S.; Fukuda, N.; and Sudoh, K.: Effect of a metallic interfacial layer on peel strength deterioration between a Cu thin film and a polyimide substrate. J. Vac. Sci. Technol. B **15(1)**, 53 (1997)
11. Gjostein, N. A.: Diffusion in Metals. Westerville, OH, ASM (1973)
12. Holloway, K.; Fryer, P. M.; Cabral. C.; Harper, J. M.; Bailey, P. J.; and Kelleher, K. H.: Tantalum as a diffusion barrier between copper and silicon: Failure mechanism and effect of nitrogen additions. J. Appl. Phys. **71(11)**, 5433 (1992)
13. Choe, H. S. and Danek, M.: MOCVD TiN diffusion barriers for copper interconnects. Proc. IEEE Int. Interconnect Technology Conference 62 (1999)
14. Olowolafe, J.; Mogab, C.; Gregory, R.; and Kottke, M.: Interdiffusions in Cu/reactive-ion-sputtered TiN, Cu/chemical-vapor-deposited TiN, Cu/TaN, and TaN/Cu/TaN thin-film structures: Low temperature diffusion analyses. J. Appl. Phys. **72(9)**, 4099 (1992)
15. Ko, D.; Park, B.; Kim, Y.; Ha, J.; and Park, Y.: Advanced Metallization and Interconnect Systems for ULSI Applications in 1995, Mater. Res. Soc., Pittsburgh, 257 (1996)
16. Bai, G.; Wittenbrock, S.; Ochoa, V.; and Bohr, M.: Effectiveness and reliability of metal diffusion barriers for copper interconnects. Mater. Res. Soc. Symp. Proc. **403**, 501 (1996)
17. Kim, K.: Advanced Metallization and Interconnect Systems for ULSI Applications. Mater. Res. Soc., Pittsburgh, 281 (1995)
18. Min, K.-H.; Chun, K.-C.; and Kim, K.-B.: Comparative study of tantalum and tantalum nitrides (Ta_2N and TaN) as a diffusion barrier for Cu metallization. J. Vac. Sci. Technol. B **14(5)**, 3263 (1996)
19. Eustathopoulos, N.; Nicholas, M. G.; and Drevet, B.: Wettability at High Temperatures, Pergamon, Amsterdam (1999)
20. Naidich, Yu. V.: Wettability of solids by molten metals. In Progress in Surface and Membrane Science. Cadenhead, D. A. and Danielli, J. F., Eds. Academic Press, New York, **14**, 353 (1981)
21. Chatain, D.; Rivollet, I.; and Eustathopoulos, N.: Thermodynamic adhesion in nonreactive liquid metal-alumina systems. J. Chem. Phys. **83**, 561 (1986)
22. Naidich, Yu. V.: Contact phenomena in molten metals. Naukova Dumka; Kiev (1972)
23. Naidich, Yu. V.: Wettability of halides with molten metals, Physico-chemical and practical aspects. Powder Metallurgy and Metal Ceramics **39**, 355 (2000)
24. Naidich, Yu. V. and Taranets, N. Y.: Wettability of aluminum nitride by tin aluminum melts. J. Mater. Sci. **33**, 3993 (1998)
25. Ljunberg, L. and Warren, R.: Wetting of silicon nitride with. selected metals and alloys. Ceram. Eng. Sci. Proc. **10**, 1655 (1989)
26. Nicholas, M. G.; Mortimer, D. A.; Jones, L. M.; and Crispin, R. M.: Some observations on the wetting and bonding of nitride ceramic. J. Mater. Sci. **25**, 2679 (1990)
27. Ramqvist, L.: Wetting of Metallic Carbides by Liquid Copper, Nickel, Cobalt and Iron. Int. J. Powder Metall. **1(4)**, 2 (1965)

28. Sinke, W.; Frijlink, P.; and Saris, F.: Oxygen in titanium nitride diffusion barriers. Appl. Phys. Lett. **47(5)**, 471 (1985)
29. Park, K. and Kim, K.: Comparative Study on the Titanium Nitride (TiN) As a diffusion Barrier Between Al/Si and Cu/Si: Failure Mechanism and Effect of `Stuffing. Mater. Res. Soc. Symp. Proc. **391**, 211 (1995)
30. Doussin, L. and Omnes, J.: Technical report (report 1/1259 M), Office National d'Etudes et de Recherches Aerospariales, Direction des Materiaux, Chatillon, France (1967)
31. Nicholas, M. and Poole, D. M.: The influence of oxygen on wetting and bonding in Cu-W sys-. Tem. J. Mater. Sci. **2(3)**, 269 (1967)
32. Lane, M. W.; Liniger, E. G.; and Lloyd, J. R.: Relationship between interfacial adhesion and electromigration in Cu metallization. J. Appl. Phys. **93(3)**, 1417 (2003)
33. Rossnagel, S. M.: Sputter deposition for semiconductor: Manufacturing. IBM J. Res. Develp. **43**, 163 (1999)
34. Cuomo, J. J. and Rossnagel, S. M.: Hollow Cathode Enhanced Magnetron Sputtering. J. Vac. Sci. Technol. A **4**, 393 (1986)
35. Liu, D.; Dew, S. K.; Brett, M. J.; Janacek, T.; Smy, T.; and Tsai, W.: Experimental study and computer simulation of collimated sputtering of titanium thin films over topographical features. J. Appl. Phys. **74(2)**, 1339 (1993)
36. Mayo, A. A.; Hamaguchi, S.; Joo, J. H.; and Rossnagel, S. M.: Across-wafer nonuniformity of long throw sputter deposition. J. Vac. Sci. Technol. B **15**, 1788 (1997)
37. Smy, T.; Tang, L.; Chan, K.; Tait, R. N.; Broughton, J. N.; Dew, S. K.; and Brett, M. J.: A simulation study of long throw sputtering for diffusion barrier deposition into high aspect vias and contacts. IEEE Trans. Electron Devices **45**, 1414 (1998)
38. Rossnagel, S.; Mikalsen, D.; Kinoshita H.; and Cuomo, J. J.: Collimated magnetron sputter deposition. J. Vac. Sci. Technol. A **9(2)**, 261 (1991)
39. Motegi, N.; Kahimoto, Y.; Nagatani, K.; Takahashi, S.; Kondo, T.; Mizusawa, Y.; and Nakayama, I.: Long-throw low-pressure sputtering technology for very large-scale integrated devices. J. Vac. Sci. Technol. B **13(4)**, 1906 (1995)
40. Broughton, J. N.; Brett, M. J.; Dew, S. K.; and Este, G.: Titanium sputter deposition at low pressures and long throw distances. IEEE Trans. Semiconduct. Manufact. **96**, 122 (1996)
41. Yamashita, M.: Sputter Type High Frequency Ion Source for Ion Beam . . . Sputtering Apparatus. J. Vac. Sci. Technol. A **7**, 151 (1989)
42. Rossnagel, S. M. and Hopwood, J. H.: Magnetron sputter deposition with levels of metal ionization. Appl. Phys. Lett. **63**, 3285 (1993)
43. Hamaguchi, S. and Rossnagel, S. M.: Liner conformality in ionized magnetron sputter metal deposition process. J. Vac. Sci. Technol. B **14**, 2603 (1996)
44. Sugiyama, K.; Pac, S.; Takahashi, Y.; and Motojima, S.: LowTempera-ganometallic. Compounds. J. Electrochem. Soc. **122**, 1545 (1975)
45. Fix, R.; Gordon, R.; and Hoffman, D.: Chemical vapor deposition of vanadium and tantalum nitride thin films. Chem. Matter. **5**, 614 (1993)
46. Tsai, M.; Sun, S.; Tsai, C.; Chuan, S.; and Chiu, H.: Comparison of the diffusion barrier properties of chemical-vapor-deposited TaN and sputtered TaN between Cu and Si. J. Appl. Phys. **79(9)**, 6932 (1996)
47. Kim, H.: Atomic layer deposition of metal and nitride thin films: Current research efforts and applications for semiconductor device processing. J. Vac. Sci. Technol. B **21(6)**, 2231 (2003)
48. Krishnamoothy, A.; Chanda, K.; Murarka, S. P.; Ryan, J.; and Ramanath, G.: Self-assembled near-zero-thickness molecular layers as diffusion barriers for Cu metallization. Appl. Phys. Lett. **78(17)**, 2467 (2001)
49. Ganesan, P. G.; Gamba, J.; Ellis, A.; Kane, R. S.; and Ramanath, G.: Polyelectrolyte nanolayers as diffusion barriers for Cu metallization. Appl. Phys. Lett. **83(16)**, 3302 (2003)
50. Mikami, N.; Hata, N.; Kikkawa, T.; and Machida, H.: Robust self-assembled monolayer as diffusion barrier for copper metallization. Appl. Phys. Lett. **83(25)**, 5181 (2003)

51. Ding, P. J.; Lanford, W. A.; Hymes, S.; and Murarka, S. P.: Room-temperature continuous-wave operation of a single-layered 1.3 μm quantum dot laser. Appl. Phys. Lett. **75(21)**, 3267 (1994)
52. Lanford, W. A.; Ding, P. J.; Wang, W.; Hymes, S.; and Murarka, S. P.: Low-temperature passivation of copper by doping with Al or Mg. Thin Solid Films **62(1–2)**, 234 (1995)
53. Smithells Metals Reference Book, 7th Ed., Brandes, E. A.; and Brook, G. B., eds. Butterworth Heinemann (1992)
54. Frederick, M. J. and Ramanath, G.: Kinetics of interfacial reaction in Cu–Mg alloy films on SiO_2. J. Appl. Phys. **95(1)**, 363 (2004)
55. Hino, M.; Nagasaka T.; and Takehama, R.: Activity measurement of the constituents in liquid Cu-Mg and Cu-Ca alloys with mass spectrometry. Metall. Mater. Trans. **31B**, 927 (2000)
56. Jacob, K. T.; Priya, S.; and Waseda, Y.: A thermodynamic study of liquid Cu-Cr alloys and metastable liquid immiscibility. Z. Metallkd. **91(7)**, 594 (2000)
57. Lewin, K.: Thermodynamic study of the Cu-Mn system. Scan. J. Metall. **22**, 310 (1993)
58. Oyamada, H.; Nagasaka, T.; and Hino, M.: Activity measurement of the constituents in liquid Cu-Al alloy with mass-spectrometry. Mater. Trans. **12**, 1225 (1998)
59. Witusiewicz, V; Arpshofen, I; and Sommer, F.: Thermodynamics of liquid Cu-Si and Cu-Zr alloys. Z. Metallkd. **91**, 594 (2000)
60. Katayama, I.; Shimatani, H.; and Kouzuka, Z.: Thermodynamic Study of Solid Cu-Ni and Ni-Mo Alloys by E. M. F. Measurements using the solid electrolyte. J. Jpn. Inst. Metall. **37(5)**, 509 (1973)
61. Azakami T. and Yazawa. A.: Activity measurements of liquid copper binary alloys. Can. Metall. Quart. **15**, 111 (1976)
62. Hondros, E. D. and Seah, M. P.: In Physical Metallurgy. Cahn, R. W. and Haaser, P., Eds. North-Holland, Amsterdam 855 (1983)
63. Landolt-Bornstein: Numerical Data and Functional Relationships in Science and Technology, New Series, Group III: Crystal and Solid State Physics, Vol. 26, Diffusion in Solid Metals and Alloys, ed. by H. Mehrer, Springer, Berlin, 185 (1990)
64. Koike, J. and Wada, M.: Self-forming diffusion barrier layer in Cu–Mn alloy metallization. Appl. Phys. Lett. **87(4)**, 041911 (2005)
65. Koike, J.; Haneda, M.; Iijima, J.; Otsuka, Y.; Sako, H.; Neishi, K.: Growth kinetics and thermal stability of a self-formed barrier layer at Cu-Mn/SiO_2 interface. J. Appl. Phys. **102(4)**, 043527 (2007)
66. Usui, T.; Nasu, H.; Takahashi, S.; Shimizu, N.; Nishikawa, T.; Yoshimaru, M.; Shibata, H.; Wada, M.; and Koike, J.: Highly reliable copper dual-damascene interconnects with self-formed MnSixOy barrier Layer. IEEE Trans Electron Devices **53(10)**, 2492 (2006)

Chapter 22
Low-*k* Dielectrics

Yoshihiro Hayashi

22.1 Introduction

As CMOS transistors were scaled, interconnects to link them are also shrunk to reduce the line pitches [1–10]. As shown in Fig. 22.1, the interconnect pitches have been shrunk from 180 nm, 140 nm, and 100 nm for 65 [4], 45 [32], and 32 nm nodes [10] LSIs, respectively. To eliminate the interconnect parasitic capacitance, low-*k* dielectric films which have lower permittivity than the conventional silica (SiO_2) dielectrics have been introduced. Figure 22.2 shows the technology trend of the *k*-value and the deposition process, in which the low-*k* films are deposited by spin-on-dielectric (SOD) method or plasma-enhanced CVD. In the case of SOD precursor solution is poured on a rotated wafer, and the precursor film is heated to vaporize the solvent followed by reaction and densification to make a low-*k* film. In the case of PECVD [36, 42], on the other hand, precursor solution is vaporized with inert carrier gas such as He, and the precursor gas is introduced into PECVD chamber with RF power. The vaporized precursor gas is exited from plasma, depositing a low-*k* film on a wafer heated in high vacuum. The SOD method is advantageous to decrease the *k*-value, while PECVD method is superior in the adhesion strength due to the possibility of in-suite plasma surface treatment in vacuum just before the low-*k* deposition.

Irrespective of the deposition methods [11], the permittivity had been reduced by replacing oxygen atoms (O) of high polarity with low-polarity ones such as fluorine (F), hydrogen (H), and carbon (C) from 180 nm node to 90 nm node. The compositional modification accomplished the *k*-value from 4.2 to 3–2.8, which was a lower limit in silica systems. Further reduction below $k = 2.7$ requires the introduction of pores with $k \sim 1$ into rigid SiOCH films, or so-called as "porous films," of which the film properties depend on the porosity and the pore size distribution.

Y. Hayashi (✉)
ULSI Fundamental Research Laboratory, Microelectronics Research Laboratories,
NEC Electronics Corporation, 1120, Shimokuzawa, Sagamihara, Kanagawa 229, Japan
e-mail: hayashi@mel.cl.nec.co.jp, y-hayashi@az.jp.nec.com

Y. Shacham-Diamand et al. (eds.), *Advanced Nanoscale ULSI Interconnects: Fundamentals and Applications*, DOI 10.1007/978-0-387-95868-2_22,

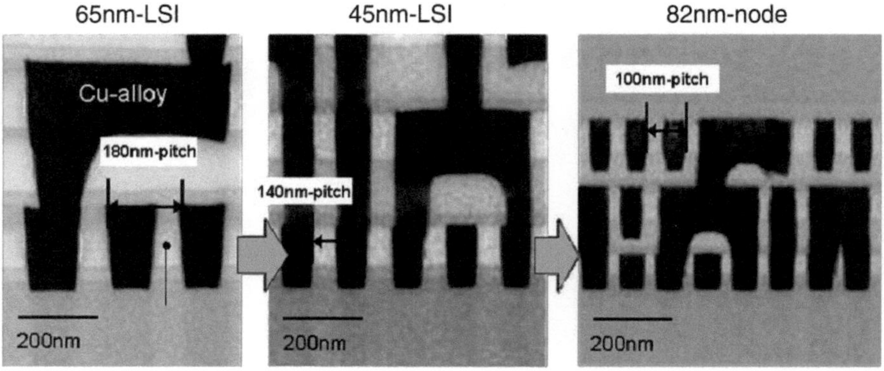

Fig. 22.1 TEM micrographs of ULSI interconnects of 65, 45, and 32 nm nodes in which the interconnect pitches are reduced from 180, 140, and 100 nm, respectively. Here, the inter-metal dielectrics (IMD) are isolated by porous SiOCH film [4], molecular-pore-stacked (MPS) SiOCH film [29, 32], and density-moderated MPS film [10]

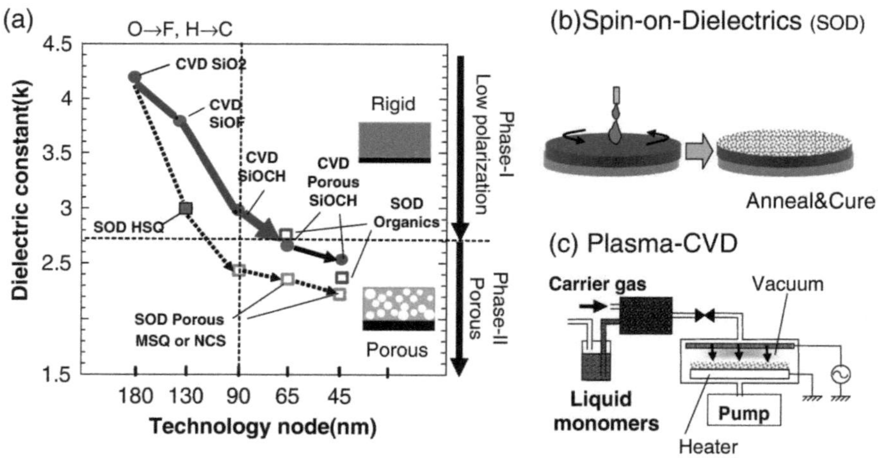

Fig. 22.2 Technology trends of low-*k* dielectric film: (**a**) the *k*-value as a function of LSI technology node and the low-*k* deposition processes of (**b**) spin-on-dielectric method and (**c**) plasma-enhanced CVD method

Figure 22.3 illustrates the potential issues of introduction of the porous low-*k* films into ULSI multi-layer interconnects [12]. The technical issues are summarized in four items: (1) moisture adsorption, (2) plasma process damages, (3) CMP defects, and (4) interfacial defect between the Cu lines increasing the leakage currents. All of these issues are related to the physical and chemical properties of low-*k* films. The main technical topics in the leading-edge ULSI devices are how to deposit and/or control the porous films. Typical analytical methods for porous low-*k* film are also listed in Fig. 22.3, such as modified edge lift-off test (m-ELT), 4-point

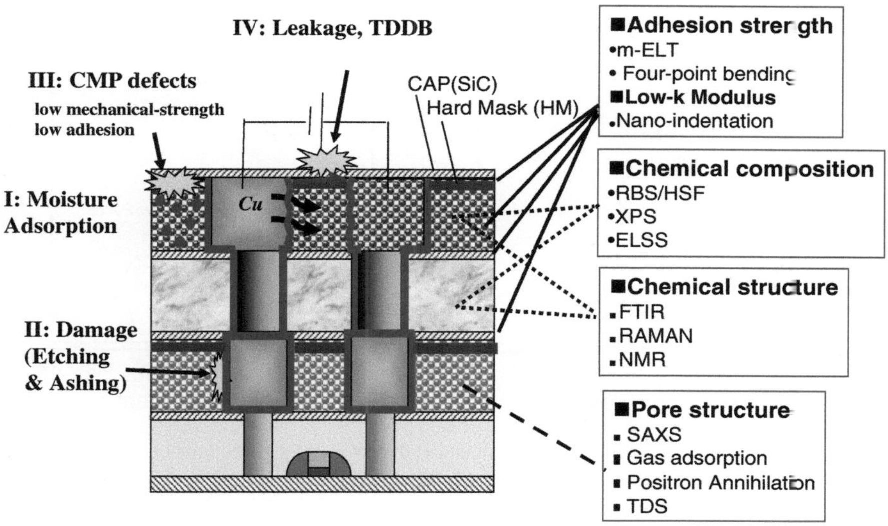

Fig. 22.3 Issues on ULSI interconnects by introducing porous low-*k* film and the analytical methods to understand and control the film structure [12]

bending [13], and nano-indentation for the mechanical evaluation [14]; RBS/HFS, XPS ELSS, FTIR, RAMAN [15], NMR [16] for the chemical evaluation; and SAXS [17], gas adsorption, positron annihilation [18], and TDS [19] for the pore structure evaluation.

In this section, we focus mainly on porous low-*k* films as well as non-porous low-*k* films. First, the basic characteristics of porous films and the technical issues are described, next the technical innovation to overcome issues such as post-curing method to harden the porous films and nanometer-scale pore structure control techniques is presented.

22.2 Basic Properties of Low-*k* Films

By the chemical composition modification, such as addition of SiO_2 to non-porous SiOCH, the *k*-value is reduced to 3.0–2.7. By reducing the *k*-value further, pores are introduced into the non-porous low-*k* films. Figure 22.4 shows the pore size distributions as a function of the *k*-value [12]. It is clearly seen that in order to reduce the *k*-value, large amount of pores have to be introduced, resulting in the pore agglomeration in the films. For example, a low-*k* film of $k = 2.2$ had the average pore size of 2.7 nm with the broad distribution to 6 nm as the maximum. The pore introduction degrades the mechanical strength of the low-*k* films such as the modulus and the adhesion strength as shown in Fig. 22.5 [20]. Here the film modulus and the adhesion strength of PECVD SiO_2 are mentioned to be 70 GPa and >0.3 $MPam^{0.5}$,

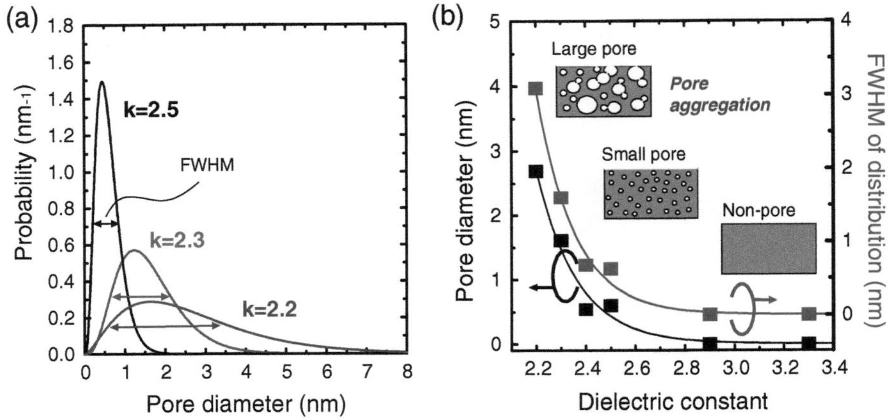

Fig. 22.4 Pore size and *k*-value of low-*k* films: (**a**) the pore size distributions observed by SAXS and (**b**) relation between the pore size and the *k*-value [12]

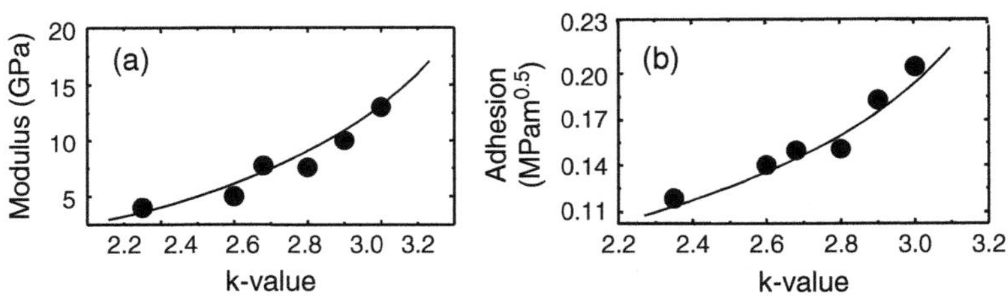

Fig. 22.5 Mechanical properties of low-*k* films: (**a**) the mechanical strength (modulus) by nano-indentation method and (**b**) the adhesion strength by m-ELT method as a function of the *k*-values [20]

respectively. The mechanical strengths are degraded even in a non-porous SiOCH film of $k = 3.0$, having the film modulus and the adhesion strength as 13 GPa and 0.21 $MPam^{0.5}$, respectively. When the *k*-value is reduced to 2.35, the mechanical strengths are lowered very much to 3 GPa and 0.12 $MPam^{0.5}$, respectively.

These degradations of the mechanical properties result in serious damages to ULSI interconnects such as the film de-lamination during chemical–mechanical polishing (CMP) [21] and/or the low-*k* cracking due to the thermal stress from packaging mold resin [22] as shown in Fig. 22.6. The hard-mask (HM) de-lamination was frequently observed during barrier metal CMP due to the lack of adhesion strength between the low-*k* film and the HM. The cracks were penetrated mostly at the interface between the porous low-*k* film and the barrier dielectrics such as SiCN during the thermal cycle test of packaging module. The mold resin with a large thermal expansion coefficient is immersed into the film stacks of the porous low-*k* film from the diced edges, giving rise to thermal stress. The other issues are moisture adsorption and plasma damages, where water vapor or process gas such as the dry-etching

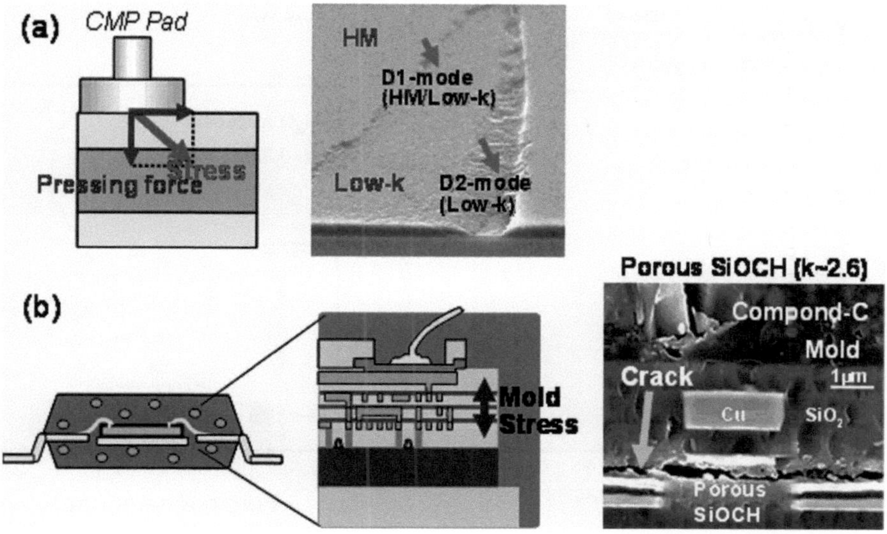

Fig. 22.6 Defects in porous low-*k* films: (**a**) hard-mask (HM) de-lamination on the porous low-*k* film during Cu/TaN-CMP process [21] and (**b**) cracks in low-*k* films in a packaged LSI chip (QFP) [22]

gas or O_2-ashing gas penetrates through channels of pores connected from the film surface or the etched surface.

Figure 22.7a shows the effects of the water adsorption on the *k*-value [12]. Here assume that the initial porous film with $k = 2.2$ has 22% porosity. When the water with $k{\sim}80$ penetrates into the total pore volumes of 50 and 85%, the *k*-values increase to $k = 3.1$ and 4.2, respectively, which are almost the same values as in the case of a non-porous SiOCH film and SiO_2. To prevent the porous low-*k* film from water adsorption, it is effective to reduce the pore size as shown in Fig. 22.7b [12]. The diffusion coefficient (*Dp*) of water in the pore is a function of the pore diameter of "*d*" as follows [23]:

$$Dp = 2/3\, d \sqrt{(2RT)} \,/\, (\pi M)$$

where *R* is gas constant, *T* the temperature, and *M* the molecular weight of water. It is quite obvious that the diffusion rate of water is retarded by decreasing the pore diameter. When the pore diameter was smaller than 1 nmϕ which was three times as large as the water molecular size, the water adsorption was not detected. Of course, when each of the pores was isolated completely, no water adsorption occurred even though the diameter was approximately 10 nmϕ [12].

Therefore, the control of the mechanical strength and pore size is crucial for realizing porous low-*k* films into the ULSI interconnects.

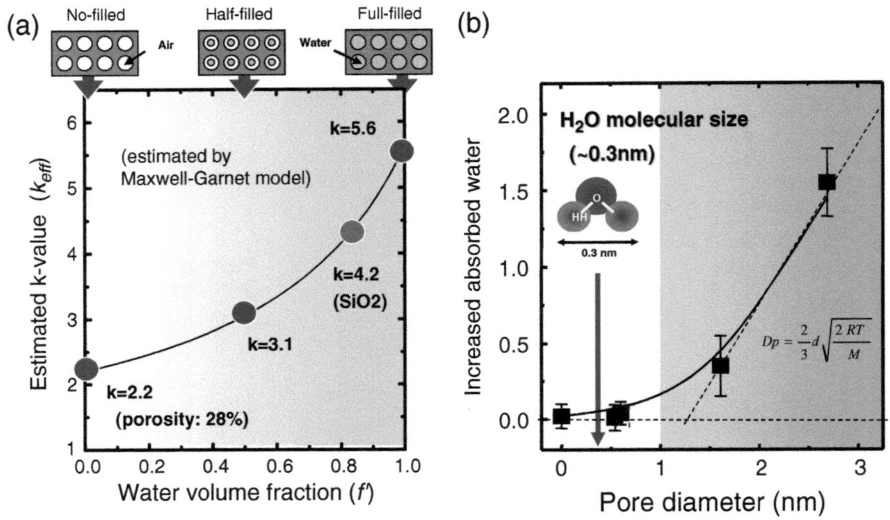

Fig. 22.7 Water adsorption into the porous low-*k* films: (**a**) the effects of the water adsorption on the *k*-value and (**b**) the relation between the pore diameter and the amount of the water adsorption [12]

22.2.1 Post-curing Techniques for Low-k Hardening

To improve the mechanical properties of porous low-*k* film, EB [24] or UV [25, 38] radiation is the practical technique as shown in Fig. 22.8. For example [20], when a porous SiOCH film with $k = 2.6$ was irradiated by UV ray with $\lambda = 100$–400 nm, the modulus was found to increase monotonically with the irradiation time period. The *k*-value decreased initially and then increased again. The initial decrease in *k*-value is related to the evaporation of water from the porous low-*k* film inside. An important point regarding these post-curing techniques is the control of the *k*-value and modulus by changing the irradiation conditions. In addition, the adhesion strength of the under-layered SiCN was increased.

Figure 22.9 shows the relation between the *k*-values and mechanical properties such as the modulus and the adhesion strength of the porous low-*k* films as well as those of post-cured. By post-curing, the modulus and the adhesion strength were increased more than the standard relation between the *k*-values and the mechanical properties of as-deposited films. By post-curing, it is possible that a porous SiOCH film of $k = 2.6$ can be converted to a porous film of $k = 2.7$ with the same mechanical properties as those of a rigid as-deposited SiOCH film. The film hardening is caused by chemical and physical structural changes in the porous low-*k* film using the EB or UV irradiation [20].

Figure 22.10 shows the FTIR analysis of the porous SiOCH films cured by UV irradiation [20]. When the UV was irradiated, the relative intensity of Si–CH_3 decreased, while that of the Si–H increased, indicating that the methyl (CH_3) was released and then hydrogen was terminated. The change in the chemical

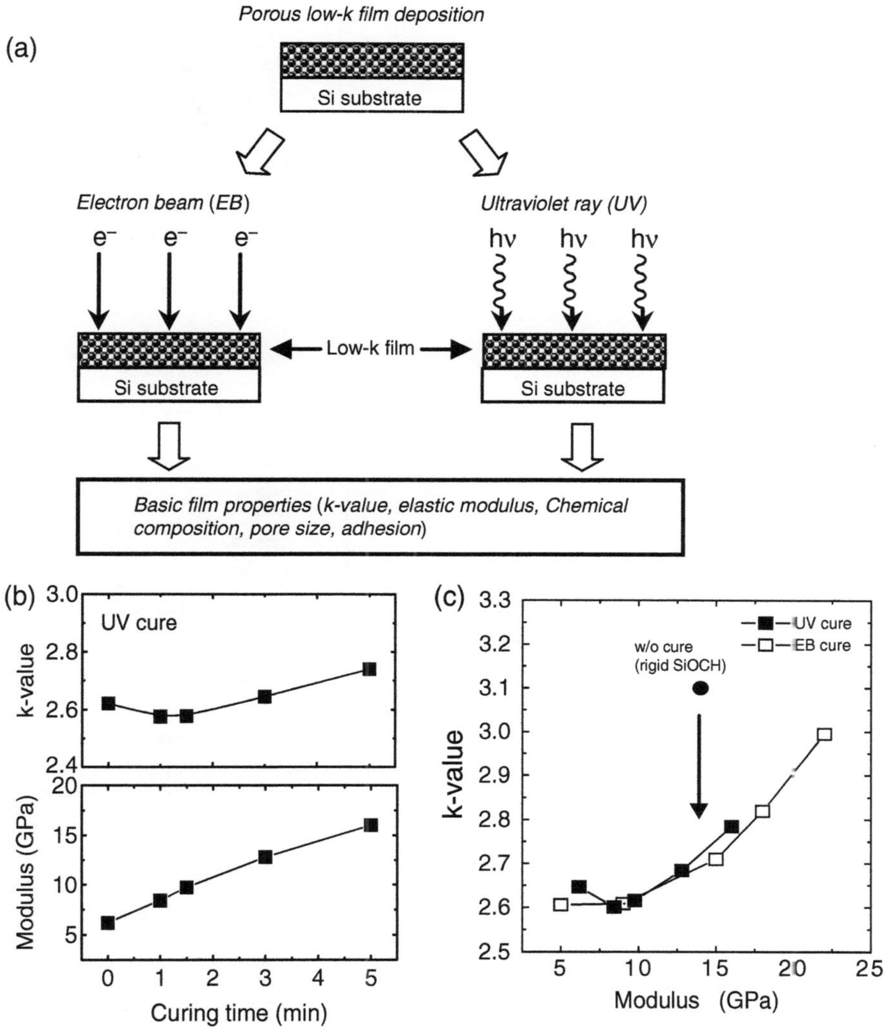

Fig. 22.8 Post-curing techniques of (**a**) EB or UV radiations for low-*k* hardening, (**b**) the changes in *k*-values and modulus as a function of UV curing time period, and (**c**) the relation between the *k*-value and modulus by EB or UV curing [20]

structure because of EB or UV irradiation corresponds to the modulus increment. Figure 22.11 shows the change in the pore structure of the porous SiOCH films during the UV curing measured by SAXS [20]. The pore diameter in the films was increased by UV curing while the modulus increased. Based on the experimental results from FTIR and SAXS, the structural evolution of the cured film occurs as follows. Densification of SiOCH occurs due to the UV irradiation by releasing steric-hindered CH3-groups thus increasing the film modulus, At the same time some of the pores are agglomerated enlarging their size.

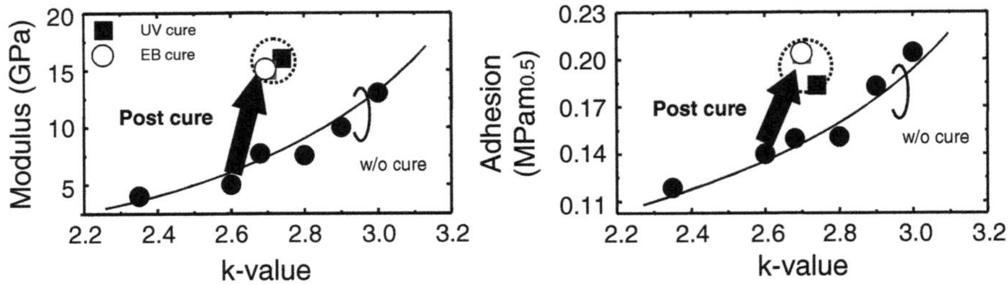

Fig. 22.9 Relation between the *k*-value and the mechanical properties such as (**a**) the film modulus by nano-indentation and (**b**) adhesion strength [20]

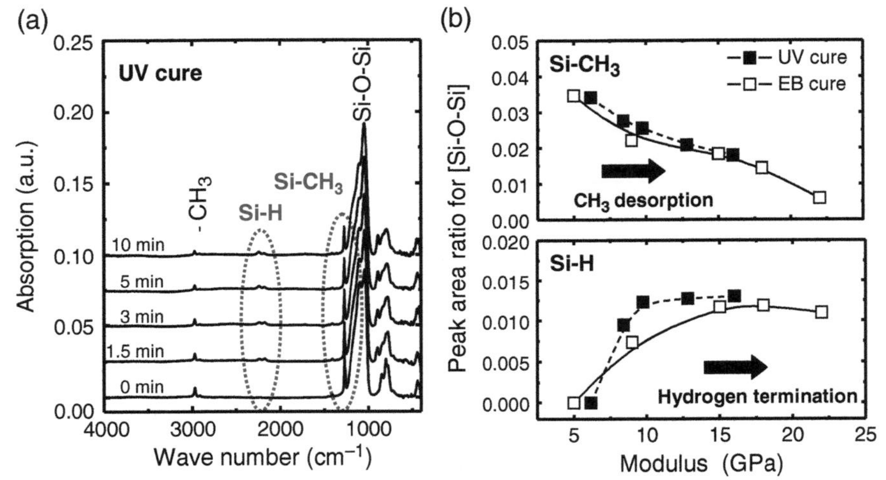

Fig. 22.10 FTIR analysis of the porous SiOCH films cured by UV irradiation: (**a**) FTIR spectroscopy by UV irradiations from 0 to 10 min, and (**b**) the relative FTIR peak intensities of Si–CH3 and S–H bonds as a function of the film modulus [20]

The post-curing technique by EB or UV irradiation is useful to improve the mechanical properties of the porous low-*k* films, applicable for the dielectric films used for ULSI interconnects.

22.3 Innovations of Low-*k* Material and Process

Recently, the chemical and physical structures of porous low-*k* films are controlled to improve the mechanical structure and the stability by novel innovations based on the material science as shown in Fig. 22.11. The conventional technique for porous low-*k* fabrication, irrespective of SOD [26, 39] or PECVD [27, 37] method, is based on the precursor film deposition containing porogen followed by thermal or UV/EB-assisted decomposition of the porogen to create pores as the out gas channels. The porogen usually is organic chemicals, which are decomposed at lower temperature

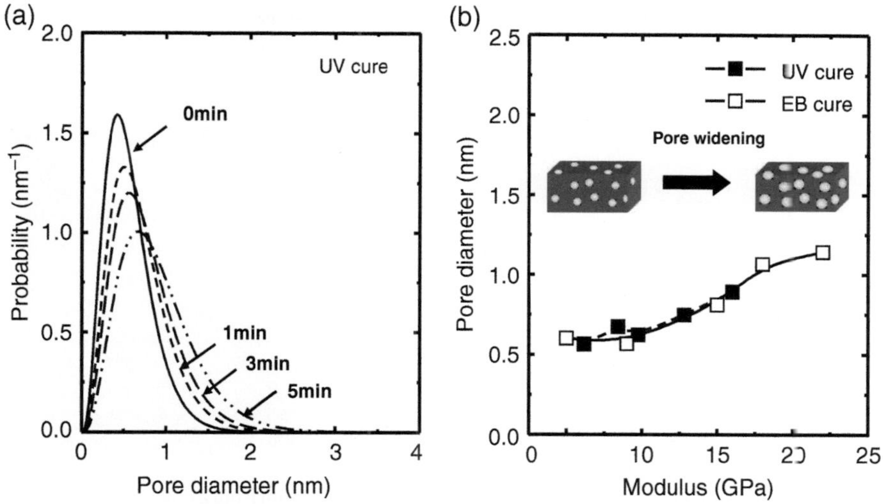

Fig. 22.11 Change in the pore structure by UV curing: (**a**) pore diameter distributions of the porous SiOCH film cured from 0 to 5 min and (**b**) the average pore diameter as a function of modulus of the cured films [20]

than the matrix material of low-*k* film itself. By this technique, the pores agglomerate each other, connecting to the film surface just like channels which act as the pathway of moisture and/or process gases. The serious issue is that it is very difficult to control the pore size with its sharp distribution.

Newly innovated porous low-*k* films are based on pre-determined pores in precursor materials without porogen chemicals irrespective of SOD and PECVD methods. For example, nano-crystalline silica (NCS) film has been developed by SOD method, in which a special siloxane precursor with special additives made self-organized pores [28]. The film with $k = 2.35$ had the average pore diameter of 1 nmϕ with relatively narrow distribution of 0.5–2 nmϕ. The mechanical stability was improved to have the modulus of 10 GPa, which is introduced into 65 nm node and 45 nm node ULSI interconnects.

In the case of PECVD method, a new idea has been innovated such as "molecular pore stacking (MPS) method" based on plasma-polymerization reaction [29–32], in which ring-type organo-silica molecules with unsaturated hydrocarbon side chains are introduced into PECVD chamber with low-power He plasma, and polymerized on a wafer to preserve the original ring-type molecular backbone as the molecular pore as shown in Fig. 22.12. The pore size is pre-determined as the original ring size, or essentially the number of silica atoms ringed, of the precursor molecules. For example, 6-membered ring of Si and O atoms with the molecular pores of 0.35 nmϕ resulted in the pores of 0.35 nmϕ in the film deposited, while 8-membered ring with the molecular ring of 0.45 nmϕ derived the pore of 0.5 nmϕ. The original ring structure was confirmed in the film deposited by solid-state NMR and RAMAN spectroscopy. Note that the pore diameter distribution, especially from

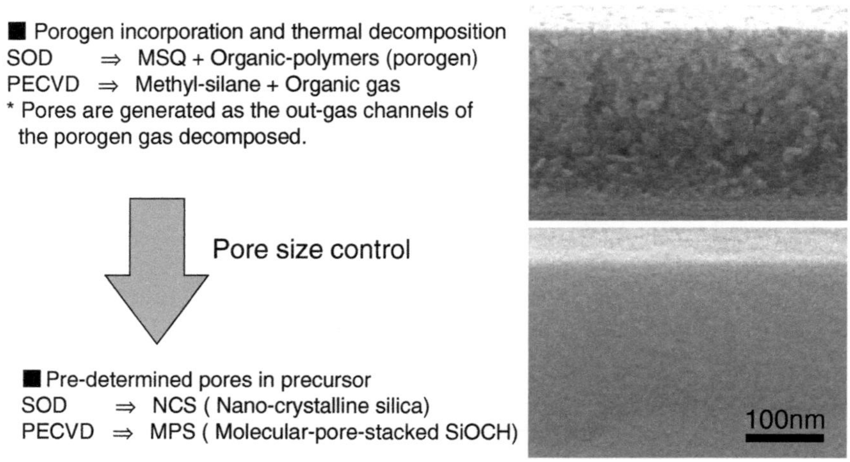

Fig. 22.12 Technology trend for the fabrication of porous low-*k* film from a conventional porogen-aided process to no porogen addition process with pre-determined pores in precursor chemicals such as a nano-crystalline silica (NCS) film or a molecular-pore stacked (MPS) SiOCH

the 6-membered ring precursors, is very sharp within sub-nanometer ranges due to the pre-determination of the pore diameter by the precursor molecular structure.

The controls of the pore diameter and the chemical structure provide great impact on the stability of porous low-*k* films. Figure 22.11a shows the effect of pore size on the *k*-value increment by pressure-cooker test (PCT). No *k*-value variation was observed for the porous low-*k* film with the pore diameters less than 0.8 nmϕ even though the PCT was severe with 100% humidity at 125°C. The chemical structure as well as the pore diameter affects the plasma damage by process gases such as O_2 plasma for the resist ashing process. Figure 22.11b shows cross-sectional SEMs of Cu interconnects isolated by a porous SiOCH film with the chemical composition of Si:O:C=1:1.58:0.86 and the carbon-rich MPS SiOCH film with the chemical composition of Si:O:C= 1:0.81:2.96. Here, O_2 plasma was exposed more intensively than the normal condition to clarify the effects of the physical and chemical compositions on the plasma damages. The samples were dipped into BHF solution to remove the SiO_2-like component. It is clearly seen that the carbon-rich MPS SiOCH film had high durability to the process gases during Cu interconnect fabrication. In addition to the deep sub-nanometer-sized pores, the carbon-rich composition of the MPS SiOCH film is believed to respond with high durability against the process plasma gases [43, 44].

Most recently, a plasma co-polymerization technology has been implemented to control the physical and chemical structures precisely in the low-*k* stacks in the Cu interconnects [10]. The technology concept is quite simple to use in two kinds of precursor molecules such as a molecule of ring-type siloxane for a carbon-rich porous SiOCH film and a molecule of chain-type siloxane for a non-porous SiOCH reinforcing the film structure as shown in Fig. 22.13. By changing the mixing ratio of these vaporized precursor gases, the *k*-value and the mechanical properties such as

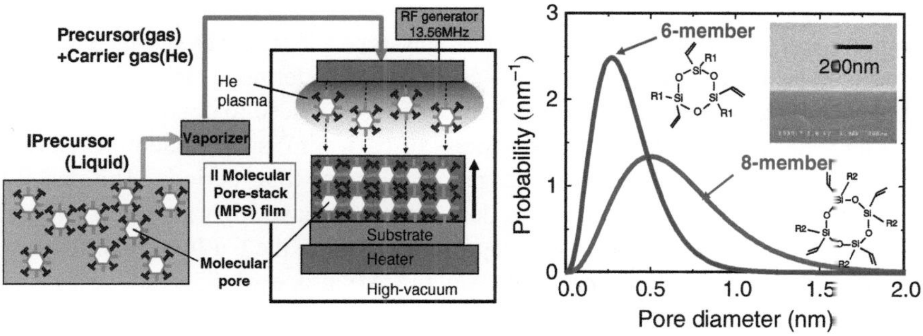

Fig. 22.13 Change in the pore structure by UV curing: (**a**) pore diameter distributions of the porous SiOCH film cured from 0 to 5 min and (**b**) the average pore diameter as a function of modulus of the cured films

the film modulus and the adhesion strength can be adjusted. For example, a non-porous SiOCH film ($k = 3.1$) with extremely high modulus of more than 20 GPa was deposited by 100% supplement of the chain-type precursor gas, while a carbon-rich MPS SiOCH film with $k = 2.45-2.5$ was grown by 100% supplement of the ring-type precursor gas. By changing the mixing ratio, these values were controlled to adequate ones. The adhesion strength was also controlled by changing the mixing ratios. Note that the non-porous SiOCH film from the chain-type siloxane had 1.2 times larger adhesion strength than a conventional SiOCH with the same *k*-value.

By applying the plasma co-polymerization reaction, a density-moderated SiOCH film was obtained by a continuous deposition process without vacuum break as shown in Fig. 22.14. At the early part of the deposition for the via-isolation, a mixed

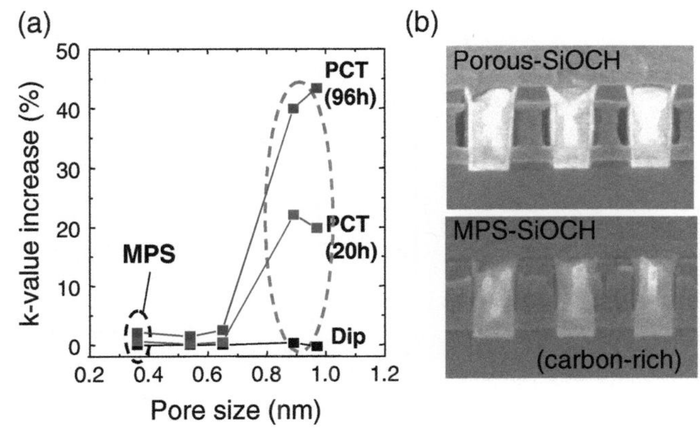

Fig. 22.14 Durability of porous low-*k* films: (**a**) the effect of pore size on the *k*-value increment by pressure-cooker test (PCT) with 100% humidity at 120°C and (**b**) cross-sectional SEMs of Cu interconnects isolated by a porous SiOCH film with Si:O:C=1:1.58:0.86 and the carbon-rich MPS SiOCH film with Si:O:C= 1:0.81:2.96. Here, O_2 plasma was exposed more intensively than the normal condition. The samples were dipped into BHF solution to remove the SiO_2-like component

precursor gas was supplied to make a SiOCH film ($k = 2.8$) with high adhesion strength, and then a precursor gas of 100% ring-type siloxane was supplied to make a carbon-rich MPS SiOCH film ($k = 2.45$–2.5) for the intermetal dielectrics. Finally, a precursor gas of 100% chain-type siloxane was introduced to deposit a high modulus hard-mask on the MPS. The thickness of each film is controlled precisely in nanometer scale by coordinating the gas supplement time periods. The density-moderated SiOCH stack realizes an ideal feature to have high adhesion strength to the lower layer, the low parasitic capacitance among the Cu lines, and the high mechanical protection withstanding the Cu/Ta-CMP process followed.

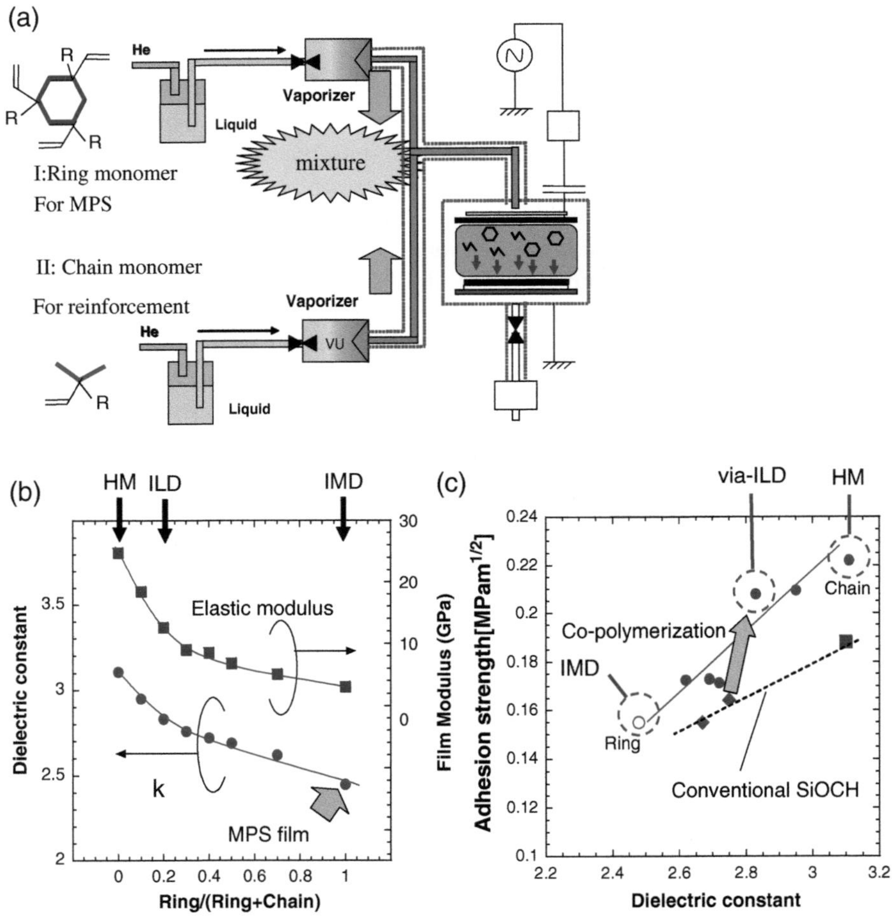

Fig. 22.15 Plasma co-polymerization technology for controlling the chemical and physical properties of the low-k film [10]: (**a**) illustration of the deposition system, (**b**) the k-values and the film modulus as functions of the gas mixing ratio between the chain-type and ring-type siloxane precursor molecules, and (**c**) the adhesion strength of the under-layered SiCN as a function of the k-values

Therefore, the precise control of the physical and chemical structures is crucial for realizing the porous low-k films with high durability against the process gases as well as the operational circumstances.

22.4 Future Trends on CAP Dielectrics

Figure 22.15 illustrates the technology on low-k materials implemented from 180 nm node to 32 nm node. For the 180 nm node, the inter-metal dielectric (IMD) film and the barrier dielectrics or cap dielectric (CAP) film were SiO_2 of $k = 4.2$ and SiN of $k{\sim}7$. The parasitic capacitance of the fine-pitched interconnects is determined mainly by the k-values of both IMD and CAP, which have been reduced by introducing rigid SiOCH ($k{\sim}3$) and SiCN ($k{\sim}4.5$) at 90 nm node, respectively. For 65 nm node, most of the semiconductor industries kept the IMD structure unchanged, while some of them introduced porous SiOCH film with $k{\sim}2.6$ (Fig. 22.16).

For 45–32 nm nodes, the main development target is the reduction of the k-value of CAP, which strongly affects the interconnect reliabilities such as EM [33] and TDDB [34, 35] among the narrow-pitched lines. The interface control between the CAP and Cu or HM is primarily important to keep the reliabilities. SiC(O) films with $k = 3.5-3$ was introduced as replacement for SiCN CAP [40], and the pre-deposition treatment before the CAP deposition is one of the key factors [41]. A big challenge is to implement new materials for CAP films with $k<3$ and high barrier property against Cu diffusion such as plasma-polymerized BCB with $k = 2.7$ (Fig. 22.17).

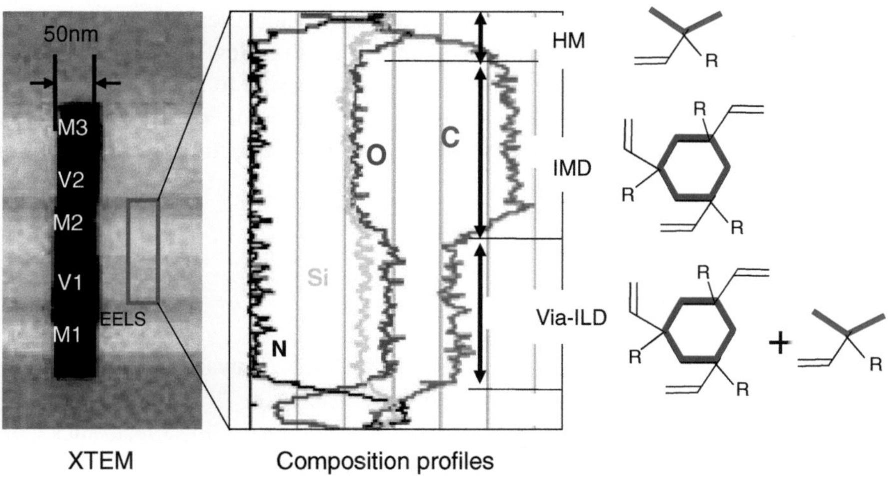

Fig. 22.16 Cross-sectional TEM and the compositional profiles in the density-moderated SiOCH film obtained by plasma co-polymerization technology. The density-moderated SiOCH stack realizes an ideal feature to have the high adhesion strength to the lower layer, the low parasitic capacitance among the Cu lines, and the high mechanical protection withstanding the Cu/Ta-CMP process followed [10]

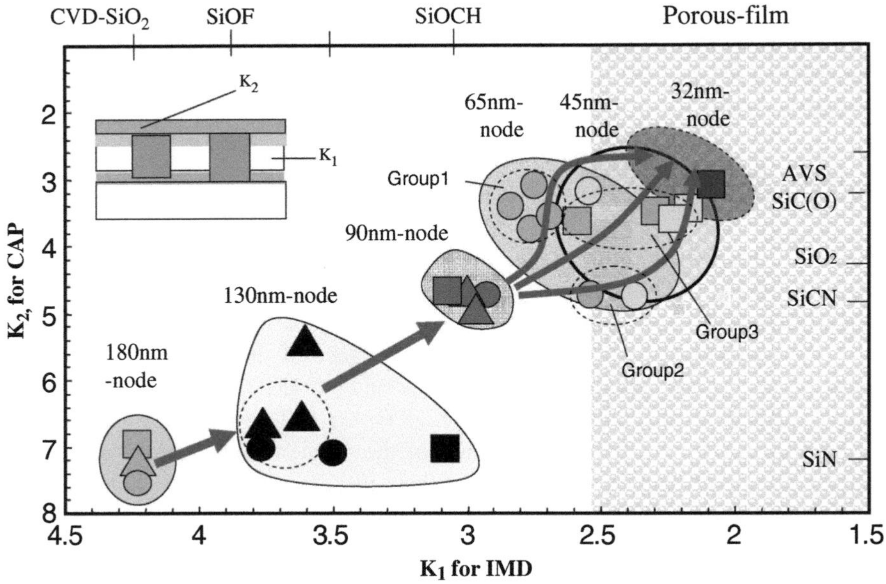

Fig. 22.17 Technology trends on the CAP film as well as on the IMD film

22.5 Summary

Introduction of low-*k* dielectric film, especially porous film, is a key factor to reduce the parasitic capacitance (C_{int}) among multi-layer interconnects in leading-edge ULSI devices. The mechanical properties as well as the electrical reliability depend on the physical and chemical structures of the porous film. New innovations such as EB/UV post-curing process and molecule structural design to stabilize the porous films have been implemented. Reduction in the *k*-values of barrier dielectrics or capping films becomes important for decreasing C_{int} effectively, especially for 45/32 nm nodes.

22.6 Acknowledgments

The author would like to acknowledge Drs. F. Ito, J. Kawahara, N. Inoue, M. Tagami, M. Ueki, K. Hijioka, H. Yamamoto, T. Takeuchi, T. Onodera, S. Saito, and N. Furutake of LSI Fundamental Research Laboratory, NEC Electronics Corporation and Dr. Tada, Device Platforms Research Laboratories, NEC Corporation, through researches on Advanced LSI BEOL technologies. A part of research on the plasma co-polymerization technology was supported by New Energy and Industrial Technology Development Organization (NEDO), Japan, through MIRAI Project from 2001 to 2006. The author also acknowledges Dr. Y. Mochizuki, NEC and H. Watanabe, NEC Electronics Corporation for the research support.

References

1. Hayashi, Y.: Impacts of low-k film on sub-100 nm-node, ULSI devices. IEEE Intl. Interconnect Tech. Conf., San Francisco, USA 145 (2002)
2. Takahashi, S.; Edahiro, M.; and Hayash, Y.: Interconnect design strategy: Structures, repeaters and materials with strategic system performance analysis (S 2 PAL) model. IEEE Trans. Electron Devices **48(2)** 239, (2001)
3. Davis, J. A. and Meindl, J. D.: Interconnect Technology and Design for Gigascale Integration. Kluwer Academic Publishers, Dordrecht (2003)
4. Ueki, M.; Narihiro, M.; Ohtake, H.; Tagami, M.; Tada, M.; Ito, F.; Harada, Y.; Abe, M.; Inoue, N.; Arai, K.; Takeuchi, T.; Saito, S.; Onodera, T.; Furutake, N.; Hiroi, M.; Sekine, M.; and Hayashi, Y.: Highly reliable, 65 nm-node Cu dual Damascene interconnects with full porous-SiOCH (k=2.5) films for low-power ASICs. Symposium on VLSI Technology, Digest 60 (2004)
5. Tada, M.; Harada, Y.; Tamura, T.; Inoue, N.; Ito, F.; Yoshiki, M.; Ohtake, H.; Narihiro, M.; Tagami, M.; Ueki, M.; Hijioka, K.; Abe, M.; Takeuchi, T.; Saito, S.; Onodera, T.; Furutake, N.; Arai, K.; Fujii, K.; and Hayashi, Y.: A 65 nm-node, Cu interconnect technology using porous SiOCH film (k=2.5) covered with ultra-thin, low-k pore seal (k=2.7). IEEE IEDM2003, Technical Digest 35.2.1 (2003)
6. Kajita, A.; Usui, T.; Yamada, M.; Ogawa, E.; Katata, T.; Sakata, A.; Miyajima, H.; Kojima, A.; Kanamura, R.; Ohoka, Y.; Kawashima, H.; Tabuchi, K.; Nagahata, K.; Kato, Y.; Hayashi, T.; Kadomura, S.; and Shibata, H.: Highly reliable Cu/low-k dual-Damascene interconnect technology with hybrid (PAE/SiOC) dielectrics for 65 nm-node high performance eDRAM. IEEE Intl. Interconnect Tech. Conf., San Francisco, USA, 9 (2003)
7. Nakai, S.; Kojima, M.; Misawa, N.; Miyajima, M.; Asai, S.; Inagaki, S. Iba, Y.; Ohba, T.; Kase, M.; Kitada, H.; Satoh, S.; Shimizu, N.; Sugiura, I.; Sugimoto, F.; Setta, Y.; Tanaka, T.; Tamura, N.; Nakaishi, M.; Nakata, Y.; Nakahira, J.; Nishikawa, N.; Hasegawa, A.; Fukuyama, S.; Fujita, K.; Hosaka, K.; Horiguchi, N.; Matsuyama, H.; Minami, T.; Minamizawa, M.; Morioka, H.; Yano, E.; Yamaguchi, A.; Watanabe, K.; Nakamura, T.; and Sugii, T.: A 65 nm CMOS Technology with High Performance and Low Leakage Transistor, a 0.55 μm 6T-SRAM cell and Robust hybrid-ULK/Cu Interconnects for Mobile Multimedia Applications. Tech. Dig. IEEE IEDM, 11.31 (2003)
8. Jeng, C. C.; Wan, W. K.; Lin, H. H.; Ming-Shuo L.; Tang, K. H.; Kao, I. C.; Lo, H. C.; Chi, K. S.; Huang, T. C.; Yao, C. H.; Lin, C. C.; Lei, M. D.; Hsia, C. C.; and Mong-Song L.: BEOL process integration of 65 nm Cu/low k interconnects. IEEE IITC 2004, 199 (2004)
9. Lee, K.-W.; Shin, H. J.; Hwang, J. W.; Nam, S. W.; Moon, Y. J.; Wee, Y. J.; Kim, I. G.; Park, W. J.; Kim, J. H.; Lee, S. J.; Park, K. K.; Kang, H.-K.; and Suh, K.-P.: Highly manufacturable Cu/low-k dual Damascene process integration for 65 nm technology node. IEEE Intl. Interconnect Tech. Conf., (San Francisco, USA) 57 (2004)
10. Tada, M.; Yamamoto, H.; Ito, F.; Narihiro, M.; Ueki, M.; Inoue, N.; Abe, M. Saito, S.; Takeuchi, T.; Furutake, N.; Onodera, T.; Kawahara, J.; Arai, K.; Kasama, Y. Taiji, T.; Tohara, M.; Sekine, M.; and Hayashi, Y.: Plasma Co-polymerization Technology with Molecular-level Structure Tightening in "In-situ" SiOCH Stacks for 32 nm-node Cu Interconnects. IEEE Intl. Electron Device Meeting, Tech. (San Francisco, CA,USA), Digest 351 (2006)
11. Maex, K.; Baklanov, M. R.; Shamiryan, D.; and Lacopi, F.: Low dielectric constant materials for microelectronics. J. Appl. Phys. **93(11)**, 8793 (2003)
12. Itoh, F.; Hijioka, K.; Takeuchi, T.; and Hayashi, Y.: Nanostructure Control of Porous Low-k Dielectric Films with High Water Resistance. Conference Proceedings AMC XX © 2005 Materials Research Society, 381 (2005)
13. Damayanti, M.; Widodo, J.; Sritharan, T.; Mhaisalkar, S. G.; Lu, W.; Gan, Z. H.; Zeng, K. Y.; and Hsia, L. C.: Erratum to "Adhesion study of low-k/Si system using 4-point bending and nanoscratch test" [Mater. Sci. Eng. B 121 (3) (2005) 193–198]. Materials Science and Engineering: B, **122(3)**, 249 (2005)

14. Johnson, M.; Li, Z.; Wang, J.; and Yan Y.: Mechanical characterization of zeolite low dielectric constant thin films by nanoindentation. Thin Solid Films **515(6)**, 3164 (2007)
15. Doux, C.; Aw, K. C.; Niewoudt, M.; and Gao W.: Analysis of HSG-7000 silsesquioxane-based low-k dielectric hot plate curing using Raman spectroscopy. Microelectron. Eng. **83(2)**, 387 (2006)
16. Su, Y.-C. and Chang F.: Synthesis and characterization of fluorinated polybenzoxazine material with low dielectric constant. Polymer **44(26)**, 7989 (2003)
17. Omote, K.; Ito Y.; and Kawamura S.: Small angle x-ray scattering for measuring pore-size distributions in porous low- films. Appl. Phys. Lett. **82**, 544 (2003)
18. Wang, C. L.; Weber, M. H.; Lynn, K. G.; and Rodbell, K. P.: Several issues regarding the nanopore structure in low-dielectric-constant films detected by positron annihilation lifetime spectroscopy. Radiat. Phys. Chem. **68(3–4),** 439 (2003)
19. Yanazawa, H.; Fukuda, T.; Uchida, Y.; and Katou, I.: Water sorbability of low-k dielectrics measured by thermal desorption spectroscopy. Surf. Sci. **566–568(1)**, 566 (2004)
20. Itoh, F.; Takeuchi, T.; and Hayashi, Y.: Improvement of Mechanical Properties of Porous SiOCH films by Post-cure Treatment. Conference Proceedings AMC XXI © 2006 Materials Research Society, 291 (2006)
21. Hijioka, K.; Ito, F.; Tagami, M.; Ohtake, H.; Harada, Y.; Takeuchi, T.; Saito, S.; and Hayashi, Y.: Mechanical Property Control of Low-k Dielectrics for Diminishing Chemical Mechanical Polishing (CMP)-Related Defects in Cu-Damascene Interconnects. Jpn. J. Appl. Phys., **43(4B)**, 1807 (2004)
22. Tagami, M.; Ohtake, H.; Abe, M.; Ito, F.; Takeuchi, T.; Ohto, K.; Usami, T.; Suzuki, M.; Suzuki, T.; Sashida, N.; and Hayashi, Y.: Comprehensive process design for low-cost chip packaging with circuit-under-pad (CUP) structure in porous-SiOCH film. IEEE Intl. Interconnect Tech. Conf. San Francisco, USA, Proc. 12 (2005)
23. Burganos, V. N.: Monte Carlo simulation of gas diffusion in regular and randomized pore systems. J. Chem. Phys. **98**, 2268 (1993)
24. Miyajima, H.; Fujita, K.; Nakata, R.; Yoda, T.; and Hayasaka, N.: The application of simultaneous ebeam cure methods for 65 nm node Cu/low-k technology with hybrid (PAE/MSX) structure. IEEE Intl. Interconnect Tech. Conf. San Francisco, USA, 222 (2004)
25. Furusawa, T.; Miura, N.; Matsumoto, M.; Goto, K.; Hashii, S.; Fujiwara, Y.; Yoshikawa, K.; Yonekura, K.; Asano, Y.; Ichiki, T.; Kawanabe, N.; Matsuzawa, T.; and Matsuura, M.: UV-hardened high-modulus CVD-ULK material for 45-nm node Cu/low-k interconnects with homogeneous dielectric structures. IEEE Intl. Interconnect Tech. Conf San Francisco, USA, 45 (2005)
26. Park, S. J.; Shin, J. J.; Min, S. K.; and Rhee, H. W.: Formation of nanoporous organosilicate films using cyclodextrins as a porogen. Curr. Appl. Phys. **6(4)**, 743 (2006)
27. Chapelon, L. L.; Arnal, V.; Broekaart, M.; Gosset, L. G.; Vitiello, J.; and Torres, J.: Characterization and integration of a CVD porous SiOCH ($k < 2.5$) with enhanced mechanical properties for 65 nm CMOS interconnects and below. Microelectron. Eng. **76(1–4)**, 1 (2004)
28. Nakamura, T. and Nakashima, A.: Robust multilevel interconnects with a nano-clustering porous low-k ($k<2.3$). IEEE Intl. Interconnect Tech. Conf. San Francisco, USA 175 (2004)
29. Hayashi, Y.; Itoh, F.; Harada, Y.; Takeuchi, T.; Tada, M.; Tagami, M.; ohtake, H.; Hijioka, K.; Saito, S.; Onodera, T.; Hara, D.; and Tokudome, K.: Novel molecular-structure design for PECVD porous SiOCH films toward 45 nm node, ASICs with $k=2.3$. IEEE Intl. Interconnect Tech. Conf. San Francisco, USA, 225 (2004)
30. Tada, M.; Ohtake, H.; Narihiro, M.; Ito, F.; Taiji, T.; Tohara, M.; Motoyama, K.; Kasama, Y.; Tagami, M.; Abe, M.; Takeuchi, T.; Arai, K.; Saito, S.; Furutake, N.; Onodera, T.; Kawahara, J.; Kinoshita, K.; Hata, N.; Kikkawa, T.; Tsuchiya, Y.; Fujii, K.; Oda, N.; Sekine, M.; and Hayashi, Y.: Feasibility study of a novel molecular-pore-stacking (MPS), SiOCH film in fully-scale-down, 45 nm-node Cu Damascene interconnects. Symp. VLSI Technol. (Kyoto, JPN), Digest 18, (2005)

31. Abe, M.; Tada, M.; Ohtake, H.; Furutake, N.; Narihiro, M.; Arai, K.; Takeuchi, T.; Saito, S.; Taiji, T.; Motoyama, K.; Kasama, Y.; Arita, K.; Ito, F.; Yamamoto, H.; Tagami, M.; Tonegawa, T.; Tsuchiya, Y.; Fujii, K.; Oda, N.; Sekine, M.; and Hayashi, Y.: A robust 45 nm-node, dual Damascene interconnects with high quality cu/barrier interface by a novel oxygen absorption process. IEEE Intl. Electron Device Meeting, Tech. Washington DC, USA, Digest 77 (2005)
32. Tagami, M.; Ohtake, H.; Tada, M.; Ueki, M.; Ito, F.; Taiji, T.; Kasama, Y.; Iwamoto, T.; Wakabayashi, H.; Fukai, T.; Arai, K.; Saito, S.; Yamamoto, H.; Abe, M.; Narihiro, M.; Furutake, N.; Onodera, T.; Takeuchi, T.; Tsuchiya, Y.; Oda, N.; Sekine, M.; Hane, M.; and Hayashi, Y.: High-Performance Cu-Interconnects with Novel Seamless Low-k SiOCH Stacks (SEALS) Featured by Compositional Modulation Process for 45 nm-Node ULSI Devices. Symp. VLSI Tech. Hawai, USA, 108 (2006)
33. Li, B.; Sullivan, T. D.; Lee, T. C.; and Badami, D.: Reliability challenges for copper interconnects. Microelectron. Reliab. **44(3)**, 365 (2004)
34. Konishi, N.; Yamada, Y.; Noguchi, J.; Jimbo, T.; and Inoue, O.: Influence of CMP process on defects in SiOC films and TDDB reliability. IEEE Intl. Interconnect Tech. Conf. San Francisco, USA, Proc. 123 (2005)
35. Tada, M.; Ohtake, H.; Kawahara, J.; and Hayashi, Y.: Effects of material interfaces in Cu/low-/spl kappa/ Damascene interconnects on their performance and reliability. IEEE trans. On electron devices **51(11)**, 1867 (2004)
36. McGahay, V.; Bonilla, G.; Chen, F.; Christiansen, C.; Cohen, S.; Cullinan-Scholl, M.; Demarest, J.; Dunn, D.; Engel, B.; Fitzsimmons, J.; Gill, J.; Grunow, S.; Herbst, B.; Hichri, H.; Ida, K.; Klymko, N.; Kiene, M.; Labelle, C.; Lee, T.; Liniger, E.; Liu, X. H.; Madan, A.; Malone, K.; Martin, J.; McLaughlin, P. V.; Minami, P.; Molis, S.; Muzzy, C.; Nguyen, S.; Patel, J. C.; Restaino, D.; Sakamoto, A.; Shaw, T. M.; Shimooka, Y.; Shobha, H.; Simonyi, E.; Widodo, J.; Grill, A.; Hannon, R.; Lane, M.; Nye, H.; Spooner, T.; Wisnieff, R.; and Ivers, T.: 65 nm Cu Integration and Interconnect Reliability in Low Stress K=2.75 SiCOH. IEEE Intl. Interconnect Tech. Conf. San Francisco, USA, Proc. 9 (2006)
37. Sankaran, S.; Arai, S.; Augur, R.; Beck, M.; Biery, G.; Bolom, T.; Bonilla, G.; Bravo, O.; Chanda, K.; Chae, M.; Chen, F.; Clevenger, L.; Cohen, S.; Cowley, A.; Davis, P.; Demarest, J.; Doyle, J.; Dimitrakopoulos, C.; Economikos, L.; Edelstein, D.; Farooq, M.; Filippi, R.; Fitzsimmons, J.; Fuller, N.; Gates, S. M.; Greco, S. E.; Grill, A.; Grunow, S.; Hannon, R.; Ida, K.; Jung, D.; Kaltalioglu, E.; Kelling, M.; Ko, T.; Kumar, K.; Labelle, C.; Landis, H.; Lane, M. W.; Landers, W.; Lee, M.; Li, W.; Liniger, E.; Liu, X.; Lloyd, J. R.; Liu, W.; Lustig, N.; Malone, K.; Marokkey, S.; Matusiewicz, G.; McLaughlin, P. S.; McLaughlin, P. V.; Mehta, S.; Melville, I.; Miyata, K.; Moon, B.; Nitta, S.; Nguyen, D.; Nicholson, L.; Nielsen, D.; Ong, P.; Patel, K.; Patel, V.; Park, W.; Pellerin, J.; Ponoth, S.; Petrarca, K.; Rath, D.; Restaino, D.; Rhee, S.; Ryan, E. T.; Shoba, H.; Simon, A.; Simonyi, E.; Shaw, T. M.; Spooner, T.; Standaert, T.; Sucharitaves, J.; Tian, C.; Wendt, H.; Werking, J.; Widodo, J.; Wiggins, L.; Wisnieff, R.; and Ivers, T.: A 45 nm CMOS node Cu/Low-k/ Ultra Low-k PECVD SiCOH (k=2.4) BEOL Technology. IEEE Intl. Electron Device Meeting, Tech. San Francisco, USA, Digest 355 (2006)
38. Nakao, S.; Ushio, J.; Ohno, T.; Hamada, T.; Kamigaki, Y.; Kato, M.; Yoneda, K.; Kondo, S.; and Kobayashi, N.: UV/EB Cure Mechanism for Porous PECVD/SOD Low-k SiCOH Materials. IEEE Intl. Interconnect Tech. Conf. San Francisco, USA, 66 (2006)
39. Kagawa, Y.; Enomoto, Y.; Shimayama, T.; Kameshima, T.; Okamoto, M.; Kawshima, H.; Yamada, A.; Hasegawa, T.; Akiyama, K.; Masuda, H.; Miyajima, M.; Shibata, H.; and Kadomura, S.: Robust 45-nm Node Cu/LJLK Interconnects using Effective Porogen Control. IEEE Intl. Interconnect Tech. Conf. San Francisco, USA, Proc. 207 (2006)
40. Yoneda, K.; Kato, M.; Nakao, S.; Matsuki, N.; Matsushita, K.; Ohara, N.; Kaneko, S.; Fukazawa, A.; Kamigaki, Y.; and Kobayashi, N.: Robust Low-k Diffusion Barrier (k=3.5) for 45-nm Node Low-k (k=2.3)/Cu Integration. IEEE Intl. Interconnect Tech. Conf. San Francisco, USA, Proc. 184 (2006)

41. Favennec, L.; Jousseaume, V.; Zenasni, A.; Bouchu, D.; and Passemard, G.: New low k a-SiC:H dielectric barrier for advanced interconnects. IEEE Intl. Interconnect Tech. Conf. San Francisco, USA, 110 (2006)
42. Tajima, N.; Ohno, T.; Hamada, T.; Yoneda, K.; Kobayashi, N.; Shinriki, M.; Miyazawa, K.; Sakota, K.; Hasaka, S.; and Inoue, M.: Carbon-Rich SiOCH Films with Hydrocarbon Network Bonds for Low-k Dielectrics: First-Principles Investigation. IEEE Intl. Interconnect Tech. Conf. San Francisco, USA, 122 (2006)
43. Hayashi,Y.; Ohtake, H.; Kawahara, J.; Tada. M.; Saito, S.; Inoue, N.; Ito, F.; Tagami, M.; Ueki, M.; Furutake, N.; Takeuchi, T.; Yamamoto, H.; and Abe, M.: Comprehensive Chemistry Designs in Porous SiOCH Film Stacks and Plasma Etching Gases for Damageless Cu Interconnects in Advanced ULSI Devices. IEEE Trans. Semicond. Manuf. **21(3)**, 469 (2008)
44. Inoue, N.; Furutake, N.; Ito, F.; Yamamoto, H.; Takeuchu, T.; and Hayashi, Y.: Impact of Barrier Metal Sputtering on Physical and Chemical Damages in Low-k SiOCH Films with Various Hydrocarbon Content. Jpn. J. Appl. Phys. **47(4)**, 2468 (2008)

Chapter 23
CMP for Cu Processing

Manabu Tsujimura

23.1 CMP Review

23.1.1 Chemical Mechanical Planarization

CMP (chemical mechanical polisher) is now regarded as an essential process among semiconductor device manufacturing processes. This CMP was first introduced by IBM in the early 1980s for planarization of logic devices [1]. Figure 23.1 shows trends in CMP adoption relative to device trends such as finer nodes and larger wafers. As previously mentioned, in days of 200 mm wafer with 500–350 technical nodes, CMP was not adopted. Since late 250 nm nodes or since 180 nodes, CMP began to prevail primarily in the area of logic device manufacturing. In the days of 300 mm wafers and 130 nm nodes, CMP came to be regarded as essential.

23.1.2 Application of CMP

Figure 23.2 shows CMP applications such as ILD (interlayer dielectric), W-plug, STI (shallow trench isolation), and metals like CMP for Cu processing. Such applications are divided into two CMP processes: the blind process, which has no stopper layers like ILD, and the recess process, which has stopper layers such as W-plug, STI, and metals. Definitions of planarization performances are given below.

1. Step height reduction for blind process such as ILD step height reduction function is the basic performance of planarization. The initial step height is reduced in accordance with polishing amount. The ratio of residual step height and an initial step height is defined as the step height ratio, which is very useful in

M. Tsujimura (✉)
Ebara Corporation, Nissay Aroma Square, 5-3-7 Kamata, Ohta-ku, Tokyo 144-8721, Japan;
4-2-1 Honfujisawa, Fujusawa-shi 251-8502, Japan
e-mail: tsujimura.manabu@ebara.com, tsujimura@fuj.ebara.co.jp

Y. Shacham-Diamand et al. (eds.), *Advanced Nanoscale ULSI Interconnects: Fundamentals and Applications*, DOI 10.1007/978-0-387-95868-2_23,

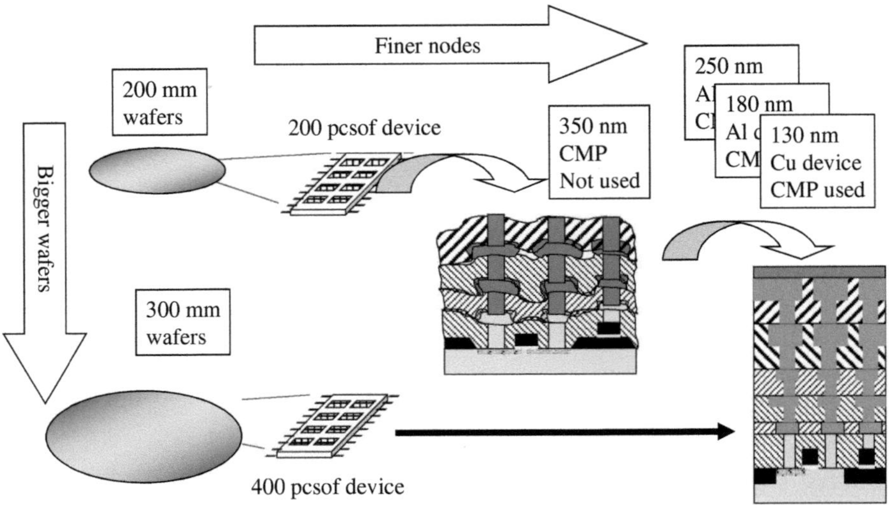

Fig. 23.1 CMP in relation to finer nodes and larger wafers

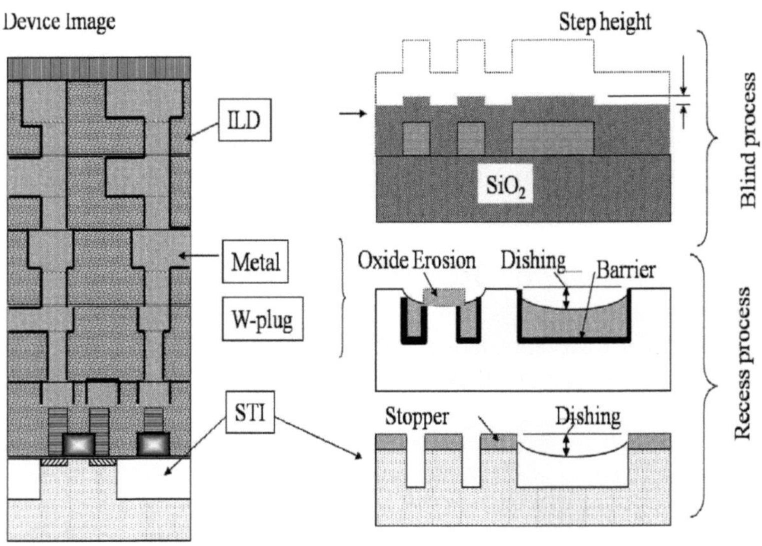

Fig. 23.2 Application of CMP

comparing planarization performances. Harder pads provide better step height reduction performance.

2. Dishing, erosion, and total loss for recess processes such as STI, W-plug, and metals

In the case of metals such as Cu, dishing is defined as the recess amount from the surface, while oxide erosion is defined as the amount eroded from the surface. These values depend on the extent of overpolishing.

23.2 CMP System

Figure 23.3 shows subsystems and related materials in a CMP system. In addition to a polisher and post-CMP cleaning, a CMP system encompasses a slurry supply, waste treatment, monitors, slurry, and pad.

Figure 23.4 shows bird's-eye view of CMP. Figure 23.5 shows the outline of CMP.

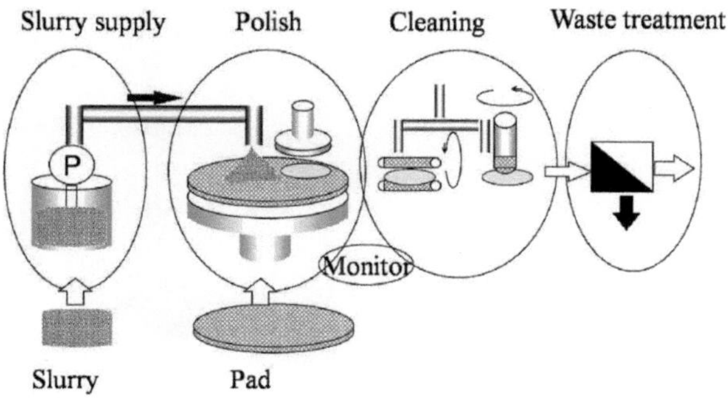

Fig. 23.3 CMP system

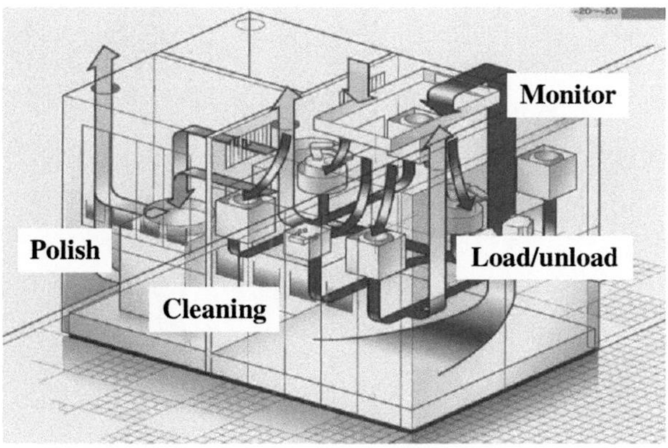

Fig. 23.4 Tool bird's-eye view

23.3 Cu-CMP

23.3.1 Application

Figure 23.6 shows an MPU with five layers of Cu metal for 130 nm nodes in which Cu is adopted and planarized.

It is easily surmised that Cu and low-k materials will often be used in several layers for finer nodes such as 250, 180, 130, and 90 nm. At a same time, it is also

Fig. 23.5 Photo of tool

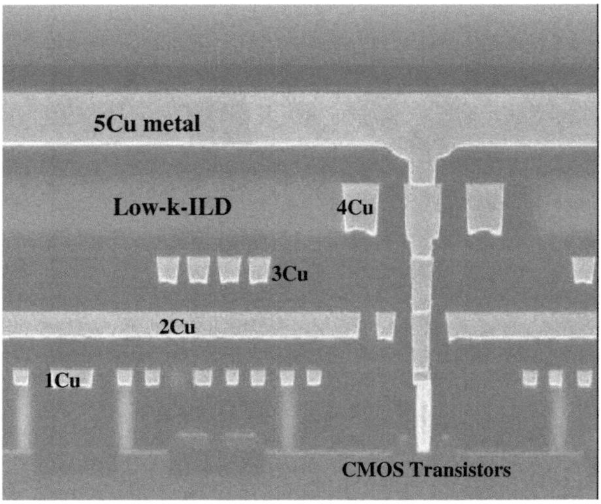

Fig. 23.6 MPU with five layers Cu metal for 130 nm node (H. Shibata of Toshiba:ADMETA2002 Tutorial session)

clear that CMP technology is essential for such device design. Figure 23.7 provides a definition of planarization performance for metal CMP in which dishing, erosion, and metal (Cu) recess (total Cu loss) are defined. For example, dishing is evaluated in 100 ?m line and space pattern. Erosion is evaluated in 90% dense area. Although those evaluation methods are not standard, it is strongly required to make an evaluation standard in order to compare apple to apple condition.

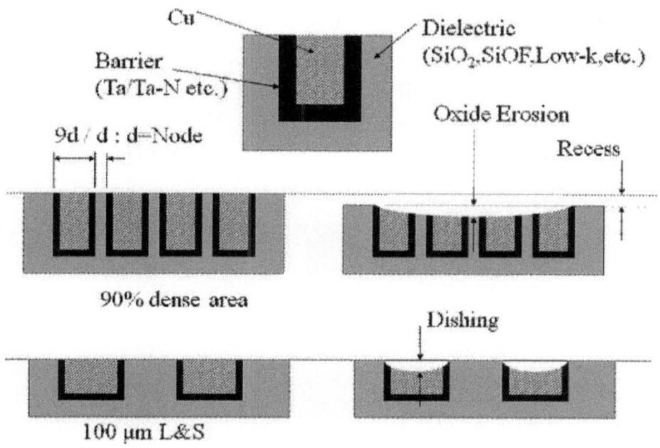

Fig. 23.7 Definition of planarization performances

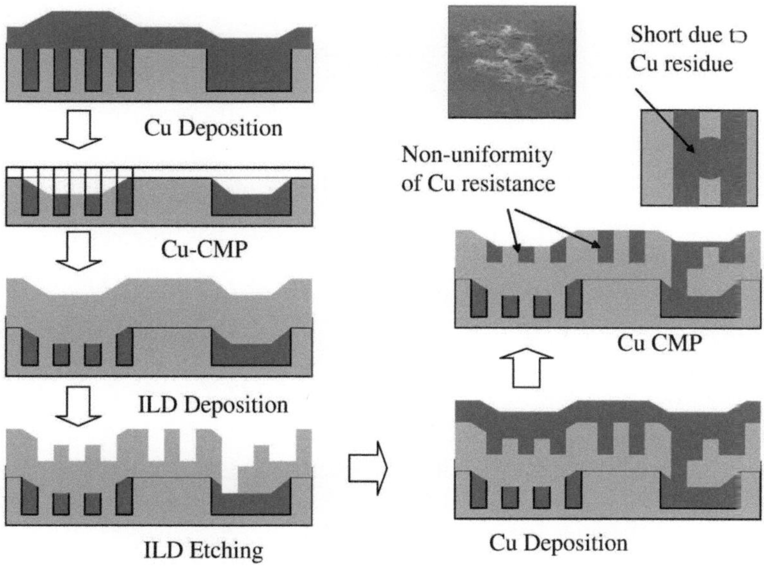

Fig. 23.8 Problems induced by dishing and erosion

It is explained in Fig. 23.8 that dishing and erosion would be a cause of Cu residue in the next layer.

If there are dishing and erosion in the first layer, this would be caused by the interconnect short by Cu residue and the reliability of interconnects, and the dishing and erosion values would affect the next layer. Therefore, requirement for planarization would be more severe in finer technical nodes.

23.3.2 Cu Polish Principle and Performances

The principle of Cu-CMP is described in Fig. 23.9. Cu slurry consists of particles like Al_2O_3 and chemicals. Chemical consists of (1) oxidizer, (2) complexing, and (3) pH control agent.

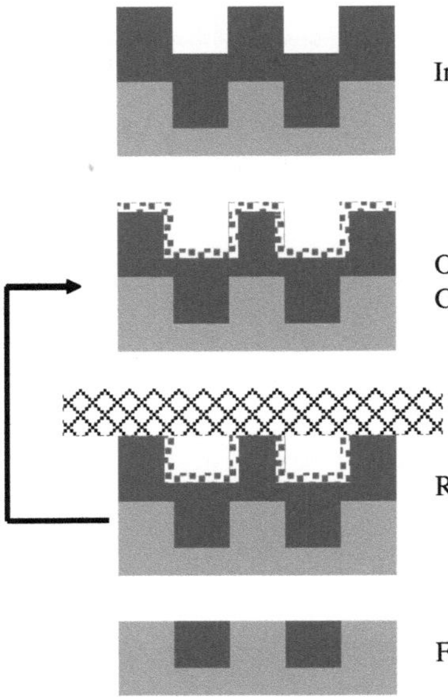

Fig. 23.9 The principle of Cu-CMP

The polish sequence is as follows:

1. Cu is oxidized
   $Cu \rightarrow Cu^{2+} + 2e^-$
   $O_2 + 2H_2O + 4e^- \rightarrow 4OH^-$
2. Oxidized Cu becomes Cu(Complex), like Cu(BTA)
3. Cu(Complex) is removed mechanically
4. Exposed Cu surface is again oxidized. = >(1)

There are two general ways for approaching the best Damascene interconnects shown in Fig. 23.10: (1) The usual deposition film by plating has step heights at chip level or wafer level. CMP is required to planarize such step heights. (2) An advanced plating technology gives less or no step height deposition. In this case, CMP planarization becomes easy.

23.3.3 Slurry Requirements

Figure 23.11 is the basic data of CMP, i.e., PVT dependence of slurry. All slurry manufacturers must supply this data first. In accordance with this basic data, CMP

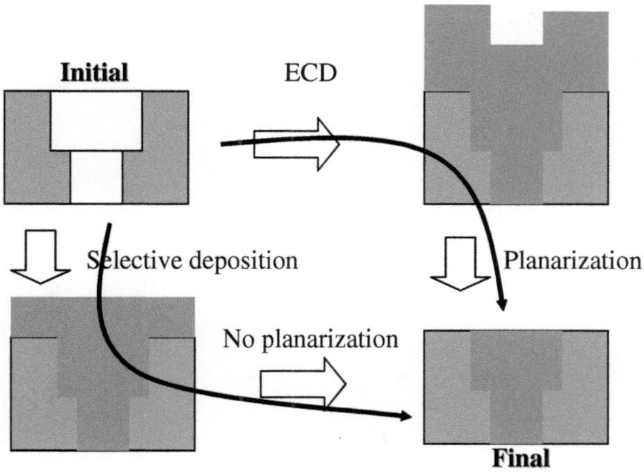

Fig. 23.10 How to approach the best Damascene interconnects

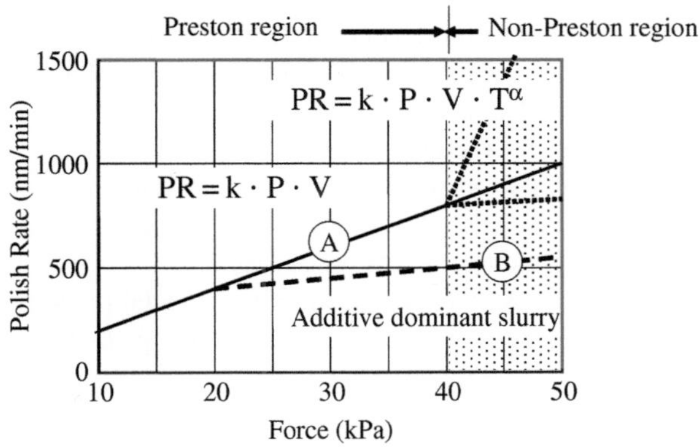

Fig. 23.11 PVT dependence of slurries

operation conditions are designed. There are several slurry types, such as: (1) Preston slurry which follows Preston's Law, (2) Arrhenius slurry which follow PVT law, and (3) additive slurry which is different from those (1) and (2).

23.3.4 Monitors

As described before, CMP processes fall into one of two categories: one for blind processes such as ILD, the second for recess processes such as metals. In the case of Cu recess process, it is possible to detect a different signal on a different material by friction changes, vibration changes, measure of Cu thickness by eddy current, and so on. The several methods are listed herein as follows:

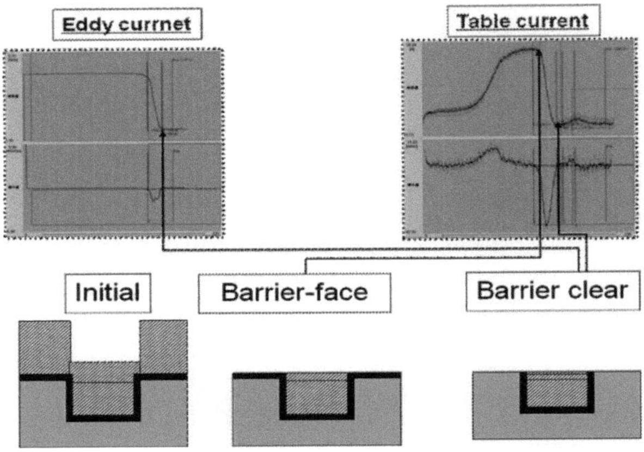

Fig. 23.12 Table current and Eddy current

1. Friction detect by motor current: This is one of the most popular recess process methods. In the example of metal CMP, friction forces differ for metals, barrier metals, and substrates. These friction differences can be detected by the motor current of a wafer carrier or a table.
2. Vibration detect by vibration sensor: Carrier vibration changes when different materials are polished or when different topographies are polished. This method can be used for both blind and recess processes.
3. Eddy current by magnetism: This is used only for metal polishing. When metal thickness changes, the eddy currents induced by magnetism also change. This system also measures metal thickness.
4. Photo detection: Different reflections are detected by different materials. This method is applicable to recess processes.

Eddy current monitors are effective and easily metal polished. Table current monitor or photo monitor may also be used. Figure 23.12 compares table current and eddy current monitors.

23.3.5 Cleaning

The surface image of post-Cu-CMP is shown on Fig. 23.13. In ILD post-cleaning, the surface to be cleaned is only ILD film such as SiO_2. However, in post-Cu-CMP, there are several materials on the surface to be cleaned such as Cu, barriers such as Ta/Ta-N, dielectric layer. Although Cu is cleaned up, Cu exists on the same surface to be cleaned.

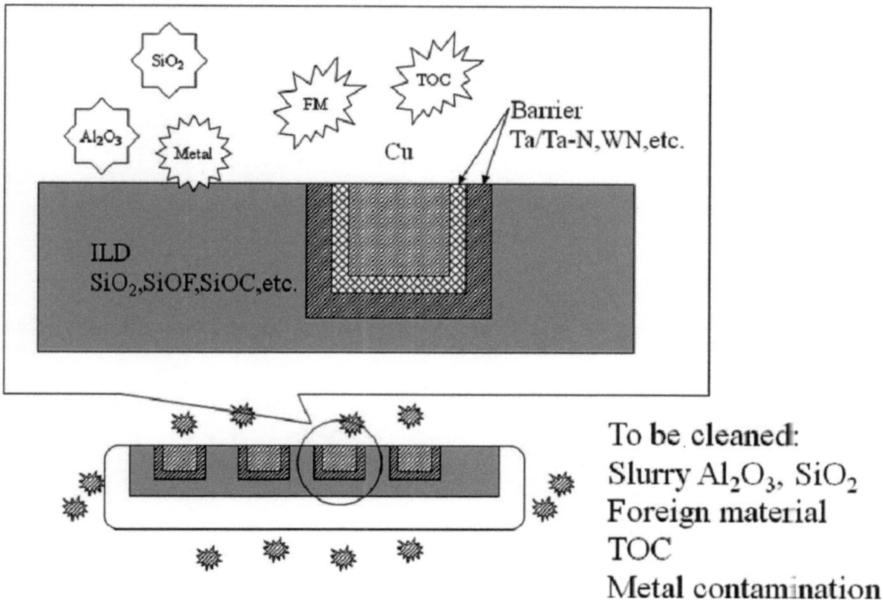

Fig. 23.13 Cleaning surfaces and contaminations

23.4 General Principle of Cu-CMP

23.4.1 Several Planarization Technologies

Although several low-*k* materials have been developed to reduce *RC* delay and power consumption, actual progress has lagged significantly behind the original schedule. As the lower-k materials feature lower strength (ex. lower Young's Modulus), lower down-force planarization technology is also required. Several technologies have been developed, such as soft CMP (chemical mechanical polishing), ECP (electro chemical polishing) [2], ECMP (electro chemical mechanical polishing) [3], CE (chemical etching) [4], and combinations of these technologies [5]. In this chapter, the general principle of planarization governing CMP, ECP, ECMP, and CE is introduced.

Figure 23.14 shows some typical planarization technologies such as CMP, ECMP, ECP, and CE. The respective principles are as follows:

The principle of Cu polishing is as shown in Fig. 23.15, as follows:

1. Chemical reaction: Cu oxidized by oxidizer. $Cu \rightarrow Cu^{++}$
2. Chemical reaction: Oxidized Cu becomes Cu complex. $Cu^{++} \rightarrow Cu(Complex)$
3. Mechanical reaction: Cu complex is removed by slurry and pad.

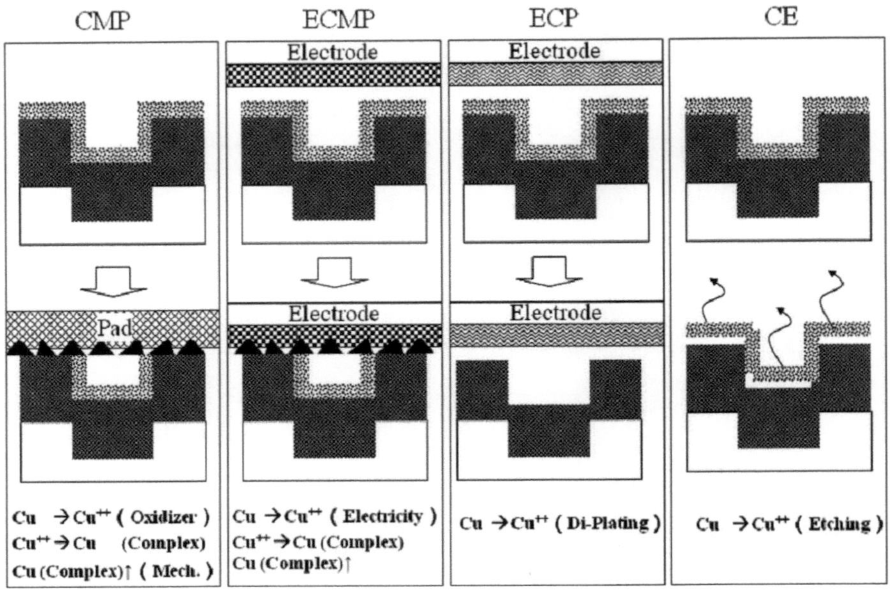

Fig. 23.14 Several planarization technologies

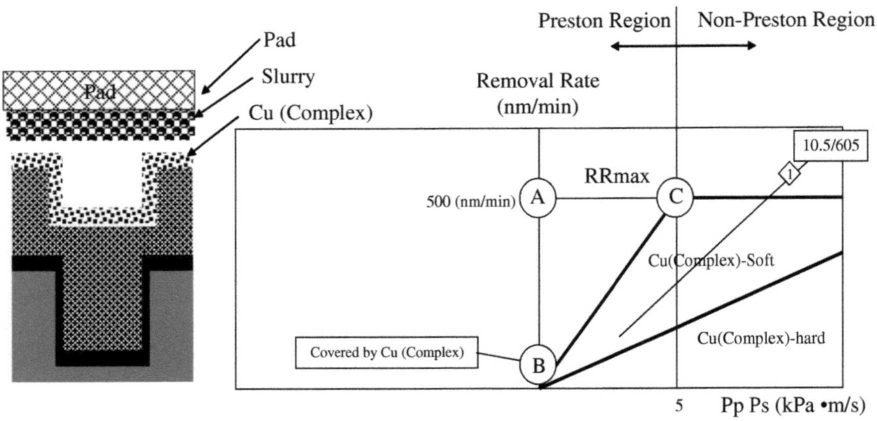

Fig. 23.15 The principle of polishing

Removal rate is governed by Preston's Law. RR=k1 Pp Ps

Where RR : Removal rate (nm/min)
k1 : Coefficient of Preston's Law
Pp : Polishing pressure (kPa)
Ps : Polishing speed (m/s)

23.4.2 CMP (Chemical Mechanical Polisher): Preston's Law

Just after the Cu surface is first oxidized by chemical oxidizer, the oxidized Cu surface is immediately changed to Cu(Complex) by the complex agent. This Cu(Complex) is removed mechanically. This represents the competing reaction between chemical reactions (1) and (2) and mechanical reaction (3), in accordance with Preston's Law. Point A shows the maximum removal rate determined by the chemicals used in reaction (1). Point B shows the minimum removal rate determined by the chemicals used in reaction (2). Point C is the intersection of the maximum removal rate and Preston's curve. The removal rate is governed by Preston's Law if Pp × Ps is situated to the left side of point C. The left side of point C is called the Preston region and the right side is the non-Preston region. The removal rate curve depends on the hardness of the Cu(Complex). The removal rate of hard Cu(Complex) is lower than that of soft Cu(Complex).

23.4.3 ECP (Electro Chemical Polishing): Faraday's Law

The principle underlying ECP is shown in Fig. 23.16. ECP is the reverse reaction of plating. It is thus referred to as di-plating. Here, the removal rate is governed by Faraday's Law.

1. Chemical reaction when Cu is oxidized by electricity $Cu \rightarrow Cu^{++}$

   Removal rate is governed by Faraday's Law. RR=k2 Cd

   Where RR : Removal rate (nm/min)
   k2 : Coefficient of Faraday's Law
   Cd : Current density (A/cm2)

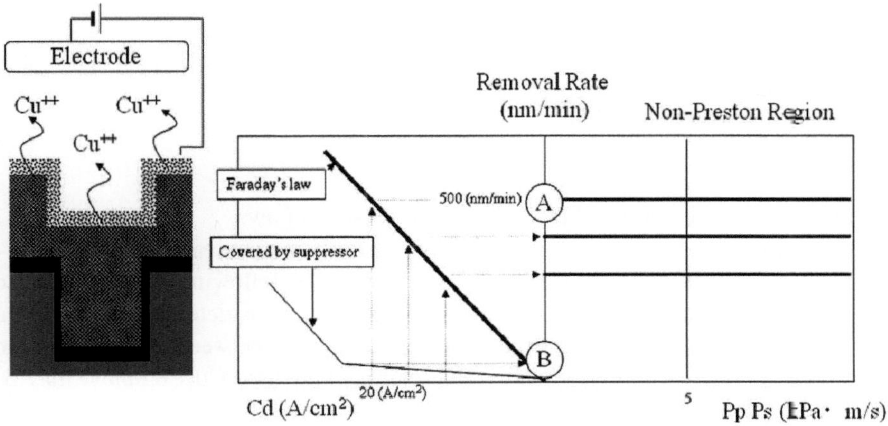

Fig. 23.16 The principle of ECP

Normally ECP does not offer the planarization performance required in semiconductor devices, even though ECP is well known to feature a smoothing function of several surfaces. However, if current can be concentrated only on the convex portion of the device pattern using a chemical agent, ECP may provide the planarization.

23.4.4 ECMP (Electrical Chemical Mechanical Polisher): Preston and Faraday's Laws

The principle of ECMP is as shown in Fig. 23.17, as follows:

1. Chemical reaction when Cu is oxidized by electricity Cu $\rightarrow$ Cu^{++}
2. Chemical reaction in which oxidized Cu becomes Cu(Complex) $Cu^{++}\rightarrow$Cu(Complex)
3. Mechanical reaction in which Cu(Complex) is removed by slurry and pad.

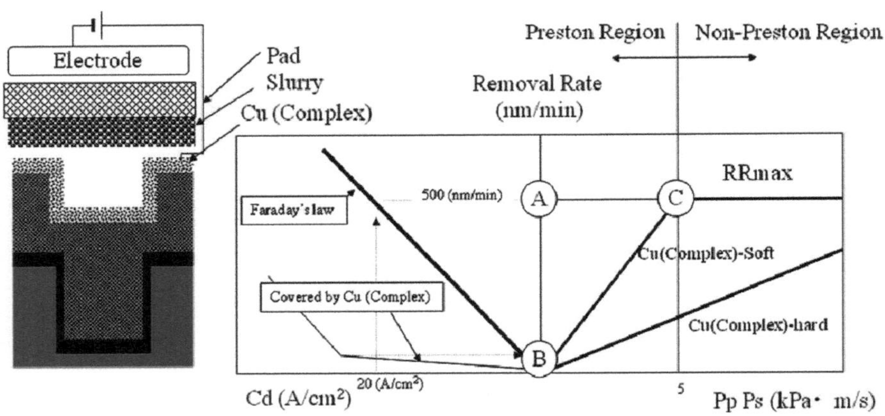

Fig. 23.17 The principle of ECMP

Removal rate is governed by Faraday's & Preston's Laws.

The difference between CMP and ECMP lies only in the means of oxidization – using chemicals or using electricity – shown in (1). The following reactions (2) and (3) are the same as in CMP. The maximum removal rate is determined by current density. It is important to determine the intersection C between the Preston and non-Preston regions complex agent, which means how hard Cu(Complex) may be formed by this complex agent.

23.4.5 CE (Chemical Etching): Dissolution Law

The principle of CE is shown in Fig. 23.18. Here, the removal rate is governed by the Dissolution Law.

1. Chemical reaction in which Cu is dissolved by chemical Cu → Cu^{++}
2. Chemical reaction in which Cu is covered by suppressor Cu → Covered
3. Mechanical reaction in which suppressor is removed

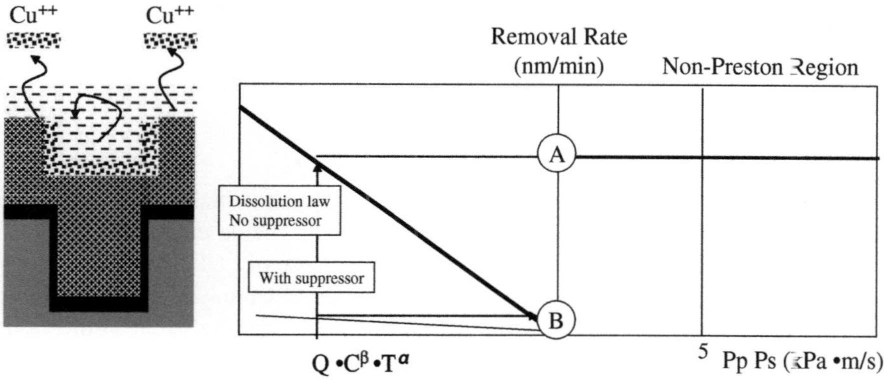

Fig. 23.18 The principle of etching polishing

Removal rate is governed by Dissolution Law $RR = k3\ Q\ C^{\beta}\ T^{\alpha}$

Where RR : Removal rate (nm/min)
k3 : Coefficient of Dissolution Law
Q: Flow rate of etching chemical
C^{β} : Concentration of etching chemical
T^{α} : Temperature of etching chemical

Although etching without a suppressor does not provide planarization, CE with a suppressor provides a certain degree of planarization.

23.5 General Principle of Planarization

The general principle of planarization is as shown in Fig. 23.19, although in the figure several of the above-mentioned planarization technologies are combined. All Cu planarization is completed through a process similar to the following:

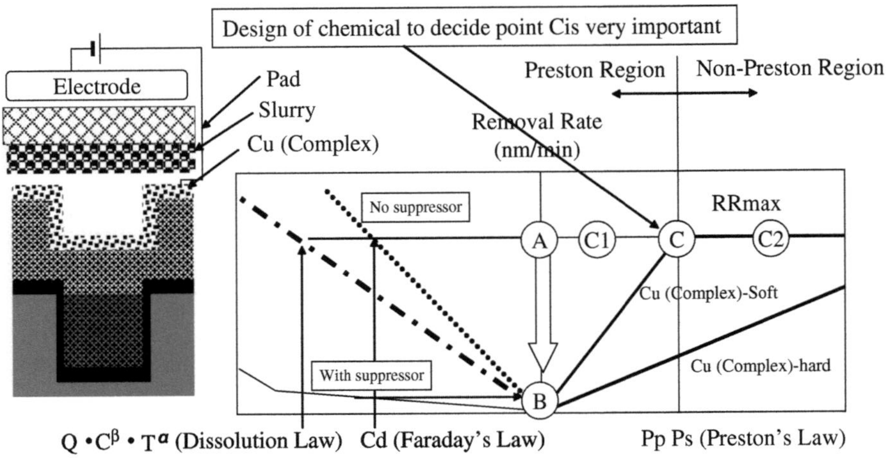

Fig. 23.19 General planarization principle diagram

Point A shows the maximum removable rate, RRmax, which is determined by flow rate Q, concentration C^{β}, and temperature T^{α} in the cases of CMP and CE (Dissolution Law), and by current density Cd in the case of ECP and ECMP (Faraday's Law). Point B shows the minimum removal rate, RRmin, which is determined by a complex agent in CMP and ECMP, and by a suppressing agent in the cases of ECP and CE. The line passing through point C shows the removal rate in accordance with Preston's Law. The side to the left of point C is called the Preston region and the side to the right is the non-Preston region. This intersection C is determined by the composition of the complex agent. The extent of hardness of Cu(Complex) is critical in determining which range – in the Preston or non-Preston regions – is used in this planarization work [5]. As shown in Fig. 23.19, this intersection point C may be changed C1 softer side and C2 harder side. It is thus easy to understand the planarization mechanism with reference to this diagram of the general principle involved.

The general principle of planarization technologies unifying CMP, ECP, ECMP, and CE is introduced. A general principle diagram enables easy understanding of the planarization mechanism and contributes to the development of new technologies combining these planarization technologies.

References

1. Kashiwagi, M.: CMP Science, 26 (1977)
2. Shin-Chieh, C.; et al.: Pattern on Planarization Efficiency of Cu Electropolishing. Jpn. J. Appl. Phys. **41**, 7332 (2002)
3. Economicos, L.; Wang, X.; Sakamoto, X.; Ong, P.; Naujok, M.; Knarr, R.; Chen, L.; Moon, Y.; Neo, S.; Salfelder, J.; Duboust, A.; Manens, A.; Lu, W.; Shrauti, S.; Liu, F.; Tsai, S.; and Swart, W.: Integrated Electro-Chemical Mechanical Polishing(Ecmp) for Future Generation. IEEE IITC, 233 (2004)

4. Pallinti, J.; Lakshminarayanan, S.; Barth, W.; Wright, P.; Lu, M.; Reder, S.; Kwak, L.; Catabay, W.; Wang, D.; and Ho, F.: An Overview of Stress Free Polishing of Cu with Ultra Low-*k*(k<2.0) Films. IEEE IITC, 83 (2003)
5. Sato, S.; Yasuda, Z.; Ishihara, M.; Komai, N.; Ohtorii, H.; Yoshio, A.; Segawa, Y.; Horikoshi, H.; Ohoka, Y.; Tai, K.; Takahashi, S.; and Nogami, T.: Newly Developed Electrochemical polishing process of copper inlaid in frangile low-*k* dielectrics. IEEE IEDM, 4.4.1 (2001)

Chapter 24
Electrochemical View of Copper Chemical–Mechanical Polishing (CMP)

D. Starosvetsky and Y. Ein-Eli

24.1 Introduction

Copper is the metal of choice, replacing aluminum in integrated circuit interconnections [1]. This switch was emerged and stimulated due to copper advantage characteristics, such as low resistivity and high immunity to electromigration, which in turn result in greater circuit reliability and markedly higher clock frequency. Copper dual-Damascene technology includes two main electrochemical steps. First step is the electrochemical copper deposition (or copper electroplating) into trenches and vias. Second electrochemical step in copper Damascene technology utilizes chemical–mechanical polishing or planarization (CMP) aiming at the removal of overburden copper after its electrochemical deposition. CMP primer objective is to achieve a global planarization of patterned surface. CMP appears to be the most promising pattern delineation technique [2–6]. Planarization via CMP process is achieved by simultaneous actions of both mechanical abrasion and electrochemical dissolution of the planarized metal.

It is absolutely clear that CMP is a quite complicated technological process, in which its efficiency is beyond the scope of only metal planarization. It depends on other various factors such as chemical composition and design of planarized structure, type of CMP polisher, pad and abrasive characteristics, and most importantly the chemical composition of the chosen slurry. The planarized structure contains a barrier layer and interlayer dielectric in addition to the metal lines, which should be planarized simultaneously with the metal layer. Most of these factors have already been discussed thoroughly in number of reviews [2–10]. In the present discussion, we would like to consider and elaborate on the electrochemical aspects of copper CMP.

Y. Ein-Eli (✉)
Department of Materials Science and Engineering, Technion-Israel Institute of Technology, Haifa 32000, Israel
e-mail: eineli@technion.ac.il

Y. Shacham-Diamand et al. (eds.), *Advanced Nanoscale ULSI Interconnects: Fundamentals and Applications*, DOI 10.1007/978-0-387-95868-2_24,

It is well known that effective CMP of metals can be obtained only when the slurry fulfills the following requirements [11]:

- Planarized metal undergoes passivation in the slurry used. Thus, metal lines which are not in a direct contact with the polishing pad region remain passive during the planarization.
- Dissolution of planarized metal occurs only at the activated upper surface sites, where passive film is mechanically removed due to a direct contact with either the pad or the abrasive.
- Rapid repassivation of previously exposed metal sites due to the mechanical removal of the oxide.

Thus, the major requirement for CMP slurry is to provide a means for self-passivation of the planarized metal; it is the essence of CMP application. Originally, CMP was developed for aluminum (Al) and tungsten (W) [11], i.e., for metals which posses enhanced ability to develop passivation (or self-passivation). The main goal of a mechanical abrasion process, in accordance with a classical description of a CMP process, is the initiation of a rapid metal dissolution of top exposed metal vicinities due to the removal of the passive film. It is important to note that mechanical removal of planarized metal is strictly discouraged. In other words, the cyclic sequential actions in CMP planarization have to be a removal of the passive film, metal dissolution of freshly exposed surface sites, and their rapid repassivation.

All the above-mentioned requirements for an efficient CMP process are valid once copper is considered. An efficient copper CMP process can be achieved only in slurry, which could provide a strong rapid passivity of copper. However, unlike aluminum and tungsten, copper is not characterized with an enhance capability to develop a self-passivation and thus, efficiency copper CMP does not seem so obvious. Potential-pH (Pourbaix) diagrams predict that copper passivity could be expected in pH region between 7 and 13, i.e., in neutral and alkaline solutions [12]. However, the protective layer covering copper surface in this pH region is characterized as being resemble to precipitations of coarse insoluble products (such as copper oxides and hydroxides, Cu_2O, CuO, and $Cu(OH)_2$ [12–16]) rather than a thin protective (passive) oxide film.

It is generally believed that an efficient copper CMP process can be achieved by the addition of different compounds to the slurry. These compounds include oxidizers, inhibitors, buffering, and complexing agents. The role of both inhibitors and buffering agents in a slurry composition is abundantly clear: to prevent corrosion damage of the planarized surface (inhibitor) and to maintain a pH value of the slurry (buffering agent) during the CMP operation. Oxidizer is usually added to slurries in order to accelerate the formation of a passive film at the copper surface due to a shift in its corrosion potential to more positive values. Complexing agent role is to prevent formation of insoluble copper compounds and their deposition at the planarized wafer surface.

Different types of slurry were suggested for copper CMP application. Most prominent slurries compositions in CMP application are the once based on ammonium hydroxide, nitric acid, and peroxide solutions. These solutions are markedly differing from each other in theirs pH values. Whereas, ammonium hydroxide-based slurry is characterized with a strong alkalinity (pH between 11 and 12 for a recommended concentration), the pH values of nitric acid-based slurry is very low ($pH \leq 2$) [17]. Peroxide-based slurry pH can be chemically related to weak acid solutions (pH from 3.5 to 4) [18]. However, evaluation of Pourbaix diagram predicts that copper can be actively dissolved in solutions below pH value of 7 [19]. It is obvious that acidic solutions cannot provide copper passivity and therefore, their use in CMP slurries is problematic. In the next section we consider these three types of Cu CMP slurries.

24.2 Ammonium Hydroxide-Based CMP Slurry

Ammonium hydroxide-based solutions were practically the first to be developed and recommended for copper CMP applications. In addition to ammonium hydroxide, other ammonia-based slurries, such ammonium nitride and ammonium chloride, have been investigated extensively by Murarka, Steigerwald, Gutman et al. [19–23], Osseo-Asare and Mishra [24], and Luo et al. [18, 25].

An effort to develop effective copper CMP slurry on the basis of ammonia-containing solutions may seem to be highly reasonable since:

Ammonia-based slurries are either neutral (ammonium nitride, ammonium chloride) or basic (ammonium hydroxide) media. These slurries are associated to a pH region where copper passivity is to be expected [12].

Ammonia is known as a strong complexing agent which is capable of transferring copper ions into soluble cations, such as $Cu(NH_3)_2^+$ and $Cu(NH_3)_4^{2+}$, thus preventing precipitation of insoluble Cu (1) compounds (hydroxides and/or oxides) on the planarized copper surface.

In these studies, copper planarization was conducted at open circuit potential (OCP). The highest planarization rate in oxidant-free solutions was achieved in a slurry based on ammonium hydroxide, compared with ammonium chloride and nitride solutions [19–23]. Slight increase in copper CMP rate in ammonium hydroxide-based slurries was achieved due to the addition of oxidants, such as $Cu(NO_3)_2$ [20], NH_4NO_3 [22, 25], $NaClO_3$ [18], and MnO_4 [26]. Acceleration in dissolution rate in these studies was achieved due to increase in copper dissolution by shifting the corrosion potential in a positive direction. However, it should be noted that copper does not develop self-passivation in ammonium hydroxide.

Thus, the main requirement of CMP slurry (the passivity of treated metal) is not being fulfilled once copper in ammonium hydroxide is considered. Copper actively dissolves in ammonium hydroxide and passive oxide film does not form at the copper surface during its exposure at OCP. It is the main reason for the equivalent copper dissolution rate obtained from non-abraded (0.91 nm/min) and abraded copper (0.92 nm/min, against a scour pad), found in NH_4OH by linear polarization measurements (LPR) method [24].

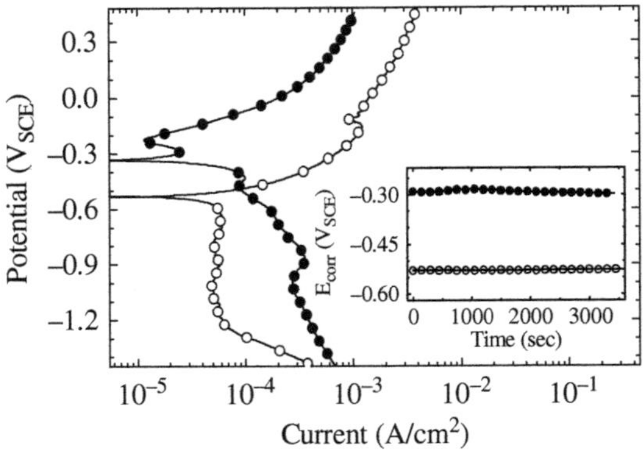

Fig. 24.1 Polarization curves of copper obtained in 2.35 g/l (•) and 30 g/l (o) NH_3 (25°C) after 1 h exposure at OCP at scan rate of 1 mV/s. Inserted graph: E_{Corr} transient of copper in 2.35 and 30 g/l NH_3 solutions at 25°C

Typical anodic potentiodynamic curves of copper in NH_4OH solution are shown in Fig. 24.1 [27, 28]. The onset of an anodic current was observed at –0.34 V in a solution containing 2.35 g/l NH_3 and at –0.54 V with a solution containing 30 g/l NH_3. Anodic current gradually increased during potential sweep in the positive direction. Anodic current peak was detected in curves obtained in both NH_3 concentrations. In a solution containing 2.35 g/l NH_3 the anodic peak appears at a potential of −0.28 V, while in a solution containing 30 g/l NH_3 the anodic peak is located at much higher potential (−0.15 V).

This shape of the anodic curve is associated with an active dissolution of copper. It was mentioned above that copper dissolves in the presence of ammonium via the formation of soluble ammonium complexes. As can be seen, the parameters of dissolution processes strongly depend on ammonia concentration. In a solution with lower ammonia concentration (2.35 g/l NH_3), the possibility for dissolved Cu (1) ions to be transformed into ammonium complexes ($Cu(NH_3)_2^+$) is limited. This causes the formation of excess Cu (1) ions at the electrode/solution interface, resulting in a formation of insoluble oxide and hydroxide compounds of Cu (1), which precipitate at the copper surface, leading to reduction in the recorded anodic current, presented in the appearance of an anodic peak. In a solution containing higher NH_3 concentration (30 g/l), the amount of Cu (1), which can be transformed into ammonium complexes, significantly increased and the formation of Cu (1) oxides and hydroxides would be expected at higher anodic currents and, consequently, at more positive potentials, as shown in Fig. 24.1.

This explanation is supported by exploring the effect of potential scan rate on the anodic profile. It was established [27, 28] that increase in scan rate causes disappearance of the anodic peak as a result of a decrease in the amount of Cu (1) ions transferred to the solution and, consequently, the amount of copper oxides and hydroxides precipitations is minimized.

Etching rate of copper in ammonium hydroxide solution was measured to be quite low at both studied ammonium concentrations (2.35 and 30 g/l). The results, measured by linear polarization, are shown in Table 24.1. As can be seen, copper corrosion rate in the studied ammonium solutions was approximately 1 nm/min.

Table 24.1 Corrosion potential, corrosion current, and corrosion rate of copper in different ammonium hydroxide solutions at OCP

Concentration NH_3 (g/l)	E_{corr} (V_{SCE})	I_{corr} ($\mu A/cm^2$)	Corrosion rate (nm/min)
2.35	−0.315	29.76	1.313
30	−0.509	51.93	2.29

Taking into account that copper removal rate during CMP in ammonium hydroxide is more than two order of magnitude higher (approximately 500 nm/min copper removal was reported by Strengewald [19–22]), we can conclude that copper CMP process in ammonium hydroxide is practically a process of mechanical removal type.

Copper surface morphology subsequent to short and extended OCP exposure in different ammonium hydroxide solutions was studied by SEM, shown in Fig. 24.2 [27, 28]. One can easily observe features of copper etching in 2.35 and 30 g/l ammonium hydroxide and realize that copper corrosion in ammonium hydroxide is a non-uniform process. Severe localized corrosion attacks can be observed at the copper surface subsequent to both short and extend exposure periods. Strong attacks through grain boundaries and shedding of whole sub-grains can be clearly seen in these micrographs. Figure 24.2a even reveals dissolution

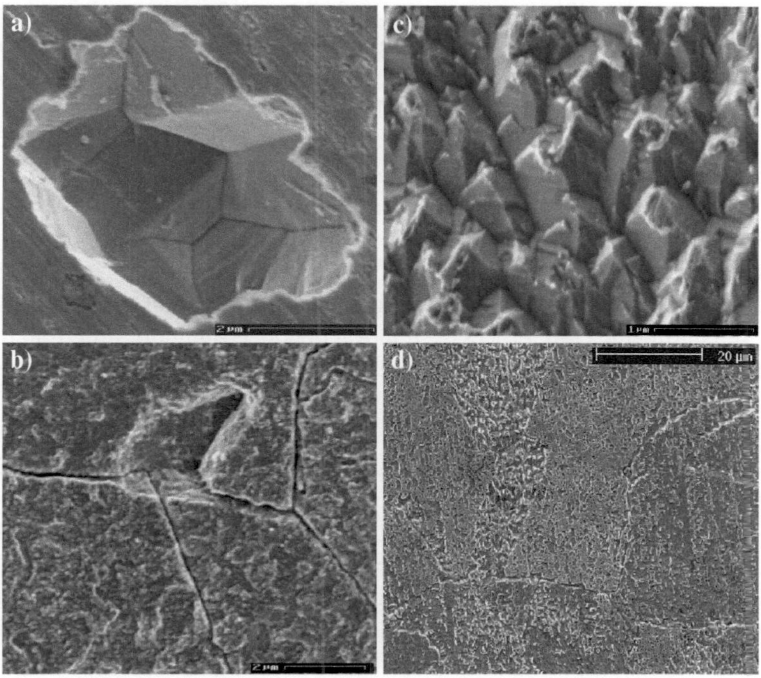

Fig. 24.2 Different fragments of copper surface after different exposure at OCP in 2.35 g/l NH_3 solution (**a** – 1 min, **b** – 1 h) and in 30 g/l NH_3 (**c** – 1 min, **d** – 1 h)

of entire sub-grain, exposing high oriented crystallographic plains, indicating of anisotropic dissolution process. Corrosion propagation through grain boundaries and sub-grain boundaries can lead to high rate of copper removal during CMP, since these grains or sub-grains can be detached mechanically from the copper surface.

Thus, the electrochemical behavior of copper in ammonium hydroxide solutions is far from complying with the main requirements for an efficient copper CMP process. Therefore, the use of ammonium hydroxide-based solutions in CMP slurries is questionable and should be reconsidered, since it does not comply all copper CMP slurry requirements, mainly due to an active copper dissolution during exposure at OCP. This may damage thin Cu layer, resulting in severe fracture of the copper interconnect. Addition of oxidants results in a shift of the corrosion potential in a positive direction, leading to an increase in copper dissolution which would not provide copper passivity.

24.3 Nitric Acid (HNO_3)-Based CMP Slurry

As was mentioned earlier, copper is actively being dissolved in nitric acid [12]. The active dissolution and high rate of cathodic depolarization in nitric acid (nitric acid is considered as a strong oxidizer) provide sufficient conditions for high etching rate. Under these conditions, abrasion of actively dissolved Cu surface could only decrease the dissolution rate due to transport limitations of reactants or dissolution products provided by the abrasive pad [27]. Thus, solutions based on nitric acid seem as unsuitable for a conventional CMP.

In order to fully comply with the outlined CMP requirements, the addition of inhibitors (e.g., benzotriazole, BTA) in acidic solutions was suggested [22, 26, 29]. Addition of BTA results in the formation of tightly adherent Cu–BTA protective layer film at the copper surface [30, 31], which blocks copper surface removal, leading to a decrease in the dissolution rate of the metallic copper. Many authors [22, 26, 29, 32] indicated that this protective layer may have the same role as a passive oxide film during the CMP process: a removal of copper-inhibitor film from the surface of the copper during CMP would result in an active dissolution of bare sites. Recovering and reformation of a protective film at these sites during further exposure would decrease the dissolution rate. It was suggested that repetitive activation and passivation of exposed surface sites may result in surface polishing [1, 22, 26].

Practically, the studies involving nitric acid slurry were in a search for an optimum BTA concentration [26, 29, 32, 33]. As was shown by Wang et al. [33], copper corrosion rate was decreased with increasing BTA concentration up to 0.01 M. Further increase in BTA concentration saturates the inhibiting effect. The authors indicated that the use of 0.01 M BTA solution would have an effective copper inhibition once applied in low HNO_3 concentration (volume percentage $<10\%$). It was shown that higher polishing quality could be obtained uniformly by adding 0.01 M BTA into HNO_3 ($1-3$ vol.%) slurry [33].

Steigerwald et al. [22] also praised the same advantages in HNO_3–BTA slurry: formation of a Cu–BTA monolayer, which blocks the etching action of copper in the slurry, and therefore prevents dissolution of the recessed copper. Envisioning HNO_3–BTA-based slurry, one can achieve favorable conditions (such as reducing mechanical energy (low down-pressure and velocity)), leading to the removal of the Cu–BTA layer, while Cu itself is not abraded. In such a process, etching of copper by the nitric acid action would occur on the top and exposed sites, where the Cu–BTA layer is being removed. In this case, there would be only a chemical removal of the copper; therefore, the copper and underlying films would experience little or no mechanical abrasion. However, this hypothesis has not been experimentally verified to date.

Hu et al. [32] proposes a novel inhibitor, citric acid, to be used in HNO_3-based slurry, instead of BTA. The authors noted that citric acid is also capable to reduce copper etch rate in nitric acid. However, experiments conducted in our laboratory did not reveal any inhibition properties of citric acid. Moreover, strong acceleration in Cu etch rate was observed in HNO_3-based slurry once citric acid was added [34].

However, some serious questions regarding the efficiency of BTA in copper CMP are raised, since: (i) BTA was developed and applied as an inhibitor for copper/water systems, while in acidic solutions it is not considered as highly effective; [31, 35–37] (ii) BTA, as all inhibitors, is usually used in copper protection (for example, in cooling systems) for a long term (months, years), because the formation of protective inhibitor layer consumes relatively much time (from some minutes to a few hours). Therefore, protective layer formation rate during the CMP process should be a fraction of a second. Thus, there is a doubt whether the addition of any inhibitor would provide extremely high rates of protective layer formation, required for CMP.

Figure 24.3 shows the potentiodynamic curves of copper in nitric acid solutions with concentrations of 0.05, 0.2, 1, and 3 vol.% [28, 29]. Potential sweep was applied once the corrosion potential reached a steady state values during OCP exposure. Both cathodic and anodic polarization curves were obtained separately by a shift of the potential in negative and positive directions, respectively. As can be seen, anodic current is gradually increased with a positive shift in the applied potential, indicating an active dissolution. The cathodic branch of the polarization curve reveals a peak appearance subsequent to OCP exposure. In all the studied solutions, the peak was located approximately 50 mV below the corrosion potential and decreased with increase in nitric acid concentration. It is reasonable to suggest that the cathodic peak is associated with a reduction of corrosion products (e.g., $Cu(NO_3)_2$, CuO, Cu_2O, or CuOH) precipitated at the copper surface during OCP exposure. The amount of precipitants is decreased with increase in nitric acid concentration as a result of higher Cu (1) ions solubility. Values of corrosion potentials, corrosion currents, and corrosion rates obtained from nitric acid-based solutions with the use of linear polarization technique are shown in Table 24.2 [2]. Copper dissolution rate in acidic solution is much higher than the rate detected in ammonium hydroxide solutions (see Table 24.1). Corrosion rate of copper grows significantly with increase in nitric acid concentration.

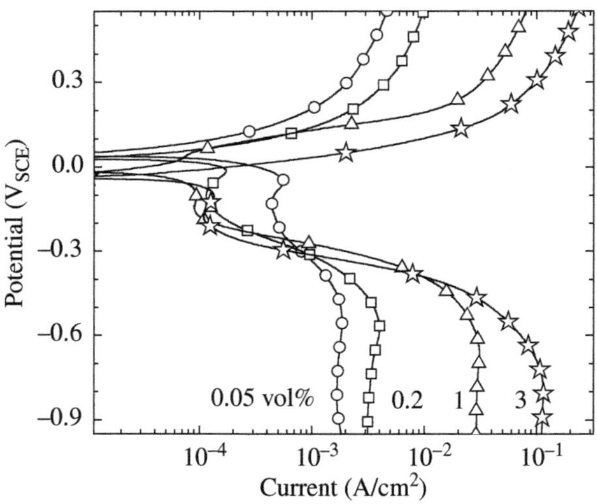

Fig. 24.3 Polarization curves of copper obtained at scan rate of 1 mV/s in solutions with different concentrations of nitric acid: 0.05 vol.% (○); 0.2 vol.% (□); 1 vol.% (△), and 3 vol.% (∗)

Table 24.2 Corrosion currents, corrosion potential, and corrosion rate of copper in nitric acid solutions

Concentration HNO_3 (vol.%)	pH	E_{corr} (V_{SCE})	I_{corr} (mA/cm^2)	Corrosion rate (nm/min)
0.2	1.78	0.02	0.604	13.3
1	1.19	0.04	1.658	36.6
3	0.9	0.052	4.468	100.45

Figure 24.4 presents the effect of BTA on copper anodic characteristic in 3 vol.% HNO_3 solution measured subsequent to 1 h exposure at OCP [27, 28]. One can see that BTA dramatically reduces the anodic current in a wide potential range (up to 0.18 V). Therefore, the attempts to realize copper CMP in nitric acid-based slurries containing BTA seem highly feasible and reasonable, in accordance with these anodic characteristics:

1. Dramatic decrease in the observed anodic current in a wide potential range.
2. Increase in etching rate of bare surface sites, where protective Cu–BTA layer is being removed by the action of a mechanical abrading.
3. An efficient protection of the recessed surface sites.

However, BTA is an efficient inhibitor once a long-term protection is required (i.e., when the protection time is weeks, months, or years).

In an effort to estimate the rate of protective Cu–BTA layer formation (in the potential region where inhibition properties of BTA were detected), we evaluated (Fig. 24.5) the anodic current transient measured at 0.1 V in nitric acid (3%) before and subsequent to an addition of 0.02 M of BTA [27, 28]. Subsequent to BTA

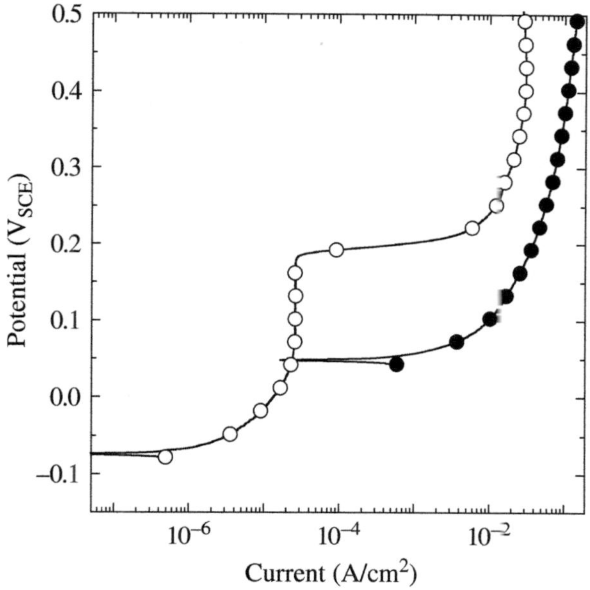

Fig. 24.4 Anodic potentiodynamic curves (1 mV/s) of copper in 3 vol.% nitric acid without (•) and with (○) 0.02 M BTA

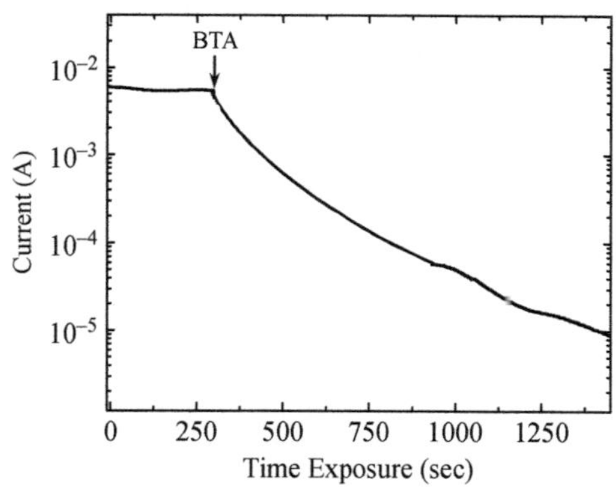

Fig. 24.5 Anodic current transient of copper measured at 0.1 V in 3 wt.% nitric acid before and after the addition of 0.02 M BTA

addition, the anodic current slowly decreased during a relatively long time period, i.e., BTA does not provide immediate copper protection in a fraction of second, as required. Protection provided by BTA occurred in time scale of minutes. Therefore, an effective CMP conducted with a rapid surface abrading strictly requires much more rapid formation of a protective layer, compared with the one formed by BTA. These results are in a good agreement with CMP experiments conducted in nitric acid containing BTA and BTA-free slurries, which showed an equal rate of Cu removal [18]. We attribute these results to the sluggish formation rate of BTA protection layer at the abraded sites.

24.4 Peroxide-Based CMP Slurries

Many types of slurry developed for copper CMP processes are based on hydrogen peroxide [33, 38–51]. Hydrogen peroxide is known as a powerful oxidizer, which can be effectively used in various solutions in a wide range of pH values. In addition, H_2O_2 is a very clean oxidizer, which its use does not require a post-CMP cleaning procedure. Its reduction is considered to be "clean"; it is accompanied only with oxygen and hydrogen evolution, hydroxide ions, and water formation. This advantage drives the development of slurries containing only water and hydrogen peroxide [43–51]. Peroxide concentrations in these solutions are varied from 1 to 30 vol.%, while the pH values of such solutions are between 4.5 and 3.5 (decreasing with increase in H_2O_2 content) [43, 44]. However, active copper dissolution should be expected in this pH range, according to Pourbaix diagram [12]. Hence, increase in copper dissolution rate with increase in peroxide concentration is expected due to increase in the solution acidity and acceleration of cathodic process as was observed in $Fe(NO_3)_3$ and nitric acid solutions [51]. However, the well-expected increase in copper dissolution rate with peroxide content was found only in slurries utilizing H_2O_2 in concentrations of 1–3%. Further increase in H_2O_2 concentration resulted in copper dissolution rate reduction [33, 41]. It was suggested by Zeidler et al. [38, 41, 43–46], Nguyen et al. [42], Wang et al. [33], and Kondo et al. [39] that this observation was a result of copper passivation. Similar explanation was suggested by Molodov et al. [50], describing copper behavior in $HClO_4$–H_2O_2 media. It was also assumed that copper passivity is provided by OH radicals absorbed at the copper surface [50]. Copper passivation due to the formation of oxide film in peroxide solutions was also reported by Hernandez et al. [41] and Hirabayashi et al. [51].

However, copper etching rate values reported in these studies could be hardly attributed to copper passivity. For example, copper dissolution rate of 10 nm/min was obtained in 15% H_2O_2 [40, 42–44]. From a corrosion science point of view, a metal having a corrosion rate higher than 0.1 mm/year is considered to be a low-corrosion resistance substance. The etching rate of copper in hydrogen peroxide solution is rather associated with an active copper dissolution than passivity, since etch rate of 10 nm/min rewritten in corrosion units corresponds to more than 5 mm/year, considered as high dissolution rate.

The anodic behavior of copper in peroxide solutions is characterized by very low anodic current values measured in the entire potential range, from OCP to 0.6 V [27]. Increase in peroxide concentration resulted in a shift of E_{corr} toward positive potentials with an increase in the anodic currents. The low anodic current values obtained from the potentiodynamic measurements were reported to be actually a result of low ionic conductivity of peroxide solutions [27, 28]. It was shown that the addition of sodium sulfate (Na_2SO_4) significantly increases solution conductivity and anodic current is gradually increased by a positive potential shift, indicating an active copper dissolution. Figure 24.6 presents anodic behavior of copper in 10 g/l Na_2SO_4 solution containing 3 vol.% H_2O_2 (pH 4) [52]. Anodic characteristics were measured potentiodynamically at different scan rates subsequent to cathodic pretreatment (1 min exposure at −1.3 V), which was conducted in order to reduce

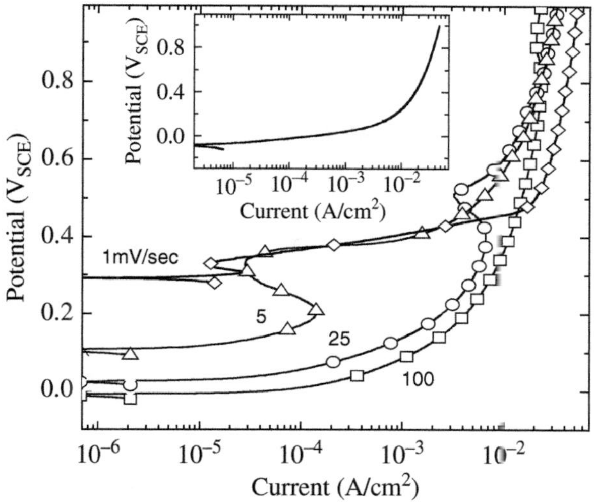

Fig. 24.6 Anodic curves of copper obtained in 10 g/l Na_2SO_4 with 3 vol.% H_2O_2 (pH 4) at different scan rates: 1 (◇); 5 (△); 25 (○), and 100 mV/s (□). Inset: anodic curves of copper obtained in 10 g/l Na_2SO_4 (pH4)

copper oxides covering the electrode surface. Cathodic pretreatment was followed immediately by a positive potentiodynamic sweep. The onset of anodic currents at a scan rate of 100 mV/s occurred at a potential of approx. 0.0 V. Positive sweep of the applied potential at this scan rate resulted in a gradual increase in the anodic current, similar to the results obtained in Na_2SO_4 peroxide-free solution (Fig. 24.6, inset), associated with an active copper dissolution. At a scan rate of 25 mV/s, the onset of anodic currents shifted in a positive direction and was observed at 0.03 V. During the potential sweep an anodic current peak can be clearly seen in a potential region between 0.4 and 0.5 V. Onset of anodic current increased with a decrease in scan rate: the onset of 0.1 V and 0.3 V was measured at 5 and 1 mV/s, respectively. Current value related to the anodic current peak markedly decreased and the peak narrowed with a decrease in scan rate (Fig. 24.6). Further shift of the voltage in a positive direction involves a rapid acceleration in anodic current at potentials above 0.3 V.

The decrease in anodic current along with anodic peak formation detected at slow scan rate was attributed to precipitation of copper oxides/hydroxides. This is shown by potentiostatic measurements conducted at different potentials related to the anodic current peak range (Fig. 24.7) [52, 53]. As can be seen, copper exposure at potentials below 0.4 V (region of current peak) is accompanied with a significant current decrease, which can be attributed to the formation of a protective layer of copper oxides or hydroxides, covering the copper surface. However, at potentials above 0.4 V and within a short exposure period (40 s as can be seen in Fig. 24.7), the current decreases, while during further exposure the anodic current increases, indicating the initiation of other electrochemical processes.

It was established by SEM studies that copper surface exposed at potentials below 0.3 V was covered with a thick deposit layer composed of copper oxide or hydroxide precipitations. Subsequent to an exposure in potentials higher than 0.3 V,

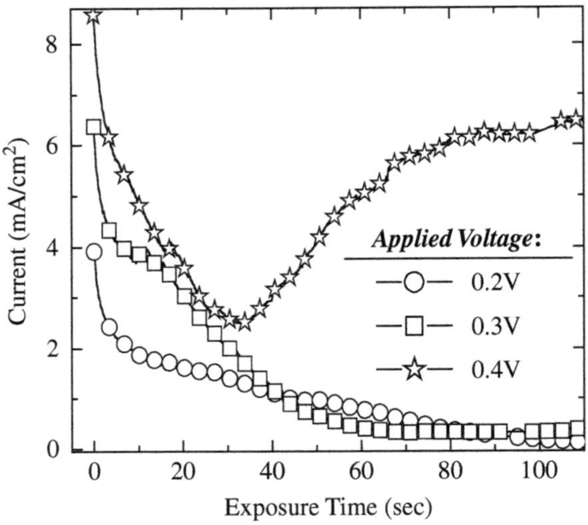

Fig. 24.7 Current–time profiles obtained by potentiostatic exposure of copper electrode at different applied voltages (0.2, 0.3, and 0.4 V) in 10 g/l Na_2SO_4 solution with 3 vol.% H_2O_2 peroxide (pH 4)

the electrode surface suffered from a localized corrosion attack. Numerous small pits (Ø 10–50 nm) were detected at copper surface [52, 53].

The formation of a dense deposition film of copper oxides at potentials below 0.3 V may be attributed to an increase in the solution pH at the electrode–electrolyte interface to values above pH 5.5, where copper oxides remain thermodynamically stable. The increase in the pH value can occur as a result of peroxide cathodic reduction ($H_2O_2 + H^+ + 2e^- \rightarrow 2H_2O$).

It was reported that the addition of inhibitors, such as benzotriazole (BTA), to peroxide solutions is a route for achieving a rapid copper passivation and formation of thin protective film instead of a thick deposition layer of copper oxides and hydroxides [54, 55]. In order to suppress copper oxides and hydroxides deposition, it was also recommended to add glycine to the slurry composition as a complexing agent.

The effect of BTA and glycine on the anodic behavior of copper in peroxide containing solutions is illustrated in Fig. 24.8 [52, 53]. The addition of BTA to Na_2SO_4 solution significantly decreases the anodic current at potentials below 0.2 V, indicating the formation of a protective layer. During further positive sweep, one can observe a slow increase in anodic current up to ~0.4 V and a rapid current increase at potentials above 0.4 V. The addition of peroxide to the Na_2SO_4 – BTA solution results in a pronounced shift in the onset of the anodic current, up to a potential of 0.4 V. The positive shift in the potential from OCP is accompanied with a gradual increase in the anodic current (Fig. 24.8a). Unlike BTA, the addition of glycine to Na_2SO_4 solutions containing BTA and H_2O_2 did not affect copper electrochemical behavior (Fig. 24.8b).

The effect BTA addition of to Na_2SO_4 solution on copper protection in different potential regions is shown in Fig. 24.9, presenting potentiodynamic curves obtained during positive potential scan and backscan [52]. No hysteresis between positive

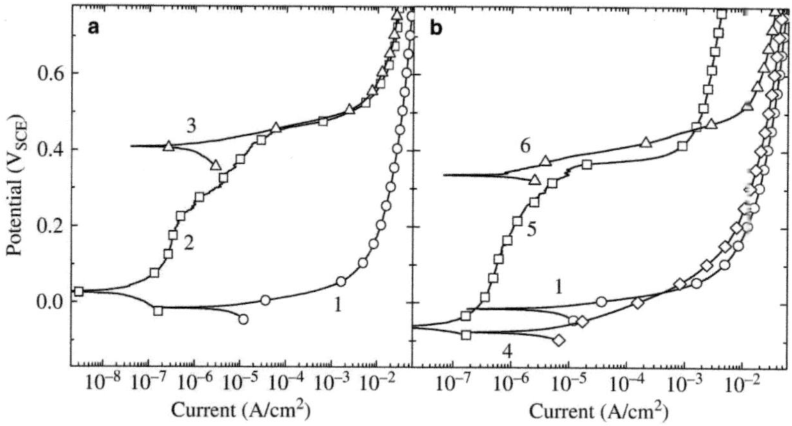

Fig. 24.8 Potentiodynamic (1 mV/s) profiles of copper in Na_2SO_4 solution (pH 4) with BTA (**a**) and glycine (**b**): (1) 10 g/l Na_2SO_4; (2) Na_2SO_4 + 0.01 M BTA; (3) Na_2SO_4 – 0.01 M BTA + 3 vol.% H_2O_2; (4) Na_2SO_4 + 0. 1 M glycine; (5) Na_2SO_4 + 0.01 M BTA+0.1 M glycine; (6) Na_2SO_4 + 0.01 M BTA + 0.1 M glycine + 3 vol.% H_2O_2

Fig. 24.9 Potentiodynamic profile (1 mV/s) of copper in 10 g/l Na_2SO_4 solution containing 0.001 M BTA

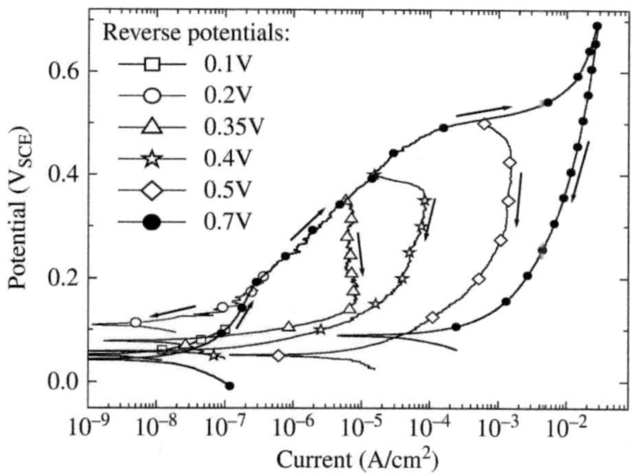

and backscan anodic curves was detected at potentials below 0.2 V, indicating a protective film formation at the copper surface in this potential region. Marked hysteresis was detected at reverse potentials above 0.2 V, which can be attributed to a degradation of the protective film at the copper surface and breakdown initiation, once applied potential exceeds potential values above 0.2 V.

It should be noted that the corrosion potential of Cu in 10 g/l Na_2SO_4 solution containing 0.001 M BTA is located in a potential range (< 0.2 V) where copper surface is covered with a protective layer. Corrosion attack in this solution is initiated under anodic polarization at potentials above 0.2 V. However, once peroxide addition is being considered, the corrosion potential of copper is situated at potentials

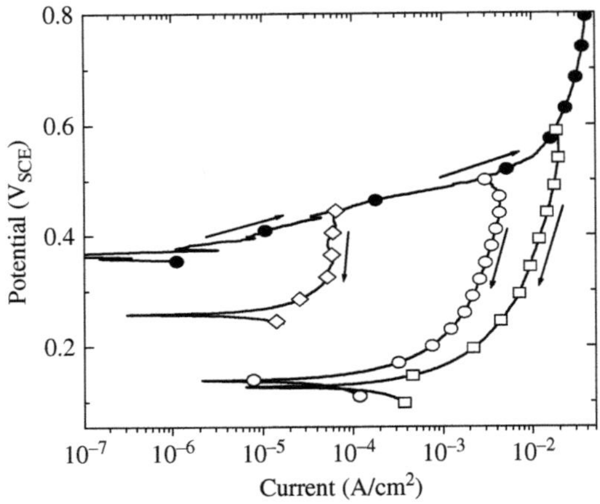

Fig. 24.10 Potentiodynamic (1 mV/s) profiles of copper in 10 g/l Na_2SO_4 containing both 0.001 M BTA and 3 vol.% H_2O_2, which were obtained at positive scan (•) and reverse scan performed at 0.45 V (◇), 0.47 V(○), and 0.52 V(□)

above 0.2 V, i.e., it undergoes a localize corrosion attack during exposure in this new OCP. This can be clearly observed in Fig. 24.10, presenting anodic curves obtained potentiodynamically during positive potential scan and backscan of copper in 3 vol.% H_2O_2 solution containing 10 g/l Na_2SO_4 and 0.001 M BTA [52, 53].

Reverse scan was set to be performed at potentials of 0.45, 0.47, and 0.52 V. It is evident that the gradual increase in the anodic current observed during positive potential scan and the pronounced hysteresis detected at reverse scan are the result of a localized corrosion attack, initiated due to a poor protective characteristic of BTA layer covering copper during OCP exposure in the presence of peroxide.

Thus, the main CMP requirement for slurry, to provide Cu passivity, is not being complied in all of the discussed slurries, so far. Cu is being actively dissolved in all of them. In ammonium hydroxide and nitric acid active copper dissolution proceeds non-uniformly, with a deep intergranular penetration. This may lead to a damage of thin Cu layer, resulting in severe dents and fractures in the copper interconnect.

Copper protection with the addition of an inhibitor addition (BTA) is proven to be *ineffective* under rapid surface abrading conditions occurring during CMP process. On the other hand, dissolution rate of copper in the studied solutions is relatively low, compared with Cu mechanical removal rate and, of course, cannot affect this process. Therefore, mechanical abrading is the dominant process in Cu CMP in these solutions.

24.5 Carbonate- and Sorbate-Based Solutions

Our earlier discussion points that the main CMP requirement from slurries, to provide Cu passivity, is not met. It was mentioned earlier that one of the main objectives of a CMP process is to provide a global planarization by removing copper overburden through a rapid dissolution. It means that copper dissolution should be the dominant process in CMP. It is driven from basic principles of CMP [11], which are

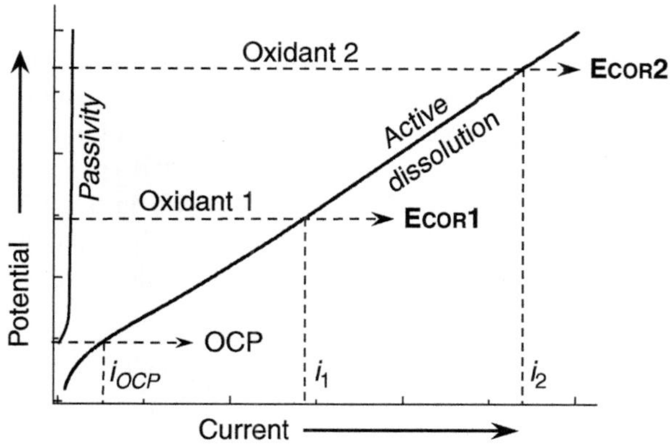

Fig. 24.11 Schema of CMP principles

clarified in Fig. 24.11. Two anodic characteristics of a metal, which are related to its passive state and active dissolution (when passive film is completely removed from the surface), are shown schematically in Fig. 24.11. It is known that active dissolution rate increases with a positive (anodic) shift in the applied potential. Therefore, dissolution rate of surface sites, in which the passive (protective) film is removed during CMP, can be accelerated by positively shifting the applied potential.

In certain solutions, which are capable of preserving the planarized metal passivity in a wide potential range, this situation can be achieved by the addition of an oxidizer to the slurry composition. The oxidizer role is to positively shift the corrosion potential (Fig. 24.11). Now, we should answer whether such situation can be achieved in nitric- and peroxide-based solutions with and without the presence of BTA in the slurry.

As was shown earlier, copper actively dissolves in nitric acid while dissolution rate at OCP was relatively low (Table 24.2). In particular, this value was much lower compared with mechanical copper removal rate. Increasing nitric acid concentration leads to increase in copper dissolution rate as well. However, this acceleration in copper dissolution rate is insufficient, since it still remains far below the rate of a mechanical removal. The addition of BTA does not result in a positive shift of the corrosion potential (Fig. 24.4), and therefore, acceleration in copper dissolution rate in the presence of BTA is hardly to be expected. Moreover, it should be noted that BTA provides copper protection in nitric acid solutions only in a narrow potential range above corrosion potential (Fig. 24.4). This implies that even a small and seemingly insignificant shift in the corrosion potential can disintegrate the protective Cu–BTA film in nitric solution, resulting in a corrosion attack. This scenario is also valid once considering peroxide–BTA-based slurries. Copper actively dissolves in peroxide solution while Cu–BTA film provides an effective copper protection only at potentials below 0.2 V. Above this threshold potential, the protective film is disintegrated and thus copper is not protected.

We should move to the next stage in our slurry design, a design of a particular slurry solution which is capable of providing copper the electrochemical conditions, as schematically illustrated in Fig. 24.11. Our recent studies establish that carbonate- and sorbate-based solutions could provide condition for copper passivation in a similar manner as aluminum and tungsten develop a passivity state [56–59]. Figure 24.12 presents anodic polarization curve of Cu in 4 g/l K_2CO_3 solution obtained during positive and reverse potential scans [57]. Scans were reversed at various potentials, related to different regions in the anodic curve.

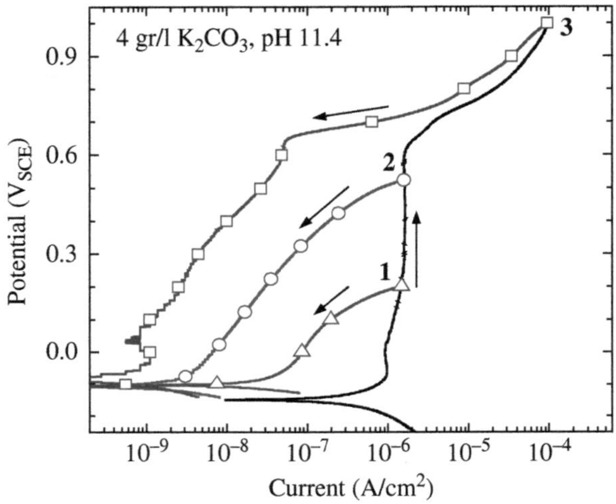

Fig. 24.12 Potentiodynamic (1 mV/s) positive scan curve and reverse scan profiles obtained from positively polarizing copper in 4 g/l K_2CO_3 solution. Scan direction was reversed at potentials of 0.2 V (Δ), 0.5 V (○), and 1.0 V (□)

As can be seen, copper remained passive in a wide potential range up to 1.0 V. Subsequent to the application of a reverse scan the anodic current also gradually decreased, indicating enhanced copper passivity. It was also established that repassivation rate of exposed surface sites was rapid, and fresh protective film was instantly formed subsequent to a mechanically surface rupture [56–58].

These experiments were conducted during potentiostatic exposure of copper electrode in carbonate-studied solution under different applied potentials (Fig. 24.13). The potentials were applied upon immersion of the copper electrode in the carbonate-based solution. Abrupt anodic current signals appeared in the current–time curve is a result of a repeated surface scratching. Anodic current rapidly reached a maximum value subsequent to a surface scratching and instantly decreased to the initial low current values, indicating on high repassivation rate of mechanically exposed surface sites covered initially with a protective film.

Similar results were obtained in sorbate-based solutions, as well. Similarly to carbonate-based solution, copper is characterized with a wide region of passivity up to 1.0 V [59]. Sorbate solutions provide enhanced copper passivity, compared with a well-known copper corrosion inhibitor, benzotriazole (BTA). Such study was performed in equal-molar solutions, containing 70 mM of each inhibitor and 1000 ppm of sulfate ions, as illustrated in Fig. 24.14.

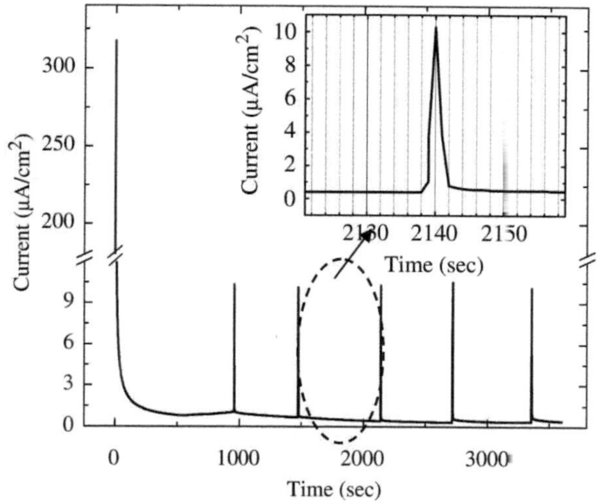

Fig. 24.13 Current transient of copper in 4 g/l K_2CO_3 solution during exposure at 0.2 V. Repeated mechanical rupturing of the protective film was simulated by glass scrubber surface scratching. Inset displays a "zoom-in" of an individual current–time transient

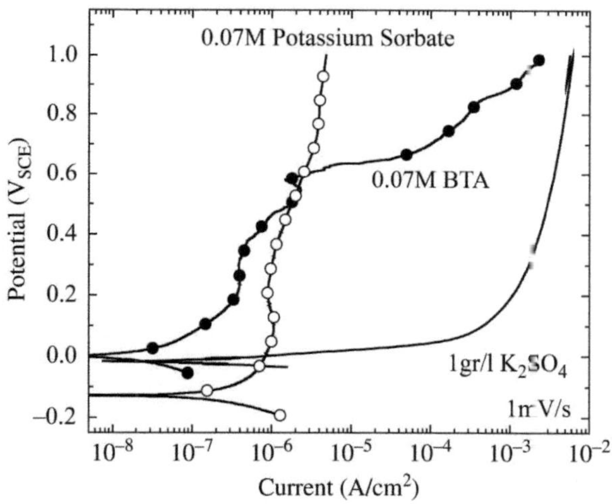

Fig. 24.14 Anodic polarization of copper obtained in sulfate solutions (Na_2SO_4, 1 %wt.) in the presence of either 0.07 M benzotriazole (BTA) or K-sorbate

As can be seen, the anodic potentiodynamic curves of copper measured in both solutions reveal that while both materials passivate copper surface, the potential range of copper passivation is considerably extended with the use of sorbate. Passive film formed on copper in Na_2SO_4/BTA electrolyte is stable up to 0.3 V, while passive film formed in Na_2SO_4/K-sorbate is stable up to 1 V [59].

The search for copper passivity in CMP process conditions is an ongoing study. The combination between superior passivating agents (such as carbonate, sorbate, and other fatty acid salts) and a strong oxidizer for copper CMP applications would be the next step in slurry design.

References

1. Murarka, S. P.; Verner, I. V.; and Gutman, R. J.: Copper – fundamental mechanisms for microelectronic applications, John Wiley & Sons, Inc., New York, 337 (2000)
2. Sethuraman, A. R.; Wang, J.-F.; and Cook, L. M.: Review of planarization and reliability aspects of future interconnect materials. J. Electron. Mater. **25(10)**, 1617 (1996)
3. Shinn, G. B.; Korthuis, V.; and Wilson, A. M.: Chemical-mechanical polishing. In. Handbook of semiconductor manufacturing technology. Nishi, Y. and Doering, R., Eds. Marcel Dekker, Inc., New York, 415 (2000)
4. Li, S. H. and Miller, R.: Chemical mechanical polishing in silicon processing. Academic Press, New York, Semiconduct. Semimet **63**, (2000)
5. Zantye, P. B.; Kumar, A.; and Sikder, A. K.: Chemical mechanical planarization for microelectronics applications, Materials Since and Engineering: Reports, Elsevier B.V. **45(3–6)**, 89 (2005)
6. Steigerwald, J. M.; Murarka, S. P.; and Gutman, R. J.: Chemical mechanical planarization of microelectronic materials, John Wiley & Sons, Inc., New York, (1997)
7. Singh, R. K. and Bajaj, R.: Advances in chemical-mechanical-planarization. MRS Bulletin **27(10)**, 743 (2002)
8. Babu, S. V.; Li, Y.; and Jindal, A.: Chemical mechanical planarization of Cu and Ta: Role of different slurry constituents. JOM **53(6)**, 50 (2001)
9. Thakurta, D. G.; Schwendeman, D. W.; Gutmann, R. J.; Shankar, S.; Lei J.; and William, G.: Three-dimensional wafer-scale copper chemical–mechanical planarization model. Thin Solid Films **414(1)**, 78 (2002)
10. Chen, K. W.; Wang, Y. L.; Liu, C. P.; Yang, K.; Chang, L.; Lo, K. Y.; and Liu, C. W.: Evaluation of advanced chemical mechanical planarization techniques for copper damascene interconnect. Thin Solid Films **447–448**, 531 (2004).
11. Kaufman, F. B.; Thompson, D. B.; Broadie, R. E.; et al.: Chemical-mechanical polishing for fabricating patterned tungsten metal features as chip interconnects. J. Electrochem. Soc. **138(11)**, 3460 (1991)
12. Pourbaix, M.: Atlas of Electrochemical Equilibria in Aqueous Solutions. 2nd US ed. NACE, Houston, TX, (1974)
13. Maurice, V.; Strehblow, H. H.; and Marcus, P.: In situ STM study of the initial stages of oxidation of Cu(111) in aqueous solution. Surface Science **458**, 185 (2000)
14. Chan, H. Y. H.; Takoudis, C. G.; and Weaver, M. J.: Oxide film formation and oxygen adsorption on copper in aqueous media as probed by surface-enhanced raman spectroscopy. J. Phys. Chem. B **103**, 357 (1999)
15. Tromans, D. and Sun, R. J.: Anodic behavior of copper in weakly alkaline solutions. J. Electrochem. Soc. **139(7)** 1945 (1992)
16. Kautec, W. and Gordon II, J. G.: XPS studies of anodic surface films on copper electrodes. J. Electrochem. Soc. **137** 2672 (1990)
17. Zeidler, D.; Stavreva, Z.; Plötner, M.; and Drescher. K.: Characterization of Cu chemical mechanical polishing by electrochemical investigations. Microelectron. Eng. **33**, 259 (1997)
18. Luo, Q.; Campbell, D. R.; and Babu, S. V.: Chemical–mechanical polishing of copper in alkaline media. Thin Solid Films **311**, 177 (1997)
19. Steigerwald, J. M.; Zirpoli, R.; Murarka, S. P.; Price D.; and Gutman R. J.: Pattern geometry effects in the chemical-mechanical polishing of inlaid copper structures. J. Electrochem. Soc. **141**, 2842 (1994)
20. Steigerwald, J. M.; Murarka, S. P.; Gutmann, R. J.; and Duquette, D. J.: Effect of copper ions in the slurry on the chemical-mechanical polish rate of titanium. J. Electrochem. Soc. **141**, 3512 (1994)
21. Steigerwald, J. M.; Duquette, D. J.; Murarka, S. P.; and Gutmann, R. J.: Electrochemical potential measurements during the chemical-mechanical polishing of copper thin films. J. Electrochem. Soc. **142**, 2379 (1995)

22. Steigerwald, J. M.; Murarka, S. P.; Gutmann, R. J.; and Duquette, D. J.: Chemical processes in the chemical mechanical polishing of copper. Mater. Chem. Phys. **41**, 217 (1995)
23. Sainio, C. A.; Duquette, D. J.; Murarka S. P.; and Steigerwald, J. M.: Electrochemical effects in the chemical-mechanical polishing of copper for integrated circuits. J. Electrn. Mater. **25(10)**, 1593 (1996)
24. Osseo-Asake, K. and Mishra, K. K.: Solution chemical constraints in the chemical-mechanical polishing of copper: Aqueous stability diagrams for the Cu-H_2O and Cu-NH_3-H_2O systems. J. Electron. Mater. **25(10)**, 1599 (1996)
25. Luo, Q.; Mackay, R. A.; and Babu, S. V.: Copper Dissolution in Aqueous Ammonia-Containing Media during Chemical Mechanical Polishing. Chem. Mater. **9(10)**, 2101 (1997)
26. Carpio, R.; Farkas, J.; and Jairath, R.: Initial study on copper CMP slurry chemistries. Thin solid films **266**, (1995)
27. Ein-Eli, Y.; Abelev, E.; Rabkin, E.; and Starosvetsky, D.: The Compatibility of Copper CMP Slurries with CMP Requirements. J. Electrochem. Soc. **150(9)**, C646 (2003)
28. Ein-Eli, Y.; Rabinovich, E.; Rabkin, E.; and Starosvetsky, D.: Electrochemical view on copper chemical-mechanical planarization. Proceedings of the Electrochemical Society meeting, 2002–22 (Copper Interconnects, New Contact Metallurgies, Structures, and Low-k Interlevel Dielectrics) 211–225 (2003)
29. Kondo, S.; Sakuma, N.; Homma, Y.; Goto, Y.; Ohashi, N.; Yamaguchi, H.; and Owada, N. J.: Abrasive-Free Polishing for Copper Damascene Interconnection. J. Electrochem. Soc. **147(10)**, 3907 (2000)
30. Graham, M. J.: Reviews on Corrosion Inhibitor Science and Technology. Eds. A. Raman, P. Labine. NACE, 1-8-1 – 32 (1989)
31. Holander, O.: Structure-activity relationship of triazole copper-corrosion inhibitor: rational development of enhanced activity inhibitors- Reviews on corrosion inhibitor science and technology. Eds. A. Raman, P. Labine, NACE, Huston. ll-13-1 – 16 (1989)
32. Hu, T. C.; Chiu, S. Y.; Dai, B. T.; Tsai, M. S.; and Tung, I.-C.: Nitric acid-based slurry with citric acid as an inhibitor for copper chemical mechanical polishing. Mater. Chem Phys. **61**, 169 (1999)
33. Wang, M. T.; Tsai, M. S.; Liu, C.; Tseng, W. T.; Chang, T. C.; Chang, L. J.; and Chen, M. C.: Effects of corrosion environments on the surface finishing of copper chemical mechanical polishing. Thin solid Films **308–309**, 518 (1997)
34. Starosvetsky, D. and Ein-Eli, Y.: unpublished results
35. Poling, G. W.: Reflection Infrared Studies of Films Formed by BTA on Copper'. Corrosion Sci. **10**, 359 (1970)
36. Cotton, J. B. and Scholes, I. R.: Benzotriazole and related compounds as corrosion inhibitors for copper. Brit. Corr. J. **2**, 1 (1967)
37. Mansfeld, F.; Smith, T.; and Parry, E. P.: Benzotriazole as corrosion inhibitor for copper. Corrosion **27**, 289 (1971)
38. Stavreva, Z.; Zeidler, D.; Plotner, M.; and Drescher, K.: Chemical mechanical polishing of copper for multilevel metallization. Appl. Surf. Sci. **91(1–4)**, 192 (1995)
39. Kondo, S.; Sakuma, N.; Homma, Y.; and Ohashi, N. J.: Slurry chemical corrosion and galvanic corrosion during copper chemical mechanical polishing. Jap. J. Appl. Phys. **39**, 6216 (2000)
40. Luo, Q.; Ramarajan, S.; and Babu, S. V.: Modification of the Preston equation for the chemical–mechanical polishing of copper. Thin Solid Films **335**, 160 (1998)
41. Hernandez, J.; Wrschka, P.; and Oehrlein, G. S.: Surface chemistry studies of copper chemical mechanical planarization. J. Electrochem. Soc. **148(7)**, G389 (2001)
42. Nguyen, V.; Van Kranenburg, H.; and Woerlee, P.: Dependency of dishing on polish time and slurry chemistry in Cu CMP. Microelectronic Engineering **50**, 403 (2000)
43. Stavreva, Z.; Zeidler, D.; Plotner, M.; and Drescher, K.: Characteristics in chemical-mechanical polishing of copper: Comparison of polishing pads. Appl. Surf. Sci. **108(1)**, 39 (1997)

44. Stavreva, Z.; Zeidler, D.; Plotner, M.; and Drescher, K.: Influence of process parameters on chemical-mechanical polishing of copper. Microelectron. Eng. **37/38**, 143 (1997)
45. Zeidler, D.; Stavreva, Z.; Plotner, M.; and Drescher, K.: Characterization of Cu chemical mechanical polishing by electrochemical investigations. Microelectron. Eng. **33**, 259 (1997)
46. Zeidler, D.; Stavreva, Z.; Plotner, M.; and Drescher, K.: The interaction between different barrier metals and the copper surface during the chemical-mechanical polishing. Microelectron. Eng. **37/38**, 237 (1997)
47. Zeidler, D.; Plotner, M.; and Drescher, K.: Endpoint detection method for CMP of copper. Microelectron. Eng. **50**, 411 (2000)
48. Kou, H. S.; and Tsai, W. T.: Effects of alumina and hydrogen peroxide on the chemical-mechanical polishing of aluminum in phosphoric acid base slurry. Mat. Chem. Phys. **69**, 53 (2001)
49. Stein, D. J.; Dale, L.; Hetherington, D.; and Cecchia, J. L.: Investigation of the kinetics of tungsten chemical mechanical polishing in potassium iodate-based slurries: II. Roles of colloid species and slurry chemistry. J. Electrochem. Soc. **146(5)**, 1934 (1999)
50. Molodov, A. I.; Markosyan, G. N.; and Losev, V. V.: Laws of the Autodissolution of Copper in the Presence of H//2o//2. Soviet Electrochem. **18(9)**, 1052 (1982)
51. Hirabayashi, H.; Higuchi, M.; Kintoshita, M.; Kaneko, H.; Hayasaka, N.; Mase, K.; and Oshima, J.: Copper-based metal polishing solution and method for manufacturing semiconductor device. Proc. of the 2nd International CMP for ULSI Multilevel Interconnects, Conference (CMP-MIC) 119 (1996)
52. Ein-Eli, Y.; Abelev, E.; and Starosvetsky, D. J.: Electrochemical Behavior of Copper in Conductive Peroxide Solutions. Elecrochem. Soc. **151(4)**, G236 (2004)
53. Ein-Eli, Y.; Starosvetsky, D.; and Abelev, E.: Electrochemical behavior of copper CMP in conductive peroxide solutions, proceedings of the electrochemical society, 2003–21 (Chemical Mechanical Planirazatio (CMP-VI)) 68 (2003)
54. Lee, S. M.; Choi, W.; Cracium, V.; Jung, S.-H.; Singh, R. K.: Electrochemical measurements to understand the dynamics of the chemically modified surface layer formation during copper CMP. MRS Spring Meeting, San Francisco, Symposium I, paper No. I4. 11 (2002)
55. Singh, R. K.; Lee, S. M.; Choi, K. S.; Basim, G. B.; Choi, W.; Chen, Z.; and Moudgil, B. M.: Fundamentals of slurry design for CMP of metal and dielectric materials. MRS Bull. **27(10)**, 752 (2002)
56. Ein-Eli, Y.; Abelev, E.; Auinat, M.; and Starosvetsky, D.: Observation of extended copper passivity in carbonate solutions and its future application in copper CMP. Electrochem. Solid-State Lett. **8(12)**, B69 (2005)
57. Ein-Eli, Y.; Abelev, E.; Auinat, M.; and Starosvetsky, D.: Copper passivity in carbonate base solution and its application to chemical mechanical planarization (CMP). Proceedings of the 9th International Symposium on the Passivation of Metals and Semiconductors. Paris, France, June, 27th July, 1st, (2005)
58. Ein-Eli, Y.; Abelev, E.; and Starosvetsky, D.: Food Preservatives Serving as Nonselective Metal and Alloy Corrosion Inhibitors. Electrochem. Solid-State Lett. **9(1)**, B5 (2006)
59. Ein-Eli, Y; and Starosvetsky, D.: unpublished results

Chapter 25
Copper Post-CMP Cleaning

D. Starosvetsky and Y. Ein-Eli

25.1 Introduction

Copper on-chip interconnects Damascene technology utilizes chemical mechanical polishing (CMP) in order to remove copper overburden after its electro deposition and achieve global planarization of patterned surface. CMP is a simultaneous action of mechanical overburden metal removal and its electrochemical dissolution. It is performed with the movement of a polisher pad in acidic or alkaline aqueous CMP electrolytes (slurry) containing dispersive abrasive particles (Al_2O_3 or SiO_2), pH buffer, certain electrolyte salts to control ionic strength, oxidants, and corrosion inhibitors. Mechanical and chemical interactions with a patterned wafer surface introduce different defects and contaminations in interlayer dielectric (ILD) surfaces and copper layers. These can either be mechanical (physical) or chemical-based defects and contaminants [1–6].

Mechanical defects occurred during CMP can be classified into surface defects and particle defects [1]. Surface defects, such as scratches, voids, grooves, and pits, are initiated by a mechanical damage of interlayer dielectric and metal layer surfaces during the polishing process. Surface defects are typically situated in a damaged layer, having a thickness between 1 and 10 nm [1, 2]. The amount and intensity of surface defects depend mostly on CMP process-operating conditions. The damaged layer should be removed as it causes various undesired surface effects.

Particle contamination of wafer surface occurred during polishing (emerging from pad material, or as suspended abrasive particles from various slurries, or from polished surface materials) is one of the major issues that should be taken into account subsequent to the CMP process. The amount of particles on the wafer surface is strongly depending on the CMP conditions, slurry type, and the hardness of surface layer [7]. It is well known that particle contamination can occur due to adsorption process manifested by Van der Waals and/or electrostatic forces. In

D. Starosvetsky (✉)
Department of Material Science and Engineering, Technion-Israel Institute of Technology, Haifa 32000, Israel
e-mail: davstar@tx.technion.ac.il

Y. Shacham-Diamand et al. (eds.), *Advanced Nanoscale ULSI Interconnects: Fundamentals and Applications*, DOI 10.1007/978-0-387-95868-2_25,

addition, particles can be physically embedded into the surface substrate due to pad pressure applied by the polisher. Particles must be removed from the contaminated wafer surface as soon as possible, because the adhesion strength of particles to surface increases with time [8, 9]. It was reported that the initial and yet dominating physical adsorption of particles is being gradually transferred into a chemical adsorption type, forming chemical bonds between the contaminant particle and the metal substrate.

Chemical-based defects and contaminants are identified as metallic particles, chemicals, and corrosion damages (defects initiated by corrosion attack and surface contamination by corrosion products) [1–6]. One of the main dangerous chemical contaminations is recognized to be the presence of metallic particles on the wafer surface [1, 2, 5]. Metallic contaminants are present on the wafer surface as adsorbed metal clusters and/or metal-containing compounds, such as oxides, hydroxides and salts. Typically, the size of metallic contaminations of wafer surface is between 10^{11} and 10^{12} atoms/cm^2. Fast diffusion of metallic ions (copper, sodium, and potassium) onto the surface and into silicon and dielectrics could have detrimental effects on characteristics of wafer surface, generating shorts and leaky paths between interconnects [1].

Most critical process for wafer surface characteristics is corrosion of plugs and interconnects. Corrosion attacks would be detrimental for thin ($\leq$ 100 nm) features as interconnects and plugs. Corrosion process is usually accompanied with formation of insoluble corrosion products, which are accumulated on the wafer surface as metallic contaminants. Thus, it is clear that corrosion effect due to CMP slurry or post-CMP cleaning must be completely excluded.

Since high-quality multilayer structure cannot be manufactured using CMP with accumulated mechanical and chemical-based defects, the post-CMP cleaning process became a mandatory step subsequent to CMP process. Post-CMP cleaning process should comply with different requirements. The process eventually should be capable of totally removing enormous variability of defects and contaminations from the planarized wafer surface. For example, post-CMP cleaning must be capable of dissolving mechanical defects from interlayer dielectric (SiO_2) and copper layer surfaces to dissolve metallic contaminants, such as oxides, hydroxides, and salts, and finally to remove abrasive particles from the surface. In addition, this process should totally eliminate copper corrosion during the cleaning, providing copper passivity for further wafer storage. These requirements determine the cleaning procedure and chemical composition of the cleaning solutions.

It is assumed that solutions with highly complicated chemical composition may answer all the requirements, considering only one cleaning operation step. However, two cleaning operation steps are also developed. Each of the two steps solves a definite set of problems and thus the chemical composition of the solutions for each step sounds less complicated. In this chapter we will consider electrochemical approaches in post-CMP cleaning, in order to eliminate different defects and clean contaminations generated at the wafer surface from the CMP process.

Initially, we will discuss the methods and solutions which are typically used in removing particles from planarized surface. It should be noted that the large portion of particles, which contaminate the wafer surface after CMP, adsorbed selectively on the surface of the copper interconnects rather than SiO_2 ILD surface [4]. The simplest approaches of surface cleaning from particles are based on using hydrodynamic forces for particle lifting and removal. These methods, such as liquid spray, spinning of wafers in liquid, techniques based on acoustic streaming and cavitations effect, could provide hydrodynamic lifting force of particles [6, 11, 12]. However, the methods based on hydrodynamic forces are not efficient enough. The major problem in particle removing is their strong adhesion to the surface due to the physical and chemical adsorption, or alternatively embedding of particles onto the planarized surface by the CMP pad [1, 13]. Particles which are strongly adsorbed to the surface or embedded onto the surface cannot be effectively removed by simple water rinsing even at high pressure spray. Efficient removal of particles can be achieved by overcoming all the adhesion forces between the adhered particles and the substrate. One simple and relatively effective technique of particle removal is based on brush scribing process [1, 8]. Similar process is the buffing process, which is conducted similarly to CMP process by action of a polishing pad in abrasive-free solutions [14].

Scribing is usually conducted with polyvinyl brushes on a wet substrate surface. In this case, the mechanical forces for particles removal, which provided by the brushes, can be increased significantly due to hydrodynamic drags. In brush scribing process, the solutions decrease the adhesion forces between the particles and the substrate, hindering the development of chemical bridging bonds between the particles and the surface [1, 4]. It is usually achieved with the addition of different surfactants to the solution composition. A frequently used surfactant in post-CMP cleaning is Triton B. The wetability of wafer surface can be efficiently modified in the presence of Triton B, especially subsequent to CMP (or buffing process) in Benzotriazole (BTA)-containing solution, once the surface is covered with a Cu^+–BTA layer [4]. The decrease in the adhesion forces can also be achieved by a decrease in the electrostatic forces developed between adhered particles and planarized wafer surface. Surface charging during exposure in electrolyte occurs as a result of ions adsorption and accompanied with the formation of a double electric layer in the solution. The polarity and magnitude of a surface charge is characterized by the polarity and magnitude of the measured zeta potential. Since ion adsorption is depended on surface nature, the polarity and magnitude of zeta potential at the surface of different materials (for example, abrasive particles and copper layer) could also be different, resulting in development of attractive forces. Decrease in attractive electrostatic forces and even transition from attractive to repulsion forces is usually achieved by manipulating the pH value, increase in the solution ionic force (by modifying solution composition and electrolyte concentration), increase in dielectric constant, and addition of surfactant [1, 4, 7]. It was found that adhesion force can be decreased with an increase in pH value (up to a pH value of 10) and also by a decrease in the pH, below a value of 2. This is usually achieved by the addition of NH_4NO_3 or citric acid, respectively [1, 3]. These chemicals can be also used in variation of the solution ionic force. Distilled water is frequently used in post-CMP cleaning procedure as medium with high dielectric constant [10].

However, fulfilling all the above-mentioned parameters by the brush scribing process is helpful in removing physically or chemically adsorbed solid particles. For embedded particles the brush scribing in such solutions is inefficient. In this case, under-etching removal process is being used. The main controlling parameter of this process is the etching rate. The solution selected for under-etching removal process should be capable of meeting different requirements: to under-etch particles embedded both into copper layers, which mostly contaminated by particles, and into interlayer dielectric; to dissolve different metallic contaminants present at the wafer surface during CMP; and to remove the damage zones (scratches, pits, etc.) formed on metal and interlayer dielectric surface during the CMP [1]. In addition to the ability to etch metals and dielectric, these solutions should be capable of preventing re-adhesion of just-liberated particles. Thus, the solution composition has to provide repulsion forces between just-liberated particles and the surface.

The solution for post-CMP cleaning has a very complicated chemical composition, which strongly depends on the type of materials presented at the planarized substrate surface, as well as post-cleaning procedure (number of steps at post-CMP cleaning). Generally, all the etching processes are conducted in acidic solutions containing different additives, such as surfactants, complexing agents, and inhibitors.

Etching of planarized surface containing silicon oxide layer is frequently conducted in diluted HF solutions [5]. Etching in these solutions allows the removal of damages presented at the silicon oxide surface. At the same time, these etchants can electrochemically dissolve both copper and other metallic contaminants. During dissolution in HF the dissolved Cu^{2+} ions are being complexized by F^{-} ions [15]. Removal of copper oxide and hydroxides layer from the copper surface can also be effectively performed in HF-containing solutions. Solution based on ammonium hydroxide and peroxide ($H_2O/H_2O_2/NH_4OH$) is also used for removal of alumina and silica particles from interlayer dielectric surface (SiO_2) [1].

HNO_3-based solutions are very effective in etching of copper layers. The buffing process in HNO_3 solution, containing BTA as inhibitor, was suggested for colloidal silica abrasive removal (first post-CMP cleaning step) from wafer surface, having Cu interconnects [4]. However, copper etching in this solution has a practical disadvantage; it was shown by many reports that copper dissolution in the presence of BTA is accompanied by a formation of protective Cu–BTA layer, which completely covered copper surface [16, 17]. In acidic solutions this layer is very thick, approximately some hundreds of nanometers (nm). Thus, it is concluded that copper surface during cleaning in this solution may be contaminated by organic compounds, which must be removed at the next steps of post-CMP cleaning.

It should be noted that etching process may constitute a threat to the surface of the copper layer, since it can lead to surface contamination and even to surface damage. It is well known that etched copper surface is frequently contaminated by corrosion products, such as oxides, hydroxides, cuprous salts, and Cu(1) ions. The formation of cuprous oxides and/or hydroxides during copper exposure was detected in alkaline- and peroxide-containing solutions [18, 19]. In chloride solutions (neutral and acidic), copper dissolution is accompanied by the formation of cuprous chloride (Cu(1)-Cl) [20, 21]. Copper dissolution in sulfuric acid solutions results in a sludge formation, which covers the copper surface as a thick black layer [22]. Sludge formation is usually accompanied with an accelerated copper dissolution process through grain boundaries. Since post-CMP cleaning must be performed to avoid degradation of the electrical characteristics of copper interconnects, selection of appropriate etchant for copper surface cleaning should be performed, taking into consideration these phenomena.

In this section we present electrochemical behavior of copper in some acidic solutions, which are commercially recommended for a post-CMP cleaning process. Two cleaning solutions were evaluated: Electroclean (Airproduct) and CX-100 (WAKO chemical).

The results obtained are shown in Figs. 25.1 and 25.2. Figure 25.1 presents anodic curves of copper in post-cleaning solutions of CX-100 (pH 2.0) and Electroclean (pH 3.8) subsequent to an exposure at open circuit potential. The difference in Cu corrosion potential, E_{corr}, as a function of time (shown in Fig. 25.1 inset) was approx. 0.1 V.

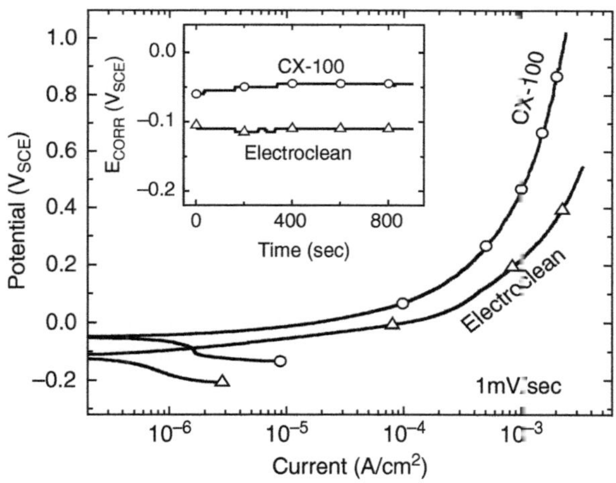

Fig. 25.1 Anodic potentiodynamic (1 mV/s) characteristics of Cu in two post-CMP cleaning solutions (CX-100 and Electroclean) obtained subsequent to 15 min exposure at OCP (Inset: corrosion potential, E_{corr}, transient of Cu during OCP exposure in studied post-cleaning solutions)

The value of E_{corr} remained almost constant during the entire exposure time. During a positive potential sweep, the anodic currents gradually increased, indicating an active copper dissolution in both solutions. Thus, anodic characteristics point that both solutions seem as suitable for Cu etching and appropriate in a post-CMP cleaning. However, further studies of copper cathodic characteristics in these solutions reveal the appearance of a cathodic peak (Fig. 25.2) subsequent to a short-time OCP exposure of Cu in the solutions. As can be seen in Fig. 25.2, the cathodic peak is positioned close to the copper corrosion potential in these solutions. Thus, the cathodic peak detected in the potentiodynamic profile is unambiguously associated with a deposition of a corrosion product at the copper surface during etching in the solutions. Unfortunately, the nature of these depositions cannot be determined only by electrochemical measurements, while the chemical composition of CX-100 and Electroclean solutions is undisclosed.

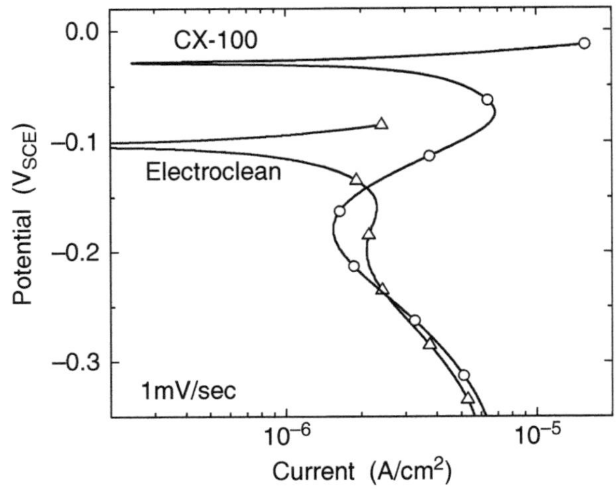

Fig. 25.2 Cathodic potentiodynamic profiles of Cu measured in two post-CMP cleaning solutions (CX-100 and Electroclean) subsequent to a 5 min exposure at OCP

Following an exposure in the cleaning solutions, copper surface was studied by SEM and energy dispersive spectroscopy (EDS). Figure 25.3 presents SEM micrographs obtained from copper surface subsequent to a 5 min exposure in CX-100 cleaning solution. In order to eliminate the effect of surface relief and elucidate the effect of chemical composition, surface study and analysis was conducted by backscattering electrons. Heterogeneous distribution of colors through the SEM micrograph obtained by backscattering electrons can be considered as evidence of deposits presented at the copper surface. Copper oxide deposition was identified at the Cu surface by EDS. At present, it is difficult to assess how critical these contaminations are and their influence on the quality of subsequent production steps. However, it can be state, with assurance, that deposition of corrosion products during post-CMP cleaning does not improve the quality of the finished product.

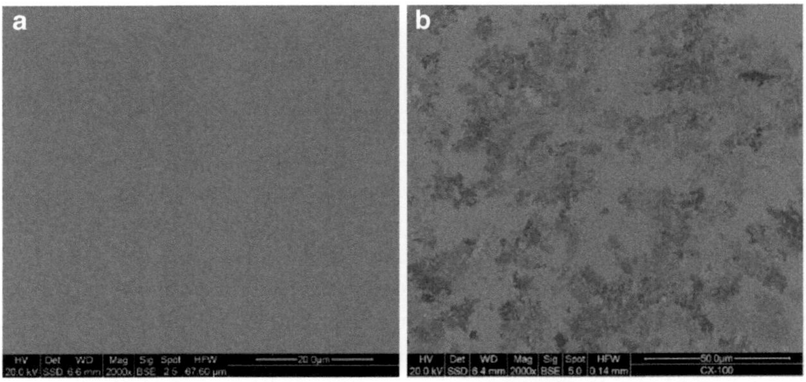

Fig. 25.3 SEM micrograph of copper surface before (**a**) and after (**b**) exposure in CX-100 cleaning solution

25.2 Summary

Post-CMP cleaning is an important step in Damascene technology utilizing copper on-chip interconnects. This process can significantly improve wafer surface quality by removing different defects and contaminations from copper layers and ILD. Brush scribing technique in association with wet chemical treatment and etching is being routinely used. An effective post-CMP cleaning can also be achieved in a combination of different techniques, for example, buffing and brush scribing in different cleaning solutions. Etching of wafer surface plays an important role in abrasive particles removal from the copper surface layer and interlayer dielectric. Copper layers are mostly contaminated with abrasive particles which determine strategy and procedure of post-CMP cleaning.

HNO_3-based solutions are highly effective in copper layers etching. Simultaneous etching of silicon oxide and copper layers can be efficiently conducted in diluted HF solutions. The composition of etching solution usually includes inhibitors and surface active compounds.

However, copper etching in aqueous solutions is usually accompanied with a deposition of corrosion products (Cu(1) compounds). Formation of such deposition is being determined electrochemically subsequent to an etching process (at OCP) in some of the recommended post-CMP cleaning solutions. These results imply that one should be very cautious in choosing the most appropriate etching solution for Cu post-CMP cleaning.

References

1. Zantye, P. B.; Kumar, A.; and Sikder, A. K.: Chemical mechanical planarization for microelectronics applications. Materials Since and Engineering: Reports, Elsevier B.V. **45(3–6)**, 89 (2005)
2. Li, S. H. and Miller, R.: Chemical mechanical polishing in silicon processing. Academic Press, New York, Semiconduct. Semimet. **63**, (2000)
3. Park, J.-G. and Busnaina, A.: Copper Post CMP cleaning and the Effect of additives. Semiconduct. Int. **28**, 39 (2005)
4. Chen, P.-L.; Chen, J.-H.; Tsai, M.-S.; Dai, B. T.; and Yeh, C. F.: Post-Cu CMP cleaning for colloidal silica abrasive removal. Microelectron. Eng. **75**, 352 (2004)
5. Biverina, A.; Bernard, H.; Palleau, J.; Torres, J.; and Tardif, F.: Copper Photocorrosion Phenomenon during Post CMP Cleaning. Electrochem. Solid-State Lett. **3(3)**, 156 (2000)
6. Zhang, L.; Raghavan S.; and Weling, M.: Minimization of chemical-mechanical planarization (CMP) defects and post-CMP cleaning. J. Vac. Sci. Technol. B **17(5)**, 2248 (1999)
7. Liu, C.-W.; Dai, B.-T.; and Yeh, C.-F.: Post cleaning of chemical mechanical polishing process. Appl. Surf. Sci. **92**, 176 (1996)
8. Carpio, R.; Farkas, J.; and Jairath, R.: Initial study on copper CMP slurry chemistries. Thin Solid Films **266(2),** 238 (1995)
9. Burdick, G. M.; Berman, N. S.; and Beaudoin, S. P.: A theoretical analysis of brush scrubbing following chemical mechanical planarization. J. Electrochem. Soc. **150**, G140 (2003)
10. Ramachandran, S.; Busnaina, A. A.; Small, R.; Shang, C.; and Chen, Z.: Non contact post-CMP cleaning of thermal oxide wafers using chelating basic chemistry, Internet web site: http// www.ekctech.com/post-cmp.htm

11. Olim, M.: A theoretical evaluation of megasonic cleaning for submicron particles. J. Electrochem. Soc. **144**, 3657 (1997)
12. Busnaina, A. A. and Dai, F.: Megasonic cleaning. Semicond. Int. **20**, 85 (1997)
13. Isrealachvili, J.: Intermolecular and surface forces. 2nd ed., Academic, San Diego, (1994)
14. Shen, J. J.; Costas, W. D.; Cook, L. M.; and Farber, J.: The effects of post chemical mechanical planarization buffing on defect density of tungsten and oxide wafers. J. Electrochem. Soc. **145(12)**, 4240 (1998)
15. Tardif, F.; Beverina, A.; Bernard, H.; Constant, I.; and Robin, F.: Photo-Corrosion effects during copper interconnect cleaning. The Electrochemical Society and the Electrochemical Society of Japan Meeting, Abstracts, Honolulu, HI **196**, (1999)
16. Holander, O. and Industrial, B.: Reviews on corrosion inhibitor science and technology. Ed. A. Raman, P. Labine, NACE, Huston. ll-13-1-16 (1989)
17. Graham, M. G.: Reviews on corrosion inhibitor science and technology. Ed. A. Raman, P. Labine, NACE, Huston. 1-8-1-32 (1989)
18. Thomas, J. G. N. and Tiller, A. K.: Formation and breakdown of surface films on copper in sodium hydrogen carbonate and sodium chloride solutions. I. Effects of anion concentrations. Br. Corrosion J. **7**, 256 (1972)
19. Ehlers, C. B.; Villegas, I.; and Stickney, J. L.: The surface chemistry of copper(100) in hydrochloric acid solutions as a function of potential: a study by LEED, Auger spectroscopy and depth profiling. J. Electroanal. Chem. **284(2)**, 403 (1990)
20. Gomez Becerra, J.; Salvareza, R. C.; and Arvia, A. J.: Electrochemical behaviour of copper in aqueous solution containing potassium ethylxanthate. J. Appl. Electrochem. **17**, 779 (1987)
21. Elsner, C. I.; Salvareza, R. C.; and Arvia, A. J.: The influence of halide ions at submonolayer levels on the formation of oxide layer and electrodissolution of copper in neutral solutions. Electrochim. Acta **33(12)**, 1735 (1988)
22. Frankel G. S.; Schrot, A. G.; Isaacs, H. S.; Horkans, J.; and Andricacos, J.: Behavior of Cu(P) and oxygen free high conductivity cu anodes under electrodeposition conditions. J. Electrochem. Soc. **140(4)**, 959 (1993)

Part VII
Electrochemical Processes and Tools

Part VII is about electrochemical processing tools for ULSI interconnect technology development. Chapter 26 provides the introductory remarks and a brief discussion of the impact of Electrochemical Processing Tools in the semiconductor industry particularly in dual Damascene plating for copper chip metallization. New challenges in further miniaturization of interconnects provide ample opportunities for the development of novel electrochemical processes and tools and to make significant contribution toward continuation of Moore's Law. The following chapters of this part on Processes and Tools cover in detail some of the topics mentioned in this introductory note. Semiconductor equipment industry experts have been invited to write on these topics. In Chapters 27, 28 and 29 Tom Ritzdorf and his team from Semitool provide a detailed description of advanced electrodeposition processes and tools for copper interconnects technology and control of electrodeposition processes; electroless deposition processes for copper, nickel and gold, and tools; and equipments for monitoring and control of bath components, respectively. In Chapter 30 Bill Lee and Igor Ivanov of Blue29 describe the evolving electroless Co and CoW capping technology. This chapter includes Co alloy capping applications, film properties and requirements, process sequence and integration, and deposition chemistry for Co alloy capping along with the Co alloy capping system. Finally, in Chapter 31 Bulent Basol of ASMNutool describes advanced planarization techniques including electrochemical polishing methods, novel electropolishing approaches, electrochemical mechanical deposition, and electro chemical mechanical planarization.

Chapter 26
Electrochemical Processing Tools for Advanced Copper Interconnects: An Introduction

Madhav Datta

26.1 Dual Damascene Process for Electroplated Copper Interconnects

The change from vacuum-deposited aluminum to electroplated copper in 1997 brought about a paradigm shift in interconnect technology and in chip making [1]. Since then, most of the leading chip manufacturers have converted to electroplated Cu technology for chip interconnects. Cu interconnects are fabricated by dual Damascene process which is referred to a metallization patterning process by which two insulator (dielectric) levels are patterned, filled with copper, and planarized to create a metal layer consisting of vias and lines. The process steps consist of laying a sandwich of two levels of insulator and etch stop layers that are patterned as holes for vias and troughs for lines. They are then filled with a single metallization step. Finally, the excess material is removed, and the wafer is planarized by chemical mechanical polishing (CMP). While finer details of exact sequence of fabrication steps vary, the end result of forming a metal layer remains the same in which vias are formed in the lower layer, and trenches are formed in the upper layer. Electroplating enables deposition of Cu in via holes and overlying trenches in a single step thus eliminating a via/line interface and significantly reducing the cycle time. Due to these reasons and due to relatively less expensive tooling, electroplating is a cost-effective and efficient process for Cu interconnects [2, 3]. Compared with vacuum deposition processes, electroplated Cu provides improved super filling capabilities and abnormal grain growth phenomena. These properties contribute significantly to improved reliability of Cu interconnects. With the proper choice of additives and plating conditions, void-free, seam-free Damascene deposits are obtained which eliminates surface-like fast diffusion paths for Cu electromigration.

Development of a defect-free dual Damascene electroplating process required an understanding of the formation and elimination of voids [2, 4]. Elimination of

M. Datta (✉)
Cooligy, Inc., 800 Maude Avenue, Mountain View, CA 94043, USA
e-mail: mdatta@cooligy.com

Y. Shacham-Diamand et al. (eds.), *Advanced Nanoscale ULSI Interconnects: Fundamentals and Applications*, DOI 10.1007/978-0-387-95868-2_26,

voids is extremely critical to minimize electromigration issues and to ensure that the Cu vias and lines are free of trapped electrolytes. Formation of voids is very common, particularly in narrow trenches, where top ridges build up first thus creating a void in the middle of the structure. Even in conformal plating, formation of seams is commonly observed that can be as disastrous as the voids. The only way to achieve defect-free trench filling is to obtain a condition where higher deposition rate is achieved in less accessible bottom of the trenches as compared with the easily accessible flat top surface and upper side walls, similar to the leveling mechanism discussed in the literature [5–8]. Based on the above principle, different proprietary additives have been developed by different chip manufacturers and electroplating tool vendors. These aspects are described in detail in Chapter 3 in this book.

Along with a defect-free dual Damascene electroplating process, the semiconductor equipment industry's drive to develop electroplating and CMP tools was an essential factor for the implementation of copper interconnect technology. Indeed, the development and availability of high volume manufacturing electroplating and CMP tools that are compatible with ultra-clean fab standards represented a major shift in the semiconductor industry's strategy that was hitherto focused on vacuum processing.

26.2 Tooling Requirements

The plated interconnect structure has to meet all the requirements of high yielding, reliable microprocessor manufacturing. These requirements include excellent adherence of Cu lines/vias to the dielectric, low resistivity, and resistance to electromigration. In order to meet these criteria, the integration scheme must include attention to proper selection of a barrier/adhesion/seed layer between dielectric and the plated Cu, and a reliable CMP process. Furthermore, since the electroplating process optimization is specific to a chosen electroplating tool, different chip manufacturers have their unique combination of tools, tailored plating bath, monitoring and control system, and integration scheme [3].

Electroplating tools for semiconductor wafer processing must meet some basic requirements that are essential in a fab-processing environment [9]. Automated wafer processing equipment must provide facilities to unload dry wafers from a cassette or a front opening unified pod (FOUP), and deliver dry, processed wafers back to the same carrier. These requirements necessitate the integration of multiple processing cells in a single set of automated process equipment. The plating tools and their fixture must be compatible with other thin film vacuum processing equipments in the fab. Robust hardware and processes must ensure the high yields that are required in a semiconductor manufacturing. Electrical contact to the wafer, terminal effect, edge exclusion zone management, back side contamination, and thickness uniformity are some of the key issues that dictate the design of an electroplating tool for semiconductor wafer processing. While the reactor design is generally specific to a given set of process chemistry and conditions, it must take into account all the above issues.

The CMP process for Cu metallization and the understanding of its interaction with the plating process played an important role in implementation of Cu interconnect technology [10]. Electroplating/CMP interactions are at both local and global levels and create challenges in process integration. The CMP process must possess the qualities of high rate of polishing, high selectivity, very low dishing ,and no corrosion. Overfill effect of dense lines during electroplating is an example of such interactions. Overfill effects are known to leave residual Cu in dense lines during CMP resulting in electrical failures due to shorting. Overpolishing of wafers to clear the Cu residuals leads to excessive dishing in wide Cu lines. Another example is the need for significant excess plating to fill up the wide lines for geometric leveling during plating. However, such "overburden copper" places heavy burden on the CMP process. Optimization of these processes, therefore, involves a closer interaction between CMP and electroplating teams.

Attention to the above issues together with an aggressive integration scheme has enabled the development and manufacturing of advanced interconnects that have been serving several generations of chip manufacturing. A significant part of these advances is due to the efforts of equipment manufacturing companies that provided advanced electroplating and CMP tools. Indeed, the defect-free fabrication of complex interconnect structures ranging down to 65 nm node demonstrates the exceptional strength of electrochemical processing technologies in advanced chip interconnections.

26.3 45 nm Node and Beyond: Novel Processes and Tools

As the semiconductor industry prepares for the 45 nm node and beyond, some material and processing changes are imminent. The industry is currently engaged in addressing issues related to integration of advanced interlayer dielectric (ILD) material into finer Cu lines. Accordingly, novel electrochemical processing methods are evolving that address issues related to planarization of the fragile Cu/ILD structure and electromigration. These developments are discussed below.

Low permittivity (low-*k*) dielectric materials are being developed to provide a reduced capacitance and a reduced crosstalk between conductors. These materials are typically SiLK or carbon-doped silicon oxide materials [11, 12]. Efforts are ongoing to make these materials highly porous to further decrease their permittivity value. The dielectric materials are expected to have effective k values less than 2.5. The key concern about the low-*k*/ultra low-*k* (ULK) dielectric materials is their hardness, cohesive strength, and fracture toughness. In other words, these materials are extremely fragile so that integrating ULK dielectric materials pose several issues. First of all, there is interaction between ULK porosity and the barrier film's coverage, especially on via sidewalls. Selective pore sealing of side walls without deposition on the bottom of the copper via and without significantly changing the dielectric constant of ULK is an important process development consideration. Another more important issue is the planarization of Cu interconnects (with hardness ~130 GPa) embedded in the fragile porous interlayer dielectric (ILD) materials (with hardness of ~1–2 GPa). Due to such tremendous hardness mismatch between adjacent layers, the current chemical mechanical polishing (CMP) technology damages the Cu lines and produces delamination of low-*k*/ULK dielectric. While very low-pressure CMP has been somewhat successful, it is an extremely low throughput process. In the case of ultra low-*k* dielectric materials, even "soft" CMP leads to deformation and delamination of Cu/ULK dielectric interface. The semiconductor equipment industry is currently engaged in the development of planarization methods with no physical contact or minimal contact pressure to the wafer.

26.3.1 Electropolishing for Planarization

The main objectives of Damascene Cu planarization are twofold. On the one hand emphasis is on development of electroplating process that can provide minimized overburden without compromising Cu conductivity. On the other hand novel approaches are sought to remove overburden Cu at a fast rate and polish with minimum dishing. Electropolishing is a cost-effective method of Cu overburden removal. However, electropolishing does not have significant planarization capability, especially for large features with low aspect ratio. A combination of electrochemical and mechanical means of Cu removal is evolving as a novel planarization technique. Currently the industry is engaged in the development of mainly two types of planarization techniques. In one approach planarization is achieved during electroplating using electrochemical mechanical deposition (ECMD) which is followed by electropolishing to remove the overburden, while in the other approach electroplating is followed by electrochemical mechanical polishing (ECMP). Both approaches emphasize electropolishing as the key metal removal method, thus positioning electropolishing at the center stage in the planarization of interconnect structures.

ECMD involves simultaneous electrochemical metal deposition and planarization [13]. Mechanical action during plating is accomplished by sweeping action of a pad on the wafer surface. Sweeping of the pad influences the additive content at the top surface thus allowing Cu growth rate to be higher in the features. Chemical modifications of the plating bath in combination with ECMD may make the planarization action more effective without the need to apply pressure to the pad.

In ECMP metal removal and planarization are accomplished by a combination of aspects of electropolishing and CMP. Compared to conventional CMP, ECMP allows 10× lower down force for planarization. A pad is used in conjunction with an electrolyte for electropolishing. As in ECMD, no abrasive slurry is used in ECMP as well. At the surface where the pad is in contact with the wafer, more Cu removal takes place due to both electrochemical and mechanical action. On the other hand, only electrochemical action removes material from locations of the wafer where contact with the pad is absent. This differential in material removal can be made more effective by selecting conditions whereby surface passivation plays a major role in electrochemical reaction.

A serious limitation of the above two planarization approaches is their inability to remove the barrier layer. In fact, electropolishing in ECMD as well as ECMP leaves behind a thin layer of Cu and the underlying barrier layer which need to be removed either by soft CMP (also known as kiss polishing) or by dry processing.

26.3.2 Electroless Processes for Damascene Capping

With the increasing trend toward finer lines and finer pitch, electromigration issues are becoming the most serious reliability concern in chip making [14]. The copper line is encapsulated on the sides and bottom by barrier layers and on the top by barrier/etch stop dielectric layers. The copper/dielectric interface has weaker adhesion than the copper/barrier metal interface, consequently copper diffusion takes place predominantly at the top surface. Under high current densities, copper atoms move in the direction of the electron flow, and vacancies accumulate in the opposite direction into voids that cause the device to fail. Consequently, there is a need to develop barrier and cladding layers for passivation and adhesion promotion.

The PVD-sputtered Ta(N) liner and PECVD Si(C)N cap technologies have been used as diffusion barriers for copper interconnects from the 0.25 μm to 65 nm process nodes [1, 15, 16]. The PVD Ta(N) liner is a relatively high resistance film and it accounts for approximately 15% of the metal area on lower metal layers. It is crucial to rapidly scale down the liner thickness while maintaining the proper diffusion and adhesion properties, but it is difficult to form a thin, conformal layer over complex dual Damascene topography using a PVD process due to the directional nature of the deposition process. New liner technologies using atomic layer deposition of metal nitride alloys provide an immediate solution to the wire resistance problem but add new challenges for wire current density scaling and integration with porous low-*k*/ULK ILD materials.

The key problem with the PECVD Si(C)N dielectric capping is that it forms a relatively weak chemical bond with copper that allows for excessive copper migration at the Cu–Si(C)N interface thus giving rise to electromigration problems that limit the maximum current density in the underlying wire [17]. The PECVD Si(C)N cap film also has a high dielectric constant ($k \sim 7$). There is, therefore, a tremendous need for the search of an alternative capping layers.

The important criteria for the selection of an alternative capping layer is its ability to provide better adhesion at the copper–cap interface since it leads to reduced copper mobility at the interface which, in turn, reduces copper electromigration and improved interconnect current density capability [1, 17]. Selective electroless deposition of cobalt alloys offers a novel approach for forming self-aligned metallic cap layers. Such films have better adhesion to copper than dielectric films. An electroless metal cap deposition process is especially appealing because of low cost, intrinsic selectivity, and superior film properties. The electroless CoWP cap has been shown to have excellent adhesion to copper, good corrosion, and diffusion properties, as well as good selectivity and higher EM resistance [17–20]. The semiconductor equipment industry is currently involved in the development and qualification of both Pd activation-based and Pd-free CoWP processes.

A Pd-free and alkaline metal-free electroless Co alloy deposition process has been described in a recent publication by Intel authors [19]. The electroless cobalt deposition process involves catalytic oxidation of reducing agent on copper surface and reduction of cobalt ions. The plating solution contains Co ions, reducing agent, complexing and buffer agents, and alkaline metal-free pH adjuster. The electroless Co bath was developed to be autocatalytic on copper with no activation step required prior to electroless Co. A complete sequence of dielectric deposition, dual Damascene patterning and metal fill, and planarization steps were used to fabricate multi-layer structures with line width/space combinations down to 100 nm and electroless CoWP cap of about 20 nm thick. The implementation of proper surface pre-treatment before CoWP deposition caused the leakage current to be reduced by several orders of magnitude and the high leakage tails to be eliminated. No line resistance increase was observed for Cu line capped with electroless Co after annealing at 400°C for 2 hours. A SIMS depth profile showed no interdiffusion between CoWP and Cu layers. The resistances of Cu lines capped with electroless CoWP as well as via chain resistance are equivalent to those without the CoWP cap. The metal cap layer helped to block the interfacial diffusion pathway of Cu at the copper–cap interface leading to a marked improvement in electromigration performance. Indeed, the electromigration median-time-to-failure (EMMTTF) of CoWP capped wafer has been shown to be 10 times better than that of wafers with undoped copper and a Si(C)N dielectric cap layer [19]. The increase of electromigration resistance with an electroless Co cap can be explained by the complete passivation of the top interface resulting in increased adhesion and interfacial bonding strength. These results indicate that electroless Co (alloy) capping will enable the ITRS roadmap for interconnect current density for several more generations.

Another way of improving copper electromigration performance is to add a dopant to the copper that will segregate to the Cu–cap interface and inhibit copper interfacial migration. Tin appears to be a potential candidate for copper doping although achieving a large EM benefit results in a relatively large (>10%) increase in copper resistivity [21, 22]. Yet another approach being explored is the use of redundant ruthenium (Ru) liner or Ru direct plate application [23]. The interest in Ru is because of its ability to plate Cu directly onto Ru without the need for Cu seed layer. The Cu–Re interface has low electromigration, the adhesion of Cu to Ru is good, and the solubility of Ru in Cu is low, so that net effect on Cu resistivity is minimal. However, all of these are still in the conceptual stage and require a thorough examination of integration issues before they can be considered as viable approaches.

26.4 Concluding Remarks

The introductory remarks presented above provided a brief discussion of the impact of electrochemical processing tools in the semiconductor industry particularly in dual Damascene plating for copper chip metallization. As the industry moves to 45 node technology, qualification and implementation of electropolishing-based planarization and electroless-capped layer in manufacturing are the key challenges for the next generation of interconnects. Future planarization efforts are expected to focus on the development of integrated polishing method for barrier layer removal and elimination of final CMP process. Several different approaches are expected to evolve to address electromigration issues. Search for novel plated metallization with improved conductivity and electromigration property is expected to be a future topic of interconnects development. In summary, new challenges in further miniaturization of interconnects provide ample opportunities for the development of novel electrochemical processes and tools and to make significant contribution toward continuation of Moore's Law.

The following chapters on processes and tools cover in detail some of the topics mentioned in this introductory note. Semiconductor equipment industry experts have been invited to write on these topics. Tom Ritzdorf and his team from Semitool provide a detailed description of advanced electrodeposition processes and tools, electroless deposition processes and tools, and equipments for monitoring and control of electrodeposition processes. Bill Lee and Igor Ivanov of Blue29 describe the evolving electroless Co and CoW capping technology. The topic of CMP process tools is treated in detail by Peter Burke of LSI Logic (now On Semi). Finally, Bulent Basol of ASMNutool describes advanced planarization techniques including electropolishing methods.

References

1. Edelstein, D. C.; Heidenreich, J.; Goldblatt, R.; Cote, W.; Uzoh, C.; Lustig, N.; Roper, P.; McDevitt, T.; Motsiff, W.; Simon, A.; Dukovic, J.; Wachnik, R.; Rathore, H.; Schulz, R.; Su, L.; Luce, S.; and Slattery, J.: Full Copper Wiring in a Sub-0.25 μm CMOS ULSI Technology. Tech. Dig. IEEE Intl. Eletron. Devices Conference 773 (1997)
2. Andricacos, P. C.; Uzoh, C.; Dukovic, J. O.; Horkans, J.; and Deligianni, H.: Damascene copper electroplating for chip interconnections. IBM J. Res. Dev. **42(5)**, 567 (1998)
3. Datta, M.: Electrochemical processing technologies in chip fabrication: Challenges and opportunities. Electrochim. Acta **48(20–22)**, 2975 (2003)
4. Moffat, T. P.: Bonewich, J. E.; Huber, W. H.; Stanishevsky, A.; Kelly, D. R.; Stafford, G. R.; and Josell, D.: Superconformal electrodeposition of copper in 500–90 nm features. J. Electrochem. Soc. **147**, 4524 (2000)
5. Kardos, O.: Current distribution on microprofiles, Part I, II, III. Plating, 61, 129, 229, 316 (1974)
6. Kruglikov, S. S.; Kudriavtsev, N. T.; Vorobiova, G. F.; Antonov, A. Ya.: On the mechanism of levelling by addition agents in electrodeposition of metals. Electrochim. Acta **10(3)**, 253 (1965)
7. Dukovic, J. O.; Tobias, C. W.: Simulation of Leveling in Electrodeposition. J. Electrochem. Soc. **137(12)**, 3748 (1990)
8. Madore, C.; Matlosz, M.; and Landolt, D.: Blocking inhibitors in cathodic leveling. J. Electrochem. Soc. **143**, 3927 (1996)
9. Ritzdorf, T.: In New Trends in Electrochemical Technology, Microelectronic Packaging, Datta, M.; Osaka, T.; Schultze, J. W., Eds. CRC Press, New York, **3**, 471 (2005)
10. Watts, D. K.; Kimura, N.; and Tsujimura, M.: In New Trends in Electrochemical Technology, Microelectronic Packaging, Datta, M.; Osaka, T.; Schultze, J. W., Eds. New York, **3**, 437 (2005)

11. Bohr, M.: Technology Challenges from Transistor to Packages. Intel Development Forum, Fall (2004)
12. Martin, S. J.; Godschalsx, J. P.; Mills, M. E.; Shaffer, E. O.; and Townsend, P. H.: Development of a low-dielectric-constant polymer for the fabrication of integrated circuit interconnect. Adv. Mater. **12(23)**, 1769 (2000)
13. Basol, B. M.: Mechanically induced superfilling of low-aspect-ratio cavities in an electrochemical mechanical deposition process. J. Electrochem. Soc. **151**, C765 (2004)
14. Dubin, V. M.; Simka, H. S.; Shankar, S.; Moon, P.; Marieb, T.; and Datta, M.: In New Trends in Electrochemical Technology, Microelectronic Packaging, Datta, M.; Osaka, T.; Schultze, J. W., Eds. CRC Press, New York, **3**, 31 (2005)
15. Chou, S.: Extending Moore's Law in the Nanotechnology era. Intel technology and Manufacturing Briefing. Intel Development Forum, February (2004)
16. Hu, C. K.; Gignac, L.; Rosenberg, R.; Liniger, E.; Rubino, J.; Sambucetti, C.; Domenicucci, A.; Chen, X.; and Stamper, A. K.: Reduced electromigration of Cu wires by surface coating. Appl. Phys. Lett. **81**, 1782 (2002)
17. Hu, C. K.; Gignac, L.; Liniger, E.; Herst, B.; Rath, D. L.; Chen, S. T.; Kaldor, S.; Simon, A.; and Wang, W.-T.: Comparison of Cu electromigration lifetime in Cu interconnects coated with various caps. Appl. Phys. Lett. **83**, 869 (2003)
18. Padhi, D.; and Dixit, G.: Key Process Parameters for Copper Electromigration. Solid State Tech. **46(11)**, (2003)
19. Moon, P.; Dubin, V.; Johnston, S.; Leu,J.; Raol, K.; and Wu, C.: Process Roadmap and Challenges for Metal Barriers. Proc. IEDM 141 (2003)
20. Lee, B.: Electroless CoWP Boosts Copper Reliability, Device Performance. Semiconductor International, July (2004)
21. Tonegawa, T.; Hiroi, M.; Motoyama, K.; Fujii, K.; and Miyamoto, H.: Suppression of bimodal stress-induced voiding using high-diffusive dopant from Cu-alloy seed layer. IEEE Intl. Interconnect Tech. Conf. San Francisco, USA, 216 (2003)
22. Padhi, D.; Gandikota, S.; McGujrk, C.; Ngyuen, H. B.; Ramanathan, S.; S. Parikh, and Dixit, G.: Investigation of Electromigration Issues in Copper Interconnects. Proc. Adv. Metall. Conf., San Diego, 337 (2002)
23. Rossnagel, S.: The Latest in Ru-Cu Interconnect Technology. Solid State Technology, online February (2005)

Chapter 27
Electrochemical Deposition Processes and Tools

T. Ritzdorf

27.1 Introduction

Beginning in the early 1990s, automated wafer processing equipment was adapted for electrochemical deposition (ECD) applications on semiconductor wafers and similar substrates. Integrating the automation of mainstream wafer processing equipment with ECD processes to produce equipment consistent with semiconductor processing was a significant engineering challenge. In the years since, several application-specific types of electrochemical processing equipment, each designed to meet its own specific requirements, have been produced [1, 2].

Wafer processing equipment configurations are dependent on the intended process sequence for the wafer, throughput requirements, safety guidelines, and fab-specific requirements. The first thing most people notice about automated wafer processing equipment is its cost. The multi-million dollar per tool equipment cost quickly adds up to result in wafer fabrication facilities that cost several billion dollars. This is a direct result of the strict design requirements and the error-avoidance complexities incorporated into the equipment, coupled with the third-party inspection requirements and fab-specific special options typically added to the equipment design and fabrication.

27.2 Electrochemical Processing Equipment

27.2.1 Equipment Automation and Wafer Handling

The intended use of the processing tool will dictate the type of automation required within the tool. Simple R&D tools in lab environments may have no wafer-handling automation if manual loading of the process chambers is acceptable. For tools in

T. Ritzdorf (✉)
Semitool Inc., 655 W. Reserve Dr., Kalispell, MT 59901, USA
e-mail: tritzdorf@semitool.com

Y. Shacham-Diamand et al. (eds.), *Advanced Nanoscale ULSI Interconnects: Fundamentals and Applications*, DOI 10.1007/978-0-387-95868-2_27,

manufacturing environments though, wafers are typically handled by robots that have very strict requirements on cleanliness, where they can touch the wafers, and their impact on tool operation and throughput. Tools in production environments may also need to incorporate an appropriate interface into automated intra-fab wafer transfer systems. The tool automation usually consists of WIP (work in process) automation, transfer robots, and automation of individual process chambers. All of these components must be designed to work together, as well as to interface with the carriers that are used to bring the wafers to the equipment.

The size of the wafer or substrate, fab protocol, and cleanliness requirements determine the type of wafer carriers used in a fab. Automated process equipment must interface with the carrier in order to introduce wafers into the equipment. Wafers that are 200 mm or smaller in diameter have multiple options for carriers. These range from various styles of open cassettes to environmentally controlled SMIF (Standard Mechanical InterFace) pods [3]. In some instances fabs have even adopted proprietary carrier designs. Equipment for processing 300 mm wafers utilizes SEMI (Semiconductor Equipment and Materials International) standard FOUP (Front Opening Unified Pod) carriers [4]. In the case of SMIF pods and FOUPs the interface between the WIP automation and the carriers is well defined, and the tool must open the pod in order to extract the wafers. There are also certain instances when the hardware is configured for "bridge" WIP modules that allow the processing tool to accept different styles of carriers or different size wafers. The complexity that can be introduced in the equipment design by considering all these factors is considerable and only involves getting the wafers into, and out of, the process chambers. Some of the configurations, such as those associated with handling pod carriers, also add significant cost to the equipment.

WIP automation handles the job of transferring a carrier of wafers from an ergonomic, operator-load position to a location where the main transfer robot can move the wafers into the process stations. Wafers are typically moved one at a time within the equipment between process stations in ECD equipment, which tends to utilize single-wafer processing. The loaders typically incorporate wafer-mapping capability to inventory the wafers and detect wafers that may be improperly located in the carrier, as well as openers for the standardized pod carriers described above.

Some tool architectures incorporate an additional wafer-handling robot within the WIP area to transfer wafers to and from the main processing tool in order to remove the burden for these transfers from the main process transfer robot(s). Each additional wafer-transfer robot adds cost, complexity, and footprint for the processing tool, so the need must be weighed against these factors when designing the tool.

In most semiconductor processing equipment the wafers are handed off from the wafer-handling robot(s) to some sort of automation associated with a process chamber. This may be as simple as placing the wafer on a locating mechanism and closing a door to the chamber after the robot is extracted or as complex as having the robot pass the wafer to a robotic mechanism that accepts the wafer and moves it into the appropriate processing position(s). The complexity required must be weighed against the benefits gained from having additional capability or flexibility in the processing chamber.

27.2.2 Processing Chambers

The electrochemical reactor contains the wafer and controls the process by controlling its interaction with the chemistry and power supply. It includes the wafer fixture with electrical contact and the power supply and provides both the fluid recirculation path and mounting for the counter electrode(s). Beyond these basic requirements, an electrochemical reactor often provides a variety of other functions, including wafer fixture automation, wafer fixture spinning, fluid agitation to affect the hydrodynamic boundary layer at the wafer surface, features to locally affect the hydrodynamic boundary layer, electric field shaping elements (shields), auxiliary electrodes (current thieves), process monitoring sensors, evolved-gas mitigation features, wafer-contact maintenance features, etc. Often a secondary process section is incorporated into the reactor to allow for pre- and/or post-deposition surface treatment of the product wafer. The specific process to be performed in the reactor typically determines the features that are incorporated into the reactor.

Of course, the reactor used for processing the wafer is a critical part of the tool configuration. The reactor affects the automation design and is the key piece of hardware that influences the process on the wafer. It is important to understand the strengths and limitations of a particular reactor to optimize its performance for a particular application. The same reactor does not work equally well for all electrochemical applications. Reactors that can be utilized for the electrochemical deposition of metals include rack platers, paddle cells, fountain reactors, and other unique configurations designed for single-wafer or batch processing.

The reactors that are typically used for ULSI-related processes are in the fountain reactor category. They have flow coming generally from the bottom of the reactor toward the top, where the wafer is usually rotated in a face-down position for the deposition process. The anode(s) are near the bottom of the reactor, which is cylindrically shaped. Variations occur in the anode-to-cathode spacing, number of anodes, and how many of the additional functions and features mentioned above are included in the hardware. A computer controller is used to control the reactor functions, the fluid flow to the reactor, and the function of the power supply, which may be capable of delivering direct current, pulsed waveforms, or pulse reverse waveforms.

There are several other types of reactors that are sometimes used for microelectronic processing in special applications. There have been some reactors developed that fixture the wafer in a face-up position, which have been used for either ULSI interconnect applications or for through-mask deposition processes. These are usually unique and are designed to fit a particular application. There is also a variant of deposition called electrochemical mechanical deposition, or ECMD, that requires a very unique processing chamber configuration. This process and equipment are described in Chapter 31. Paddle cells and variants of these have also been used for microelectronic processing, usually associated with the deposition of magnetic alloys used in magnetic recording or solder alloys in through-mask applications [5]. These chambers incorporate a reciprocating paddle, or similar structure, near the wafer surface to provide fluid agitation. They may come in wafer face-up, face-down, or vertical configurations.

27.2.3 Tool Configuration

The configuration of a tool for performing ECD processes on wafers is determined by taking the above considerations and combining them with the knowledge of the processes that will occur in the tool. The number of process chambers of each type that are incorporated depends on the expected range of process times for each process step and the desired tool throughput. It is typical to configure tools for copper Damascene ECD processes with throughputs of 50–100 wafers/h, while tools used for 3D interconnect applications may only be capable of achieving 5–10 wafers/h. Reducing process bottlenecks, wafer scheduling, and automation speed become increasingly important to tool throughput as the process sequence complexity increases. Additionally, there are many details that go into the design of the fluid handling components of the tool. While not considered as high-tech as the reactor design, these details are critically important in determining the successful processing of the wafers. Subtle differences in fluid flow design can have large impacts on the yield of wafers processed with the equipment.

Safety (of people and wafers) is an important consideration in tool design. SEMI standards S2 [6] and S8 [7] establish guidelines for safe and ergonomic tool design, and many microelectronics manufacturers have additional internal design protocols. Third-party verification of these standards is expensive and the time for these verifications must be accommodated in the tool manufacturing and test schedule. However, despite the extensive nature of these guidelines, there are still areas of ambiguity that customers and equipment suppliers need to discuss early in the tool design process to avoid misperceptions.

Safety guidelines can also significantly influence the choice of materials of construction of a processing tool. Materials considered fire-safe, which are required by some standards and insurance companies, are significantly more expensive than materials typically used in the industry, and chemical compatibility of approved materials must be carefully considered, especially with some of the reactive chemical mixtures used in semiconductor processing. The use of chemistries that are not compatible with each other within a single tool requires segregating drains, exhausts, and containment areas, if the tool is capable of handling them at all. In some cases physically separating sections of the tool to prevent mixing of incompatible chemicals may be necessary.

27.3 Electrochemical Processes

Generally, the metal ECD processes utilized in microelectronic manufacturing can be divided into *through-mask-deposition* and *blanket-deposition* types. Each has its own specific design requirements and operational characteristics, and there are specific applications within each category that have narrower sets of requirements. The Damascene copper process, which is the focus of Section 27.3.1, falls into the blanket-deposition category [5].

Blanket metal deposition is a class of processing in which there is no significant masking of the substrate to be plated. When one desires a fully metallized surface, this is the type of deposition used. Examples include backside metallization of GaAs wafers, or where a subtractive method, such as CMP (Chemical Mechanical Planarization) or masking and etching will be used to define the areas where metal is removed. The copper Damascene interconnect application has become the most prominent example of this type of process [8–12].

27.3.1 Damascene Copper Electrodeposition

Copper Damascene interconnects have been utilized in the semiconductor industry since the mid-1990s [13–15]. In this application, trenches and vias are etched into a dielectric to define intralevel and interlevel interconnects. Barrier and seed layers are applied to this patterned substrate and copper is electroplated to fill the features. This process requires extremely good repeatability and good uniformity control across the wafer. This means that the contact to the wafer must be extremely repeatable and circumferentially uniform, and the reactor should allow for uniform deposition throughout the entire process. Reactors commonly employed for Damascene copper ECD processes are of the fountain style, and typically have a rotating wafer with the plated surface down.

The copper Damascene fill processes are driven in large part by the adsorption of organic additives on the surface of the wafer in the initial part of the process. For this reason, and to prevent formation of defects caused by non-uniform wetting or air entrapment, it is critical for equipment used for Damascene processes to have tight control over the wafer wetting, or solution entry, process. This means that the equipment must be able to control the entry speed and wafer position throughout the entry process, and it must also maintain control of the current and voltage applied to the wafer as it enters the solution. If these things are not done properly, a defect signature will be produced on the wafer as it begins processing.

Because of the relatively thin (less than 2 μm) deposition, it is possible to use an exposed electrical contact to the wafer. The deposited metal layer covers the contact, as well as the wafer surface, and provides excellent electrical conductivity. This contact is pulled away from the wafer at the end of the process. It is also possible (because of the reversible nature of copper deposition in acid solutions) to eliminate the need for a seal to prevent deposition on the contacts by reversing the polarity of the current on the contacts after removing the wafer and stripping the copper off of the contacts before plating the next wafer [16, 17]. Sealed contacts may also be used for this process, but much care is needed to maintain repeatable processing of large numbers of wafers, especially as copper seed layers become thinner than

50 nm. Some of the considerations associated with fluid seals in this application include the complexity of removing a wet seal from the substrate without pulling liquid behind the sealing surface through capillary forces, trapping of residual acid solution during a rinse cycle, and intrusion of the seal and contact elements into the usable area of the semiconductor wafer. Failure to provide a robust hardware design can lead to etching or electrolytic removal of the copper seed layer locally near the contact points, or to oxidation or corrosion of the freshly plated surface. These issues typically manifest themselves as increased non-uniformity and/or increased defectivity near the edge of the wafer. Also, the added profile of a seal protruding beyond the wafer surface can make immersing the wafer in solution and releasing trapped air, which leads to defects on the wafer surface, more difficult [18]. Semiconductor device manufacturers are increasingly pushing toward thinner copper seed layers as the feature dimensions shrink. They are also increasingly pushing toward smaller edge exclusions to increase the number of yielding die at the edge of the wafer. These trends make the contact and its interaction with the PVD seed layer more and more critical.

Another important characteristic of the copper Damascene process is that the barrier and seed-layer resistance is appreciable at the beginning of the process, and the total film resistance decreases as the process proceeds. This causes a voltage drop between the contact terminals (at the edge of the wafer) and the center of the wafer. This is referred to as the terminal effect [19–21]. The terminal effect leads to a variation in the applied current density across the wafer in conventional reactors, which causes variation in fill capability, additive incorporation, grain size and, of course, thickness.

In addition to this spatial variation, the terminal effect is time dependent in a typical copper Damascene process because the conductivity of the film stack increases during the deposition [22]. To compensate for the changing terminal effect, the reactor must be designed so that the radial potential-field distribution can be adjusted over time throughout the deposition process [23]. Since the potential-field distribution on the wafer changes dramatically as the film conductivity changes, the potential-field in the electrolyte at the wafer surface must be changed accordingly in order to maintain a uniform current density distribution across the wafer. The potential-field curves for a typical Damascene copper process for the 45 nm technology node are shown in Fig. 27.1. These curves determine what the capability of the ECD reactor should be in order to maintain a uniform deposition throughout the process.

Various techniques have been employed by ECD equipment manufacturers to attempt to vary the potential-field distribution of the reactor in response to the potential-field distribution within the metallization layer on the wafer surface. These include the use of multiple anodes with dynamic current (or voltage) control, current shielding that can be moved (relative to the wafer) during the process, and auxiliary cathodes (current thieves) that can be adjusted to vary the current distribution near the edge of the wafer. Each of these methods must be considered with respect to its ease of implementation, the spatial control compared to the potential-field distribution across the wafer surface, and reactor maintenance impacts. In addition, it

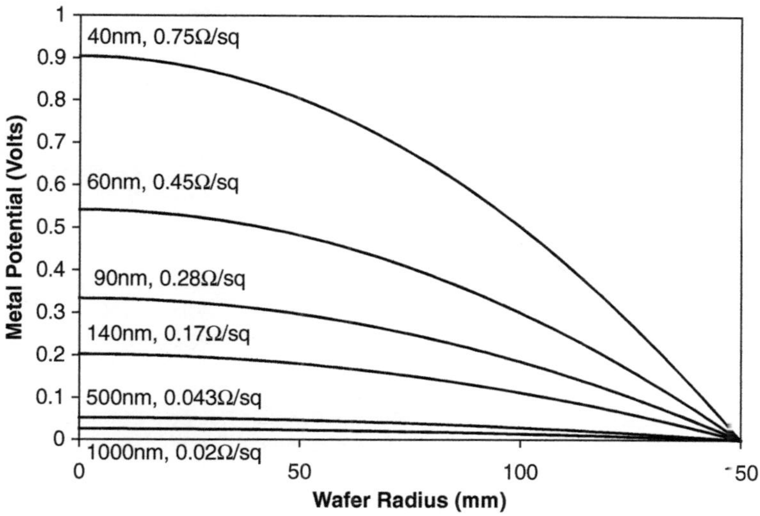

Fig. 27.1 Potential distributions in the copper film as a function of film thickness (sheet resistance) in a copper Damascene process

is important to consider any particulate contamination impacts due to mechanical motion or deposits forming near the wafer surface. The impact of utilizing one of these methods (dynamic current distribution control) can be seen in Fig. 27.2.

Recently, several industrial semiconductor equipment manufacturers have introduced systems that utilize membrane elements to separate some components of the plating chemistry from the anode and/or to separate the cathode electrolyte, or catholyte, from the anode electrolyte, or anolyte [24]. The main purpose of these systems is to extend the bath life and/or reduce the consumption of some of the bath components over the bath life. The systems that have been introduced range from those utilizing a filter membrane to effect some separation, to those using a semi-permeable, ion-selective membrane. In the example of acid copper baths used for copper Damascene deposition, the purpose of the membrane is to try to keep the organic additives from contacting the copper anode, where they are broken down, and to reduce additive usage and the associated buildup of generally unwanted additive breakdown products. Ion-selective membranes can also be used to prevent the anodic oxidation of these compounds when using noble metal anodes, sometimes referred to as inert or dimensionally stable anodes.

ECD reactors incorporating ion-selective membranes are typically designed with separate fluid flow systems for the catholyte (which the wafer is processed in) and the anolyte. This arrangement allows the separation of the electrolytes (to varying degrees of effectiveness, depending on design choices) and more effective utilization of the chemistry, especially the organic additives. Although these systems are usually described as limiting the degradation of the organic additives at the anode, the subtleties of the membrane systems are somewhat more complex. In reality, the equipment design must take into account such elements as osmosis, electrophoretic

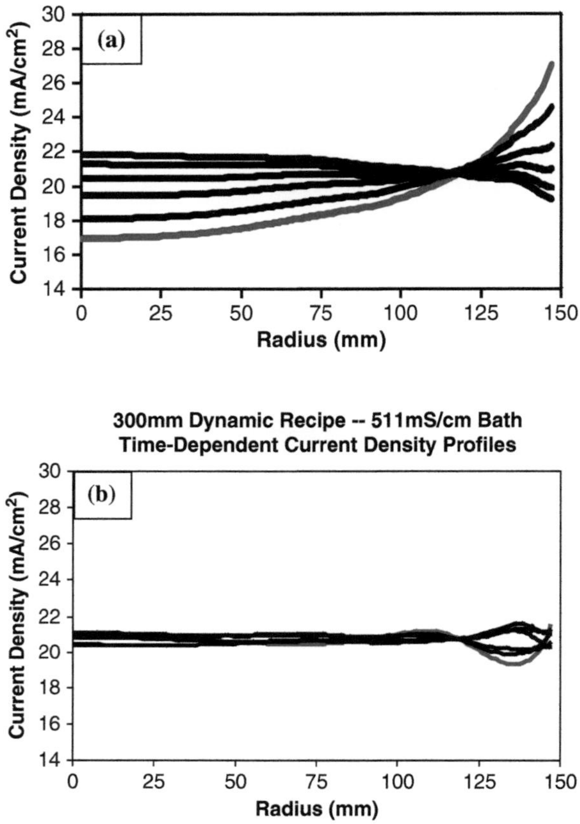

Fig. 27.2 Current density profile evolution during 300 mm blanket wafer deposition. *Grey curves* indicate initial current density profiles. (**a**) Constant electric potential distribution throughout deposition (**b**) Dynamic current density control providing uniform current density throughout deposition process. (bath conductivity: 510 mS/cm)

transfer of components across the membrane, chemical concentrations and relative bath volumes, and the transport of water of hydration that is carried across with ionic charge carriers (electro-osmosis). The effectiveness of the system at reducing additive consumption is also impacted by the concentrations of the supporting electrolytes on each side of the membrane and the waveform used for deposition.

Reactors designed with less effective membranes, such as filters, do not have completely separated anolyte and catholyte. These systems have some flow of solution through the filter which makes it important to balance the flows of electrolytes into, and out of, each section of the fluid flow system. Flow must be maintained in the preferred direction if it is to be controlled, which necessitates more consumption of chemistry by the equipment, regardless of whether wafers are being run or not. Filter-element separated systems simply slow down the mass transfer of additives to the anode, so they are not as effective as ion-selective membranes at reducing chemical consumption.

Rinsing of the wafer after deposition is also important in order to prevent surface corrosion and provide a uniform film surface. Many process chambers incorporate rinse capability into a secondary portion of the ECD reactor to guarantee fast, efficient rinsing. Additionally, wafers are usually rinsed in a subsequent process to ensure complete removal of plating chemistry. The wafer may also be rinsed with an acid to remove any oxides and corrosion products that have formed. These processes may be incorporated into a process chamber that provides backside metals removal and/or etching of the copper film on the outer perimeter of the wafer.

27.3.2 Through-Mask Electrodeposition

Many semiconductor and microelectronic processes make use of photoresist-patterned, or through-mask, electrodeposition [25, 26]. These processes may be associated with plating submicrometer gold interconnects on GaAs wafers, plating copper coils or magnetic alloys for thin-film recording heads, plating copper conductors for redistribution or integrated passive applications, or plating PbSn or lead-free solders for flip chip connection [27]. All of these processes have a blanket conductive seed layer, or conductive plating base, and a patterned dielectric template, into which metal is deposited. There are many variations, and many additional processes and applications that have not been mentioned here, but for practical purposes these processes may be classified together. We will not discuss these processes in significant detail, but they are worth mentioning as a general class of ECD processes (see Fig. 27.3).

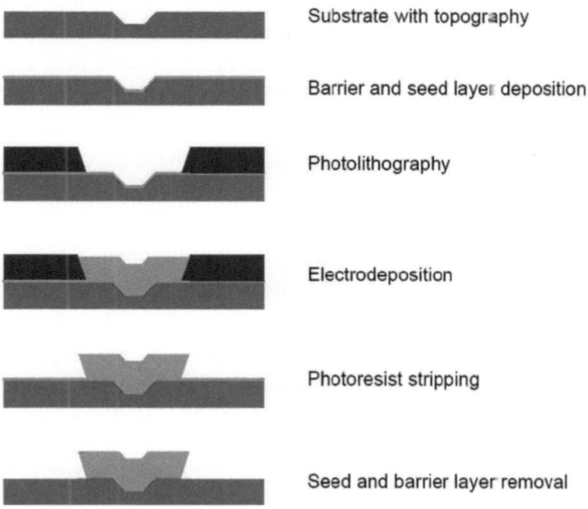

Fig. 27.3 Typical through-mask ECD process sequence

Generally, these processes may be described as having a conductive base layer stack that is deposited across the entire substrate. The substrate may be planar or may have topography at this stage of processing. The base layer typically consists of at least two metal layers: an adhesion (or diffusion barrier) layer and a conductive seed layer to provide current flow across the surface. The first layer usually provides adhesion to underlying metal and/or dielectric layers and may also prevent diffusion of the ECD layers into the underlying materials (or vice versa). The seed layer is typically a thin vacuum-deposited layer of the material that will be plated through the resist mask. It is also possible to use other materials as the seed layer, and there are many variants that provide alternate metallurgical structures for specific applications.

Photoresist is usually used as the plating template material. It is coated, exposed, developed, and baked to provide a stable photoresist template for the plating process. The photoresist usually undergoes a plasma de-scum process to remove any residual polymer that was not removed in the develop process. Metal is then deposited to a thickness near the photoresist thickness. Depending on the process flow and application, the ECD thickness may be slightly below, or even above, the photoresist thickness.

After the deposition process, the photoresist is usually stripped from the wafer, and the seed and barrier layers are removed. These processes may be vacuum and/or wet processes. Sometimes there are additional cleaning steps involved at various points in the process sequence. If the deposit is a solder material for flip chip attachment, the solder is usually reflowed to create a hemispherical solder bump after the under bump metallurgy, or UBM, is removed. Again, there are many variants that accommodate specific applications and integration schemes.

Through-mask ECD processes have different requirements than the blanket plating processes described above. Sealed electrical contacts are usually preferred for through-mask applications. While the terminal effect is sometimes important for these processes, many times it is constant, or nearly so, throughout the entire process. This is especially true if the open area in the photoresist mask is very low. There is additional complexity introduced due to the local-scale variations associated with current density variation induced by pattern distribution effects [28–30]. The mass transfer characteristics of these processes, both across the wafer surface and through the depth of the photoresist mask, can also be very different from that in a blanket plating application. It is important to understand the electrolyte characteristics and how they impact the process behavior in a through-mask ECD process. In addition, there are additional chemistry compatibility, pH, and temperature limitations associated with deposition when photoresist is present in order to avoid such problems as resist attack, underplating (depositing metal under the photoresist at pattern edges), resist swelling, and resist cracking. These problems can all be avoided by thoughtful attention to process design and process integration.

27.3.3 Electrografting

Recently, a process termed “electrografting” has achieved some notoriety [31–33]. This process is essentially the functionalization of a surface with an organic material, similar to a self-assembled monolayer (SAM). While electrografting has received more publicity, a process based more on the SAM functionalization approach has also been investigated in some process sequences, such as the deposition of metals on insulators [34, 35]. Both versions of the process, as applied to microelectronic processing, involve functionalizing a surface with an organic molecule with near-monolayer coverage, and using the functional groups or chelated metal on the organic molecule to provide activation sites for subsequent electrolytic or electroless deposition processes. In some cases, the Electrografting process may involve the electrochemical polymerization of organic species to form higher molecular weight materials which are deposited and bonded to the surface of the electrode [36].

The electrografting and chemical grafting processes are interesting in that they provide a mechanism to deposit organic, or functionalized organic, molecules on metallic or insulating surfaces. They may also be used to deposit selectively to one exposed material, while not depositing on a chemically or electrochemically dissimilar area. These processes have been used to help deposit copper seed layers directly to barrier materials, to activate and functionalize copper surfaces for subsequent deposition of self-aligned barriers by electroless deposition, and to functionalize dielectric surfaces to activate them for the electroless deposition of a liner barrier layer.

27.3.4 Electrophoretic Deposition

Electrophoretic deposition has also been used to deposit materials associated with microelectronic processes. The two processes that have been utilized in microelectronic processing are polymer deposition (photoresist) [37, 38] and metal deposition (solder alloys) [39] Electrophoresis is the transfer of charged ions, molecules, or particles due to an imposed potential gradient. An applied potential in an ECD reactor can be used to move materials to an electrode, where they can be deposited.

Electrophoretic photoresist deposition has been used in the printed circuit board industry for many years and has recently been applied to wafer processing [38]. It has been applied to both positive-tone and negative-tone photoresist materials, and the deposition may occur on the anode or the cathode, depending on the chemistry being used. The organic components of the photoresist (polymer, photo-active compounds, and solvents) are usually contained in micelles in an aqueous emulsion. Electrophoresis causes the negatively charged, or positively charged, micelles to be driven to the anode or cathode, respectively. At the appropriate electrode, the micelles are destabilized by the hydrolysis of water and a change in the localized pH of the solution. This causes the materials within the micelle to deposit out of the emulsion onto the electrode. The electrophoretic photoresist deposition processes allow very conformal coatings of relatively thick (5–20 μm) photoresist even over large 3D structures. Of course, the surface must be conducting to allow it to be charged as an electrode to promote the deposition process.

Electrophoretic deposition has also been used for depositing metallic particles on wafers. This has mainly been used for depositing solder alloy particles, which may then be reflowed to produce hemispherical solder bumps for chip connection. The metallic particles are suspended in an aqueous solution and migrate to the electrode under an applied potential gradient. Once they reach the electrode they accumulate, or deposit, at the electrode surface. This process can be carried out with a photoresist-patterned wafer as the electrode, and the photoresist openings will define where the metallic particles are deposited. These processes are limited by the particle size distribution, deposition speed, and the ability to keep the particles suspended in solution.

The equipment used for electrophoretic deposition is very similar to electrodeposition equipment. There may be some specific modifications that allow processing of slurries or solvent-containing solutions, but the equipment is essentially the same.

27.4 Summary

Much has been learned about ECD processes since the 1990s. These processes are now a mainstream part of semiconductor processing, and there are several options for manufacturing-grade equipment. Several equipment suppliers provide sophisticated equipment that has been optimized for ULSI interconnect processing. Suppliers are now focusing on reducing process costs, as opposed to trying to make the processes work. There are also other processes and equipment that are designed to perform them being used in microelectronic manufacturing, and being developed for the future.

27.5 Biographical Sketches

T. Ritzdorf *Semitool, Inc., Kalispell, MT 59901 (tritzdorf@semitool.com).*
Mr. Ritzdorf is the director of ECD Technology at Semitool, Inc. in Kalispell, MT. He received a B.S. in Chemical Engineering from Montana State University in 1986, and a M.S. in Chemical Engineering from the University of Minnesota in 1989. Mr. Ritzdorf spent 6 years working in the magnetic recording industry at Control Data, Seagate, and Storage Technology. He is an author or coauthor of at least 20 patents and over 30 technical papers, including chapters for two books on electrochemical processing. Mr. Ritzdorf is a member of the Electrochemical Society and the American Electroplaters and Surface Finishers Society.

References

1. Ritzdorf, T. L.; Wilson, G. J.; McHugh, P. R.; Woodruff, D. J.; Hanson, K. M.; and Fulton, D.: Design and modeling of equipment used in electrochemical processes for microelectronics. IBM J. Res. Dev. **49(1)**, 65 (2005)

2. Powers, J. V. and Romankiw, L. T.: US Patent #3,652,442, Electroplating Cell Including Means to Agitate the Electrolyte in Laminar Flow (1972)
3. SEMI E19 Standard Mechanical Interface (SMIF), SEMI (Semiconductor Equipment and Materials International), 30381 Zanker Road, San Jose, CA, 95134 (2002)
4. SEMI E47.1-0305 Provisional Mechanical Specification for FOUPs Used to Transport and Store 300 mm Wafers. SEMI (Semiconductor Equipment and Materials International), 30381 Zanker Road, San Jose, CA, 95134 (2005)
5. Ritzdorf, T. and Fulton, D.: Electrochemical deposition equipment, In Datta, M.; Osaka, T., and Schultze, J. W. Eds. New Trends in Electrochemical Technology, Microelectronic Packaging, Francis and Taylor (CRC), London **3**, (2005)
6. SEMI S2 Environmental, Health, and Safety Guidelines for Semiconductor Manufacturing Equipment, SEMI (Semiconductor Equipment and Materials International), 30381 Zanker Road, San Jose, CA, 95134 (2003)
7. SEMI S8 Safety Guidelines for Ergonomics Engineering of Semiconductor Manufacturing Equipment, SEMI (Semiconductor Equipment and Materials International), 30381 Zanker Road, San Jose, CA, 95134 (2003)
8. Braun, A. E.: Copper Electroplating Enters Mainstream Processing. Semiconductor International 58 (1999)
9. Andricacos, P. C.; Uzoh, C.; Dukovic, J. O.; Horkans, J.; and Deligianni, H.: Damascene copper electroplating for chip interconnections. IBM J. Res. Dev. **42(5)**, 567 (1998)
10. Pricer, T. J.; Kushner, M. J.; and Alkire, R. C.: Monte Carlo Simulations of the Electrodeposition of Copper, Part II: Acid Sulfate Solution with Blocking Additive. J. ECS **149(8)**, C406 (2002)
11. Vereecken, P.; Long, J.; Cooper, E.; Deligianni, H.; Andricacos, P.; Binstead, R.; Wu, J.; Mikkola, R.; and Calvert, J.: Effect of Differential Additive Concentrations in Damascene Copper Electroplating. Meeting Abstracts of the 203rd meeting of the Electrochemical Society, abst. 606 (2003)
12. Josell, D.; Wheeler, D.; and Moffat, T. P.: Superconformal Electrodeposition in Vias. Electrochem. Solid-State Lett. **5(4)**, C49 (2002)
13. Edelstein, D.; Heidenreich, J.; Goldblatt, R.; Cote, W.; Uzoh, C.; Lustig, N.; Roper, P.; McDevitt, T.; Motsiff, W.; Simon, A.; Dukovic, J.; Wachnik, R.; Rathore, H.; Schulz, R.; Su, L.; Luce, S.; and Slattery, J.: Full copper wiring in a sub-0.25 μm CMOS ULSI technology. Proc. IEEE IEDM. 773 (1997)
14. Venkatesan, S.; Gelatos, A. V.; Misra, V.; Smith, B.; Islam, R.; Cope, J.; Wilson, B.; Tuttle, D.; Cardwell, R.; Anderson, S.; Angyal, M.; Bajaj, R.; Capasso, C.; Crabtree, P.; Das, S.; Farkas, J.; Filipiak, S.; Fioralice, B.; Freeman, M.; Gilbert, P. V.; Herrick, M.; Jain, A.; Kawasaki, H.; King, C.; Klein, J.; Lii, T.; Reid, K.; Saaranen, T.; Simpson, C.; Sparks, T.; Tsui, P.; Venkatraman, R.; Watts, D.; Weitzman, E. J.; Woodruff, R.; Yang, I.; Bhat, N.; Hamilton, G.; and Yu, Y.: A high performance 1.8 V, 0.20 mm CMOS technology with copper metallization. Proc. IEEE IEDM 769 (1997)
15. Zielinski, E. M.; Russell, S. W.; List, R. S.; Wilson, A. M.; Jin, C.; Newton, K. J.; Lu, J. P.; Hurd, T.; Hsu, W. Y.; Cordasco, V.; Gopikanth, M.; Korthuis, V.; Lee, W.; Cerny, G.; Russell, N. M.; Smith, P. B.; O'Brien, S.; and Havemann, R. H.: Damascene Integration of Copper and Ultra-Low-*k* Xerogel for High Performance Interconnects, Proc. IEEE IEDM 936–938 (1997)
16. Graham, L. W.; Ritzdorf, T. L.; and Turner, J. I.: US Patent # 6,599,412, In-situ cleaning processes for semiconductor electroplating electrodes (2003)
17. Graham, L. W.; Hanson, K.; Ritzdorf, T. L.; and Turner, J. I.: US Patent # 6,270,647 Electroplating system having auxiliary electrode exterior to main reactor chamber for contact cleaning operations (2001)
18. Law, S. B.; Loh, S.; Ng, H. W.; Zhou, B. B.; Zhang, H.; Tan, W. L.; Sudijono, J.; and Hsia, L. C.: Impact of the additives and the current density of copper electroplating process on the backend-of-line metallization of ULSI, Meeting Abstracts of the 203rd meeting of the Electrochemical Society, abst. 675 (2003)

19. Lanzi, O. and Landau, U.: Terminal effect at a resistive electrode under tafel kinetics. J. ECS **137(4)**, 1139 (1990)
20. Mehdizadeh, S. and Dukovic, J. O.: Transient Plating-Rate Distribution on Resistive Thin Films with Point Contact Terminals. Extended Abstracts of the 184th meeting of the ECS, 93-2, IBM-J, Abst. 210 (1997)
21. Deligianni, H.; Dukovic, J. O.; Walton, E. G.; Contolini, R. J.; Reid, J.; and Patton, E.: Model of wafer thickness uniformity in an electroplating tool. Electrochemical Processing in ULSI Fabrication and Semiconductor/Metal Deposition II, ECS PV **99(9)**, 83 (1999)
22. Jorne, J.: Uniformity of copper electroplating on wafers. Meeting Abstracts of the 193rd meeting of the Electrochemical Society, abst. 256 (1998)
23. Klocke, J.; McHugh, P.; Wilson, G.; Ritari, K.; Roberts, M.; and Ritzdorf, T.: Overcoming Terminal Effects During Electrochemical Deposition of Copper Films for 300 mm Damascene Interconnect Applications. Advanced Metallization Conference (AMC) 2002, San Diego, CA (2002)
24. Hafezi, H.; Huang, Y. C.; Zhang, A.; Singh, S.; and Ngai, C.: Enhanced Copper ECP for 45 nm Devices. Semiconductor International (2005)
25. Datta, M.: Applications of electrochemical microfabrication: An introduction. IBM J. Res. Dev. **42(5)**, 563 (1998)
26. Saenger, K. L.; Costrini, G.; Kotecki, D. E.; Kwietniak, K. T.; and Andricacos, P. C.: Submicron Pt Electrodes by Through-Mask Plating. Meeting Abstracts of the 199th meeting of the Electrochemical Society, abst. 317 (2001)
27. Datta, M.; Shenoy, R. V.; Jahnes, C.; Andriacos, P. C.; Horkans, J.; Dukovic, J. O.; Romankiw, L. T.; Roeder, J.; Deligianni, H.; Nye, H.; Agarwala, B.; Tong, H. M.; and Totta, P.: Electrochemical Fabrication of Mechanically Robust PbSn C4 Interconnections. J. ECS **142(11)**, 3779 (1995)
28. Mehdizadeh, S.; Dukovic, J. O.; Andricacos, P. C.; Romankiw, L. T.; and Cheh, H. Y.: The influence of lithographic patterning on current distribution: A model for microfabrication by electrodeposition. J. ECS **139(1)**, 78 (1992)
29. Mehdizadeh, S.; Dukovic, J.; Andricacos, P. C.; Romankiw, L. T.; and Cheh, H. Y.: The influence of lithographic patterning on current distribution in electrodeposition: Experimental study and mass-transfer effects. J. ECS **140(12)**, 3497 (1993)
30. DeBecker, B. and West, A. C.: Workpiece, pattern, and feature scale current distributions. J. ECS **143(2)**, 486 (1996)
31. Haumesser, P. H.; Giblat, F.; Ameur, S.; Cordeau, M.; Maitrejean, S.; Mourier, T.; Bureau, C.; and Passemard, G.: Electro-grafting: a New Approach for Copper Seeding or Direct Plating. In Advanced Metallization Conference 2003. Ray, G. W.; Smy, T.; Ohta, T.; and Tsujimura, M., Eds. MRS, 575 (2004)
32. Shih, C. H.; Su, H. W.; Lin, C. J.; Ko, T.; Chen, C. H.; Huang, J. J.; Chou, S. W.; Peng, C. H.; Hsieh, C. H.; Tsai, M. H.; Shue, W. S.; Yu, C. H.; and Liang, M. S.: Direct Plating of Cu on ALD TaN for 45 nm-node Cu BEOL Metallization. IEDM 2004, Proc. IEEE IEDM (2004)
33. Haumesser, P. H.; Cordeau, M.; Maitrejean, S.; Mourier, T.; Gosset, L. G.; Besling, W. F. A.; Passemard, G.; and Torres, J.: Copper metallization for advanced interconnects: the electrochemical revolution. In Proceedings of the 2004 IITC, IEEE 3 (2004)
34. Osaka, T.; Yoshino, M.; Yokoshima, T.; Hashimoto, A.; and Hagiwara, Y.: Formation of Diffusion Barrier Layer on Low-*k* Material Using Wet Fabrication Process. In Electrochemical Processes in ULSI and MEMS. Deligianni, H.; Moffat, T.; Mayer, S. T.; and Stafford, G., Eds. 205th ECS- **PV-17**, 251 (2004)
35. Decorps, T.; Haumesser, P. H.; Peyne, C.; Cordeau, M.; Raynal, F.; Rabinzohn, P.; and Bureau, C.: Chemical Grafting to Improve Electroless Deposition of Self-Aligned Barriers. In Advanced Metallization Conference 2004. Erb, D.; Ramm, P.; Masu, K.; and Osaki, A., Eds. MRS 823 (2005)
36. ChemPhysChem 2004, www.chemphyschem.org, Wiley-VCH Verlag GmbH& Co. **5**, 1468 (2004)

37. Hendriks, H.; Tajadod, J.; and Klocke, J.: Photoresist application for 3D features on wafer surfaces. Compound Semiconductor (2003)
38. Klocke, J. and Steeper, J.: Patterning Three-Dimensional Structures on Wafers with Electrophoretic Photoresist. Pan Pacific Microelectronics Symposium, Maui, HI 35 (2002)
39. Buetow, M.: Industry Seen Open Stance from Congress in China. (Industry News) Circuits Assembly 6 (2005)

Chapter 28
Electroless Deposition Processes and Tools

Z. Hu and T. Ritzdorf

28.1 Introduction

Electroless deposition, which was introduced in 1946 by Brenner and Riddell [1], is a type of electrochemical deposition that is gaining interest in semiconductor and related applications. Electroless deposition utilizes complementary electrochemical reactions to cause metal deposition. The oxidation of a reducing agent supplies the electrons needed for reducing metal ions to their metallic state [2]. This encompasses both immersion (displacement) deposition reactions and autocatalytic reactions. One should use electroless deposition in cases where it is desirable to deposit metal on non-conducting surfaces or to deposit metal selectively to certain underlying materials, especially if there is no possibility to have a continuous conductive underlayer. Current applications of interest in the microelectronics industry include copper deposition for seed layer and interconnect metallization, nickel and gold depositions for contact metallurgy in microelectronics packaging and related applications, and cobalt–tungsten alloys as diffusion barrier for copper interconnects.

Immersion deposition occurs when a metal on the substrate surface is oxidized (and solvated) to provide electrons that enable the reduction and deposition of a more noble metal from the solution. The deposition ceases when the less noble metal on the substrate is completely covered. As a result, the maximum thickness of deposits produced this way is limited.

Autocatalytic deposition utilizes a reducing agent in the solution as the electron source and continues as long as the substrate remains in the deposition solution. Once the deposited metal covers the activated surface, it functions as a catalyst to promote its own deposition.

Z. Hu (✉)
Semitool Inc., 655 W. Reserve Dr., Kalispell, MT 59901, USA
e-mail: zhu@semitool.com

Y. Shacham-Diamand et al. (eds.), *Advanced Nanoscale ULSI Interconnects: Fundamentals and Applications*, DOI 10.1007/978-0-387-95868-2_28,

28.2 Electroless Copper Deposition

In comparison with aluminum, copper has a lower electrical resistivity and higher resistance to stress-induced voiding. As a result, copper has been widely adopted for metallization in microelectronic devices. Copper can be deposited by vacuum-based techniques, including physical vapor deposition and chemical vapor deposition, and by electrochemical deposition techniques, including electrodeposition and electroless deposition. Compared with vacuum-based deposition techniques, electrochemical deposition techniques have the advantages of low tool cost and low processing temperature. Electrochemical deposition techniques, particularly electrodeposition, also provide fast deposition rate and superconformal via/trench filling capability. While electrodeposition of copper has been widely used in microelectronic manufacturing, electroless deposition of copper may be used in applications where electrodeposition faces processing difficulties. For instance, as electrodeposition requires a continuous seed layer that is conductive enough to carry current uniformly across the wafer without significant potential drop, electroless deposition may be used in cases where seed layer is absent or the seed layer is not conductive enough.

28.2.1 Copper Deposition Chemistry

Electroless copper deposition is realized though the reduction of copper ions in solution by a reducing agent. The reducing agents that may be used for electroless copper deposition include formaldehyde [3–7], glyoxylic acid [8–12], and hypophosphite [13, 14]. Among them, formaldehyde is by far the most commonly used reducing agent for this purpose, as it produces copper deposits at higher deposition rates with excellent mechanical properties. Commercial electroless copper baths that use formaldehyde as reducing agent are available. Glyoxylic acid is similar to formaldehyde in nature, but with a lower vapor pressure. This is advantageous over formaldehyde as the toxicity of formaldehyde is exacerbated by its low vapor pressure in working places. Glyoxylic acid reducing agent has been reported to produce smooth and conformal copper deposits compatible with ULSI applications [9–11]. Hypophosphite has also been investigated as a reducing agent for electroless copper deposition [13, 14]. However, as copper is not a catalyst for the oxidation of hypophosphite, other catalytic metals such as nickel must be present in the bath to keep the deposition going as desired. In the process, nickel is codeposited with copper, thus catalyzing the oxidation of hypophosphite and causing the continuous deposition of copper and nickel on the substrate. Phosphorous is codeposited as well, when hypophosphate is used as the reducing agent. The codeposition of both nickel and phosphorous would increase the resistivity of the deposit.

In a formaldehyde-based electroless deposition system, copper deposition only occurs at high pH conditions. This is believed to be due to the involvement of methanediolate anion, $H_2C(OH)O^-$, in the deposition reactions [15–17].

Methanediolate anion is an intermediate formed from the reaction of formaldehyde and hydroxide (Eq. 28.1).

$$HCHO + OH^- \leftrightarrow H_2C\,(OH)\,O^- \tag{28.1}$$

The intermediate methanediolate anion adsorbs on the surface of a catalytic substrate to which copper is to be deposited and dehydrogenates, resulting in hydrogen radical (Eq. 28.2). The dehydrogenated species reacts further with a hydroxide ion, producing an electron (Eq. 28.3) [15–17].

$$\left[H_2C\,(OH)\,O^-\right]_{ads} \rightarrow \left[HC\,(OH)\,O^-\right]_{ads} + H_{ads} \tag{28.2}$$

$$\left[HC\,(OH)\,O^-\right]_{ads} + OH^- \rightarrow HCOO^- + H_2O + e^- \tag{28.3}$$

The hydrogen radicals combine together to form hydrogen gas. The anodic half reaction can thus be written as:

$$2HCHO + 4OH^- \rightarrow 2HCOO^- + 2H_2O + H_2 \uparrow + 2e^- \tag{28.4}$$

At alkaline conditions, copper ions tend to form copper hydroxide and precipitate out. To prevent copper ions from precipitating, complexing agents must be used. The complexing ability of the complexing agent used for this purpose and its concentration must be carefully considered. Apparently, the complexing agent must be strong enough to dramatically reduce free cupric ion concentration $[Cu^{2+}]$, so that $[Cu^{2+}][OH^-]^2 < 4.8\times10^{-20}$, which is the solubility product of copper hydroxide. On the other handif copper ions were too strongly complexedinsufficient free copper ions would be available for deposition and the deposition rate would lower than desired. The most common complexing agent used for formaldehyde-based electroless copper deposition is ethylenediaminetetraacetic acid (EDTA). In the presence of the complexing agent EDTA, copper deposition is the coupling of the anodic half reaction (Eq. 28.4) and the cathodic half reaction (Eq. 28.6) where the free copper ion concentration is determined by the dissociation of copper–EDTA complex (Eq. 28.5). The overall reaction is represented by Eq. 28.7.

$$CuEDTA^{2-} \leftrightarrow CU^{2+} + EDTA^{4-} \tag{28.5}$$

$$Cu^{2+} + 2e^- \rightarrow Cu \tag{28.6}$$

$$2HCHO + CuEDTA^{2-} + 4OH^- \rightarrow 2HCOO^- + Cu + 2H_2O + H_2 \uparrow + EDTA^{4-} \tag{28.7}$$

28.2.2 Electroless Copper for Seeding Barriers

The reduction of copper by formaldehyde only takes place at a catalytic substrate, which can be Cu, Au, Pt, PdRh, or Ag [18, 19]. Therefore, for deposition on a substrate seeded with copper, there is no need to activate the surface of the substrate prior to copper deposition. On other substrates, such as dielectrics and diffusion/adhesion barriers (e.g., Ti/TiN and Ta/TaN), however, the surface of the substrate generally needs to be activated.

Electroless copper deposition on diffusion/adhesion barriers is currently, and has been for quite some time, an area of significant interest [3, 4, 7, 8, 10, 11, 20–26]. As copper is mobile in silicon and dielectrics, a diffusion barrier is required to prevent copper from penetrating into dielectrics and the silicon substrate. In a Damascene metallization process, the diffusion barrier (e.g., Ta/TaN bilayer) is first applied on the wafer that has been patterned. On top of the diffusion barrier, a thin layer of copper is then deposited by sputtering, or PVDto provide a low-resistance seed for the subsequent electrodeposition of copper. Sputtering produces copper seeds with excellent adhesion. The step coverage of sputtering, although limited has also been able to meet the need for the technology nodes up till now. However, as feature dimensions continue to scale down, the step coverage of sputtered copper seed is expected to become the primary factor that limits gap fill in the subsequent copper electrodeposition process. Electroless deposition has been recognized as a viable alternative for this purpose. Electroless copper deposition has the potential for conformally seeding fine and high aspect ratio features because of its excellent step coverage. In addition, electroless copper deposition has the potential for being used for filling features along with the seeding in a single process step, which would certainly provide a cost-saving advantage.

Diffusion barriers commonly utilized in microelectronics are not catalytic to electroless copper deposition. To enable electroless copper deposition on barriers (e.g., TiN/Ti and TaN/Ta), the surface of the barriers have to be first activated either by copper or by another catalytic metal such as palladium. In fact, surface activation is often the key step that affects the adhesion and resistivity of the resulting copper film and the selectivity of the deposition process.

28.2.3 Surface Activation Using Palladium

Generally, activation of TiN/Ti and TaN/Ta barriers is accomplished by depositing small amounts of palladium on the barrier surface from solutions containing Pd(II) ions (e.g., from $PdCl_2$) and fluoride (e.g., from NH_4F and/or HF). The driving force behind the activation is the displacement reaction between Pd(II) ions and Ti (or Ta) metal at its metallic state. The displacement reaction may also take place between Pd(II) and Ti (or Ta) species at its lower oxidation states, e.g., Ti(I). Taking Ti metal at its metallic state as an example, the activation process in the presence of fluoride can be expressed by the following overall reaction between Pd(II) ions and metallic Ti. Fluoride ions in the activation solution facilitate the displacement reaction between Pd(II) and Ti through their complexation with Ti(III) (Eq. 28.8)

$$2Ti + 3Pd^{2+} + 12F^- \rightarrow 2{TiF_6}^{3-} + 3pd \tag{28.8}$$

It can be deduced from the above discussion that if there were no oxidizable metal species on the surface, nucleation of palladium would not occur. Both Ti and Ta can be easily oxidized by air, forming metal oxides that will not react with Pd(II) ions. To enable palladium nucleation, the surface oxide must be removed. In the presence of fluoride ions in the palladium activation solution, the surface is cleaned by the solution through the reaction of fluoride with the surface oxides. In the case of a Ti substrate, the cleaning process can be represented by Eq. (28.9).

$$Ti_2O_3 + 12F^- + 6H^+ \rightarrow 2{TiF_6}^{3-} + 3H_2O \tag{28.9}$$

With the palladium activation solution that is usually acidic, the etching of Ti and Ta would also occur according to the following redox reactions.

$$2Ti + 6H^+ + 12F^- \rightarrow 2{TiF_6}^{3-} + 3H_2 \uparrow \tag{28.10}$$

$$2Ta + 10H^+ + 14F^- \rightarrow 2{TaF_7}^{2-} + 5H_2 \uparrow \tag{28.11}$$

28.2.4 Surface Activation Through Sensitization

Sensitization with Sn(II) species can be used to facilitate palladium activation of barriers for the subsequent electroless copper deposition. This is particularly true when the barrier under investigation is essentially composed of metal nitride that is in stoichiometric ratio of nitrogen to metal and has no metallic Ti (or Ta) for Pd(II) to react with. In a sensitization-preceded palladium activation process, the barrier substrate is first immersed in a solution that contains Sn(II) species (e.g., $SnCl_2$ in HCl solution) for sensitization and then in a palladium solution for activation, where Pd(II) ions in solution react with Sn(II) species adsorbed on the barrier substrate from the sensitization step, leaving palladium nuclei on the barrier. The use of palladium activation in conjunction with Sn(II) sensitization for electroless copper deposition has been demonstrated with TiN and TaN barriers [4, 7].

28.2.5 Direct Surface Activation Using Copper

Direct activation of TiN/Ti and Ta surface with copper, without using palladium, has also been reported [3, 26]. In the seeding process, Cu(II) ions in solution react, in the presence of fluoride, with metallic Ti (or Ta) on the substrate, yielding copper nuclei on the substrate. The substrate was then transferred into an electroless copper solution for building up a continuous copper film. It can be expected that because of the markedly lower oxidation ability of Cu(II) compared with that of Pd(II), the concentration of fluoride needed to make the displacement reaction between Cu(II) and Ti (or Ta) feasible would be significantly higher. As a consequence, the etching of Ti (or Ta), the substrate material, in the process may be significantly higher. The etching of the substrate material is certainly an issue that needs to be considered as barriers employed in microelectronics fabrication are very limited in thickness. A lower pH would exacerbate the etching (cf. Eqs. 28.10 and 28.11). In addition, hydrogen evolution resulting from the etching would adversely affect seeding. Indeed it was found that copper films deposited following seeding with copper in a HF solution show significantly poorer adhesion compared with those deposited after copper seeding in a NH_4F solution [26]. For the copper seeding process, ethanol and ascorbic acid were added to the seeding solution and inert gas purging were employed to reduce surface oxide formation and thus improve adhesion.

28.2.6 Electroless Copper for Gap Fill

Besides direct deposition on barriers for seeding, electroless copper deposition has been shown to be promising for bottom-up filling of features [5, 6, 10], where additives such as dipyridyl and polyethylene glycol (PEG), which suppresses copper deposition, and additives such as bis((3-sulfopropyl)disulfide (SPS) and 3-*N*,*N*-dimethylaminodithiocabamoyl-1-propanesulfonic acid (DPS), which accelerate copper deposition, were used to achieve bottom-up deposition. The approach is analogous to that in electrodeposition where the combination of these types of suppressors and accelerators are used to achieve superconformal (or bottom-up) deposition of copper. Electroless copper deposition for selectively filling of features has also been reported [3].

28.2.7 Copper Deposition Process Summary

Electroless copper depositions are generally performed at elevated temperatures. For instance, electroless copper depositions with formaldehyde or glyoxylic acid as the reducing agent were reported to be performed in the temperature range of 50–75°C [4–6, 8–11, 22, 25]. As is true with other electroless depositions, increasing temperature would boost copper deposition rate. On the other hand increasing temperature would destabilize the bath and may cause the bath to decompose spontaneously. The threshold temperature at which the onset of spontaneous decomposition occurs depends on bath composition, including copper, complexing agent, and reducing agent concentrations as well as stabilizer concentration. It also depends on the history of the bath (e.g., the unintentional introduction of particles). The presence of fine particles in an electroless bath would facilitate the onset of bath decomposition, as they act as catalytic centers for the reduction of copper ions by the reducing agent used. Lastly, it depends on the operating parameters such as agitation and bath pH.

As Eq. 28.1 indicates, the concentration of methanediolate anion depends on hydroxide concentration, or pH. For methanediolate anion to be present in sufficiently high concentration, solution pH must be at least 11. Furthermore, as shown in Eq. 28.7 the thermodynamic driving force for copper deposition becomes greater as pH increases. However, bath pH must be well controlled. When bath pH is too high ($pH > 13$), the bath becomes unstable, which may lead to spontaneous decomposition of the bath. In general, electroless copper depositions using formaldehyde as the reducing agent are performed at a pH of around 12.5.

When formaldehyde is used as the reducing agent, the oxidation product is formate. As a result, there would be no non-metal codeposition in the electroless copper deposition process, which is highly desired for interconnect applications. The same is true for the deposition using glyoxylic acid since the oxidation product of glyoxylic acid is oxalate. However, there will be *inclusions* of organic species in the resulting copper deposit. In addition, as can be seen from the copper deposition reaction (Eq. 28.7), hydrogen is one of the products of the redox reaction. The evolution of hydrogen on the substrate surface impairs the adhesion of the deposited copper with the substrate. Hydrogen inclusion also degrades the deposited copper film, causing film density to decrease and resistivity to increase. Care must be taken to prevent hydrogen inclusion. The use of surfactants in copper deposition baths would mitigate hydrogen inclusion in the deposit [3].

In general, the resistivity of the as-deposited copper film is significantly higher than the bulk resistivity of copper, making post-deposition annealing necessary. Annealing also improves the adhesion of copper to the substrate. Annealing for this purpose is typically performed in an inert atmosphere (e.g., nitrogen or argon) or in a reducing environment, e.g., a forming gas with a composition of 5% hydrogen and 95% nitrogen.

Electroless copper deposition is a promising alternative to electrodeposition for applications in microelectronics. On the other hand there are still critical technical challenges, including adhesion (to barriers in particular), deposit surface morphology, and deposit density, which need to be further addressed to make the approach a generic metallization technology for the fabrication of microelectronic devices.

28.3 Electroless Nickel Deposition

Since the work by Brenner and Riddell in 1946 [1], electroless nickel deposition has found widespread use in corrosion protection, wear resistance improvement, and metallization of nonconductive materials such as ceramics, glass, and plastics. Electroless nickel deposition has also found applications in microelectronics, particularly for contact metallurgies with which microelectronic devices are connected to printed circuit boards. For example, nickel is electrolessly deposited on aluminum prior to the deposition of palladium and/or gold to form bonding pads. Nickel is also electrolessly deposited on copper to form a diffusion barrier between copper and the subsequently deposited solder materials such as palladium, gold, tin-lead alloy, and tin-silver alloy to limit adverse intermetallic compound formation.

28.3.1 Nickel Deposition Chemistry

Electroless nickel deposition can be realized by using various reducing agents, including borohydride, DMAB, hydrazine, and hypophosphite [2, 27, 28]. The deposition processes using these reducing agents are all autocatalytic. In other words, nickel is a catalyst for the oxidation of the reducing agents and the accompanying reduction of nickel ions.

Borohydride tends to undergo fast decomposition through hydrolysis in acidic and neutral solutions. To mitigate its hydrolysis, an electroless nickel deposition with borohydride has to be carried out at highly alkaline conditions (pH ~12.5), which necessitates the use of strong complexing agents such as ethylenediamine and EDTA to keep nickel ions in solution. Since one of the oxidation products of borohydride is boron, the resulting nickel deposit is a nickel–boron (NiB) alloy, with its composition being dependent on deposition conditions. In comparison with borohydride, DMAB is a weaker reducing agent. Although DMAB decomposes in acidic conditions, it is significantly more stable than borohydride. Generally, DMAB is used for electroless nickel deposition at solution pH close to neutral, particularly in slightly alkaline conditions (pH~9). Similar to the use of borohydride, boron is codeposited with nickel in the deposition process using DMAB, resulting in a NiB alloy. The reduction of nickel ions by hydrazine, on the other hand produces a deposit of nearly pure nickel, because the oxidation products of hydrazine are nitrogen and water. Generally, deposits are 97 to 99% nickel, with the balance being oxygen, nitrogen, and other trace elements [2, 29]. Electroless nickel depositions using hydrazine are performed under alkaline conditions (pH ~11). Although electroless nickel deposition using hydrazine is capable of producing a purer nickel deposit, the process has found little practical application, as the resulting nickel deposit tends to be brittle and dark in color, often with high stress and poor corrosion resistance [27].

Hypophosphite is the most commonly used reducing agent for electroless nickel deposition. By using hypophosphite, well-behaved and stable baths can be formulated in a wide pH range, spanning from pH 4 to pH 10. Electroless deposition of nickel using hypophosphite can be generally expressed by the overall reaction (Eq. 28.12).

$$2H_2PO_2^- + Ni^{2+} + 2H_2O \rightarrow Ni + 2H_2PO_3^- + 2H^+ + H_2 \uparrow \qquad (28.12)$$

The exact mechanism of nickel deposition involving hypophosphite on a catalytic surface is still a subject that fascinates researchers. However, it is generally believed that the deposition process involves the generation of hydrogen radicals and the subsequent reduction of nickel ions by the hydrogen radicals on the catalytic surface. The partial reactions involved include the following (Eqs. 28.13, 28.14, 28.15 and 28.16),

$$H_2PO_2^- + H_2o \rightarrow H_2PO_3^- + 2H_{ads} \quad (28.13)$$

$$Ni^{2+} + 2H_{ads} \rightarrow Ni + 2H^+ \quad (28.14)$$

$$2H_{ads} \rightarrow H_2 \quad (28.15)$$

$$H_2PO_2^- + H_{ads} + H^+ \rightarrow 2H_2O + P \quad (28.16)$$

As indicated in the above partial reactions, the generation of hydrogen radicals and their subsequent reaction with hypophosphite ions lead to the generation and incorporation of phosphorous. The content of phosphorous in the resulting nickel–phosphorous (NiP) alloys ranges from 3 to 13 wt.% [27], depending on deposition conditions such as nickel concentration in solution and bath pH. Increasing nickel concentration in solution reduces phosphorous content in the deposit, whereas lowering solution pH favors the codeposition of phosphorous in the deposit. Phosphorous content in NiP deposits affects the microstructure of the deposits. The microstructure differs for different phosphorous contents, from microcrystalline at lower phosphorous contents to amorphous at higher phosphorous contents. Phosphorous content also determines the physical and chemical properties of the deposits. In general, high-phosphorous deposits offer excellent resistance to acidic corrosive environments, whereas low-phosphorous deposits are more resistant to alkaline corrosion. Compared with high-phosphorous nickel deposits, low-phosphorous nickel deposits exhibit lower electrical resistance and better solderability.

28.3.2 Substrate Activation

While electroless nickel deposition on non-conductive substrates such as dielectrics and silicon has been investigated for its use in microelectronics [30], most of its applications in this area involve copper and aluminum substrates [31, 32]. Since neither copper nor aluminum is a catalyst for the oxidation of hypophosphite and the concurrent reduction of nickel, electroless nickel deposition on both copper and aluminum substrates using hypophosphite must be preceded with surface activation of the substrates.

28.3.3 Palladium Activation of Copper Substrate

For electroless nickel deposition on copper, palladium activation is the most common approach employed. In the activation process, Pd(II) ions in solution react with metallic copper on the substrate that is in contact with the Pd(II) solution, generating palladium nuclei on the copper substrate (Eq. 28.17).

$$Pd^{2+} + Cu \rightarrow Cu^{2+} + Pd \tag{28.17}$$

The resultant palladium nuclei on the copper substrate serve as catalytic centers where hypophosphite ions reduce nickel ions through dehydrogenation (Eqs. 28.13 and 28.14). Once nickel is deposited on the substrate the deposition process proceeds autocatalytically, with film thickness being controllable by deposition conditions such as deposition time, temperature, and bath composition.

28.3.4 Self-Activation of Copper Substrate

In contrast to hypophosphite, DMAB is able to reduce nickel ions directly on copper. In other words, copper is a good catalyst for the reduction of nickel ions by DMAB. The resultant deposit is NiB alloy. Once the copper substrate under investigation is seeded with nickel (or NiB alloy), the reduction of nickel ions by DMAB on the substrate proceeds autocatalytically. Alternatively, the NiB seeded copper substrate is further deposited with NiP alloy in a hypophosphite-based electroless nickel bath [33]. The approach of self-activation using DMAB-based electroless nickel solutions is particularly beneficial in situations where the incorporation of palladium needs to be avoided.

28.3.5 Activation of Aluminum Substrates

For electroless deposition of nickel on aluminum, because of the strong tendency of aluminum to form a dense oxide layer on its surface, it is essential to remove the oxide layer and prevent the re-oxidation of the cleaned surface prior to electroless nickel deposition. Various methods have been proposed for this purpose, including both zincate-free and zincate-based processes [27, 32, 34–38]. The general approach of zincate-free process is to mitigate the reoxidation of the aluminum surface by using oxygen-deficient media (organic solvent or de-aerated water) to rinse after surface cleaning and protect the cleaned aluminum surface. The displacement reaction between Ni(II) ions and metallic aluminum is then utilized to seed the aluminum surface with nickel, which catalyzes the subsequent deposition of nickel in a continuous manner. For instance, an aluminum surface was first etched in a 3% HF solution to remove surface oxide. The substrate was then rinsed with de-aerated water that contains less than 2 ppm oxygen. The rinsed substrate was quickly transferred into a Ni(II)-containing solution for seeding with nickel, which was then followed by electroless deposition of nickel in another solution containing both nickel and hypophosphite [32]. In another case, a mixture of NH_4OH and H_2O_2 was used to etch aluminum surface. Following the rinse with isopropanol, the substrate was transferred directly into an electroless nickel deposition solution where the seeding of nickel and the subsequent autocatalytic deposition of NiP alloy took place [34].

Zincation processes, however, are certainly the prevailing approach for this purpose, as zincation of aluminum surfaces provides a protective layer that may be exposed to air and oxygen-containing solutions during processing and is able to react with Ni(II) in solution to initiate autocatalytic deposition of nickel. Because of the effective protection of the cleaned aluminum surface, electroless nickel deposits of high adhesion to aluminum substrates could be obtained. A variety of zincation solutions that all contain Zn(II) species but various auxiliary components at different pH values have been proposed for this purpose [27, 35–38]. Among them, highly alkaline zincation solutions, where zinc exists in the form of zincate, $Zn(OH)_4{}^{2-}$ (or $ZnO_2{}^{2-}$), are the most widely used. This type of zincation solution is generally made from zinc oxide and sodium hydroxide.

A zincation process using a highly alkaline zincate solution removes aluminum oxide by forming aluminate through the reaction of hydroxide ions with aluminum oxide (Eq. 28.18).

$$Al_2O_3 \cdot xH_2O + 2OH^- \rightarrow 2AlO_2{}^- + (x + 1)\,H_2O \qquad (28.18)$$

The zincate ions in the solution then react with aluminum in accordance with the following simplified overall reaction (Eq. 28.19), depositing a thin layer of zinc on the aluminum substrate.

$$3ZnO_2{}^{2-} + 2Al + 2H_2O \rightarrow 2AlO_2{}^- + 3Zn + 4OH^- \qquad (28.19)$$

The rate of the displacement reaction is determined by the composition of the zincate solution, the type of the aluminum substrate, and the operating conditions such as temperature. In the presence of oxygen, which is usually the situation, metallic aluminum is also etched by the highly alkaline zincate solution through the oxidation of aluminum by oxygen. As a result, excessive zincation should be avoided to mitigate aluminum etching, particularly when the thickness of the aluminum film involved is limited and/or undercutting at locations where another material overlaps the aluminum must be minimized.

The zinc film formed on aluminum not only protects aluminum from being oxidized but also serves as a sacrificial seed layer for electroless nickel deposition that follows. When a zincated aluminum substrate is introduced into an electroless nickel bath at operating conditions (e.g., at an elevated temperature) a displacement reaction between Ni(II) ions in solution and zinc on the substrate takes place immediately, yielding nickel nuclei that concurrently initiate autocatalytic deposition of nickel on the substrate.

To ensure uniform seeding of zinc, a multiple-zincation process is usually preferred over single-zincation processes. In a multiple-zincation process, the aluminum substrate zincated in the previous zincation step is de-smutted in a nitric acid solution (e.g., 30% nitric acid by volume) to strip off granular zinc deposits and then re-zincated in the zincation solution prior to electroless nickel deposition. The process significantly improves the uniformity of zinc seeding and thus the uniformity and surface morphology of the subsequent nickel deposit, even when the

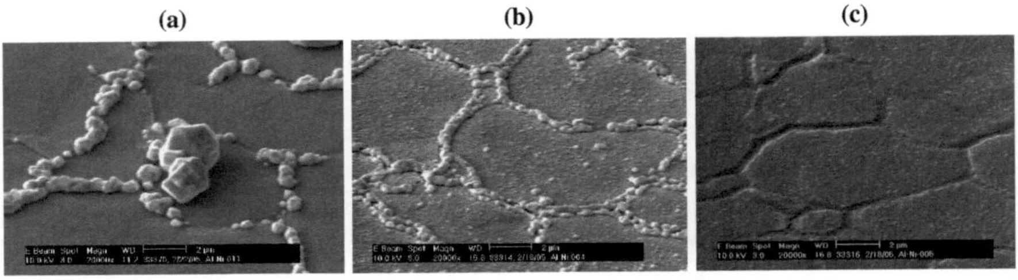

Fig. 28.1 SEM micrographs of aluminum surface after (**a**) single zincation for 30-s, (**b**) double zincation: 15-s first zincation, 30-s de-smutting in a nitric acid solution, and 15-s second zincation, and (**c**) triple zincation: 10-s for each zincation, 30-s de-smutting after the first and second zincation. Zincation and de-smutting at room temperature

zincation time used in a single-zincation process equals the total zincation time used in a multiple-zincation process.

The effect of zincation on the microscopic uniformity of zinc seeding and the surface morphology of the resulting nickel deposit can be seen from Figs. 28.1 and 28.2, respectively. The SEM micrographs shown in Fig. 28.1 were obtained with aluminum pads on wafers following (A) a single-zincation process, (B) a double-zincation process, and (C) a triple-zincation process, with the total zincation time being the same. The process conditions are detailed in the caption of the figure. As is evident, a multiple-zincation process significantly improves the microscopic seeding uniformity, which would impact the result of the subsequent electroless nickel deposition. Shown in Fig. 28.2 are the SEM micrographs observed after electroless nickel deposition on the aluminum pads that were zincated (A) before the optimization of zincation process conditions and (B) after the optimization of zincation process conditions, with the same electroless nickel deposition process. It is expected that uniform seeding is especially important for applications where thin

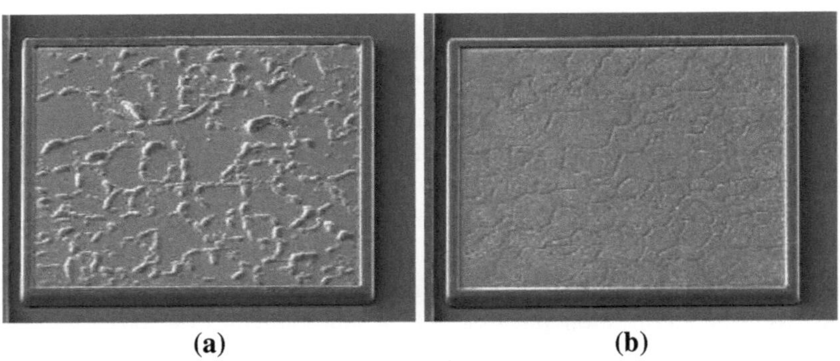

Fig. 28.2 SEM micrographs showing the surface morphology of the electroless nickel deposits on aluminum pads obtained at identical nickel deposition condition following zincation under different conditions. (**a**) Prior to zincation process optimization and (**b**) after zincation process optimization

nickel deposits with smooth surface morphology are required as the impact of seeding uniformity on the subsequent nickel deposits is particularly significant when the thickness of the nickel deposits is very limited.

28.3.6 *Nickel Deposition Process Summary*

Generally, electroless deposition only occurs at elevated temperatures. For electroless nickel deposition with hypophosphite as reducing agent, the deposition is usually performed at 60–90°C. The rate of the deposition increases exponentially with the increase in temperature, as is true with other electroless depositions. Increasing temperature appears to be the easiest way to meet the requirement for a higher deposition rate. However, as previously discussed increasing temperature too much would destabilize the bath, which may trigger spontaneous decomposition of the bath.

The stability of an electroless nickel bath can be markedly enhanced by the addition of a small amount of stabilizers or suppressors. Stabilizers are chemical species that preferentially adsorb on the catalytic surfaces of particles in solution, thus preventing the homogenous reaction that triggers the spontaneous decomposition. Stabilizers also preferentially adsorb at the catalytic centers on the substrate onto which nickel is to be deposited leading to the suppression of electroless deposition on the substrate. Stabilizers such as Pb(II) ions and thiourea are used to stabilize electroless nickel baths and improve deposit surface morphology. By judicially choosing the chemical nature of stabilizers and their concentrations, the deposition behavior of electroless nickel baths can be well controlled and the spontaneous decomposition of the baths can be virtually eliminated.

28.4 Electroless Gold Deposition

In addition to electroless copper and nickel, a variety of other metals and their alloys can be deposited using electroless deposition processes, including gold that is primarily used for contact metallurgy. Electroless gold deposition is commonly used in electronic packaging applications [39–43]. For instance, following electroless nickel deposition on copper or aluminum substrates, gold is usually deposited as the final finish of the contact metallurgy to provide a low contact-resistance surface and enhance solderability.

28.4.1 *Cyanide-Based Gold Deposition*

Because of the high nobility of gold, aqueous solutions of its ions can be obtained only by using complexing agents. In other words, to form a gold bath, one or more complexing agents must be employed to stabilize gold ions in solution. Cyanide forms strong complexes with gold ions. The stability constant (or formation constant) of Au(I)–cyanide complex, $Au(CN)_2^-$, is 10^{38}, which is at least one order of magnitude higher than those of the gold complexes formed with other complexing agents such as thiosulfate (10^{26}) and sulfite (10^{10}). As a result, cyanide has been traditionally used for electrolytic and electroless deposition of gold.

28.4.2 Immersion Deposition

Cyanide-based gold baths can be used to deposit gold on metals (e.g., nickel) and metal alloys (e.g., NiP and NiB) through a galvanic displacement reaction between gold ions in solution and the metal or metal alloy under investigation. The deposition process is called an immersion process. In this case, the electrons needed to reduce gold ions, e.g., Au(I) species, in solution are provided by the oxidation of the metal. A typical electroless gold bath for this purpose is made from $KAu(CN)_2$, which is the source of gold and KCN, which functions as a stabilizer. In the case that the substrate metal is nickel, the displacement reaction is represented by Eq. 28.20.

$$2Au\,(CN)_2^- + Ni + 2CN^- \rightarrow 2Au + Ni\,(CN)_4^- \qquad (28.20)$$

The deposition process is self-limiting. In other words, the deposition rate slows down with the advance of the deposition. Once all the available nickel on the substrate is covered with gold the deposition process ceases. The maximum thickness that can be achieved with the immersion process is about 0.3 μm [41, 43]. In addition, as the deposition is based on metal displacement, the process does not always produce pore-free deposits.

28.4.3 Autocatalytic Deposition

A variety of reducing agents such as hypophosphite, hydroxylamine, thiourea, borohydride, and DMAB have been investigated and reported in the literature for formulating electroless gold baths in the presence of cyanide [40–42, 44]. While there are some questions as to whether the deposition processes using the above reducing agents can all be categorized as autocatalytic and the formulated baths are stable enough for practical use, the electroless gold deposition systems based on the reducing agents borohydride and DMAB are autocatalytic and have found practical applications.

Borohydride and DMAB can autocatalytically reduce gold ions on a gold substrate, depositing more gold on the substrate. Such an autocatalytic gold deposition also takes place on a substrate of other metals such as nickel and copper, since gold nuclei will be generated on the substrate through a displacement reaction once the substrate is brought in contact with the gold deposition solution at operating temperature. It is believed that borohydride functions as the reducing agent by first producing BH_3OH^- through a hydrolysis reaction (Eq. 28.21) and DMAB as the reducing agent by generating the same intermediate BH_3OH^- through a substitution reaction (Eq. 28.22) [41, 45, 46].

$$BH_4^- + H_2O \rightarrow BH_3OH^- + H_2 \uparrow \quad (28.21)$$

$$(CH_3)_2NH \cdot BH_3 + OH^- \rightarrow BH_3OH^- + (CH_3)_2NH \quad (28.22)$$

The resultant BH_3OH^- ions then reduce gold ions in accordance with the following reaction (Eq. 28.23).

$$BH_3OH^- + 3Au(CN)_2^- + 3OH^- \rightarrow 3Au + 6CN^- + BO_2^- + 3/2H_2 \uparrow + 2H_2O \quad (28.23)$$

Electroless gold depositions using either borohydride or DMAB as the reducing agent are generally carried out at highly alkaline conditions (pH >13). Borohydride is a powerful reducing agent. At highly alkaline conditions, the reactivity of borohydride is moderated. More importantly, the hydrolysis decomposition of borohydride is mitigated at highly alkaline conditions. For DMAB, increasing pH minimizes its hydrolysis decomposition and in the mean time, increases its reducing ability. In other words, the thermodynamic driving force for the reduction of gold ions by DMAB becomes greater as pH increases (Eqs. 28.22 and 28.23). In a highly alkaline solution, the concentration of hydroxide is substantial and the pH of the solution is relatively stable. However, care should still be taken to monitor and control bath pH, since hydroxide ions in solution are not only consumed by the deposition reaction (Eq. 28.23) but also by the dissolving of carbon dioxide from air into the bath. Carbon dioxide is readily soluble in highly alkaline solutions, forming carbonate by consuming hydroxide.

Although gold–cyanide complexes are highly stable at alkaline conditions, the autocatalytic baths are per se thermodynamically unstable. The compromise in the stability of the baths may be exacerbated by the introduction of impurities, e.g., nickel and cobalt ions. It has been noted that even though the introduction of copper ions caused no adverse effect, the introduction of nickel and cobalt ions into the borohydride-based gold bath is highly detrimental, causing the bath to spontaneously decompose even when the impurities are present at a concentration as low as 0.1 M.$^-$ As a result, the autocatalytic baths are not suitable for electroless gold deposition directly on substrates of these metals. Striking with a galvanic displacement (immersion) gold bath prior to the autocatalytic deposition can be used to remedy the problem and extend the useful life of the autocatalytic gold baths. The addition of some complexing agents in the autocatalytic gold baths would alleviate the problem as well. Organic complexing agents such as EDTA and ethylenediamine that complex strongly with metal contaminants are found to be effective in stabilizing these baths.

To improve deposit surface morphology, inorganic grain refiners such as Pb(II) and thallous ions are commonly used [40, 41]. The presence of trace amounts of the grain refining species that function as depolarizers also increases deposition rate. By using either borohydride or DMAB as the reducing agent, soft gold deposits of high purity, suitable for bonding and soldering, have been obtained.

28.4.4 Substrate (Nickel)-Catalyzed Deposition

In addition to the aforementioned galvanic displacement deposition and autocatalytic deposition, substrate (nickel)-catalyzed deposition has also been reported [47–49]. This is a process that lies in between the autocatalytic deposition process and the displacement deposition process and is exclusively applicable to nickel substrates. In this deposition process, hydrazine is used as the reducing agent. Unlike borohydride and DMAB, the anodic oxidation of hydrazine takes place only on the nickel substrate; the deposition of gold ceases once the nickel substrate is completely covered with a layer of gold. Under the catalysis of nickel, the overall redox reaction can be written as:

$$N_2H_4 + 4Au\,(CN)_2^{\;-} + 4OH^- \rightarrow 4Au + N_2 \uparrow + 8CN^- + 4H_2O \qquad (28.24)$$

By using the substrate-catalyzed deposition process, uniform and compact gold deposits on nickel substrates could be obtained and the corrosion of nickel substrates (e.g., NiP and NiB) common in displacement-based gold deposition processes could be mitigated. On the other handthe maximum obtainable thickness of gold deposit with the process is intrinsically limited. It was reported that with the substrate-catalyzed process the obtainable thickness of gold deposit was limited to $\sim 0.7\ \mu m$ [49]. In addition, it was found that such a substrate-catalyzed process yields different results on NiB and NiP substrates: on NiB substrates the process produced uniform and adherent gold deposits, whereas on NiP substrates acceptable gold deposits were obtained only when the phosphorous content in the substrate was low and the surface of the substrate was pretreated to have the surface oxide removed [48].

28.4.5 Non-cyanide Gold Deposition

Because of the toxicity of cyanide and its incompatibility with photoresists used to delineate circuit patterns and bonding pads, non-cyanide gold electroless deposition is highly desired. A great deal of effort has been made to develop non-cyanide electroless gold deposition processes [2, 41, 50–53]. Sulfite forms Au(I) complex sufficiently stable for electrodeposition of gold. Sulfite has also been investigated for electroless gold deposition with the use of such reducing agents as hypophosphite, formaldehyde, hydrazine, borohydride, and DMAB. In addition, chloride was used to complex Au(III) ions for electroless gold deposition using borane reducing agents [2]. EDTA was used to complex Au(I) ions at a pH close to neutral for electroless gold deposition using polyphenols [52]. In general, the gold electroless baths formed with the above-mentioned non-cyanide complexing agents and reducing agents are not as stable as their cyanide-based counterparts.

Electroless gold deposition in the presence of both thiosulfate and sulfite has been investigated [50–51]. In this deposition system, thiosulfate functions as the complexing agent, whereas sulfite acts as both complexing agent and reducing agent. On NiB substrates, the deposition proceeds through the combination of galvanic displacement deposition and substrate-catalyzed deposition, with the latter playing a major role. Thiosulfate forms a stronger complex with Au(I) ions than sulfite does. However, the presence of thiosulfate would lead to the codeposition of sulfur, which is the decomposition product of thiosulfate [54, 55], making the deposited gold harder. Soft gold is desired for bonding applications in microelectronics.

Due to the limited complexing ability of non-cyanide complexing agents, the stability of non-cyanide electroless gold baths, in which traditional reducing agents are used to provide electrons for the reduction of gold is fundamentally limited. This certainly restricts their use for practical applications. There continues to be a substantial effort to develop non-cyanide autocatalytic electroless gold deposition processes. At present, galvanic displacement-based deposition is the primary means for non-cyanide electroless gold deposition.

28.5 Electroless Deposition Equipment

The equipment used for electroless deposition processes typically consists of immersion reactors, with attention paid to flow-control or solution handling and process sequencing. Critical factors affecting electroless deposition and the design of reactors for this process include temperature control and uniformity, flow uniformity, and equipment configuration complexity. It is extremely important to maintain elevated temperatures (as high as 90°C) without producing localized hot spots that may lead to autocatalytic deposition on the reactor components. Because of the inherent instability of electroless deposition solutions it is also extremely important to pay close attention to the reactor and plumbing layout within the process tool. Subtle changes in the configuration can have dramatic effects on process and chemistry stability. This instability has even led some to consider adding reducing agents immediately before use and designing systems for single use of the deposition chemistry. Electroless deposition process sequences typically consist of several cleaning, activation, and deposition steps; therefore, the issues described in the equipment design section relating to system configuration and throughput are important [2, 56–65].

Electroless deposition may be performed in equipment ranging from simple batch immersion systems to application-specific single wafer processing systems designed for a single electroless deposition process. In addition to the process-specific design criteria mentioned above, electroless deposition equipment is oftentimes designed with fluid flow components to handle chemistries for cleaning or etching of the primary fluid flow components in the case of plate-out of the metals from the solutions. This adds to equipment complexity and cost.

28.6 Biographical Sketches

T. Ritzdorf *Semitool, Inc., Kalispell, MT 59901 (tritzdorf@semitool.com).* Mr. Ritzdorf is the Director of ECD Technology at Semitool, Inc. in Kalispell, MT. He received a B.S. in Chemical Engineering from Montana State University in 1986, and a M.S. in Chemical Engineering from the University of Minnesota in 1989. Mr. Ritzdorf spent 6 years working in the magnetic recording industry at Control Data, Seagate, and Storage Technology. He is an author or coauthor of at least 20 patents and over 30 technical papers, including chapters for two books on electrochemical processing. Mr. Ritzdorf is a member of the Electrochemical Society and the American Electroplaters and Surface Finishers Society.

Zhongmin Hu *Semitool, Inc., Kalispell, MT 59901 (zhu@semitool.com).* Dr. Hu is a Sr. Process Development Engineer in ECD Technology at Semitool, Inc. in Kalispell, MT. He received a B.S. in Chemistry from Zhejiang University of Technology in 1982, a M.S. in Electrochemistry from Shanghai University in 1987, and a Ph.D. from the University of Cincinnati in 1998. Before joining in Semitool, Inc., he spent two years working as a postdoctorate research associate in Ames Laboratory of DOE at Iowa State University. He is an author or coauthor of several patents and more than 30 peer-reviewed publications. Hu is a member of American Chemical Society and the Electrochemical Society.

References

1. Brenner, A. and Riddell, G. E.: Temperature coefficients for proving rings. J. Res. Natl. Bur. Stan. **37**, 31 (1946)
2. Mallory, G. O. and Hajdu, J. B.: Electroless Plating: Fundamentals and Applications, American Electroplaters and Surface Finishers Society, Orlando, FL (1990)
3. Dubin, V. M.; Shacham-DiamandY.; Zhao, B.; Vasudev, P. K.; and Ting, C. H.: Selective and blanket electroless copper deposition for ultralarge scale integration. J. Electrochem. Soc. **144(3)**, 898 (1997)
4. Hsu, H.-H.; Lin, K.-H.; Lin, S.-J.; and Yeh, J.-W.: Electroless Copper Deposition for Ultralarge-Scale Integration. J. Electrochem. Soc. **148(1)**, C47 (2001)
5. Lee, C. H.; Lee, S. C.; and Kim, J. J.: Improvement of Electrolessly Gap-Filled Cu Using 2,2'-Dipyridyl and Bis-(3-sulfopropyl)-disulfide (SPS). Electrochem. Solid-State Lett. **8**, C110 (2005)
6. Lee, C. H.; Cho, S. K.; and Kim, J. J.: Electroless Cu bottom-up filling using 3-N,N-dlmethylaminodithiocarbamoyl-1-propanesulfonic acid. Electrochem. Solid-State Lett. **8(11)**, J27 (2005)
7. Hsu, H.-H.; Teng, C.-W.; Lin, S.-J.; and Yeh, J.-W.: Sn/Pd catalyzation and electroless Cu deposition on TaN diffusion barrier layers. J. Electrochem. Soc. **149(3)**, C143 (2002)
8. Kim, Y.-S.; Bae, D.-L.; Yang, H.; Shin, H.-S.; Wang, G.-W.; Senkevich, J. J.; and Lu, T.-M.: Direct copper electroless deposition on a tungsten barrier layer for ultralarge scale integration. J. Electrochem. Soc. **152(2)**, C89 (2005)

9. Shacham-DiamandY.: Electroless copper deposition using glyoxylic acid as reducing agent for ultra-large-scale-integration metallization. Electrochem. Solid-State Lett. **3(6)**, 279 (2000)
10. Shingubara, S.; Wang, Z.; Yaegashi, O.; Obata, R.; Sakaue, H.; and Takahagi, T.: Bottom-up fill of copper in deep submicrometer holes by electroless plating. Electrochem. Solid-State Lett. **7(6)**, C78 (2004)
11. Wang, Z.; Ida, T.; Sakaue, H.; Shingubara, S.; and Takahagi, T.: Electroless plating of copper on metal-nitride diffusion barriers initiated by displacement plating. Electrochem. Solid-State Lett. **6(3)**, C38 (2003)
12. Honma, H. and Kobayashi, T.: Electroless copper deposition process using glyoxylic acid as a reducing agent. J. Electrochem. Soc. **141(3)**, 730 (1994)
13. Li, J. and Kohl, P. A.: The acceleration of nonformaldehyde electroless copper plating. J. Electrochem. Soc. **149(12)**, C631 (2002)
14. Li, J. and Kohl, P. A.: The deposition characteristics of accelerated nonformaldehyde electroless copper plating. J. Electrochem. Soc. **150(8)**, C558 (2003)
15. Schumacher, R.; Pesek, J. J.; and Melroy, O. R.: Kinetic analysis of electroless deposition of copper. J. Phys. Chem. **89(20)**, 4338 (1985)
16. Wiese, H. and Weil, K. G.: On the mechanism of electroless copper deposition. Ber. Bunsenges. Phys. Chem. **91**, 619 (1987)
17. Weber, C. J.; Pickering, H. W.; and Weil, K. G.: Surface development during electroless copper deposition. J. Electrochem. Soc. **144(7)**, 2364 (1997)
18. Bindra, P. and Roldan, J.: Mechanisms of electroless metal plating. J. Electrochem. Soc. **132(11)**, 2581 (1985)
19. Dubin, V. M.: Selective electroless Ni-Cu(P) deposition for via hole filling and conductor pattern cladding in VLSI multilevel interconnection structures. J. Electrochem. Soc. **139(2)**, 633 (1992)
20. Hsu, H.-H.; Hsieh, C.-C.; Chen, M.-H.; Lin, S.-J.; and Yeh, J.-W.: Displacement activation of tantalum diffusion barrier layer for electroless copper deposition. J. Electrochem. Soc. **148(9)**, C590 (2001)
21. Hong, S. W.; Shin, C.-H.; and Park, J.-W.: Palladium activation on TaN_x barrier films for autocatalytic electroless copper deposition. J. Electrochem. Soc. **149(1)**, G85 (2002)
22. Hong, S. W. and Park, J.-W.: Effect of nitrogen content in TaN_x ($x = 0$–1) barrier substrates on electroless copper deposition. Electrochem. Solid-State Lett. **5(12)**, C107 (2002)
23. Hong, S. W.; Lee, Y. S.; Park, K.-C.; and Park, J.-W.: Nucleation and Growth of Electroless Palladium Deposition on Polycrystalline TiN Barrier films for Electroless Copper Deposition. J. Electrochem. Soc. **150(1)**, C16 (2003)
24. Hsu, H.-H.; Yeh, J.-W.; and Lin, S.-J.: Repeated 3D nucleation in electroless Cu deposition and the grain boundary structure involved. J. Electrochem. Soc. **150(11)**, C813 (2003)
25. Oh, Y.-J.; Cho, S. M.; and Chung, C.-H.: Control of topographical selectivity in palladium-activated electroless copper metallization. Electrochem. Solid-State Lett. **8(1)**, C1 (2005)
26. Wang, Z.; Li, H.; Shodiev, H.; and Suni, I. I.: Immersion/electroless deposition of Cu onto Ta. Electrochem. Solid-State Lett. **7**, C67 (2004)
27. Riedel, W.: Electroless Nickel Plating, Finishing Publications, Stevenge, England (1991)
28. Kohl, P. A.: In Modern Electroplating, 4th edition. Schlesinger, M.; and Paunovic, M., Eds. John Willey & Sons, New York (2000)
29. Djokic, S. S.: Electroless Deposition of Cobalt Using Hydrazine as a Reducing Agent. J. Electrochem. Soc. **144(7)**, 2358 (1997)
30. Takano, N.; Hosoda, N.; Yamada, T.; and Osaka, T.: Mechanism of the Chemical Deposition of Nickel on Silicon Wafers in Aqueous Solution. J. Electrochem. Soc. **146(4)**, 1407 (1999)
31. Lin, K.-L. and Wu, C.-H.: Structural evolution of electroless nickel bump. J. Electrochem. Soc. **150(5)**, C273 (2003)
32. Watanabe, H. and Honma, H.: Fabrication of nickel microbump on aluminum using electroless nickel plating. J. Electrochem. Soc. **144(2)**, 471 (1997)
33. Xu, H.; Brito, J.; and Sadik, O. A.: Mechanism of stabilizer acceleration in electroless nickel at wirebond substrates. J. Electrochem. Soc. **150(11)**, C816 (2003)

34. Rohan, J. F.; Murphy, P. A.; and Barrett, J.: Zincate-free, electroless nickel deposition on aluminum bond pads. J. Electrochem. Soc. **152(1)**, C32 (2005)
35. Heiman, J.: Deposition of metals on aluminum by immersion from solutions containing fluorides. J. Electrochem. Soc. **95(5)**, 205 (1949)
36. Lashmore, D. S.: Immersion coatings on aluminum. Plat. Surf. Finish. **67(1)**, 37 (1980)
37. Datta, M.; Merritt, S. A.; and Dagenais, M.: Electroless remetallization of aluminum bond pads on CMOS driver chip for flip-chip attachment to vertical cavity surface emitting lasers (VCSEL's). IEEE Transactions on Components and Packaging Technology **22**, 299 (1999)
38. Wei-Chin, N.; Tze-Man, K.; Chen, W.; and Guo-Jun, Q.: The Effects of Immersion Zincation to the Electroless Nickel Under-bup Materials in Microelectronic Packaging. 1998 IEEE/CPMT Electronics Packaging Technology Conference 89 (1998)
39. Harman, G. G.: Wire Bonding in Microelectronics. McGraw-Hill, New York (1997).
40. Ali, H. O. and Christie, I. R. A.: A review of electroless deposition processes. Gold. Bull. **17(4)**, 118 (1984)
41. Simon, F.: Deposition of gold without external current source. Gold. Bull. **26**, 14 (1993)
42. Sargent, A. and Sadik, O. A.: Probing the mechanism of electroless gold plating using an EQCM: II. Effect of bath additives on interfacial plating processes. J. Electrochem. Soc. **148(6)**, C413 (2001)
43. Krasopoulos, A. V.; Li, J.; Josowicz, M.; and Janata, J.: Rapid substitution of gold for aluminum metallization on integrated circuits. J. Electrochem. Soc. **144(3)**, 1070 (1997)
44. Sargent, A.; Sadik, O. A.; and Matienzo, L. J.: Probing the mechanism of electroless gold plating using an electrochemical quartz crystal microbalance I. Elucidating the nature of reactive intermediates in dimethylamine borane. J. Electrochem. Soc. **148(4)**, C257 (2001)
45. Okinaka, Y.: Electroless gold deposition using borohydride of dimethylamine borane as reducing agent. Plating **57(9)**, 914 (1970)
46. Okinaka, Y.: An electrochemical study of electroless gold-deposition reaction. J. Electrochem. Soc. **120(6)**, 739 (1973)
47. Iacovangelo, C. D. and Zarnoch, K. P.: Substrate-catalyzed electroless gold plating. J. Electrochem. Soc. **138(4)**, 983 (1991)
48. Osaka, T.; Misato, T.; Sato, J.; Akiya, H.; Homma, T.; Kato, M.; Okinaka, Y.; and Yoshioka, O.: Evaluation of substrate (Ni)-Catalyzed electroless gold plating process. J. Electrochem. Soc. **147(3)**, 1059 (2000)
49. Shaigan, N.; Ashrafizadeh, S. N.; Bafghi, M. S. H.; and Rastegari, S.: Elimination of the corrosion of Ni-P substrates during electroless gold plating. J. Electrochem. Soc. **152(4)**, C173 (2005)
50. Sato, J.; Kato, M.; Otani, H.; Homma, T.; Okinaka, Y.; Osaka, T.; and Yoshioka, O.: Substrate (Ni)-Catalyzed electroless gold deposition from a noncyanide bath containing thiosulfate and sulfite. J. Electrochem. Soc. **149(3)**, C168 (2002)
51. Kato, M.; Sato, J.; Otani, H.; Homma, T.; Okinaka, Y.; Osaka, T.; and oshioka, O.: Substrate (Ni)-Catalyzed electroless gold deposition from a noncyanide bath containing thiosulfate and sulfite. J. Electrochem. Soc. **149(3)**, C164 (2002)
52. Ohtani, Y.; Horiuchi, A.; Yamaguchi, A.; Oyaizu, K.; and Yuasa, M.: Non-Cyanide electroless gold plating using polyphenols as reducing agents. J. Electrochem. Soc. **153(1)**, C63 (2006)
53. Sato, Y.; Osawa, T.; Kaieda, K.; and Kobayakawa, K.: Cyanide-free electroless gold plating from a bath containing disulfitoaurate and thiourea of its derivatives. Plat. Surf. Finish. **81(9)**, 74 (1994)
54. Osaka, T.; Kodera, A.; Misato, T.; Homma, T.; Okinaka, Y.; and Yoshioka, O.: Electrodeposition of soft gold from a thiosulfate-sulfite bath for electronics applications. J. Electrochem. Soc. **144(10)**, 3462 (1997)
55. Osaka, T.; Kato, M.; Sato, J.; Yoshizawa, K.; Homma, T.; Okinaka, Y.; and Yoshioka, O.: Mechanism of sulfur inclusion in soft gold electrodeposited from the thiosulfate-sulfite bath. J. Electrochem. Soc. **148(10)**, C659 (2001)
56. Lopatin, S.; Shacham-DiamandY.; Dubin, V.; and Vasudev, P. K.: Selective Electroless CoWP Deposition onto Pd-Activated In-Laid Cu Lines, 1997 VMIC, 219 (1997)

57. O'Sullivan, E. J.; Schrott, A. G.; Paunovic, M.; Sambucetti, C. J.; Marino, J. R.; Bailey, P. J.; Kaja, S.; and Semkow, K. W.: Electrolessly deposited diffusion barriers for microelectronics. IBM J. Res. Dev. **42(5)**, 607 (1998)
58. Itabashi, T.; Nakano, H.; and Akahoshi, H.: Electroless Deposited CoWB for Copper Diffusion Barrier Metal, IITC (2002)
59. Petrov, N.; Sverdlov, Y.; and Shacham-DiamandY.: Electrochemical Study of the Electroless Deposition of Co(P) and Co(W,P) Alloys, JECS **149(4)**, C187 (2002)
60. Aly, I. H. M.; Younan, M. M.; and Nageeb, M. T.: Autocatalytic (Electroless) Deposition of Ternary Nickel-Cobalt-Phosphorus Alloy, Plat. Surf. Fin. 37 (2003)
61. Wirth, A.; Mourier, T.; Turek, P.; Mayer, D.; and Moussavi, M.: Evaluation of Novel Electrolessly Deposited Diffusion Barriers for Copper Interconnects, IITC? JECS?
62. Lopatin, S.; Shacham-DiamandY.; Dubin, V.; Vasudev, P. K.; Kim, Y.; and Smy, T.: Characterization of electroless Cu, Co, Ni and their alloys for ULSI metallization, MRS Conf. Proc. ULSI XIII 437 (1998)
63. Shacham-DiamandY.; and Sverdlov, Y.: Multi-layer deposition of electroless copper, nickel, cobalt and their alloys on silicon for MEMS and ULSI applications. MRS Conf. Proc. ULSI XIV 103 (1999)
64. Min, W. S.; Lantasov, Y.; Palmans, R.; Maex, K.; and Lee, D. N.: The Formation of Pd Seeded Copper Layer on TiN Substrates by Electroless Deposition. Advanced Metallization Conference in 1998 (AMC 1998) Materials Research Soc. (1999)
65. Maex, K.; Brongersma, S. H.; Lantasov, Y.; RichardE.; Palmans, R.; and Vervoort, I.: Integration of Electroless and Electrolytic Cu in the IC Back End of Line Technologies. Electrochemical Technology Applications in Electronics III. Madore, C.; Osaka, T.; Romankiw, L. T.; and Yamazaki Y., Eds. **PV 99(34)**, 71 (2000)

Chapter 29
Tools for Monitoring and Control of Bath Components

T. Ritzdorf

29.1 Introduction

Maintaining constant chemical constituent concentrations in electrolytes is critical to ensure consistent process results in most deposition processes. In order to maintain consistent bath concentrations, it is important to be able to monitor the concentration of each constituent and to react to incorrect concentrations. The most efficient way to eliminate variation due to human factors and ensure consistent control is to automate the chemical management process [1].

The electrolyte components can be divided into organic and inorganic components. The concentration of the metal ions that are being deposited in an electrolytic process can be maintained either by using consumable anodes or by dosing concentrates into the electrolyte bath. The depletion rate of these metal ions can typically be calculated straightforwardly through an understanding of how much metal is deposited on the substrates being processed. Organic components are typically added to the electrolyte to modify the properties (composition, step coverage, grain size, etc.) of the deposited film. The concentration analysis of these components can be done either based on the amount of particular chemical species present or on the impact that these species have on the activity of the solution. These organic components tend to have decomposition rates that are dependent on applied potential (or open area, for electroless processes) and elapsed time (bath age). In order to effectively replenish components and maintain consistent concentrations it is important to understand which parameters affect the concentration of each species in the bath. Other components of the electrolyte (complexing agents, inorganic additives, antioxidants, etc.) may decrease concentration due to incorporation into a deposited film, oxidation at the anode surface, or reduction at the cathode surface. Whatever is the cause of concentration change, its rate must be understood in order to replenish the appropriate constituent correctly.

T. Ritzdorf (✉)
Semitool Inc., 655 W. Reserve Dr., Kalispell, MT 59901, USA
e-mail: tritzdorf@semitool.com

Y. Shacham-Diamand et al. (eds.), *Advanced Nanoscale ULSI Interconnects: Fundamentals and Applications*, DOI 10.1007/978-0-387-95868-2_29,

In order to build a piece of equipment that is capable of closed-loop chemical concentration control it is necessary to have the ability to analyze the concentrations of the chemical constituents to be controlled, as well as to add them to the process chemistry. Dosing is normally based on the applied charge (amp-minutes), elapsed time, and analysis results. The system must also be able to perform automatic analysis, including sample extraction if necessary, of all critical bath components. The system controller must be able to adjust the replenishment rates based on the analysis results. More advanced systems also automatically adjust the replenishment rates for the dosing based on accumulated charge and elapsed time.

29.2 Chemical Constituent Analysis

Many methods have been developed for analyzing chemical constituents of an electrolyte. Automated analysis systems that are part of a closed-loop bath control system must require little human support in order to produce repeatable analysis results, and should be small, cheap, and use little bath sample for analysis. It is desirable to configure automated analyzers to run as often as possible to obtain close to real-time concentration data, so it is important that the analytical technique does not require large amounts of chemistry. It is also important to design the overall system to minimize the amount of chemistry that is required for flushing fluid flow components and delivering fresh chemistry to the analyzers. Components that change concentration more rapidly, or are more critical to the process, are typically analyzed more frequently than other components in order to optimize the collection of analytical results based on the amount of electrolyte consumed. The quality and quantity of analytical data that can be obtained from an automated system must be balanced against the cost of the particular technique, in initial capital equipment, operations, and in the product at risk between analyses.

Methods that have been used or considered for automated plating-bath analysis include titration (potentiometric, colorimetric, or pH), HPLC (high-performance liquid chromatography), X-ray fluorescence (XRF), electroanalytical techniques (see below), photometry, pH monitoring, conductivity monitoring, oxidation–reduction potential monitoring (ORP), and mass spectrometry [2].

29.2.1 *Titration*

Titration is one of the oldest and most reliable analytical techniques and is commonly used for analysis of inorganic components (metal ions, acid content, trace ions, etc.) in automatic systems. The advantages of titration techniques are that they can be easily automated and are usually very accurate and precise. Disadvantages are that they can require large amounts of chemical usage and may have long cycle times.

Titration techniques typically require a significant analyte sample size to be extracted from the bath. The analyte is prepared by mixing with reagents to ensure the state of the constituents to be analyzed or to provide a secondary material that is analyzed to infer the concentration of the primary constituent present. Although these methods are relatively easy to automate, the requirement to have precision control of fluid volumes and to clean the fluid streams and reagent vessels requires significant complexity in terms of fluid handling components. The rinsing requirements to ensure that fluids are not affected by previous samples also result in greatly increased chemical consumption.

29.2.2 X-ray Fluorescence

X-ray fluorescence is a relatively straightforward method of analyzing metal ion concentrations and ratios of metal ions in solutions. Although such techniques are relatively easy to automate and are very fast, they are expensive and subject to drift of the X-ray source and the detector. Therefore they require frequent calibration to provide precise results. This, combined with the limited capability of XRF techniques to analyze multiple components of industrial electrolytes and their high price, has resulted in limited use of these techniques in automated bath analysis systems.

29.2.3 Electroanalytical Techniques

Electroanalytical techniques, such as cyclic voltammetric stripping (CVS) or pulsed cyclic galvanostatic analysis (PCGA), are often used for analysis of organic bath components [3–5]. Other methods that are sometimes employed to monitor electrolyte compositions include cyclic pulsed voltammetric stripping (CPVS) and chronoamperometry. The methods described here typically utilize a potentiostat and a rotating disk electrode (RDE) as the working electrode of a three-electrode analytical system. These methods are based on changes in electrodeposition kinetics in the bath sample with respect to the concentration of organic additives rather than the direct measurement of a particular chemical species. Although this approach is not a direct measurement of the organic species, it can be quite accurate and reliable. It has similar disadvantages to titration methods; a sample must be extracted and consumed to facilitate the analysis.

29.2.3.1 Cyclic Voltammetric Stripping

Cyclic voltammetric stripping has been the standard electroanalytical method employed for many years. The method, as typically used, was developed by Haak, Ogden, and Tench [6, 7]. Essentially, the potential of the RDE, or working electrode, is scanned at a constant rate of potential change with respect to time and the current flow between the working electrode and the counter-electrode is measured.

The voltage is scanned in the cathodic direction, then in the anodic direction, with reference to the working electrode. The amount of deposition during the cathodic cycle is usually deduced by integrating the area under the anodic stripping peak. The amp-time, or charge, represented by the stripping peak is directly related to the amount of metal deposited during the cathodic portion of the cycle, for a reversible reaction. For a scan with particular end points and scan rate, this stripping charge will be related to the electrochemical activity of the solution. If suppressors are added to the bath to increase the polarization the stripping charge will be reduced, and if accelerators, or catalysts, are added to the bath the stripping charge will increase.

Because CVS analysis provides a single value that represents the concentration of the constituent of interest, it is important to understand the impact of the other constituents on the result, or the "matrix effect." In order to minimize this impact, the analysis is usually carried out in a regime where non-analyte constituents are put in a concentration range where they have little impact on the result. This is usually done through the addition of suppressors into a saturated regime or through overdosing of accelerators. Titrations of a standard solution with the analyte sample or titrations of the analyte sample with known standard reagents are then used to correlate the impact on the stripping peak areas of each measurement. These correlations are used to deduce the concentration of the bath sample [8, 9].

While there is a long history of CVS usage that represents a large knowledge base, it has some disadvantages. First, like most electroanalytical techniques, the method requires standardization to be related to "known" standard samples. When the exact nature of additive breakdown products is not known, it is impossible to standardize the technique to aged bath standards. Second, the use of a single parameter (stripping peak charge) as being representative of the concentration of the constituent of interest ignores the interactions of the various components of the electrolyte, whether added purposefully, or through some other means (additive breakdown, contamination, constituent concentration drift). Third, the typical scan rates and end points chosen for CVS analysis result in a deposition time less than 5 s. This means that the current being measured is a very dynamic function and is never at kinetic equilibrium. This is especially true when the adsorption and reaction kinetics of a particular bath occur on timescales similar to or greater than the deposition time employed.

29.2.3.2 CPVS

Cyclic pulsed voltammetric stripping is very similar to CVS analysis, except that it utilizes pulsed potential input rather than linear potential sweeps [10]. The particular pulse sequence chosen can be used to alleviate some of the issues detailed above for the CVS technique. Essentially, CPVS has similar advantages and disadvantages to CVS analysis, but with a less extensive knowledge base. The CPVS techniques may offer slightly more flexibility in discriminating between effects of different bath components, however.

29.2.3.3 PCGA

Pulsed cyclic galvanostatic analysis, or PCGA, is essentially a chronopotentiometric analysis technique [4]. It utilizes a sequence of pulses to prepare the RDE for the analysis and to nucleate deposition of metal. The key to the technique, though, is to analyze the potential response as a function of time, and to use the steady-state potential to determine the polarization of the working electrode. As in the CVS and CPVS techniques, the analysis is done as the samples are titrated and the polarization associated with the analyte sample is compared to a reference set of curves.

The use of a chronopotentiometric technique mimics typical industrial plating processes, which typically regulate the current or are galvanostatic. This causes the analysis to be fairly representative of how the process is actually behaving. Additionally, the pulse cycles and the use of the steady-state potential as the feature of interest mean that the CPVS technique has the capability to discriminate between the electrochemical impacts of matrix components in the electrolyte. In fact, it could even be possible to analyze the potential/time curve that results and determine the concentrations of multiple components of the electrolyte as long as they impact different portions of the response curve. The use of the steady-state polarization as the primary response in PCGA analysis specifically avoids the limitations described above for CVS due to the high potential scan rates and dynamic kinetics of the additive absorption [3]. The PCGA technique focuses on the steady-state electrochemical behavior of the electrolyte.

As is the case with the rest of the titration/electroanalytical techniques, PCGA is not particularly fast (15–60 min per analysis cycle) and requires extracting relatively large amounts of electrolyte for analysis, when utilized in an automated analysis system. Additionally, it correlates the electrochemical activity of the electrolyte to the concentrations of the constituents through the comparison of a single value that can be impacted by matrix effects in the electrolyte.

29.2.3.4 Chronoamperometry

Chronoamperometry is very similar to chronopotentiometry, as described above under PCGA analysis. It is simply a potentiostatic process that utilizes the current/time curve as the response. It can be implemented analogously to PCGA for additive analysis.

Chronoamperometry does not mimic industrial processes (galvanostatic) quite as well as chronopotentiometry because it is a potentiostatic technique. This means that the current (density) varies during the analysis cycle and this variation as a function of time is what is used as the response. Electrochemically, this means the adsorption and reactions occurring at the RDE surface may be more consistent over the analysis time than in other techniques, but that is not necessarily similar to an industrial process. Additionally, this technique has the same advantages and disadvantages as the PCGA technique described above.

29.2.4 Photometric Techniques

Photometric techniques are simple, fast, relatively cheap, and typically do not require sample extraction. The simple application of Beer's law with an appropriate illumination source and detector allows measurement of the concentration of species that absorb energy of a particular wavelength, assuming that no other components interfere in the same region of the spectrum. They usually require calibration or standardization to a blank or known sample to account for drift of the illumination source, but these processes are also easily automated. Because of the advantages of these techniques, they have garnered much interest in automated bath analysis.

$$A = \varepsilon\, l\, c \qquad \text{Beer's Law}$$

where

A is absorbance,
ε is molar absorptivity,
l is path length, or cell length, and
c is concentration.

Photometric analysis at a specific wavelength in the UV or visible wavelengths is the most common analytical method employed to measure components of electrolytes. These methods are straightforward to employ for analyzing metal ions with characteristic absorption in the visible spectrum.

Raman spectroscopy is another photometric technique that has been investigated for analysis of organic constituents of plating baths. Although these techniques have benefits similar to those associated with the simple photometric techniques described above, the minute concentrations typically associated with organic additives (especially in Damascene copper processes) have proven problematic.

29.2.5 Probes and Electrodes

A number of direct, real-time measurement techniques provide the advantage of no sample consumption. Using a pH probe is probably the most well known and widely used technique. Ion-specific electrodes, oxidation–reduction potential (ORP) electrodes, and conductivity probes are also very common in plating bath-monitoring applications. Spectrophotometric techniques can also be used for on-line inorganic and organic analyses. Depending on the electrolyte being analyzed and the particular technique employed, these instruments may provide direct chemical measurements or simply monitor characteristics of the electrolyte that are indicative of changes in electrolyte composition.

29.2.6 Liquid Chromatography

HPLC, ion chromatography, and various separation techniques are most often used in off-line bench-top applications [11]. They can be configured for automatic on-line analysis, but their use has been limited due to high initial capital cost and/or maintenance requirements [2]. These techniques are usually designed to directly measure the chemical constituents, which can be complex, as opposed to the electrochemical properties of the chemistry. The main advantage of these analytical techniques is that they potentially provide direct measurement of all of the chemical species in the electrolyte, without regard to electrochemical activity. However, this could also be a disadvantage, since the electrochemical activity is typically considered to be the important characteristic determined by the organic components, and small concentration changes can produce large changes in electrochemical activity.

The HPLC techniques utilize eluents that must be mixed and kept available in order to keep the system running. The columns used to separate the various constituents in the electrolyte must also be kept maintained and ready to operate. These columns require periodic maintenance and replacement. All these maintenance issues make it difficult to provide reliable automated analysis equipment utilizing HPLC techniques.

29.2.7 Mass Spectrometry

Mass spectrometry is a powerful technique that, like HPLC, can directly measure chemical constituents of an electrolyte. Mass spectrometers have the additional capability of being able to identify components according to molecular weight and charge. These systems are very expensive, but have recently been adapted to the online analysis of plating baths. While the amount of information provided by these systems can be impressive, they are expensive and may require a relatively large amount of maintenance.

Mass spectrometry equipment utilizes a small amount of electrolyte sample and prepares the sample through dilution, addition of standard reagents, and volatilization through means such as electro-spray. The sample is then ionized and electrostatically analyzed to determine its molecular mass and charge. These values can be used to identify constituents or to quantify the amount of a previously identified constituent in a bath.

29.2.8 Analytical Techniques Summary

Each of the analytical techniques outlined here has its own advantages and disadvantages. These must be carefully considered when designing and operating a closed-loop chemical control system. It is especially important to understand the details of the analytical techniques if the results of different techniques are being compared. The analytical techniques that measure the electrochemical activity variations caused by organic additives, for instance, do not all measure exactly the same effect. It is important to understand exactly what is being measured, and how it relates to the actual operation of the process, when comparing these techniques. Also, it is not at all straightforward to try to correlate the techniques that measure electrolyte activity to those that measure individual chemical species concentrations when you consider molecular weight distributions of the organic components due to manufacturing and/or breakdown occurring during processing. Effective and robust analysis and control of electrolyte can only be achieved with an in-depth understanding of the advantages and limitations of the analytical techniques being employed.

29.3 Constituent Replenishment

Replenishment of the constituents of an electrochemical processing bath may be conducted on a time, and/or volumetric, basis. In a time-based approach, constant volumes of replenishment solutions are added to the bath at varying time intervals. In a volumetric-based system, varying volumes are added at a set temporal frequency. The choice of one approach over the other, or the use of a hybrid method, is dependent on the hardware utilized for the replenishment, or dosing, function. The volumetric precision and accuracy of the fluid metering hardware, combined with the temporal resolution and multi-tasking capability of the controller, are the determining factors. As long as the accuracy and precision are good enough to control the constituent concentrations within the desired limits and the hardware is reliable, the specific scheme utilized is not important.

29.4 System Design

The simplest replenishment systems dose constituents at a constant rate, based on time or charge passed through the system. Slightly more complex systems react to analysis results from an integrated analysis system. The more sophisticated systems will utilize a parameter-based model to set the replenishment rates or volumes and will adjust the parameters algorithmically based on analysis results.

A simple replenishment system can be configured by providing a reservoir of the concentrate(s) to be dosed into the bath, and a few simple fluid flow components with a simple controller (Fig. 29.1). The simplest design would include a metering pump, and a controller to add the concentrate at a constant rate, whether that is based on time or charge (amp-minutes or coulombs) passed through the system. With such

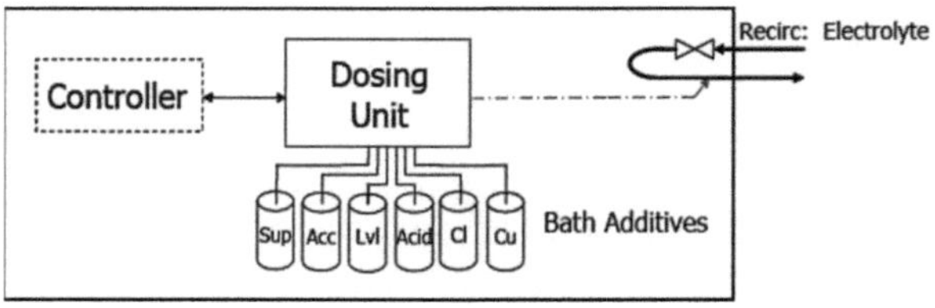

Fig. 29.1 Open-loop replenishment system

a system, it is important to periodically check the constituent concentrations in order to verify that the replenishment rates are adequate.

The replenishment system described above can be turned into a closed-loop chemical control system by adding an analysis system, as described in Section 29.2, above, and adjusting the replenishment rates based on the analysis results. The response can be as simple as adding concentrates only based on low analysis results or it could be a more sophisticated PID-style control system.

Sophisticated closed-loop chemical control systems will take into account the physics that determine constituent usage or breakdown, as well as analytical results that monitor the concentrations of the electrolyte (Fig. 29.2). Such a system would essentially implement a model-based replenishment scheme that includes parameters that are appropriate considering the constituents of the bath. These may include time, charge passed, number of wafers processed, or surface area of plated substrates. The parameters that determine the response of the replenishment rates based on these parameters may then be adjusted based on the analytical results obtained over time.

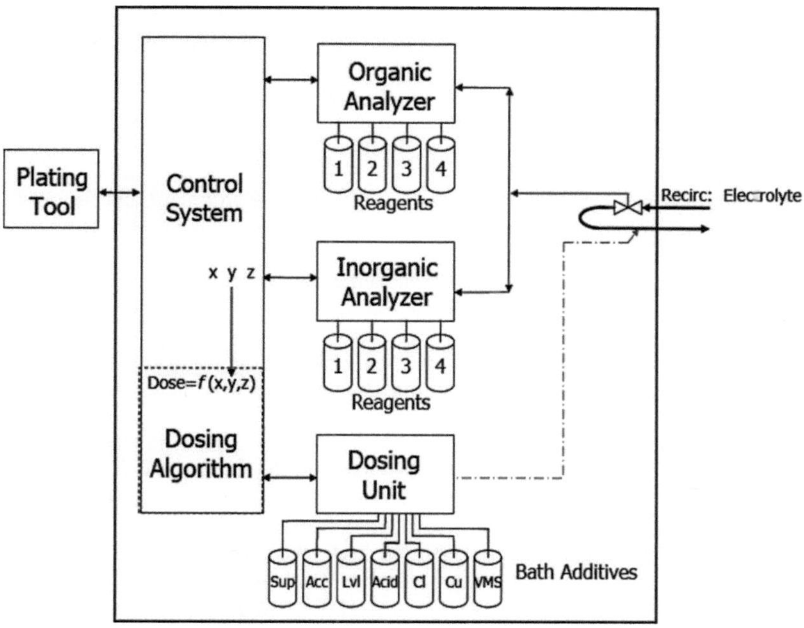

Fig. 29.2 Model-based closed-loop chemical control system

29.5 Biographical Sketches

T. Ritzdorf *Semitool, Inc., Kalispell, MT 59901 (tritzdorf@semitool.com).* Mr. Ritzdorf is the director of ECD Technology at Semitool, Inc. in Kalispell, MT.

He received a B.S. in Chemical Engineering from Montana State University in 1986, and a M.S. in Chemical Engineering from the University of Minnesota in 1989. Mr. Ritzdorf spent 6 years working in the magnetic recording industry at Control Data, Seagate, and Storage Technology. He is an author or coauthor of at least 20 patents and over 30 technical papers, including chapters for two books on electrochemical processing. Mr. Ritzdorf is a member of the Electrochemical Society and the American Electroplaters and Surface Finishers Society.

References

1. Ritzdorf, T. and Fulton, D.: Electrochemical deposition equipment, In: Datta, M.; Osaka, T.; and Schultze J. W., Eds. New Trends in Electrochemical Technology, Microelectronic Packaging, Francis and Taylor (CRC) **3**, 471 (2005)
2. Taylor, T.; Ritzdorf, T.; Lindberg, F.; Carpenter, B.; and LeFebvre, M.: Electrolyte Composition Monitoring for Copper Interconnect Applications. Electrochemical Processing in ULSI Fabrication I and Interconnect and Contact Metallization: Materials, Processes, and Reliability, ECS, Pennington, NJ, 33 (1998)
3. Graham, L.; Ritzdorf, T.; and Lindberg, F.: Steady-State Chemical Analysis of Organic Suppressor Additives Used in Copper Plating Baths. Interconnect and Contact Metallization for ULSI, ECS, Pennington, NJ, **PV 99(31)**, 143 (2000)
4. Robertson, P.; Tolmachev, Y. V.; and Fulton, D.: Galvanostatic Method for Quantification of Organic Suppressor and Accelerator Additives in Acid Copper Plating Baths. Morphological Evolution in Electrodeposition and Electrochemical Processing in ULSI Fabrication IV, ECS, Pennington, NJ, **PV2001(8)**, 309 (2004)
5. Graham, L. W.; Taylor, T. C.; Ritzdorf, T. L.; Lindberg, F. A.; and Carpenter, B. C.: US Patent # 6,365,033, Methods for Controlling and/or Measuring Additive Concentration in an Electroplating Bath (2002)
6. Haak, R.; Ogden, C.; and Tench, D.: Cyclic Voltammetric Stripping Analysis of Acid Copper Sulfate Baths, Part 1: Polyether-Sulfide-Based Additives, Plating and Surface Finishing (1981)
7. Haak, R.; Ogden, C.; and Tench, D.: Cyclic Voltammetric Stripping Analysis of Acid Copper Sulfate Baths, Part 2: Sulfoniumalkanesulfonate-Based Additives, Plating and Surface Finishing (1982)
8. Freitag, W.; Ogden, C.; Tench, D.; and White, J.: Determination of the individual additive components in acid copper plating baths. Plat. Surf. Fin. **70(10)**, 55 (1983)
9. Bratin, P.: New Developments in Use of CVS for Analysis of Plating Solutions. Proceedings of AES Analytical Methods Symposium, Chicago, IL (1985)
10. Fisher, G. L. and Pellegrino, P. J.: The Use of Cyclic Pulse Voltammetric Stripping for Acid Copper Plating Bath Analysis, Plating and Surface Finishing (1988)
11. Haak, K.: Ion Chromatography in the Electroplating Industry. Plat. & Surf. Fin., September (1983)

Chapter 30
Processes and Tools for Co Alloy Capping

Bill Lee and Igor Ivanov

30.1 Introduction

The copper Damascene process is widely established and has brought higher performance to semiconductor devices. Copper has replaced aluminum because of its lower resistivity, higher reliability, and lower cost and was expected to be better because of its higher activation energy for diffusion. However, copper also suffers from electromigration (EM) and stress migration (SM) reliability issues (Fig. 30.1) as geometries continue to shrink, and current densities increase.

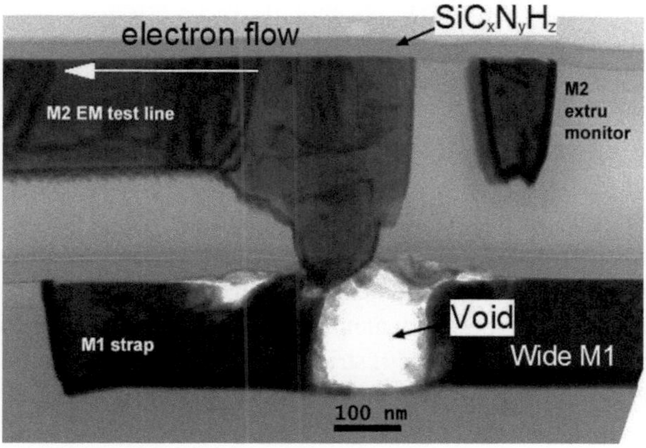

Fig. 30.1 Example of EM-induced void [1] (used with permission from C. K. Hu, IBM)

B. Lee (✉)
Blue29, 615 Palomar Avenue, Sunnyvale, CA 9408, USA
e-mail: bill.t.lee@lamrc.com

Y. Shacham-Diamand et al. (eds.), *Advanced Nanoscale ULSI Interconnects: Fundamentals and Applications*, DOI 10.1007/978-0-387-95868-2_30,

30.2 Co Alloy Capping Applications

In the copper Damascene process, the copper line is encapsulated on the sides and bottom by barrier metal (e.g., Ta/TaN), and on top by a barrier/etch stop dielectric layer. The copper to dielectric interface, typically SiN, SiCN, or SiC, has weaker adhesion than the copper to barrier metal interface, so copper diffusion occurs predominantly at the top surface. Under high current densities, copper atoms move in the direction of the electron flow, and vacancies accumulate in the opposite direction into voids that cause the device to fail as resistance increases (Fig. 30.2).

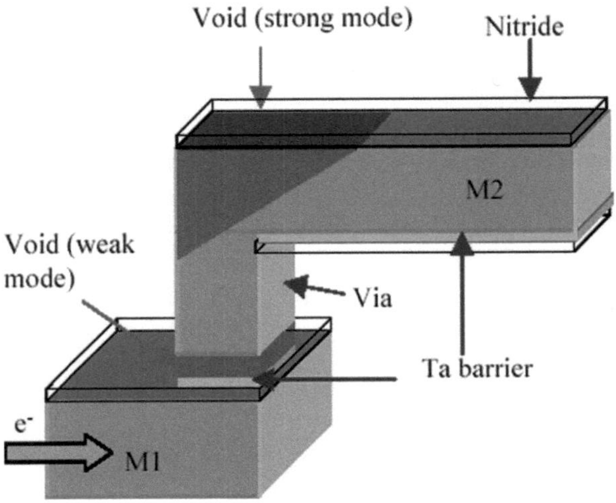

Fig. 30.2 Electromigration-induced voiding [2] (used with permission from P. Ho, U. Texas)

Attempts to improve copper to dielectric adhesion using various surface treatments, such as CVD silane to create an intermediate copper silicide layer prior to dielectric deposition provide some near-term relief, but ultimately this metal–dielectric interface must be fundamentally changed to a metal–metal interface, otherwise current densities will be restricted to the low 10^6 A/cm^2 level. Limiting current density would force drive currents to be lower or interconnect linewidths and via sizes to be larger, making devices bigger, more expensive, and slower.

Copper lines have different linewidths on different levels. The vias contacting wide lines above or below experience void formation in the wide lines caused by thermal stress cycling during wafer manufacturing or due to Joule heating caused by current crowding at the vias during device operation. Vacancies, which are induced during copper grain growth, move along grain boundaries and weak interfaces, accumulating at low stress points at the narrower vias.

These reliability problems can be overcome by adding a cobalt tungsten phosphide (CoWP) [3–6, 10, 12–17, 19, 21–23] or cobalt tungsten boride (CoWB) [7–9, 18, 20, 26] cap to the copper using selective electroless deposition after CMP. EM lifetime improvements of 10× to over 300× have been demonstrated [10–15]. EM and SM device reliability improvements are sought without affecting other reliability and electrical performance such as line resistance, line-to-line leakage, and dielectric breakdown strength.

Table 30.1 summarizes some published EM improvement results using Co alloy capping [20, 24].

Table 30.1 Sample EM improvement results using Co alloy capping

	NEC	Intel	IBM	Freescale
TTF > 50% structures (h)	> 2000*	350	316	> 6000*
Test conditions	300°C, 2 MA/cm^2	–	350°C, 1.8 MA/cm^2	350°C
Improvement	> 100×**	10×	385×	207×
Linewidth (μm)	0.2	0.12	0.18	–
Reference	SiN	Si(C)N	Si(C,H)	SiN

* Testing stopped before all die had failed. ** No failures, so this is the minimum amount of improvement.

For each device generation, geometries shrink by ~0.7× in linewidth or about half in cross-sectional area [14]. Since currents need to remain approximately constant, current density, *j*, can as much as be doubled. According to Black's law,

$$\mathrm{MTTF} = \mathrm{A}j^{-n}\exp\left(\mathrm{E_a/kT}\right)$$

EM lifetime (MTTF) is inversely proportional to j^n, where n varies between 1 and 2 depending on the mode of failure (i.e., fail during void formation or void growth). Thus, lifetime can decrease four times for each technology generation, so 4× is usually stated as the minimum lifetime improvement required and more typically 10× is specified.

Beyond reliability improvement, Co alloy capping can improve device performance by reducing the effective dielectric constant around the Cu lines. Making the cap metal act as both a Cu diffusion barrier and the etchstop layer for the via etch permits the dielectric etchstop/barrier layer to be thinned or eliminated. *RC* time constant reductions of 5–15% are anticipated by this method [15–17].

Various other applications such as image sensors [21, 28], diffusion barriers on bonding pads [27, 30], on-chip inductors [31], magnetic cladding [29], etc., may be enabled by electroless Co alloy capping [32–36].

30.3 Film Properties and Requirements

At the 65 nm and 45 nm technology nodes, the Co alloy cap (Fig. 30.3) thickness requirement is nominally 150 and 100 Å, respectively, with a within-wafer non-uniformity of < 5% 1σ and average wafer-to-wafer variability < 5% 1σ. EM improvement is relatively independent of thickness – the key is to have a good Cu:Co interface to slow the surface diffusion of Cu. SM improvement scales with thickness because film thickness affects film stress which locally modifies

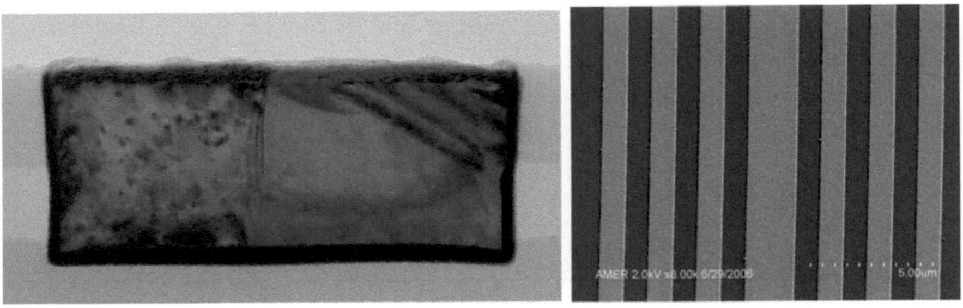

Fig. 30.3 Example of CoWP cap

stress-induced diffusion of Cu. Thickness is usually limited by line-to-line spacing because the Co alloy cap grows isotropically on the copper, so beyond the thickness of the barrier metal, further lateral growth would narrow the spacing and potentially increase line-to-line leakage.

From an electrical standpoint, the Co alloy cap must meet line-to-line leakage and TDDB requirements (which are company and device design dependent) and generally have less than 5% change in interconnect resistance. This resistance change includes contributions from both line and via resistance – from both the cross-sectional dimensional changes due to Cu loss – and from the Co alloy material resistivity and thickness which adds a parallel shunt path in the lines but adds resistance in series in the vias.

When the focus is on performance improvement through k_{eff} reduction, another requirement is for the metal cap to act as a diffusion barrier. The test criteria for this can vary widely. One method is to test annealed blanket films of cap metal on copper by surface analysis techniques such as AES, XPS, or surface SIMS to check for copper that diffused through. Another is to build an MIM capacitor stack with the bottom metal electrode capped and check the C–V properties after anneal for any copper in the dielectric.

Table 30.2 summarizes some published film results.

In addition to the film itself, it is important to create good interfaces between the Cu and the Co alloy layers, as well as between the Co alloy layer and the

Table 30.2 Sample Co alloy film properties

		Reference [17]	Reference [4]	Reference [18]	Reference [7]
Film type		CoWP	CoWP	CoWB	CoWB
Composition*	Co	90%	88–90%	87%	80%
	W	2%	1.5–2.5%	6.8%	20%
	P	8%	8–10%		
	B			4.2%	(trace)
	O			2.1%	
Resistivity*	μΩ cm	–	80 (20)	–	–

*As-deposited (after anneal)

overlying dielectric, barrier metal and seed layers. For example, the adhesion energy between Cu and Co alloy has to exceed that between Cu and etchstop layer (SiN, SiCH, SiCN, etc.) in order to achieve EM lifetime improvement by slowing Cu diffusion and void growth. Cu:SiN and Cu:SiCH interfaces have been measured to have between 10 and 20 J/m^2 debond energy with an electromigration activation energy between 0.8 and 1.1 eV while the Cu:CoWP interface exceeds 40 J/m^2 debond energy with an electromigration activation energy of 1.9–2.4 eV [13, 14, 25]. CoWP compares favorably against Cu:Ta which was measured to be 1.4 eV [14, 25] and even Cu bulk diffusion is 2.1–2.2 eV [25].

When the Co alloy film is initially deposited, it can be crystalline or amorphous depending on its composition (e.g., high P content in CoWP makes the film amorphous). However, after subsequent wafer processing (e.g., CVD), the Co alloy may undergo crystallization. CoWP film as-deposited is either amorphous or in h.c.p. phase depending on its composition. Amorphous CoWP transforms at about 290°C to the h.c.p. phase. At about 430°C, the film forms the f.c.c. phase, and at about 500°C, Co_2P and metallic Co precipitates start to form [5, 19].

30.4 Process Sequence and Integration

The deposition of Co alloys requires several steps as follows.

Surface preparation: Incoming wafers from CMP and post-CMP cleaning have to be prepared for selective deposition. This requires the Cu surface to be cleaned of corrosion and corrosion-preventing passivation material such as benzotriazole (BTA), otherwise nucleation of the Co alloy will be retarded leading to incomplete coverage or the formation of a rough film due to different nucleation times and growth rates. The dielectric surface also needs to be cleaned of any remaining copper residue that may have been embedded during the CMP process which may be below the detection limit by line-to-line leakage or TDDB testing, but sufficient nevertheless to act as unwanted nucleation sites for Co alloy nodules to form. These nodules can reduce line-to-line high voltage breakdown strength and increase leakage, so should be avoided. Typical cleaning solutions include dilute acids such as oxalic acid, sulfuric acid, citric acid etc.

Film deposition: The next step is to selectively grow the Co alloy film uniformly across the patterned wafer, with minimum pattern dependency (isolated versus dense, wide versus narrow), full coverage, and no pinhole defects. The film should have a smooth surface and be as thin as possible while satisfying the EM or SM improvement objectives. This requires the nucleation and growth to begin immediately after the surface preparation is complete, with high nucleation density and short initiation time. This can be achieved either by using a Pd activation layer to catalyze the Co alloy growth using a mild reducing agent such as hypophosphorous acid or by performing a "self-activating" Co alloy film growth using a strong reducing agent by itself such as dimethylamineborane (DMAB). A higher process temperature also reduces initiation time. Typical Co alloy deposition chemistry examples are described in the next section.

Surface finalization: After Co alloy film growth on the Cu (and to a lesser extent on the exposed barrier metal), further surface treatment may be needed. Depending on the nature of the dielectric material, a residue of Co hydroxide may remain on the dielectric surface which should be scrub cleaned using a mild acid, otherwise this residue may be reduced to conductive metal at the next process integration step (dielectric deposition) and increase line-to-line leakage. Any particles generated during the prior process steps should also be removed. The specific chemicals that are most effective in removing any remaining residue and/or particles will depend on the dielectric type and the Co alloy composition.

Process integration considerations: The Co alloy cap process is inserted between Cu CMP and dielectric (either etchstop or ILD) deposition. To be compatible with upstream integration steps, optimization may be required between the (pre-Co alloy deposition) surface treatment step and the upstream Cu plating, annealing, CMP, and post-CMP cleaning steps. To be compatible with downstream integration steps, optimization may be required between the (Co alloy) deposition plus post-dep surface finalization steps, and the dielectric CVD [20], pre-CVD reducing treatment, via etch [21], etchstop open, PR strip/clean, barrier metal deposition, sputter open, and Cu seed deposition steps.

30.5 Deposition Chemistry

Electroless Co alloys are deposited from aqueous solutions using one or more reducing agents along with several other chemicals serving specific roles. The principle of electroless deposition is as follows: [22]

Oxidation of reducing agent

$$\text{Red} --- (\text{conductive catalytic surface}) \rightarrow \text{Ox} + ne^-$$

Reduction of metal ions

$$\text{M}^{n+} + ne^- --- (\text{conductive catalytic surface}) \rightarrow \text{M}$$

Two reducing agents often used are hypophosphite and dimethylaminoborane (DMAB) – the former makes CoP and the latter CoB. Often, a secondary refractory metal such as W or Mo is co-deposited to 'stuff' the Co grain boundaries for better diffusion barrier and oxidation properties. The primary Co deposition reaction is described as follows. For the two reducing agents mentioned, the following reactions occur [22, 23]:

Using hypophosphite:

$$\text{Deposit Co: } \text{Co}^{2+} + 2\text{H}_2\text{PO}_2^{\ -} + 4\text{OH}^- = Co + 2\text{HPO}_3^{\ 2-} + \text{H}_2 + 2\text{H}_2\text{O}$$

$$\text{Co-Deposit P: } 4\text{H}_2\text{PO}_2^{\ -} = 2P + 2\text{HPO}_3^{\ 2-} + \text{H}_2 + 2\text{H}_2\text{O}$$

$$\text{Co-Deposit W}: WO_2^{2+} + 6H_2PO_2^- + 2OH^- + 2H_2O = W + 6H_2PO_3^- + 3H_2$$

Using DMAB:

$$\text{Deposit Co}: 3Co^{2+} + 2\,(CH_3)_2\,NH{:}BH_3 + 6OH^- = 3Co + 2B\,(OH)_3 + 3H_2 + 2\,(CH_3)_2\,NH$$

$$\text{Co-Deposit B}: 4\,(CH_3)_2\,NH{:}BNH_3 + 6H_2O = 2B + 2B\,(OH)_3 + 9H_2 + 4\,(CH_3)_2\,NH$$

$$\text{Co-Deposit W}: WO_2^{2+} + 2\,(CH_3)_3\,NH{:}BH_3 + 4OH^- = W + 2B\,(OH_3) + 2H_2 + 2\,(CH_3)_2\,NH$$

The deposition step is typically performed at a temperature between 75 and 95°C at a pH of 9–9.5. Higher temperature reduces the initiation time and increases nucleation density that is desirable for making the thinnest smooth (defect-free full-coverage) films needed at tight geometry nodes.

In addition to the reducing agent, the following chemicals are also needed: pH adjustor, complexant, stabilizer, pH buffer, and surfactant.

While it was historically easier to work with alkali metal salts for several of these ingredients (e.g., KOH as pH adjustor and sodium citrate as complexant), it is desirable (and often required) to eliminate all alkali metals down to trace levels in chemicals to be used in production semiconductor fabs. The following list provides examples of (alkali metal-free) ingredients used for each of the listed functions:

Cobalt source	$CoSO_4.7H_2O$ or $CoCl_2.6H_2O$
Tungsten source	WO_3 or $(NH_4)_2WO_4$
Phosphorus source*	H_3PO_2 or $NH_4H_2PO_2$
Boron source*	$(CH_3)_2NH{:}BH_3$ [DMAB] or $(CH_3)_3N{:}BH_3$ [TMAB]
Complexant	$C_6H_8O_7.H_2O$
Stabilizer	$C_3H_4O_4$ or CH_4N_2S
pH buffer	H_3BO_3
Surfactant	RE-610 or Triton X-100
pH adjustor	$(CH_3)_4NOH$ [TMAH] or NH_4OH

*Also reducing agent

30.6 Co Alloy Capping System

Integrated process: Electroless Co alloy deposition selectivity requires the surface of the Cu and dielectric to be clean. Post-CMP passivation material such as BTA has to be removed from the Cu surface, and all unwanted foreign matter has to be removed from the dielectric surface. In cases where a lower-temperature deposition step is used, an activation step (e.g., using acidic aqueous $PdCl_2$) is used to improve the selectivity of the deposition onto Cu. After Co alloy deposition, further surface treatment may be used to prepare for the next step in the integration flow. The hardware should handle multiple chemicals in an easily programmable and efficient manner. There should be no queue time between processing steps that would allow any reaction to continue unchecked the wafer surface to dry and create watermark defects, or introduce other wafer-to-wafer variability.

Closed environment: Some of the chemical formulations perform best at high temperatures near their boiling points. For example, film growth incubation time falls below 1 s above 90°C that helps to achieve thin conformal films with full coverage. This is important at the 45 nm technology node where the desired Co alloy cap thickness is 100 Å or less. The hardware should prevent evaporation losses of volatile chemical components that would change the solution composition and process performance.

Oxygen control: The focus in this application is to make the highest quality interface between the Cu and the Co alloy, and then between the Co alloy and the overlying dielectric. Once the Cu and barrier metal surface are cleaned, they should not be allowed to deteriorate before the Co alloy cap is deposited. After the Co alloy is deposited, it should not be subjected to corrosion. Exposure to air should be avoided during processing, and process chemicals and rinse water should be deoxygenated.

Chemical management: Electroless deposition relies on the interaction of several components in the solution. These should be monitored and adjusted as necessary in real time. Optical methods such as UV–Vis (Fig. 30.4) and Raman spectroscopies are preferred since they are non-invasive and can monitor multiple components simultaneously. Chemical analysis methods such as ion chromatography are useful for development but best performed in an offline manner (e.g., per shift qualification) in production due to their need for additional consumables (test chemicals).

System architecture: The first generation of stand-alone electroless deposition systems for ULSI interconnect was based on converted Cu electroplating systems. This type of system typically had one or two rows of wafer-processing stations, with each row having a bath or spray module for each of the process steps. For example, a wafer would undergo the following sequence of transfers and processing steps:

1. Face-up transfer from FOUP to wafer flipper and wafer flip
2. Face-down transfer to first spray station for pre-clean and rinse
3. Face-down transfer to second spray station for activation and rinse [optional]
4. Face-down transfer to deposition bath station for deposition and rinse
5. Face-down transfer to wafer flipper and wafer flip
6. Face-up transfer to first brush station for first brush clean, rinse, and spin dry
7. Face-up transfer to second brush station for second brush clean, rinse and spin dry
8. Face-up transfer to FOUP

Since not all steps require the same amount of time, the wafer can sit idle at some stations awaiting the next station. Although it is feasible to fill the entire system

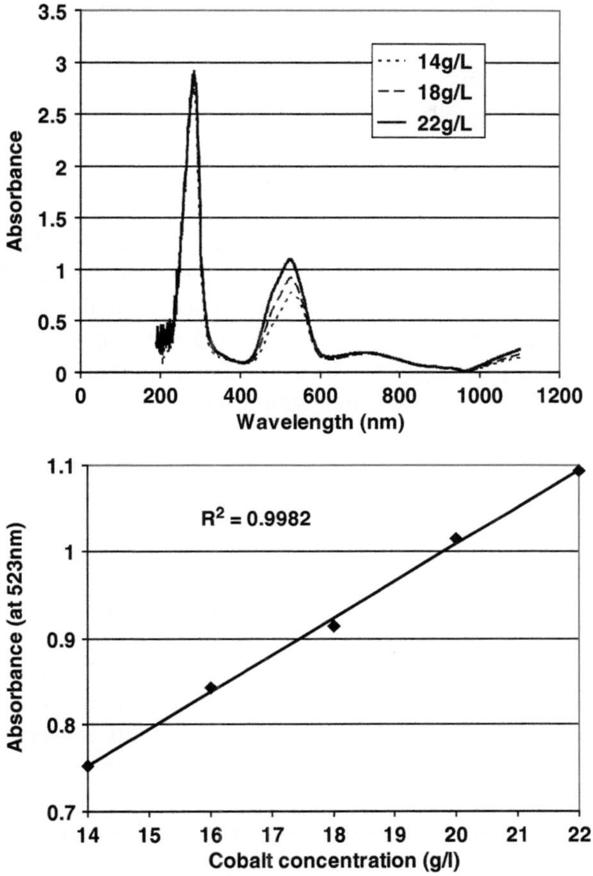

Fig. 30.4 UV–Vis spectroscopic measurement of cobalt concentration

with nitrogen, it is impractical for the low ppm oxygen levels required. Further, the conventional face-down orientation of the wafer is not preferred for electroless plating because one of the reaction byproducts, hydrogen, can form bubbles which interfere with Co alloy film nucleation and growth.

Equipment makers then developed system and chamber architectures specifically for Co alloy cap deposition because of its several different requirements than Cu electroplating. This second generation system typically has multi-step stations to minimize wafer transfer in air and keep critical processing steps together so that there are no delays between steps. For example, a wafer would undergo the following sequence of transfers and processing steps:

1. Face-up transfer from FOUP to deposition chamber, and while under N_2:
   a) Pre-clean and rinse [optionally: also activate and rinse]
   b) Deposition, rinse, and spin dry

2. Face-up transfer to brush clean chamber, and while under N_2:
    a) First clean and rinse
    b) Brush clean, rinse, and spin dry
3. Face-up transfer to FOUP

In this architecture, the wafer remains wet and under controlled ambient after the pre-clean step while the Cu is exposed. Further, film growth at the deposition step can be rapidly and reproducibly quenched. Differences in total process time between the two types of chambers can be balanced by having the appropriate ratio of chambers to balance unit process cycle times. For example, four deposition chambers each running 15 wph can be configured with two brush clean modules each running 30 wph to make a 60 wph system.

Chamber architecture: There are several types of chambers for chemical cleaning and activation (Fig. 30.5), deposition (Fig. 30.6), and brush cleaning (Fig. 30.7). In general, the process chambers should have the following attributes:

1. Small sealable volume – quick purge of air and fill with N_2 without evaporation loss
2. Face-up processing – avoid wafer flipping and H_2 bubble trapping
3. Uniform temperature – for uniform reactions and film thickness and composition
4. Consistent flux – for similar reactant and byproduct flux at every wafer location
5. Dissolved gas control – avoid metal corrosion and control chemical properties
6. Low chemical consumption – to reduce cost
7. Efficient chemical spin/rinse – to quench reaction and avoid cross-contamination
8. Effective rinse/dry – to avoid watermark defects and minimize waste volume

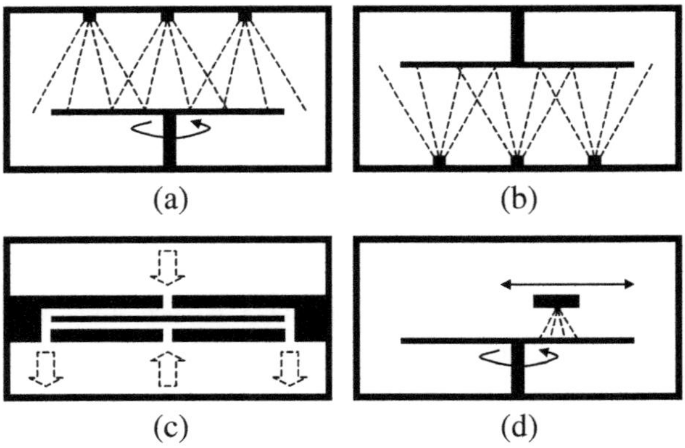

Fig. 30.5 Chemical cleaning and activation chambers: (**a**) face-up spray-down, (**b**) face-down spray-up, (**c**) face-up or face-down immersion, (**d**) face-up radially dispensed

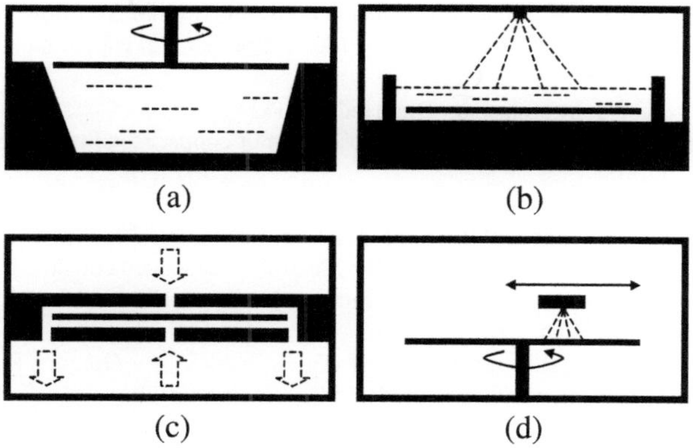

Fig. 30.6 Deposition chambers: (**a**) face-down bath, (**b**) face-up refillable reservoir, (**c**) face-up or face-down immersion, (**d**) face-up radially dispensed

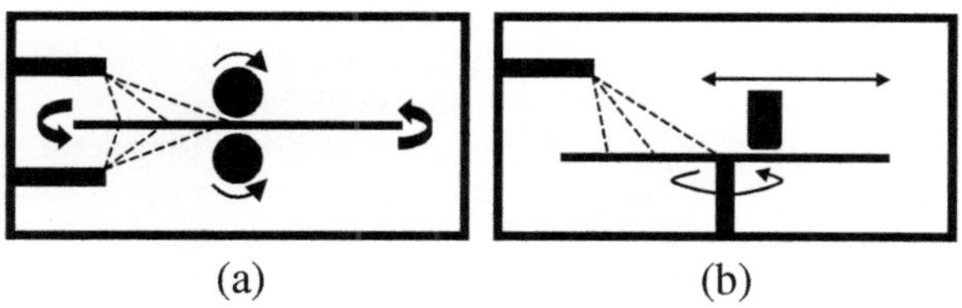

Fig. 30.7 Brush clean chambers: (**a**) face-up double roll (**b**) face-up radial pencil

30.7 Conclusion

Development of electroless cobalt cap technology is underway at the leading IC companies. Effective processes, semiconductor-grade chemicals, and production systems are now available. Cobalt cap technology will allow the semiconductor industry to overcome interconnect reliability obstacles and provide performance and cost improvements.

References

1. Hu, C. K.; Canaperi, D.; Chen, S. T.; Gignac, L. M.; Kaldor, S.; Krishnan, M.; Malhotra, S. G.; Liniger, E.; LloydJ. R.; Rath, D. L.; Restaino, D.; Rosenberg, R.; Rubino, J.; Seo, S.-C.; Simon, A.; Smith, S.; and Tseng, W.-T.: Electromigration cu mass flow in cu interconnections. Thin Solid Films **504(1–2)**, 274 (2005)
2. Ho, P. S.; Lee, K.-D.; Ogawa, E. T.; Lux; Matsuhashi, H.; Blaschke ,V. A.; and Augur, R.: Electromigration reliability of cu interconnects and effects of low K dielectrics. Proc. IEEE IEDM. 741 (2002)
3. O'Sullivan, E. J.; Schrott, A. G.; Paunovic, M.; Sambucetti, C. J.; Marino, J. R.; Bailey, P. J.; Kaja, S.; and Semkow, K. W.: Electrolessly deposited diffusion barriers for microelectronics. IBM Jour. R&D **42(5)**, 607 (1998)

4. Kohn, A.; Eizenberg, M.; Shacham-DiamandY.; Israel, B.; and Sverdlov, Y.: Evaluation of electroless deposited Co(W,P) thin films as diffusion barriers for copper metallization. Microelectron. Eng. **55(1–4)**, 297 (2001)
5. Kohn, A.; Eizenberg, M.; Shacham-DiamandY.; and Sverdlov, Y.: Characterization of electroless deposited Co(W,P) thin films for encapsulation of copper metallization. Mater. Sci. Eng. A **302(1)**, 18 (2001)
6. Petrov, N.; Sverdlov, Y.; and Shacham-DiamandY.: Electrochemical study of the electroless of Co(P) and Co(W,P) alloys. J. Electrochem. Soc. **149(4)**, C187 (2002)
7. Nakano, H.; Itabashi, T.; and Akahoshi, H.: Electroless deposited cobalt-tungsten-boron capping barrier metal on Damascene copper interconnection. J. Electrochem. Soc. **152(3)**, C163 (2005)
8. Mathew, V.; Chatterjee, R.; Garcia, S.; Svedberg, L.; Jiang, Z.-X.; Gregory, R.; Lie, K.-H.; and Yu, K.: Characterization of Electroless Plated CoWB Barrier Films. Abs. 592, 204th Meeting, The Electrochemical Society, Inc. (2003)
9. Bogush, V.; Sverdlov, Y.; Einati, H.; and Shacham-DiamandY.: Oxidation Resistance of Cu ULSI Metallization with Electroless CoWB Capping Layer. Proceedings of the Advanced Metallization Conference 2004 (AMC 2004), MRS publications, San Diego, USA. 843 (2005)
10. Ishigami, T.; Kurokawa, T.; Kakuhara, Y.; Withers, B.; Jacobs, J.; Kolics, A.; Ivanov, I.; Sekine, M.; and Ueno, K.: High Reliability Cu Interconnection Utilizing a Low Contamination CoWP Capping Layer. IEEE IITC. 75 (2004)
11. Moon, P.; Dubin,V.; Johnston, S.; Leu, J.; Raol, K.; and Wu, C.: Process Roadmap and Challenges for Metal Barriers. IEEE IEDM. **841**, 35.1.1 (2003)
12. Hu, C. K.; Gignac, L.; Rosenberg, R.; Liniger, E.; Rubino, J.; Sambucetti, C.; Domenicucci, A.; Chen, X.; Stamper, A. K.: Reduced electromigration of cu wires by surface coating. App. Phys. Lett. **81(10)**, 1782 (2002)
13. Hu, C. K.; Gignac, L.; Rosenberg, R.; Liniger, E.; Rubino, J.; Sambucetti, C.; Stamper, A. K.; Domenicucci, A.; and Chen, X.: Reduced Cu Interface Diffusion by CoWP Surface Coating. Microelectron. Eng. **70(2–4)**, 406 (2003)
14. Hu, C. K.; Canaperi, D.; Chen, S. T.; Gignac, L. M.; Herbst, B.; Kaldor, S.; Krishnan, M.; Liniger, E.; Rath, D. L.; Restaino, D.; Rosenhere, R.; Rubino, J.; Seo, S.-C.; Simon, A.; Smith, S.; and Tseng, W.-T.: Effects of Overlayers on Electromigration Reliability Improvements for Cu/Low k Interconnects. IEEE IRPS. 222 (2004)
15. Ko, T.; Chang, C. L.; Chou, S. W.; Lin, M. W.; Lin, C. J.; Shih, C. H.; Su, H. W.; Tsai, M. H.; Shue, W. S.; and Liang, M. S.: High Performance/Reliability Cu Interconnect with Selective CoWP Cap. Symposium on VLSI Technology, Digest of Technical Papers. 109 (2003)
16. Moon, P.; Dubin, V.; Johnston, S.; Leu, J.; Raol, K.; and Wu, C.: Process Roadmap and Challenges for Metal Barriers. IEEE IEDM Tech. Dig. 35.1.1 (2003)
17. Kohn, A.; Eizenberg, M.; and Shacham-DiamandY.: Copper grain boundary diffusion in electroless deposited cobalt based films and its influence on diffusion barrier integrity for copper metallization. J. App. Phy. **94(5)**, 3015 (2003)
18. Sverdlov, Y.; Bogush, V.; Einati, H.; and Shacham-DiamandY.: Electrochemical study of the electroless deposition of Co(W,B) alloys. J. Electrochem. Soc. **152(9)**, C631 (2005)
19. Armyanov, S.; Valova, E.; Franquet, A.; Dille, J.; Delplancke, J.-L.; Hubin, A.; Steenhaut, O.; Kovacheva, D.; Tatchev, D.; and Vassilev, Ts.: crystalline and amorphous electroless Co-W-P coatings. J. Electrochem. Soc. **152(9)**, C612 (2005)
20. Michaelson, L.; Mathew, V.; Gall, M.; Hauschildt, M.; Acosta, E.; and Garcia, S.: Electroless Deposition of Co-based Alloys for Selective Capping Applications. Proc. ADMETA (2005)
21. Gambino, J.; Wynne, J.; Smith, S.; Mongeon, S.; Pokrinchak, P.; and MeatyardD.: Effect of CoWP Cap Thickness on Via Yield and Reliability for Cu Interconnects with CoWP-only Cap Process. IEEE IITC. 111 (2005)
22. Dubin, V. M.; Lopatin, S.; Kohn, A.; Petrov, N.; Eizenberg, M.; and Yosi Shacham-D.: Electroless Barrier and Seed Layers for On-Chip Metallization, Microelectronic Packaging. CRC Press. 65 (2004)

23. Kolics, A.; Polyanskaya, M.; and Ivanov, I.: Initial Stages of Electroless Cap Formation on Copper Interconnects. Proc. 207th ECS. 362 (2005)
24. Gosset, L. G.; Chhun, S.; Besling, W.; Vanypre, Th.; Farcy, A.; Arnal, V.; Mellier, M.; Flake, J.; Michaelson, L.; Brun, Ph.; Ollier, E.; Ney, D.; Girault, V.; Hopstaken, M.; Jullian, S.; Kim, K.; Jiang, Z. X.; Broussous, L.; Humbert, A.; Fang, H.; Van, S.; Shanmugasundram, A.; and Torres, J.: Interest and Characterization of a Hybrid CoWP/SiCN Architecture for Sub 65 nm Technology Nodes. Proc. AMC. 587 (2005)
25. Hu, C. K. and Rosenberg, R.: Capping Layer Effects on Electromigration in Narrow Cu Lines, Stress-Induced Phenomena in Metallization. 7th International Workshop CP741. 97 (2004)
26. Einati, H.; Bogush, V.; Sverdlov, Y.; Rosenberg, Y.; Shacham-DiamandY.; et al. The effect of tungsten and boron on the cu barrier and oxidation properties of thin electroless cobalt-tungsten-boron films. Microelectron. Eng. **82(3–4)**, 623 (2005)
27. Magagnin, L.; Sirtori, V.; Seregni, S.; Origo, A.; and Cavallotti, P. L.: Electroless Co-P for diffusion barrier in Pd-free soldering. Electrochimica Acta **50(23)**, 4621 (2005)
28. Gambino, J.; Johnson, C.; Therrien, J.; Hunt, D.; Wynne, J.; Smith, S.; Mongeon, S.; Pokrinchak, P.; and Levin, T. M.: Stress Migration Lifetime for Cu Interconnects with CoWP-only Cap. Proceedings of the 12th IFPA. 92 (2005)
29. Molla, J.; D′urso, J.; Kyler, K.; Engel, B. N.; Grynkewich, G. W.; and Rizzo, N. D.: Method of Applying Cladding Material on Conductive Lines of MRAM Devices. US Patent 6,927,072, (2005)
30. Edelstein, D.; Stamper, A. K.; Rubino J. M.; and Sambucetti, C. J.: Self-Aligned Corrosion Stop for Copper C4 and Wirebond. US Patent Application 2004/0234679 (2004)
31. Gambino, J. P.; William T. M.; and Erick, G. W.: Integration of High-Performance Copper Inductors with Bond Pads. US Patent Application 2005/0160575 (2005)
32. Lee, B.: Electroless CoWP Boosts Copper Reliability, Device Performance. Semiconductor International (2004)
33. Singer, P.: The Advantages of Capping Copper with Cobalt. Semiconductor International, October (2005)
34. Lopatin, S.: Electrochemical metallization of nanostructures: Integrated circuits and micro-electro-mechanical systems. Recent Res. Devel. Electrochem. 6 (2003): 57–100, ISBN: 81-7895-107-X
35. Akolkar, R. N.; Dubin, V.; Cheng, C.; Johnston, S.; Chebiam, R.; and Fajardo, A.: Advanced Electrochemical Processes for Sub-50 nm On-chip Metallization. Proc. 208th ECS. 665 (2005)
36. Gan, C. L.; Lee, C. Y.; Cheng, C. K.; and Gambino, J.: Effect of Current Direction on the Reliability of Different Capped Cu Interconnects. Proc. MRS **863**, B 9.3.1 (2005)

Chapter 31
Advanced Planarization Techniques

Bulent M. Basol

31.1 Introduction

As the integrated circuit technology nodes reach 45 nm and beyond, growing requirement for reduced propagation delay dictates inclusion of low-*k* materials in the interconnect metallization structures. Unfortunately, mechanical properties, such as hardness and Young's modulus of the dielectric materials, deteriorate as their porosity is increased and the *k* value is reduced to 2.5 and below [1]. Reliability issues such as electromigration, stress migration, and time-dependent dielectric breakdown (TDDB) lifetimes are also becoming more challenging for multi-stack low-*k* structures. The low-*k* and ultra low-*k* materials are prone to delamination [2] and cracking [3] during CMP; risk of damage rising as the polishing pressure and time increases [4]. It has been demonstrated that delamination in low-*k* stacks was driven by the work done against the friction force during the CMP process [5]. Therefore, it is becoming more and more difficult to polish and planarize topographic copper layers, deposited on low-*k* dielectric materials, at low stress and high rate while maintaining the mechanical integrity of the overall interconnect structure. Furthermore as feature widths and depths shrink, tolerances for metal loss and line resistance variation over the wafer surface are also reduced. In advanced interconnects, adding sacrificial thickness to the dielectric layer which can then be removed during CMP overpolish step is not a good option to minimize topography because hard cap layers are often used to protect the low-*k* dielectric materials from the negative effects of CMP [6, 7] and thickness of these layers is kept to a minimum to reduce their contribution to the effective dielectric constant of the stack. Therefore, as technology nodes move beyond 45 nm, planarization steps of the interconnect manufacturing process flow are expected to offer reduced stress, higher planarization efficiency, reduced copper dishing, less dielectric erosion, better global line

---

B.M. Basol (✉)
SoloPower Inc., 5981 Optical Court, San Jose, CA 95138, USA
e-mail: bbasol@solopower.com

Y. Shacham-Diamand et al. (eds.), *Advanced Nanoscale ULSI Interconnects: Fundamentals and Applications*, DOI 10.1007/978-0-387-95868-2_31,

resistance uniformity, while at the same time maintaining high process throughput, low defectivity, and low cost.

New CMP approaches are being developed to address at least some of the above-mentioned challenges by careful selection of hardware, consumables, and process window. Low-abrasive [8, 9] or abrasive-free [6, 9–11] CMP approaches, for example, are being developed to use lower shear force and chemistries with enhanced metal removal rates. While CMP tries to keep up with the ever growing demands of the new technology nodes, use of naturally low stress electrochemical techniques is also under consideration as an alternative or a supplement to copper CMP. In the following sections we will briefly review copper electropolishing approaches with different degrees of planarization capability. Then we will concentrate on two advanced electrochemical processes, electrochemical mechanical deposition (ECMD) and electrochemical mechanical planarization (ECMP), which demonstrated excellent planarization capability for copper interconnect structures under low stress conditions.

31.2 Electrochemical Polishing

Electrochemical polishing is an anodic dissolution process that can remove and smoothen metal layers at rates higher than 1 μm/min without inducing physical stress on the surface [12, 13]. In a typical electropolishing process the metal film to be removed is placed in a polishing electrolyte across from a cathode. Upon application of the anodic voltage to the metal layer, dissolution reaction takes place on the surface with a rate primarily determined by the current density, mass transfer, temperature, and the nature of the electrolyte. For polishing action it is necessary to work in a diffusion-limited regime where a diffusion layer, a salt layer, or a passivation layer is formed on the metal surface.

31.2.1 Conventional Copper Electropolishing Processes

Because of their simplicity and extensive knowledge base [12–14], standard electropolishing methods utilizing phosphoric acid electrolytes have long been investigated for possible stress-free copper removal and planarization for interconnect applications [15–27]. Upon application of an anodic voltage in a phosphoric acid solution, polishing initiates once a diffusion layer is established on the copper surface and the process is operated at around the limiting current density, i_L, where a current plateau is observed in current–voltage characteristics (see Fig. 31.1). For currents lower than i_L preferential crystallographic etching yields a rough surface. The diffusion layer thickness for a typical copper electropolishing process in a phosphoric acid-based electrolyte is in the range of 5–30 μm [28–30] depending on the composition, viscosity, and temperature of the solution and the mass

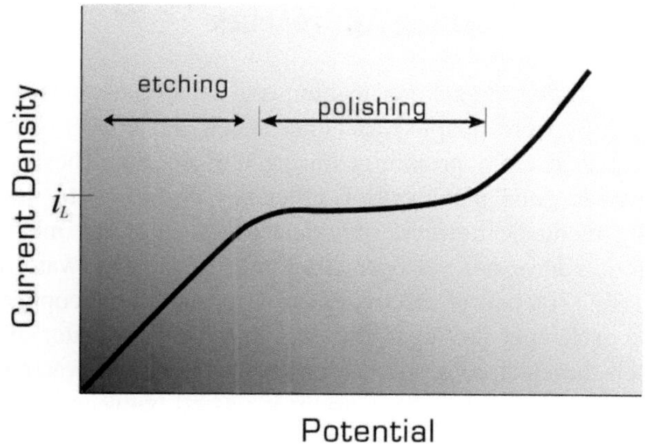

Fig. 31.1 A typical polarization curve for copper electropolishing

transfer to the wafer surface. Although electropolishing successfully planarizes copper over high aspect ratio small features, its planarization efficiency is not adequate for features with widths comparable to or larger than the diffusion length thickness. This pattern sensitivity problem is the most important shortcoming of the conventional electropolishing technique and extensive research has been done on new electrolyte compositions [21, 25] and tool designs [26, 27] to overcome this problem. Recently West et al. [28] carried out a numerical simulation of electrochemical planarization of copper and concluded that given the range of step heights that are required to be planarized, as well as the constraints on overburden thickness, conventional electropolishing processes are not likely to be effective for this application. A recent publication [25] reported near 100% planarization efficiency for 1–50 μm wide 1 μm deep copper patterns using a "super planarizing" phosphoric acid-based electrolyte containing alcohols and organic acids. Despite these encouraging reports, the ability of the standard non-contact electropolishing technique to planarize features with thin copper overburden and aspect ratios of less than 0.01 remains to be seen. Reader is referred to Suni et al. [29] and West et al. [30] for recent review of the copper electropolishing technology.

Since conventional electropolishing has limitations in planarizing low aspect ratio features, a planar copper layer delivered to the electropolishing process would eliminate the problem of pattern dependence. There have been proposals to use electropolishing after an initial CMP step which planarizes the bulk of the topographic copper layer [31]. In this approach electropolishing removes the already planarized copper in a stress-free fashion. Another proposed approach involves planarization of the copper layer during electroplating using a planar deposition or ECMD technique [32–34] and then electropolishing the resulting flat layer in a conformal manner under low-stress conditions [35].

31.3 Novel Electropolishing Approaches

In newly proposed electropolishing techniques, ion exchange materials [36, 37] or membranes [38] were adapted to either touch the copper wafer surface or hydroplane over it at close proximity during the process. These techniques are reported to provide good planarization efficiency compared to the conventional electropolishing methods; however, practical lifetimes of the membranes or ion exchange materials have not yet been disclosed. In the D.I. water electropolishing approach [36, 37] a pair of electrodes are disposed on the copper-coated wafer surface. An ion exchanger serving as the catalyst to enhance water splitting is sandwiched between the electrodes and the copper surface, and water is provided to the copper/ion exchanger interface. Current is passed between the two electrodes through the DI water and the copper coating as a relative motion is established between the electrodes and the wafer. The OH^- ions from water splitting interact with copper surface to cause removal of material across from the cathodic electrode. Since cavities are filled with water, relatively small current passes through them and material removal is primarily limited to the high points closest to the ion exchanger. Average copper removal rates of close to 400 nm/min and pattern-independent planarization of copper features have been demonstrated using this technique [36, 37].

In the membrane-mediated electropolishing method [38] a cathode and an acidic solution are confined within a half cell. The lower surface of the half cell is enclosed by a cation-selective, ion-conducting membrane that faces the copper-coated wafer surface. De-ionized water is provided to the membrane/copper interface. Once relative motion is established between the half cell and the wafer, Cu^{2+} ions formed due to anodization easily permeate the membrane and enter the half cell. Since DI water resistivity is high, etching of copper from cavities filled with DI water is reported to be much smaller than removal from the high points close to the membrane. Planarization efficiencies equivalent to CMP have been demonstrated for 100 μm wide features using this technique, and a copper removal rate of 900 nm/min is reported [38].

31.4 Electrochemical Mechanical Deposition

Standard electrochemical deposition (ECD) baths employed for interconnect fabrication are formulated to yield bottom-up growth or super-fill of copper in submicrometer size, high aspect ratio features. These electrolytes typically contain copper sulfate, sulfuric acid, chloride ions, and organic additives [39], such as suppressors, accelerators, and levelers. Suppressors are typically polyethylene glycol (PEG)-related polymers, which form a film over the copper surface increasing polarization. Accelerators are sulfur-containing compounds such as bis-(3-sulfopropyl) disulfide (SPS) or 3-mercapto-1-propanesulfonate (MPS), which de-polarize surfaces that already contain suppressors. Chloride is known to improve effectiveness of the suppressor by adsorbing to the copper surface and providing strong binding sites for suppressor adsorption. Various mechanisms have been proposed to explain the super-conformal growth phenomenon. One model assumed diffusion-limited consumption of suppressor species within the narrow features [40, 41]. Another model proposed that Cl might be consumed at the bottom of narrow features reducing suppressor adsorption or accelerator concentration might be increasing at the bottom due to rapid reduction of the internal surface area during gap fill [42]. A curvature-enhanced accelerator coverage model [43, 44] was successful in predicting super-filling phenomenon by mathematically formulating the change in the effective surface coverage of additives as a function of the varying surface curvature of the depositing copper layer.

As the brief review above demonstrates, the super-conformal deposition or bottom-up fill mechanisms require either a diffusional limitation or a change in the surface curvature within a high aspect ratio narrow cavity. Such arguments are not valid for large, low aspect ratio features, which may have widths of more than 50–100 μm. Therefore, copper deposition by conventional ECD techniques yields conformal layers over such large cavities.

Electrochemical mechanical deposition (ECMD) [32–34] is a new technique that has the ability to deposit planarized conductive layers on non-planar substrate surfaces [35, 45–48]. As applied to copper deposition on patterned wafers ECMD involves sweeping the surface of the wafer with a pad as electroplating commences. When pad action is applied to the wafer surface during plating in an electrolyte with organic additives, the process provides enhanced deposition into cavities including those with aspect ratios much smaller than 0.01 [45]. Planarization does not take place in additive-free solutions [35, 47].

31.4.1 Planarization Mechanism of ECMD

Mechanically induced super filling [46–48] (MISF) observed for ECMD of copper on patterned wafers is a result of a mechanically induced current suppression (MICS) phenomenon. Figure 31.2a–d schematically show how this mechanism planarizes a patterned wafer surface near a low aspect ratio cavity. In Fig. 31.2a the wafer surface with a thin copper layer is immersed in a plating bath containing organic accelerator and suppressor or inhibitor species, and a cathodic voltage is applied to the copper layer with respect to a counter electrode which is not shown in the figure. Under equilibrium conditions the organic additives are distributed uniformly over the top surface of the copper as well as the internal surface of the large cavity, the effective fractional surface coverage of the suppressor species given by $\theta_{no\,pad}$. Since the additive surface coverage is uniform over the whole copper surface, the current density flowing to the top surface portion, i_t, is equal to the current density flowing to the cavity, i_c, i.e., electroplating is conformal as expected for a standard ECD process. Assuming modified Tafel kinetics, under the conditions described above, the current density is given by [49]

$$i_t = i_c = i_0 \left(1 - \theta_{no\,pad}\right) exp\left(-\alpha_c FV/RT\right) \qquad (31.1)$$

Fig. 31.2 Evolution of copper layer planarization in an ECMD process: (**a**) accelerator and inhibitor are initially uniformly adsorbed; (**b**) inhibitor surface coverage is enhanced on the *top* surface swept by the pad; (**c**) copper deposition rate on the swept top surface is reduced compared to cavity; and (**d**) planar Cu structure obtained

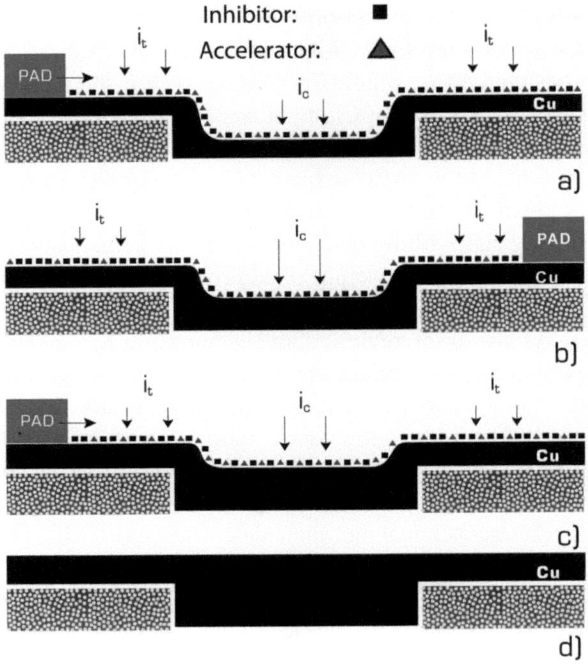

where R is the gas constant (8.314 J mol^{-1}K^{-1}), F is the Faraday's constant (96487 Coulombs/mol), T is the temperature in Kelvin, and α_c is the transfer coefficient.

Now let us assume that a pad piece sweeps the top surface of the structure shown in Fig. 31.2a in the direction of the arrow. Figure 31.2b represents the status of the wafer surface right after the pad sweep, which induces a differential between the additive populations at the top surface and within the cavity, rendering the top surface richer in suppressors. The current density at the top surface right after the pad sweep is reduced due to this MICS phenomenon and is given by [46, 48]

$$i_t = i_0\left[1 - \theta_{no\,pad} - \theta_{pad,0}\, exp\,(-kt)\right] exp\,(-\alpha_c FV/RT) \qquad (31.2)$$

where $\theta_{pad,0}$ is the additional suppressor coverage induced at the top surface due to the mechanical action of the pad right after the sweep, and k is the additive relaxation rate constant (s^{-1}). The current density going into the cavity is still governed by Eq. (31.1) since the pad action does not disturb the additive concentration within the feature, although the use of galvanostatic conditions may dictate the time variation of this current [46]. As can be seen from Eq. (31.2), the suppressed current density would rise toward the equilibrium level given by Eq. (31.1) as plating continues in the ECD mode in the absence of the pad action. This is depicted in Fig. 31.2c, where fractional surface coverage of suppressor is lower and fractional surface coverage of accelerator is higher at the top surface compared to Fig. 31.2b. Therefore, high planarization efficiency can be obtained in an ECMD process by sweeping the wafer surface at short time intervals (in the order of a few milliseconds) using a multi-strip pad and high relative motion between the pad and the wafer. The process is terminated once a planar structure shown in Fig. 31.2d is obtained.

Figure 31.3 is a schematic of a copper ECMD cell. The wafer is held by a carrier and lowered face down toward the polishing face of a multi-strip pad, which is supported on a perforated support plate. An anode is placed in an electrolyte filled cavity below the support plate and the electrolyte is pumped from the anode cavity toward the wafer surface through the perforations in the support plate. Wafer surface is pushed against the pad strips at low pressure (< 1 psi) and rendered cathodic during the process as the wafer holder is rotated (typically at 20–100 rpm) and moved in lateral direction at speeds higher than 15 cm/s. Lateral motion assures that the center of the wafer which does not have any linear velocity due to rotation is also swept by the pad. The perforations in the support plate and the channels in the pad structure [50, 51] are optimized for best deposition uniformity. After planarization the polarity of the applied voltage may be reversed as polishing by the pad continues. This way thickness of the planarized copper layer may be further reduced in a planar fashion by anodic dissolution [32, 52].

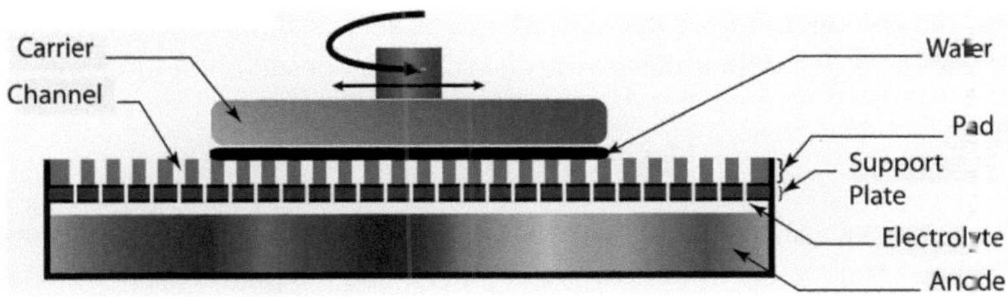

Fig. 31.3 Schematic of an ECMD cell

MICS may be due to physical removal of the adsorbed additives from the top surface by the pad action. ECMD chemistries comprise suppressor concentrations that are much larger, e.g., 100×, than accelerator concentrations. Therefore, even if both accelerators and suppressors were removed from the wafer surface by the pad, build-up of suppressor surface coverage would be much faster yielding a suppressed top surface immediately after the pad sweep. This would then be followed by the slower accelerator re-adsorption, which would take place with the rate constant k. It is also possible to explain the observed MICS by assuming that the pad action primarily removes the loosely bound accelerators from the top surface, which then start to re-adsorb with the rate constant k once the pad piece is moved away. In any case, it appears that for a given pad design and wafer movement, re-adsorption kinetics of accelerators to the swept top surface determines the achievable current density differential between the wafer top surface and the cavities. Recently, MICS phenomenon was also demonstrated on un-patterned blanket wafer surfaces by periodically sweeping part of the flat surface with a small pad during deposition [48]. Copper growth rate was found to be lower in the swept location and the resulting copper thickness profiles agreed well with the model calculations based on

Eqs. (31.1) and (31.2). Additive-free electrolytes or solutions containing only one type of additive did not show any appreciable MICS phenomenon [47, 48]. The effect was only prominent in electrolytes containing both accelerators and suppressors.

Using commercially available organic additives and galvanostatic ECMD, the additive relaxation times were found to be in the range of 1.5–5 (s^{-1}) [46, 48] depending on the condition of the copper surface before the ECMD process was initiated. Planarization efficiency in the ECMD process is a function of the chemistry, the pad design, and the initial surface coverage of additives on the copper-coated wafer surface. Typical planarization efficiency values are in the 80–95% range using commercially available copper sulfate plating baths with accelerator/suppressor additive packages and multi-strip planarization pad designs [47, 50, 51]. Levelers are detrimental to planarization efficiency in an ECMD process [47, 52] because they migrate to high current density regions (i.e., cavities) to increase polarization there.

Let us assume that the thickness of the copper film at the top surface of a wafer increases by an amount d_t during an ECMD process (e.g., going from the situation in Fig. 31.2a–c) when a total thickness of d_{plate} is deposited on the wafer surface based on the amount of the charge passed. Let us also assume that the corresponding copper thickness rise within the cavities is d_c. From conservation of charge

$$d_{plate} = (1 - D)\, d_t + D\, d_c \tag{31.3}$$

where D is the pattern density of the wafer. The step height, Δ, at the cavity edge can be written in terms of the initial feature depth, H, and the plated copper thicknesses

$$\Delta = H - (d_c - d_t) \tag{31.4}$$

Using Eqs. (31.3) and (31.4), the step height can be expressed as

$$\Delta = \left(D^{-1}\right)\left(d_t + DH - d_{plate}\right) \tag{31.5}$$

and the planarization efficiency is given by

$$PE\,(\%) = \left[100\,(\mathrm{d_c} - \mathrm{d_t})\,/\mathrm{d_c}\right] \tag{31.6}$$

Graph in Fig. 31.4 a shows the expected reduction in step height, Δ, as the total plated copper thickness, d_{plate} is increased in an ECMD process. Assuming time-independent 100% planarization efficiency, this relationship can be represented by the solid line in Fig. 31.4a. Any reduction in planarization efficiency would then move the x-axis intersection of the line to the right by an amount equal to the copper thickness plated at the top surface, d_t. Figure 31.4b is the step height data [45] obtained from ECMD of 200 mm diameter wafers with a feature depth of 0.9 μm. Incremental increases in copper thickness at the top surface and within the cavities were monitored by FIB cross sections in these experiments as a function of applied

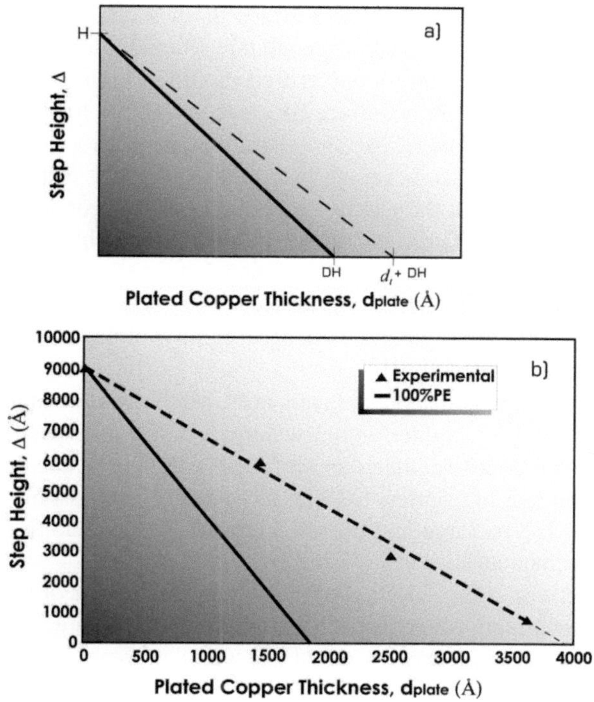

Fig. 31.4 Step height vs. plated copper thickness in an ECMD process; (**a**) theory, (**b**) experimental

charge. When 2880 Å of copper was plated, for example, the copper film thickness at the top surface and within the cavities increased by 1500 and 8000 Å, respectively, suggesting a planarization efficiency of 81%. Equation (31.4) predicts $d_c = H + d_t$ at the time when the wafer is fully planarized. Therefore, with the measured 81% planarization efficiency, the copper thickness plated on the top surface and into the cavity by ECMD would be about $d_t = 2100$ Å and $d_c = 11100$ Å, respectively, at the time of full planarization. The theoretical 100% planarization efficiency curve is also shown in Fig. 31.4b. Its intersection point with the x-axis at 1800 Å suggests that the pattern density of the wafer was about 1800/9000 = 20%.

Planarization efficiencies close to 100% can be obtained if the wafer surface is pre-treated in an accelerator containing solution and then ECMD is carried out in a bath containing only suppressors. In this case, as the pad sweeps the top surface, previously adsorbed accelerators are removed and replaced by suppressors. Since there is no accelerator available in the bath, fractional surface coverage of suppressors at the top surface is very high and the current density at the top surface stays suppressed throughout the process. As for the cavities, some of the pre-adsorbed accelerators are replaced by the suppressors with a characteristic time constant that drives the fractional surface coverage within the cavities to a new steady-state value. These results are in some way analogous to super filling achieved in high aspect ratio features by first derivatizing the wafer surface in an accelerator containing solution and then plating it in an electrolyte containing an inhibitor [53]. The degree of derivatization is very important for defect-free gap fill of high aspect ratio features, too little or too much accelerator surface coverage giving rise to voids. For the large cavities that are planarized by ECMD, however, this issue is not important and near-saturation accelerator surface coverage may be used for best planarization results. It has been shown [54] that the plating current density in a copper sulfate solution containing 300 ppm of PEG can be increased by about 30× by addition of 10 ppm of SPS. Assuming a factor of 30×, and assuming that the top wafer surface has only suppressor coverage, the planarization efficiency in such an electrolyte would be about (29/30) × 100 = 97%, which is in agreement with the ECMD experimental findings [47].

Benefits of utilizing planar copper layers in a CMP process include lower dishing and erosion [52, 55, 56], tighter within-wafer resistance distribution, and relative insensitivity of resistance to pattern density [56]. Because of its capability to provide higher deposition rate in cavities, ECMD may be especially useful for processing high level metal layers, large and deep features of 3D integration structures, and micro electromechanical systems (MEMS).

FIB cross sections of Fig. 31.5a–d compare the performance of ECD and ECMD techniques for processing wafers with 2 μm deep features. Figure 31.5a,b shows FIB pictures taken from a wafer processed by ECD out of an electrolyte with accelerator and suppressor additives. As shown in Fig. 31.5a copper overburden thickness is 3.9 μm over the dense array due to the well-known over-plating phenomenon which is prominent in leveler-free baths. Figure 31.5b is a cross section taken at a large trench location and demonstrates the conformal nature of the ECD copper, which

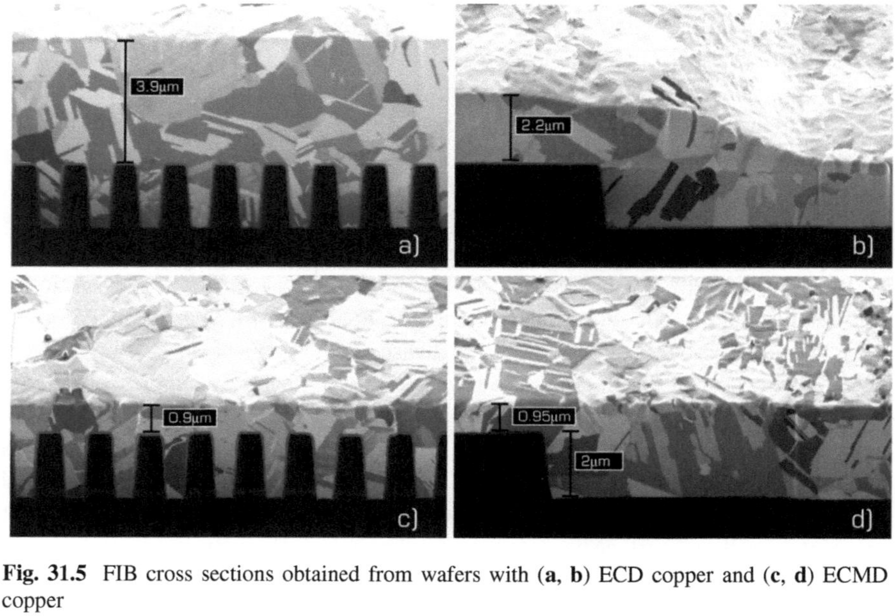

Fig. 31.5 FIB cross sections obtained from wafers with (**a, b**) ECD copper and (**c, d**) ECMD copper

produces a step near the feature edge. The step height is approximately equal to the trench depth. Cross sections in Fig. 31.5c,d were taken from another wafer, which had a copper seed layer thickness of 0.1 μm and which received a 0.25 μm thick ECD copper film before being processed by ECMD until full planarization. As can be seen from this data, the copper overburden thickness is much smaller (0.9–0.95 μm) compared to the ECD case and over-plating at the dense array region is avoided. ECMD copper film is flat and uniform irrespective of feature density or size.

31.5 ElectroChemical Mechanical Planarization (ECMP)

Electrochemical mechanical planarization or polishing (ECMP) is a technology that can be a supplement or alternative to CMP for planarization of topographic copper layers. In ECMP, electrochemical surface reactions yield a high resistance passivation layer which can be easily removed from the high points by the mechanical action of a pad at low pressures (< 1 psi). In that respect ECMP is different from the novel electropolishing approaches discussed above, where high resistance of DI water was utilized to differentiate removal rates between the high and low points on the wafer. Oxidizing agents such as H_2O_2, which are needed for the operation of CMP processes [57], do not have to be included in the ECMP chemistry since anodic potential can oxidize and dissolve copper. Although application of an anodic voltage to a wafer carrier [58] or the wafer surface [32, 59–61] while polishing it with a pad was proposed several years ago, results of integrating this technique in Back end of line (BEOL) process flow started appearing in the literature after 2004 [62–66].

31.5.1 Planarization Mechanism of ECMP

Planarization mechanism of ECMP is schematically shown in Fig. 31.6a,b. In Fig. 31.6a, a passivation film is formed on the top wafer surface as well as within the cavity in the copper layer over the low aspect ratio feature. When a pad sweeps the top surface of this structure at low pressure, it removes the passivation layer from the top surface leaving it intact within the cavity (see Fig. 31.6b). Copper etching and removal by the anodic potential and electrolyte take place only at the high points where the passivation layer is removed. The high resistivity passivation layer protects the recessed regions untouched by the pad. As a result a totally planarized copper layer (Fig. 31.6c) is formed at high planarization efficiency.

The ECMD processing cell shown in Fig. 31.3 may also be used for ECMP, with modifications, such as replacing the copper anode with a properly selected cathode material or using the support plate as the cathode. Various tool designs providing relative motion between the wafer and the perforated pad/cathode assemblies have been developed for ECMP [29]. Making electrical connection to the wafer surface is critical during the ECMP and ECMD processes since traditional contacting schemes

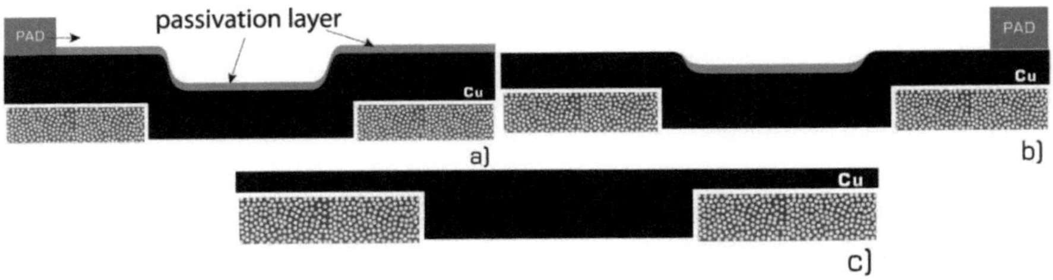

Fig. 31.6 Planarization of a topographic copper layer by ECMP: (**a**) a passivation layer is formed over the whole surface; (**b**) pad sweeping removes the passivation layer from the top surface initiating copper removal selectively from this region; (**c**) planarization is achieved since the cavity region unswept by the pad is protected by the passivation layer

used for ECD processes (such as clamp ring contacts placed at the front periphery of the wafer) would not allow full face of the wafer to be physically touched by the pad. Therefore, pad-integrated contacts have been developed [67] for use in electrochemical mechanical processes. One method of contacting the wafer surface during processing involves placing contact elements in or near the pad [67] so that the elements can slide over the wafer surface with low friction when relative motion is established between the pad and the wafer. Liquid contacts [68] and the use of a conductive polishing pad as the means to establish electrical connection with the wafer surface [67, 69] have also been proposed. Establishing low pressure electrical contact to wafers is extremely critical in electrochemical mechanical approaches especially for processing advanced node wafers with ultra low-*k* porous dielectric stacks.

For best ECMP results the passivation layer must be soft and easily abraded when contacted by the pad but must remain intact and provide high resistance in the recessed areas. Depending on the chemistry the passivation layer may be a copper complex [60] or a film composed of polymers [66]. Figure 31.7 shows the

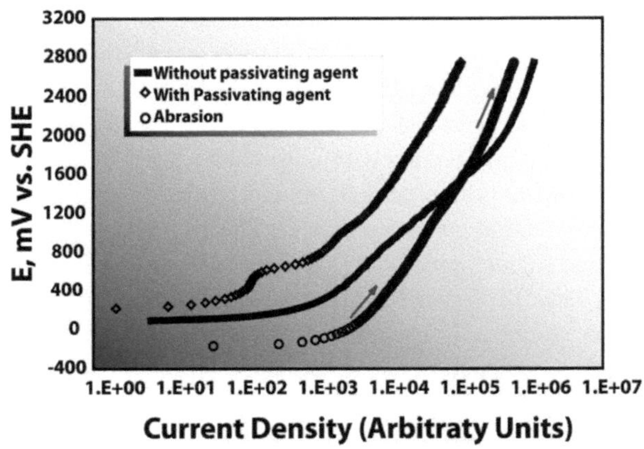

Fig. 31.7 Polarization curves of copper surface in ECMP chemistries under different conditions

polarization curves obtained from various solutions under different conditions using a modified rotating disk electrode system which allowed electrochemical measurements with and without abrasion of the copper disk surface with a pad. The data show that the current density at a given voltage got reduced by an order of magnitude upon addition of passivating agents into the virgin ECMP solution. Upon abrasion by the pad at low down force the current again increased by about an order of magnitude, demonstrating the effectiveness of the passivation layer to block current and the ease of removal of this passivating layer by the pad.

The copper removal rate in ECMP process is proportional to current density as shown in Fig. 31.8, and as reported by others [62–64]. Planarization efficiencies close to 100% have been demonstrated for both recessed regions and over-plated areas on wafers [62–66]. Figure 31.9 shows the step height reduction in an ECMP process as a function of removed copper thickness. Measurements were taken at 100 μm bond pads for this data. The second data set in Fig. 31.9 shows the reduction of copper protrusion over an array of 0.18 μm wide trenches with 50% density.

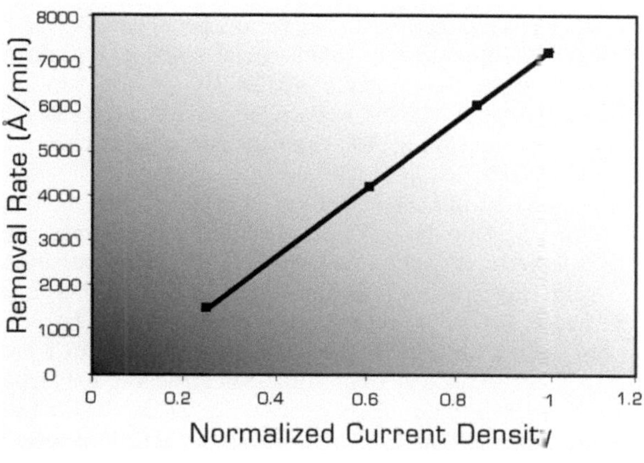

Fig. 31.8 Copper removal rate as a function of current density in Cu ECMP

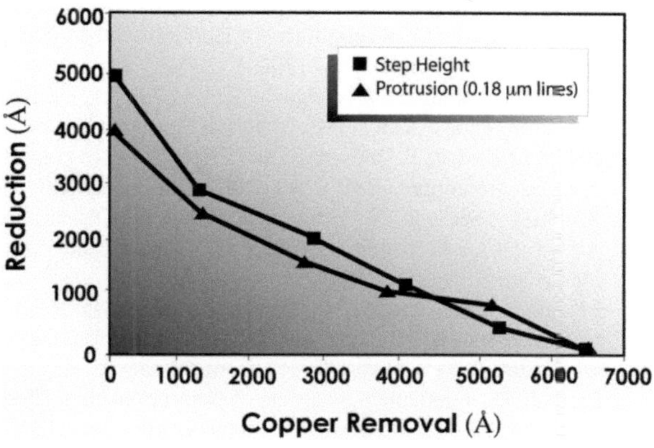

Fig. 31.9 Step height and protrusion reduction as a function of removed Cu thickness in ECMP

Results demonstrate the effectiveness of ECMP to obtain globally and locally planar copper layers.

Step height and protrusion reduction in ECMP: Using ECMP as the bulk copper removal step in a CMP tool, improved dishing results, high electrical yields, and low defect densities were reported [62, 64]. Barrier layer CMP process window was also evaluated and found to be wider for wafers processed by ECMP during the bulk copper removal step [63]. Emesh et al. [65] integrated ECMP process with the plating tool instead of the CMP tool, delivering a thin and flat copper overburden to the CMP process. Electrical characterization showed lower sheet resistance values for such wafers [65] as was the case for flat wafers processed by the ECMD process [52, 56].

References

1. Plawsky, J. L.; Gill, W. N.; Jain, A.; and Rogojevic, S.: In: Interlayer Dielectrics for Semiconductor Technologies. Murarka, S. P.; Eizenberg, M.; and Sinha, A. K.;(eds.) Elsevier Academic Press, UK, 261 (2003)
2. Kloster, G.; Scherban, T.; Xu, G.; Blaine, J.; Sun, B.; and Zhou, Y.: Porosity effects on low-k dielectric film strength and interfacial adhesion, Proc. International Interconnect Technology Conf.,IEEE, San Francisco 242 (2002)
3. Maitrejean, S.; Fusalba, F.; Patz, M.; Jousseaume, V.; and Mourier, T.: Adhesion studies of thin films on ultra low-k, Proc. International Interconnect Technology Conf., IEEE, San Francisco, 206 (2002)
4. Kondo, S.; Tokitoh, S.; Yoon, B. U.; Namiki, A.; Sone, A.; Ohashi, N.; Misawa, K.; Sone, S.; Shin, H. J.; Yoshie, T.; Yoneda, K.; Shimada, M.; Ogawa, S.; Matsumoto, I.; and Kobayashi, N. S.: Low pressure CMP for reliable porous low-k/Cu integration, Proc. International Interconnect Technology Conf., IEEE, San Francisco, 86 (2003)
5. Leduc, P.; Savoye, M.; Maitrejean, S.; Scevola, D.; Jousseaume, V.; and PassemardG.: Understanding CMP-induced delamination in ultra low-k/Cu integration, Proc. International Interconnect Technology Conf.,IEEE, San Francisco, 209 (2005)
6. Wang, X. B.; Tan, J. B.; Siew, Y. K.; Zhang, B. C.; Liu, W. P.; Zhang, F.; Leong, L. S.; Roy, R.; and Hsia, L. C.: Integration of Cu-CMP process with combination of abrasive free copper polishing and low selective barrier polishing for 90 nm Cu/low-k interconnect, AMC 2004 (Materials Research Society) 571 (2005)
7. Yamada, Y.; Konishi, N.; Watanabe, S.; Noguchi, J.; Jimbo, T.; and Inoue, O.: Study on the degradation of TDDB reliability for Cu/low-k integration caused by Cu CMP process, CMP-MIC Conference, IMIC, 567 (2005)
8. Konishi, N.; Yamada, Y.; Noguchi, J.; and Tanaka, U.: Improvement in Cu-CMP technology for 90-nm nodes, AMC 2003, Materials Research Society, 127 (2004)
9. Belov, I.; Kim, J. Y.; Moser, T.; and Pierce, K.: Novel low-abrasive slurries and abrasive-free solutions for copper CMP, CMP-MIC Conference, IMIC, 300 (2005)
10. Kondo, S.; Sakuma, N.; Homma, Y.; Goto, Y.; Ohashi, N.; Yamaguchi, H.; and Owada, N.: Abrasive-Free Polishing for Copper Damascene Interconnection, J. Electrochem. Soc. **147**, 3907 (2000)
11. Matsuda, T.; Takahashi, H.; Tsurugaya, M.; Miyazaki, K.; Doy, T. K.; and Kinoshita, M.: Characteristics of Abrasive-Free Micelle Slurry for Copper CMP, J. Electrochem. Soc. **150**, G532 (2003)
12. Landolt, D.: Fundamental aspects of electropolishing, Electrochim. Acta **32**, 1 (1987).
13. Datta, M.: Anodic dissolution of metals at high rates, IBM J. Res. Dev. **37**, 207 (1993)

14. Vidal, R.; and West, A. C.: Copper electropolishing in concentrated phosphoric acidJ. Electrochem. Soc. **142**, 2682 (1995)
15. Bernhardt, A.; and Contolini, R.: Electrochemical Planarization, US Pat. No. 5,256,565 (1993)
16. Contolini, R.; Bernhardt, A. F.; and Mayer, S. T.: Electrochemical planarization for multilevel metallization, J. Electrochem. Soc. **141**, 2503 (1994)
17. Contolini, R. J.; Mayer, S. T.; Graff, R. T.; Tarte, L.; and Bernhardt, A. F.: Electrochemical planarization of ULSI copper, Solid State Technology 155 (1997)
18. Lopatin, S.; Preusse, A.; and Cheung, R.: Interconnect and Contact Metallization for ULSI. Arita, Y.; MathadG. S.; and Rathore, H. R., Eds. Electrochemical Society, New Jersey, 221 (1999)
19. Chang, S. H.; Shieh, J. M.; Huang, C. C.; Dai, B. T.; Li, Y. H.; and Feng, M. S.: Microleveling mechanism and applications of electropolishing on planarization of copper metallization. J. Vac. Sci. Technol. B **20**, 2149 (2002)
20. Padhi, D.; Yahalom, J.; Gandikota, S.; and Dixit, G.: Planarization of copper thin films by electropolishing in phosphoric acid for ULSI applications. J. Electrochem. Soc **150**, G10 (2003)
21. Chang, S. C.; Shieh, J. M.; Dai, B. T.; Feng, M. S.; Li, Y. H.; Shih, C. H.; Tsai, H. M.; Shue, S. L.; Liang, R. S.; and Wang, Y. L.: Superpolishing for planarizing copper Damascene interconnects, Electrochem. Solid State Lett. **6**, G72 (2003)
22. Chang, S. C. and Wang, Y. L.: Effects of applied voltages on planarization efficiency of copper electropolishing. J. Vac. Technol. B **22**, 2754 (2004)
23. Huo, J.; Solanki, R.; and McAndrew, J.: Study of anodic layers and their effects on electropolishing of bulk and electroplated films of copper. J. Appl. Electrochem. **34**, 305 (2004)
24. Du, B. and Suni, I. I.: Mechanistic studies of Cu electropolishing in phosphoric acid electrolytes. J. Electrochem. Soc. **151**, C375 (2004)
25. Liu, S. H.; Shieh, J. M.; Chen, C.; Dai, B. T.; Hensen, K.; and Cheng, S. S.: Two-additive electrolytes for superplanarizing Damascene copper metals, Electrochem. Solid-State Lett. **8**, C47 (2005)
26. Huo, J.; Solanki, R.; and McAndrew, J.: A novel electroplanarization system for replacement of CMP. Electrochem. Solid-State Lett. **8**, C33 (2005)
27. Wang, H.: Methods and apparatus for electropolishing metal interconnections on semiconductor devices, US Pat. No. 6,837,984 (2005)
28. West, A. C.; Shao, I.; and Deligianni, H.: Numerical simulation of electrochemical planarization of copper overburden. J. Electrochem. Soc. **152**, C652 (2005)
29. Suni, I. I. and Du, B.: Copper planarization for ULSI processing by electrochemical methods: a review. IEEE Trans. Semicond. Manuf. **18**, 341 (2005)
30. West, A. C.; Deligianni, H.; and Andricacos, P. C.: Electrochemical polishing of interconnect metallization. IBM J. Res. Devel. **49**, 37 (2005)
31. Singer, P.: Copper challenges for the 45 nm node. Semicond. Int. **27**, 40 (2004)
32. Talieh, H.: Method and apparatus for electrochemical mechanical deposition. U.S. Patent No. 6,176,992 (2001)
33. Uzoh, C. E.; Talieh, H. B. Basol, M.; and Young, D. W.: Workpiece proximity plating apparatus. U.S. Patent No. 6,630,059 (2003)
34. Basol, B.M.: Plating method and apparatus that creates a differential between additive disposed on a top surface and a cavity surface of a workpiece using an external influence. U.S. Patent No. 6,534,116 (2003)
35. Basol, B. M.; Uzoh, C. E.; Talieh, H.; Wang, T.; Guo, G.; Erdemli, S.; Cornejo, M.; Bogart, J.; and Basol, E.C.: Planar copper electroplating and electropolishing techniques. Chem. Eng. Comm. **193(7)**, 903 (2006)
36. Noji, I.; Kobata, I.; Yasuda, H.; Izumi, T.; Kumekawa, M.; Wada, Y.; Fukunaga, A.; Tsujimura, M.; Toma, Y.; Suzuki, T.; and Saitoh, T.: Application of electro-chemical polishing in DI water to Cu Damascene wiring planarization process, AMC 2004 (Materials Research Society), 577 (2005)

37. Wada, Y.; Noji, I.; Kobata, I.; Kohama, T.; Fukunaga, A.; and Tsujimura, M.: The enabling solution of Cu/low-k planarization technology, Proc. International Interconnect Technology Conf.,(IEEE, San Francisco, 126 (2005)
38. Mazur, S.; Jackson, C. E.; and Foggin, G. W.: Membrane-mediated electropolishing of Damascene copper, Proc. International Interconnect Technology Conf., IEEE, San Francisco, 206 (2005)
39. Dow, W. P. and Huang, H. S.: Roles of chloride ion in microvia filling by copper electrodeposition. J. Electrochem. Soc., **152,** C67 (2005).
40. Andricacos, P. C.; Uzoh, C.; Ducovic, J.O.; Horkans, J.; and Deligianni, H.: Damascene copper electroplating for chip interconnections. IBM J. Res. Dev. **42**, 567 (1998)
41. West, A. C.: Theory of filling of high aspect ratio trenches and vias in presence of additives. J. Electrochem. Soc. **147**, 227 (2000)
42. ReidJ. and Mayer, S.: Factors influencing fill of IC features using electroplated copper, AMC 1999, (Materials Research Society), 53 (2000)
43. Josell, D.; Wheeler, D.; Huber W. H.; and Moffat, T. P.: Superconformal electrodeposition in submicron features. Phys. Rev. Lett. **87**, 016102-1 (2001)
44. Moffat, T. P.; Wheeler, D.; Huber, W.H.; and Josell, J.: Superconformal electrodeposition of copper, Electrochem. Solid-State Lett. **4**, C26 (2001)
45. Basol, B. M.; Uzoh, C.; Talieh, H.; Young, D.; Lindquist, P.; Wang, T.; and Cornejo, M.: ECMD technique for semiconductor interconnect applications. Microelectron. Eng. **64**, 43 (2002)
46. Basol, B. M.: Mechanically induced super-filling of low aspect ratio cavities in an electrochemical mechanical deposition process. J. Electrochem. Soc. **151**, C765 (2004)
47. Basol, B. M.; Erdemli, S.; Uzoh, C.; and Wang, T.: Planarization efficiency of electrochemical mechanical deposition and its dependence on process parameters. J. Electrochem. Soc. **153(3)**, C176 (2006)
48. Basol, B. M. and West, A. C.: Study on mechanically induced current suppression and super filling mechanisms. Electrochem. Solid-State Lett. **9(4)**, C77 (2005)
49. Cao, Y.; Taephaisitphongse, P.; Chalupa, R.; and West, A. C.: Three-additive model of superfilling of copper. J. Electrochem. Soc. **148**, C466 (2001)
50. Uzoh, C.; Basol, B.; and Talieh, H.: Pad designs and structures for a versatile materials processing apparatus. US Patent No. 6,413,388 (2002)
51. Basol, B. M.; Uzoh, C. E.; and Bogart, J. A.: Low-force electrochemical mechanical processing method and apparatus, US Patent Publication No. 2003/0064669 (2003)
52. Mourier, T.; Haxaire, K.; Cordeau, M.; Chausse, P.; DaSilva, S.; and Torres, J.: Electrochemical mechanical deposition and reverse linear planarization of copper for 45 nm node ULK integration, AMC 2004, (Materials Research Society), 597 (2005)
53. Moffat, T. P.; Wheeler, D.; Witt, C.; and Josell, D.: Superconformal electrodeposition using derivitized substrates. Electrochem. Solid-State Lett. **5**, C110 (2002)
54. Taephaisitphongse, P.; Cao, Y.; and West, A.: Electrochemical and fill studies of a multicomponent additive package for copper deposition. J. Electrochem. Soc. **148**, C492 (2001)
55. Stickney, B.; Nguyen, B.; Basol, B.; Uzoh, C.; and Talieh, H.: Topography reduction for copper Damascene interconnects. Solid State Technol. **46**, 49 (2003)
56. Vos, I.; Heylen, N.; Hernandez, J. L.; Wang, T.; Truong, T.; Basol, B.; Sprey, H.; and Vanhaelemeersch, S.: Influence of Copper Plating and Die Layout on the Copper CMP Performance, AMC 2005: Asian Session, Tokyo, Japan (2005)
57. Aksu, S. and Doyle, F. M.: The role of glycine in the CMP of copper. J. Electrochem. Soc. **149**, G352 (2002)
58. Tsai, C. S. and Tseng, P. N.: Chemical mechanical planarization apparatus and polishing methodUS Patent No. 5,575,706 (1996)
59. Uzoh, C. E. and Harper, J. M. E.: Method of electrochemical mechanical planarization. US Patent No. 5,807,165 (1998)

60. Sato, S.; Yasuda, Z.; Ishihara, M.; Komai, N.; Ohtorii, H.; Yoshio, A.; Segawa, Y.; Horikoshi, H.; Ohoka, Y.; Tai, K.; Takahashi, S.; and Nogami, T.: Newly developed electrochemical polishing process of copper as replacement of CMP suitable for Damascene copper inlaid in fragile low-k dielectrics, IEDM 2001 (IEEE), 4.4.1-4.4.4. (2001)
61. Sun, L.; Tsai, S. D.; and Redeker, F. C.: Method and apparatus for electrochemical mechanical planarization, US Patent No. 6,379,223 (2002)
62. Economikos, L.; Wang, X.; Sakamoto, A.; Ong, P.; Naujok, M.; Knarr, R.; Chen, L.; Moon, Y.; Neo, S.; Salfelder, J.; Duboust, A.; Manens, A.; Lu, W.; Shrauti, S.; Liu, F.; Tsai, S.; and Swart, W.: Integrated electro-chemical mechanical planarization for future generation device technology, Proc. International Interconnect Technology Conf., IEEE, San Francisco, 233 235 (2004)
63. Sakamoto, A.; Economikos, L.; Ong, P.; Naujok, M.; Tseng, W.; Moon, Y.; Salfelder, J.; Duboust, A.; and Nogami, T.: Electro-chemical mechanical planarization and its evaluation on BEOL with 65 nm node dimensions, CMP-MIC Conference (IMIC), 191–199 (2005)
64. Manens, A.; Miller, P.; Kollata, E.; and Duboust, A.: Advanced process control extends ECMP process consistency, Solid State Technology, February 2006.
65. Emesh, I.; Khosla, V.; Erdemli, S.; Emami, R.; and Basol, B. M.: Thin and planar copper layers for advanced interconnect fabrication, AMC 2005 (Materials Research Society), 501 (2006)
66. Duboust, A.; Wang, Y.; Liu, F.; and Hsu, W. Y.: http://www.eurosemi.eu.com. (2005)
67. Talieh, H.; Uzoh, C.; and Basol, B. M.: Device providing electrical contact to the surface of a semiconductor workpiece during metal plating, US Patent No. 6,497,800 (2002)
68. Talieh, H. and Basol, B.: Method for forming an electrical contact with a semiconductor substrate, US Patent No. 6,471,847 (2002)
69. Kondo, S.; Tominaga, S.; Namiki, A.; Yamada, K.; Abe, D.; Fukaya, K.; Shimada, M.; and Kobayashi, N.: Novel electrochemical mechanical planarization using carbon polishing pad to achieve robust ultra low-k/Cu integration, Proceedings of the International Interconnect Technology Conf., IEEE, San Francisco, 203 (2005)

Part VIII
Metrology

This part presents the integrated and thin film metrology techniques. Chapter 32 "Integrated Metrology (IM) History at a Glance" describes the utility of Im over CMP. In the early 1990s, when oxide CMP was penetrating into semiconductor manufacturing, the process suffered from significant instability and non-uniformity, thus requiring intensive process control. The penetration of the NovaScan 210 into the CMP market was very fast, since end-users realized the benefits of Integrated Metrology very quickly and the benefits are presented in the chapter. Integrated Metrology has evolved from solving the instabilities of the nascent CMP processes to providing added value to the process based on advanced metrology capabilities. These capabilities are providing information that enables the industry to continue improvement in the area of CMP and push the limits of the IC manufacturing landscape. Chapter 33 "Thin Film Metrology – X-ray Methods" describes different X-ray-based techniques. In this chapter X-ray fluorescence (XRF), X-ray reflectometery (XRR), and small-angle X-ray scattering (SAXS) methods were describedto measure concentration/thickness of the film, thickness of the film, and pore size density respectively by non-destructive means.

Chapter 32
Integrated Metrology (IM) History at a Glance

Moshe Finarov, David Scheiner, and Gabi Sharon

32.1 Introduction

In the early 1990s, when oxide CMP was penetrating into semiconductor manufacturing, the process suffered from significant instability and non-uniformity, thus requiring intensive process control. This process control was implemented by using two methods:

- The "gating" method (polish first wafer, clean, dry, and measure – then feedback the result to adjust the process for the subsequent wafers in the lot)
- Very high sampling of the oxide thickness of the polished layers (more than 5 wafers per lot). The first generation polishers were designed as "dry in – wet out," which made it necessary to clean and dry the polished wafer on separate cleaning equipment and to measure it on a stand-alone metrology tool. This procedure took many hours and caused inefficient use of polishers in production. Leading CMP users were looking for a metrology solution to shorten this process control cycle, preferably by incorporating metrology capabilities inside the polisher.

The major challenge was the incorporation of a high-accuracy, large-sized, and very sensitive optical measurement system in the process tool environment, which is characterized by mechanical vibrations, electromagnetic interference, and wet environment. Moreover, since the polished wafers were kept in water, it was necessary to carry out the optical measurements in water, something that was not practiced before. Integrating a high-precision system with the processing equipment required a metrology tool with high reliability and high throughput, so not to affect the polisher's uptime and throughput. In addition, high tool-to-tool matching of the metrology was needed in order to provide uniform metrology criteria for all polishers in the fab [1–3].

G. Sharon (✉)
Nova Measuring Instruments Ltd., Weizmann Science Park, Rehovot 76100, Israel
e-mail: gab-s@nova.co.il

Y. Shacham-Diamand et al. (eds.), *Advanced Nanoscale ULSI Interconnects: Fundamentals and Applications*, DOI 10.1007/978-0-387-95868-2_32,

At the Semicon West show in July 1995, the concept of integrated measuring instruments was first presented by Nova Measuring Instruments. It was a dedicated, compact opto-mechanical design meeting all these requirements, while keeping the wafer stationary and in water during measurement [9]. The first commercial ITM (integrated thickness monitoring) product, the NovaScan 210, was designed for measuring the thickness of dielectric films inside the polisher and integrated with Strasbaugh 6DS-SP and Westech 372 polishers. The year 1995 is considered as the beginning of the integrated metrology (IM) era and forms a paradigm shift in CMP processing in particular and in the semiconductor industry in general.

The penetration of the NovaScan 210 into the CMP market was very fast, since end-users realized the benefits of integrated metrology very quickly. These benefits included

- Increase of the polisher utilization time
- Faster response and higher sampling rate resulting in significantly better process control
- Shorter manufacturing cycle
- Exclusion of test wafers
- Saving of labor and clean room space normally required for SA metrology tools.

Integrated metrology was the right solution, in the right place, at the right time.

However, together with clear success in the marketing and technology areas, integrated metrology as a business model confronted many commercial difficulties on its way to the end-user. Inherently, IM requires very tight cooperation between three parties: the end-user, the equipment manufacturer, and the IM tool supplier. Compared to other OEM components, IM performance is strongly application dependent and requires close cooperation between end-users and IM tool suppliers in order to provide adequate performance and long-term support.

Most of the end-users were not ready for such a business model and were driven by the desire to buy the full range of equipment from one vendor with its full responsibility for both performance and support. Nevertheless, understanding the production needs, they were the main driving force for IM implementation and pushed the CMP equipment manufacturers to adopt IM solutions.

A majority of the leading process equipment manufacturers had a negative attitude regarding IM. Integrating a system that they did not manufacture into their tools, required taking overall responsibility without having the expertise and support for proper operation. Integration of the IM tools required significant engineering resources and they feared a negative impact of the IM tool on the throughput and reliability of their equipment. Additionally, the IM systems were a relatively expensive option on the processing tool and the limited additional revenue that could be obtained from customers caused the equipment manufacturers to initially resist the implementation.

It took 2–3 years before the leading process equipment manufacturers (PEMs) accepted IM and the corresponding process control capability as a critical feature of continued improvement of CMP equipment performance. Overall, successful cooperation with IM suppliers, such as Nova, Dainippon Screen, and Nanometrics, illustrates that most of the commercial and technical problems associated with the "love triangle" between end-users, PEMs, and IM equipment manufacturers have been solved.

Currently all leading polishers are offered with on-board IM tools and enable excursion control and closed-loop control (CLC) for practically all CMP processes: ILD, PMD, STI, poly-silicon, Tungsten, and Copper. The most advanced IM tools, such as NovaScan 2040, 3060, and 3090, are fast enough to measure every wafer, both before and after CMP, thus including initial thickness variations in the CLC model.

The IM systems have evolved significantly in the last decade: from simple visible light reflectometry in the first models (NovaScan 210, 420, 840) to Polarized DUV scatterometry in the latest model (NovaScan 3090). Metrology performance of the modern IM tools such as NovaScan 3090 and NanoSpec 9010 is similar to those of the advanced stand-alone metrology tools, and IM tools have largely replaced stand-alone metrology in the CMP area of the fab.

Integration between the IM tool and the CMP polisher has significantly evolved during this period. The first IM tools were installed inside the polishers but their operation was independent – recipe design, measurement runs, and data collection were separated from the polisher's computer and user interface. Currently the integration between them is more extensive: all the IM operation is carried out from the polisher's GUI through the SECS/GEM communication. The measurement results are used in real time for CLC and correcting the polishing parameters of the next wafers in the lot.

In less than one decade, integrated metrology has experienced explosive growth and resulted in major changes in the CMP area; today more than 1500 IM tools are working in all advanced semiconductor fabs. Technologically IM, together with the advances of the IC manufacturing technology, has experienced the following changes since 1995:

- From measuring thick ILD layers to measuring single Angstrom residues after STI and Copper polish
- From solid test sites measured by visible reflectometry to line arrays measured by DUV scatterometry
- From 2 min for 5 measurement points on 200 mm wafers to 13 s for 13 points on 300 mm wafers.
- From delayed statistical process control (SPC) to real-time CLC.
- IM tools have become a standard for dielectric CMP processing and continue development in order to meet new requirements of the next ITRS technology nodes.

32.2 Integrated Metrology (IM) Technology

Practically all IM tools used for CMP process control are based on spectrophotometry or reflectometry. According to this method, reflectance of any site on the wafer is measured in a broad spectral range at normal incidence. The measured spectrum is compared to a theoretical spectrum calculated in advance or in real time according to an optical model. The model describes the measured site in terms of both geometrical parameters (thickness, array period, etc.) and material optical parameters (index of refraction and extinction coefficient of each layer of measured stack as a function of wavelength). The interpretation procedure consists of iterative calculations with variation of the free parameters of the optical model until a good fit between the measured and the calculated spectra is reached. When the goodness of fit (GOF) reaches a predefined threshold, the variable parameters of the optical model are reported as the results of this measurement.

Reflectometry utilizes the well-known phenomena of light interference in thin films Fig. 32.1). Since phase shifts between light beams reflected from different boundaries of the measured layer stack depend strongly on layer thickness, the total reflectance contains information about thickness and optical properties of all layers.

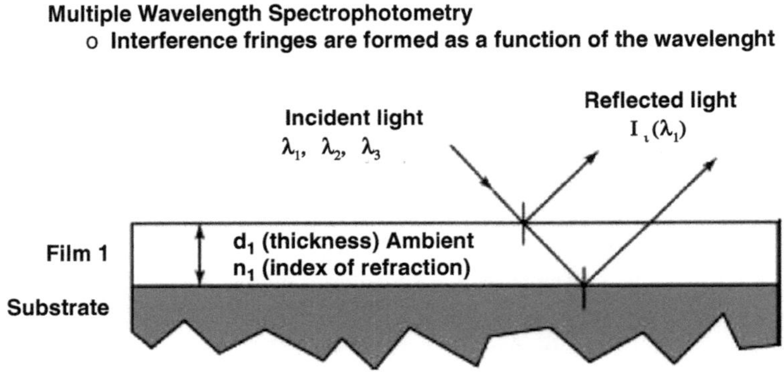

Fig. 32.1 Spectrophotometer based on refraction index

Light reflected from the top surface of the film stack interferes with light reflected from all interfaces between layers where a change in index of refraction exists, resulting in a spectrum with intensity variation as a function of wavelength. This distribution can then be used to determine the thickness of a transparent film or multiple films in a stack, if the optical film properties (i.e., index to infraction) are known.

If the number of layers and accordingly, the number of optical model variables is small, the extraction of the film thickness value is simple and unambiguous. If a multi-layer stack is measured and many of the parameters are unknown, interpretation is difficult and requires a thorough spectral analysis based on sophisticated interpretation algorithms and a very high accuracy optical measurement.

The resulting measurement accuracy of the film thickness depends on the following main factors:

- Measurement accuracy, which mainly includes dynamic repeatability and tool-to-tool matching
- Accuracy of the optical model describing the measured structure
- Sensitivity of the measured spectrum to the desired parameters such as top layer thickness.

Measurement repeatability mainly depends on the signal-to-noise ratio level (SNR) which is limited by brightness of the light source, noise, and sensitivity of the photo-detector and other instrumental factors. For measuring in small measurement sites on patterned wafers it is important to accurately locate the light spot of the measurement beam on the selected site with high positioning and focusing accuracy.

Another significant factor affecting measurement accuracy is calibration. The measured spectrum of light intensity reflected from the measurement site can be transformed to an absolute value of reflectance by normalizing the measured raw spectrum to a spectrum acquired at the same conditions from a material with a well-established and accurate optical model. Usually bare silicon is used for spectrum calibration.

The accuracy of the optical model is highly important for accurate spectral measurements. If the measured structure includes well known and stable materials such as silicon dioxide, silicon nitride, silicon, and the like, optical dispersion of the materials is well determined and can be used as known parameters. If some new material is used, it is necessary to measure its dispersion in advance or to calculate the dispersion together with the thickness during spectrum interpretation. Multiple variable parameters make interpretation less reliable and more time consuming. In practice, it is recommended that not more than 3–4 free parameters be calculated simultaneously.

Sensitivity of the measured spectrum to the desired parameters should be evaluated carefully in order to predict the level of accuracy that may be achieved with a given measurement system for a specific application (i.e., optical model). For example, in order to obtain good sensitivity to layer thickness, the correct spectral range should be applied: the thinner the measured layer, the shorter the wavelengths that should be used.

Figure 32.2 shows an optical scheme of a spectral reflectometer. It consists of three main optical channels: illumination (light source), imaging (camera), and spectrophotometer (sensor). In order to enable accurate positioning of the measurement spot on the test site, the imaging and spectrophotometric channels are combined using a pinhole mirror. Accurate placement of the spot on the test site is carried out automatically by a motion system supported by image acquisition and image processing of the pattern around the test site which is kept in the tool memory

In recent years scatterometry-based measurements have become mainstream for measurement of geometrical dimensions of patterned structures. To support this, the latest IM tools an additional polarizer in the main optical path. This enables measurement on patterned sites like line arrays, because in such cases the reflected light beam is a zero-order diffraction beam, whose properties are different for different polarizations. Rotating the polarizer to a certain position relative to the line

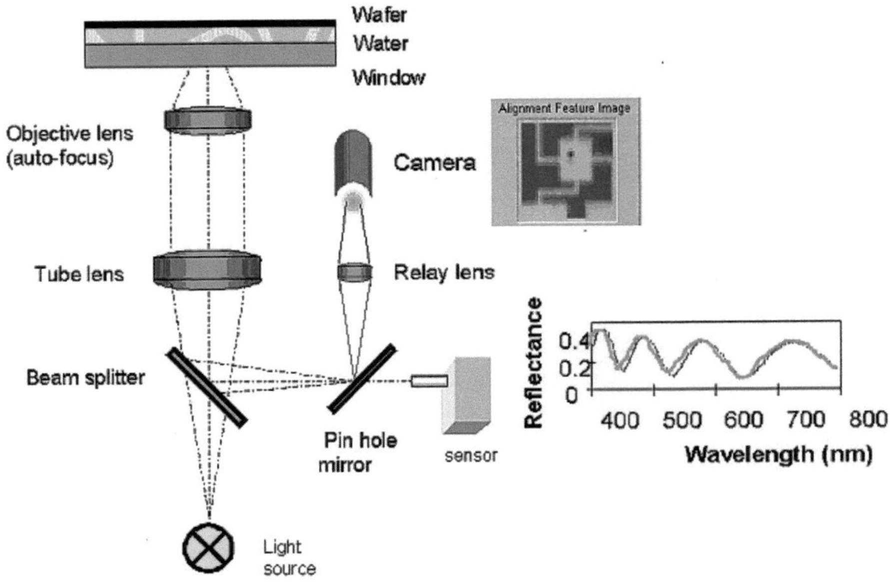

Fig. 32.2 Typical optical scheme of spectral reflectometer

array orientation, e.g., parallel and perpendicular, permits separation of these two polarizations and thus enables implementation of highly accurate optical models of diffraction such as RCWT (rigorous coupled wave theory).

One of the major components of any metrology system is a motion system that enables placement of the measurement spot on the required position on the wafer. Such a motion system should be fast and accurate, allowing single micron positioning repeatability and accuracy. There are two main motion concepts for metrology systems:

In conventional configurations, the wafer is moved in multiple degrees of freedom, while a stationary optical assembly is used for measurements.

In Nova's concept a small optical head is moved over the stationary wafer. In this case the wafer may be in any ambient, including water, allowing flexible integration with the processing equipment [6–8].

The examples of IM tools in wet integrations are presented in Figs. 32.3 and 32.4.

The first "dry in-wet out" polishers required so-called wet integrations – where the wafer was held in water during measurement. Currently most CMP polishers are of the "dry in-dry out" type, enabling easier "dry integration" where the IM tool is attached to the outside of the process tool and the wafers are measured in air after cleaning and drying, before being transferred to the output cassette [10]. However, many CMP manufacturers prefer the internal "wet integration" even for "dry in-dry out" polishers because of faster response to the process and better CLC performance. Examples of wet and dry integrations benefits are presented in Table 32.1.

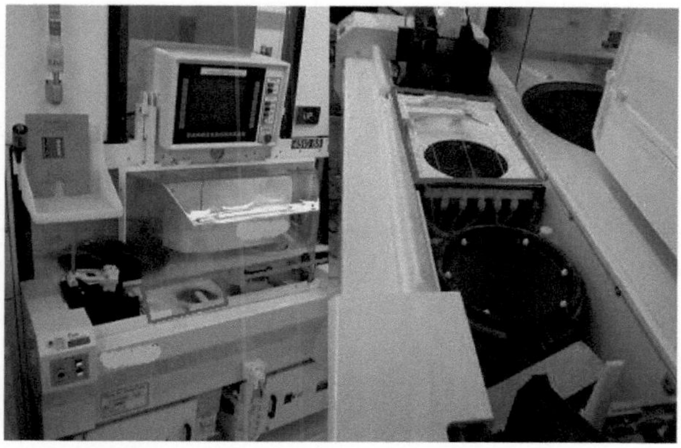

Fig. 32.3 Wet NovaScan 210 integrated on to a Westech 372

Fig. 32.4 Dry factory interface

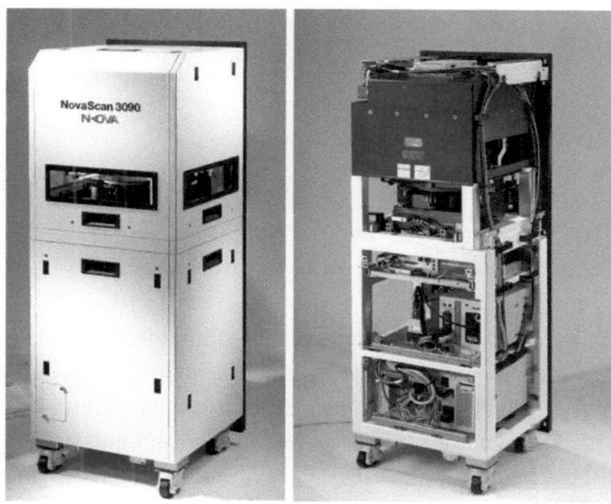

Table 32.1 Wet vs. dry IM benefits

Wet	Dry
Closer to the process	Faster and cheaper integration
Fastest CLC feedback	Easier application development
Reduces rework	Lower sensitivity to STI oxide residues
Reduces gating wafer cycle time	Recipe compatibility with stand-alone metrology tools
Fastest reactions to process drifts	

As an example, Table 32.2 presents highlights of the latest generation IM tool specification. This tool complies with requirements of the 65 nm technology node according to ITRS. The tool's main features are polarized light in a broaden spectral range – from DUV to NIR, small spot size, and very high throughput.

Table 32.2 NovaScan 3090 IM tool specification highlights

Specification requirements	
Spectral range	250–950 nm
Polarization	Rotating polarizer, full spectral range
Spot size	25×25 (μ^2)
Measurement time	15 s for 13 points per wafer one polarization

32.3 IM CLC

Enhancing the polishing process by using in-line metrology with closed-loop control IM tools enables wafer to wafer and within wafer process control. Closed-loop control (CLC), utilizing the measurement data (pre- and post-polish), provides real-time adjustment of the CMP process (Fig. 32.5).

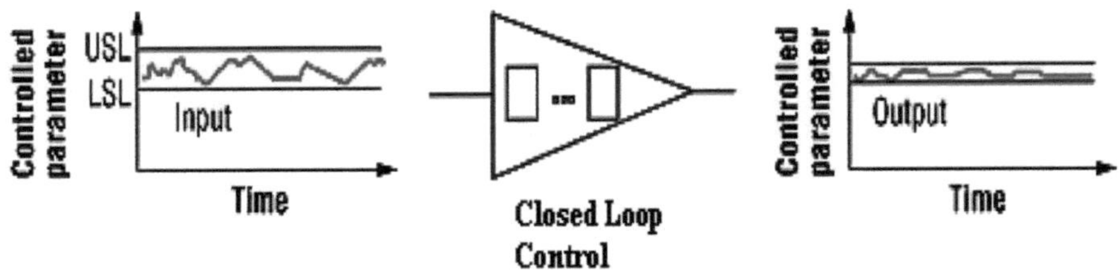

Fig. 32.5 Closed-loop control enables reduced output variation

In the production environment, maintaining stable process performance that does not drift out of specified limits is a major challenge for CMP. For example, variability in preceding deposition steps creates non-uniform pre-polish film thickness. Other effects that must be taken into account are the drift in polish rate over the lifetime of the polish pad and the influence on polish rate of various product wafer topographies.

Process control in semiconductor manufacturing has traditionally been implemented through a series of stand-alone systems that together create the control loop.

Typically, the process begins with gating wafers used to identify required process adjustments or using default polish time, the time adjustment is used for the rest of the run (see Fig. 32.6). This method is operator-intensive and time-consuming and extends overall cycle time.

CMP processing requires continuous monitoring and data feedback from each exiting wafer. The data are fed to the process controller which can adjust process conditions for the following wafer by dynamically adjusting polishing time.

The closed-loop control (CLC) delivers reliable, highly automated wafer-to-wafer uniformity over varied CMP manufacturing processes. The data of every processed wafer are measured inside the polisher, and process parameters are fed back to adjust the next wafer polish. CLC PID algorithms developed by Nova Measuring Instruments take into account a variety of different factors during the production process such as

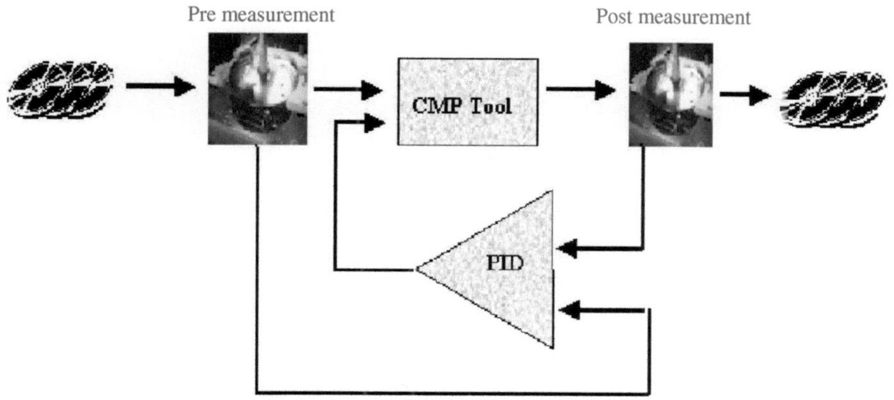

Fig. 32.6 Closed-loop control diagram

- History of the head/lot
- Post-thickness target or amount removed
- Incoming thickness
- Control single or multiple heads
- Detect and ignore abnormal removal rate change detection
- Detect polish time “Out Of Spec” (OOS)
- Detect polish time change
- Detect and ignore invalid measurements results
- Ignite Process Alarm event.

The algorithms also

- Support gating.
- Support first wafer calculation per lot or default polish time.
- Support idle time calculation.
- Support product-to-product calculation.
- Support auto rework.

The use of classical control theory and extensive accumulated field experience provides a solid basis for the CLC option. Sampling rate is high, as the ITM systems measure each wafer, and the process parameters are fed (back) for the next polished wafer.

The measurement results are used to capture process excursions and trends at early stages, which may be fed back to the polisher for automated process control (APC). In production, the overall cycle time is reduced considerably and customers achieve higher yield and better chip performance, as well as tighter consumables control.

By enabling 100% pre- and post-process control, the advantages of closed-loop control are immediate, and improve overall productivity and cost of ownership per wafer.

The benefits of using CLC are as follows [4, 5]:

- Improved outgoing CMP quality
- Reduced cycle time in CMP area
- Reduced consumables usage
- Improved polisher utilization
- Reduced labor
- Improved productivity
- Decrease test wafer use.

32.4 CLC Results

"Mean with default polish time" is an off-line calculation based on polisher performance while using the same polish time of wafer 1 for the rest of the lot (Fig. 32.7: wafers 2–25). Figure 32.8 presents the closed-loop controlled performance.

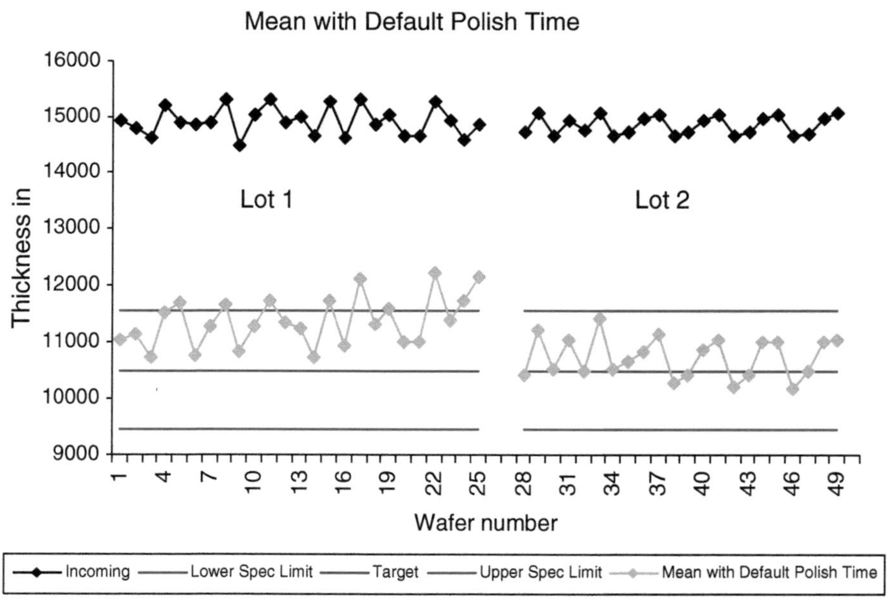

Fig. 32.7 Open-loop control using default polish time

Figures 32.9 and 32.10 show extensive data (over 3000 wafers) from different products showing the improvement in standard deviation obtained with the use of CLC.

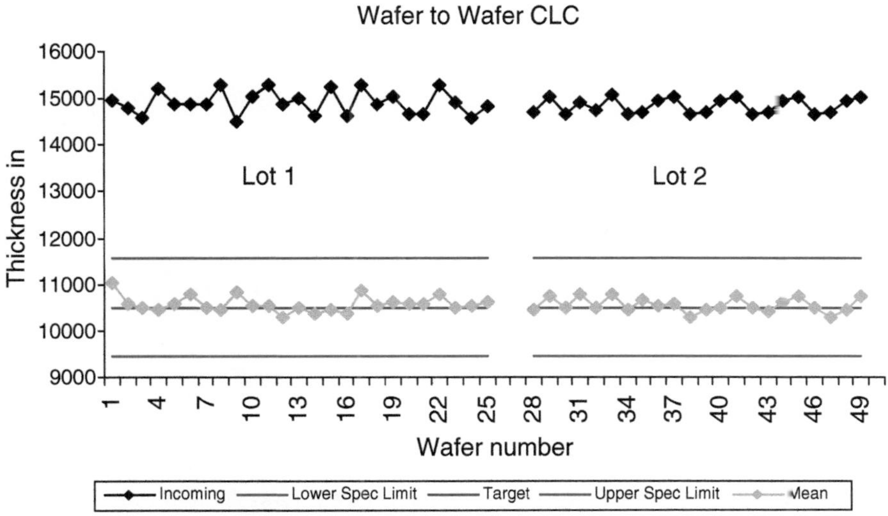

Fig. 32.8 Closed-loop control – polish time adjusting for every wafer

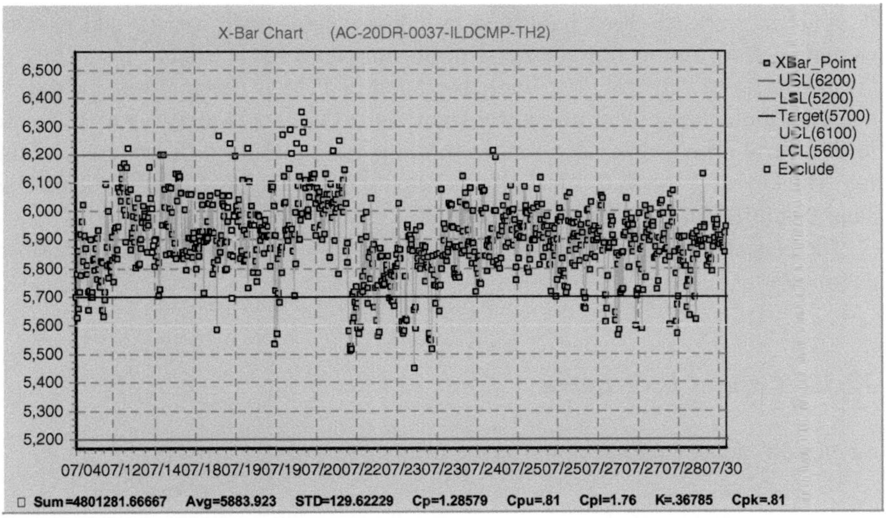

Fig. 32.9 Before using process control (STDEV = 130, CpK = 0.81)

32.5 Qualitative View for ITM Return on Investment

Qualitative considerations are used for return on investment (ROI) calculations since different IC manufacturers have specific ROI considerations and a generic model is not reliable enough for ITM since it differs a great deal with the area for implementation and the specific metrology used.

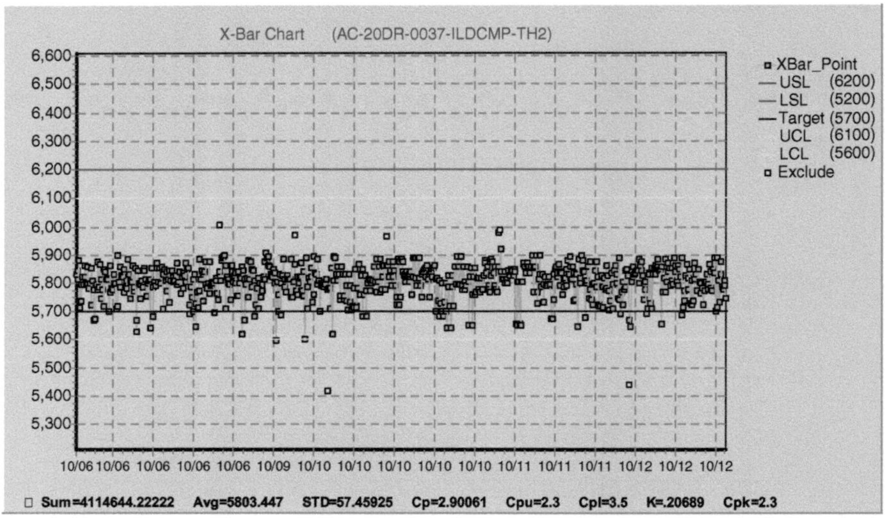

Fig. 32.10 After using process control (STDEV = 57, CpK = 2.3)

A lot of work has been published in industry conferences on the added-value stemming from run to run control. Run to run can refer to either "lot to lot" or "wafer–to-wafer" control. The implementation of ITM is playing a major role in enabling the enhancement of control from "lot to lot" to "wafer to wafer" to die level control. The added value for run to run is not equal for different areas of the fab. The basis for the ROI calculation is heavily based on these figures. Again it is clear that different areas have different ROI.

The considerations for ROI are based on the main categories that are described as follows.

32.5.1 Capital Investment

Savings in capital equipment may come from three areas

- ITM replacing SA tools
- Saving on automation needs (e.g., FOUP drops) for SA
- Clean room space needed for the stand-alone tools.

32.5.2 Labor Reduction

Operators that previously worked with stand-alone metrology tools are now redundant since the measurement is done in the process background without the need to load/unload wafers for the metrology tool. Analysis and disposition can be automated as well and save more labor with wafer costs.

32.5.3 Process

Other process savings will come from four areas:

32.5.3.1 Rework

In cases where there is a systematic problem on the process tool, it is important to correct it as soon as possible or lots will have to be reworked or scrapped. By using the ITM detecting and correcting happen and less lots are sent to rework or scrap.

32.5.3.2 Reduction of Test Wafer Usage

Use of ITM leads to a reduction of test wafer usage since the CMP tool is monitored in real time. Test wafers or "qual" wafers can be eliminated since targeting data or performance data are continuously available and therefore there is no need for periodic test wafers or targeting wafers for tool adjustments.

32.5.3.3 Excursion Detection

The on-going, real-time measurement using ITM enables early detection of excursions. This could be either at the lot level, wafer level, or die level. Early detection of excursions is an essential risk reduction factor in IC high volume manufacturing environment.

32.5.3.4 Yield

Integrated metrology has the ability to improve yield over the current control methods, which are based on stand-alone tools. Through higher density sampling plans, available without throughput hit or process delay, the end-user can have better process modeling and finer targeting. Use of APC-enabled process control, based on real-time data, real-time adjustments, and tighter C_p and C_{pk}, leads to better process control and higher yield.

32.6 Manufacturing

Integrated metrology has the ability to save costs associated with manufacturing due to the following.

32.6.1 Cycle Time Reduction

One of the inherent advantages of integrated metrology is that it saves the time associated with stand-alone measurements. Since there is no queue before the integrated metrology tool and because the measurement time takes place during the process, these times are eliminated.

The time saving components are

- Overall measurement time (embedded in the polisher cycle time).
- Queue time before measurement
- Wafer Automation, lot/Foup handling time (in 300 mm fab is a time-consuming operation)
- Working with IM eliminates these delays and the associated cost related to Foup drops and rails.

ITM in CMP has become a known practice and in many cases oxide CMP uses ITM as a default. The back-end process of copper CMP is now starting to gain momentum with ITM being incorporated into the process.

32.6.2 Reduced Downtime After PM and Faster Tool Qualification

Typically after preventative maintenance (PM) the polisher is idle until all metrology results are collected to insure it's proper working conditions. With ITM, faster qualification is achieved with less consumable usage.

With the introduction of APC, integrated metrology is moving from the historical position of being a "non-value added operation" to becoming a "added-value" operation in conjunction with APC as it helps to reduce process variation, detects excursions, and improves productivity. All the above makes a valid case for ROI and serves as an incentive for implementation.

New technologies being introduced, such as Scatterometry, are providing increased levels of process information for even better process control.

In Fig. 32.11 below, one can see that as shrinking technology nodes evolve, control schemes are getting more advanced, more measurements are needed per wafer (from 5 to over 27 points per wafer) and the throughput of the metrology tools is increasing to support this.

32.7 Shallow Trench Isolation (STI)

In STI CMP there are two main trends requiring advanced IM solutions. The first trend is that the nitride and residual oxide layers become very thin requiring the highest possible performance of the metrology tools. Thin film metrology is evolving toward the deep-UV or even the vacuum-UV spectral range, and the repeatability and tool-to-tool matching of IM tools is approaching the level of 0.1%.

Another trend is a need to carry out the measurements directly on die area, or at least on patterned measurement sites, which better correlate to the topography-dependent CMP behavior on actual die [11]. This requires sophisticated techniques like spectral scatterometry or AFM.

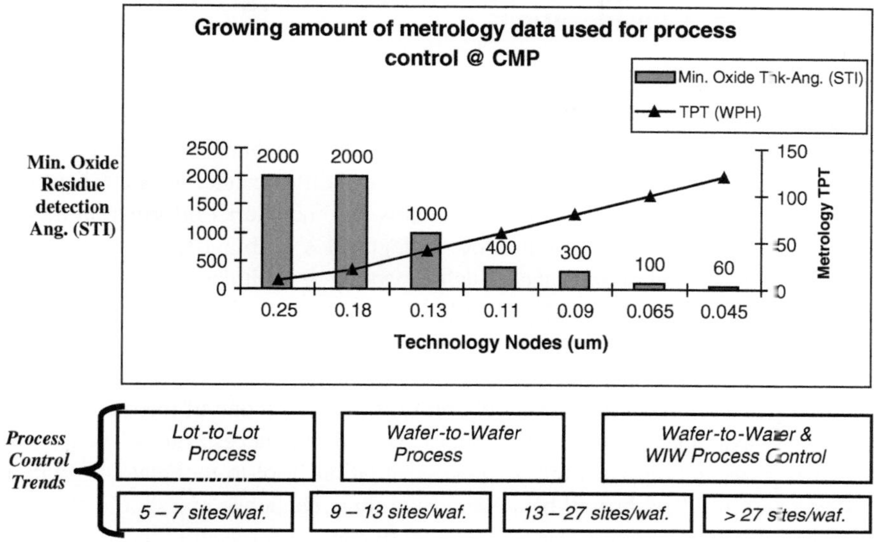

Fig. 32.11 Throughput of the metrology tools is getting higher while technology nodes are getting smaller

32.8 Pre-metal Dielectric (PMD)

Leading edge processes are evolving to new gate stack materials that include high-K dielectrics and metal gate electrodes – all of which have complex and possibly variable composition. In order to enable accurate optical metrology of these and overlying layers, optical characterization of these new materials is required. The solutions are provided either by spectral ellipsometry or by spectrophotometry based on utilization of smart algorithms used to extract the optical models.

As in many other applications, measurement on line arrays is gaining preference, leading to implementation of spectral scatterometry methods with their associated complexity of recipe creation and speed of measurements.

32.9 Copper

The main challenge of integrated metrology for copper CMP is to provide capabilities to replace off-line electrical methods for copper line thickness measurement. This being performed using spectral scatterometry with appropriate methods of accurate optical characterization of copper and barrier layers.

Other potential post-copper CMP applications are measurement of dense array erosion and detection of metal (copper and barrier) residues. Solutions for these applications are available, but their use in production requires intensive work on metrology qualification by CMP manufacturers.

32.10 Future Considerations

The integration of IM tools with CMP polishers has evolved to a high level of acceptance in part due to standardization of hardware and software interfaces. The direction in hardware integration is the BOLTS standard enabling easy connection of any IM module to the CMP front-end interface. Software interfaces are based on GEM/SECS/HSMS and other SEMI standards and include all information needed for real-time CLC. IM tools will allow e-diagnostics enabling remote monitoring tool performance and maintenance. Such capability as fault detection classification will be an inherent part of IM tools allowing prediction of the tool behavior and clear recommendation for planned preventive maintenance.

In future high volume manufacturing (HVM), there is no room for IM tool failures, HW or SW. IM performance reliability is becoming one of the main requirements, for example, an automatic pattern recognition success rate of better than 99.7% regardless of application and process variation. Tool-to-tool matching within, and across, multiple fabs will be a major metrology challenge and should reach the level of 0.1% for all applications. IM tool performance is reaching the stage where the metrology is equal to, or surpasses, the capabilities of some of the stand-alone tools.

Throughput of IM tools will continue to be one of the major challenges. With the sampling rate continuously growing, it will be necessary to measure tens of measurement sites on each wafer, to apply more time-consuming measurement methods like ellipsometry and scatterometry at continuously shorter time per wafer than now. Most applications will require both pre- and post-CMP measurements of every wafer.

The more sophisticated requirements, and therefore more expensive IM solutions, in many cases may lead to changing the business model of IM: end-users may buy the IM tools directly from IM suppliers, because buying through equipment manufacturers, with the associated markup on OEM tool, could lead to unacceptable end-users prices and ROI values.

32.11 Summary

Integrated metrology has evolved from solving the instabilities of the nascent CMP processes to providing added value to the process based on advanced metrology capabilities. These capabilities are providing information that enables the industry to continue improvement in the area of CMP and push the limits of the IC manufacturing landscape.

References

1. Dishon, G.; Finarov, M.; and Kipper, R.; Nova Measuring Instruments, Curry, J. W.; Schraub T.; and Trojan, D.; Strasbaugh, R. H.; Inc., Stambaugh, D.; Li, Y., and Ben-Jacob J.: On-Line Integrated Metrology for CMP Processing, IBM/MiCRUS Corp., CMP-MIC Conference, Feb. 22–23, ISMIC – 100P/96/0273 (1996)

2. Dishon, G.; Eylon, D.; Finarov, M.; and Shulman, A.: Dielectric CMP Advanced Process Control based on Integrated Thickness Monitoring: CIM-MIC Conference, 1998 IMIC – 300P/98/0267 (1998)
3. Fiorletta, C.: Capabilities and lessons from 10 years of APC success – Solid State Technology February (2004)
4. Liu, M.: APC from A Foundry Perspective – AEC/APC XV Symposium September (2003)
5. Moyne, J.: Benefits of Run to Run Contrail – Advanced Process Control (AEC/APC) Conference/4th European Advanced Equipment Control
6. Finarov, M.: Apparatus for optical inspection of wafers during polishing. US Patent No. 6,752,689, Issued 22-Jun (ITM) (2004)
7. Finarov, M.; Dvir, E.; Haimovich, E.; and Shulman, B.: Apparatus for optical inspection of wafers during polishing. US Patent No. 6,368,181; Issued 09-Apr (ITM with window) (2002)
8. Dvir, E.; Finarov, M.; Haimovich, E.; and Shulman, B.: Apparatus for optical inspection of wafers during polishing. April 9, US Patent # 6,368,181 (2002)
9. Finarov, M.: Apparatus for optical inspection of wafers during polishing. US Patent # 6,752,689 (2004)
10. Schiener, D. and Finarov, M.: Method and apparatus for measurements of patterned structures. US Patent # 6,100,985. August 8 (2000)
11. Finarov, M.; Ravid, A.; and Schiener, D.: Method and apparatus for monitoring a chemical mechanical planarization process applied to metal-based patterned objects. US Patent # 6,292,265

Chapter 33
Thin Film Metrology – X-ray Methods

Boris Yokhin

33.1 X-Ray Fluorescence (XRF)

X-ray fluorescence method is based on producing vacancies in the inner atomic shells with the help of external X-ray source. These vacancies then de-excite and produce the characteristic X-rays of interest. The energy of the characteristic radiation indicates the atomic number, and intensity of this radiation allows to measure concentration or thickness of the film.

Since early 1970s, when high-resolution solid state detectors appeared, the so-called energy dispersive mode of XRF became widely used. In this mode the photons coming from the sample are registered one by one, and energy of each photon is recorded by means of measuring the charge generated in the event of interaction within detector. Due to the absence of angle-dispersive crystals, the detector can capture a relatively large solid angle and then low-power X-ray tubes might be used for excitation. An example of EDXRF spectrum is shown in the Fig. 33.1.

In the semiconductor industry measuring on a small spot is often required. X-ray polycapillaries that emerged in mid-1990s allow in combination with micro-focus low-power X-ray tubes concentrating the incidence beam onto 10–30 μm spot. At the same time they allow locating Si(Li) or Si pin-diode detector very close to the sample, resulting in high output count rates. For example, the JVX6200 tool manufactured by Jordan Valley Semiconductor has a EDXRF channel with micro-focus tube/polycapillary excitation and few detectors simultaneously collecting fluorescence radiation (Fig. 33.2). The total output count rate may reach, for example, 50 Kcps from the 20 μm spot 4 kA Cu film. Accordingly, the measurement time per point goes down sometimes to 1–2 s only. The close proximity of the detectors to the wafer allows analysis of light elements like Al and P while working in regular atmosphere. Other typical applications include Cu dishing and erosion, and composition

B. Yokhin (✉)
Jordan Valley Semiconductor, Ramat Gavriel, Migdal Haeemek, Israel
e-mail: boris@jordanvalley.com

Y. Shacham-Diamand et al. (eds.), *Advanced Nanoscale ULSI Interconnects: Fundamentals and Applications*, DOI 10.1007/978-0-387-95868-2_33,

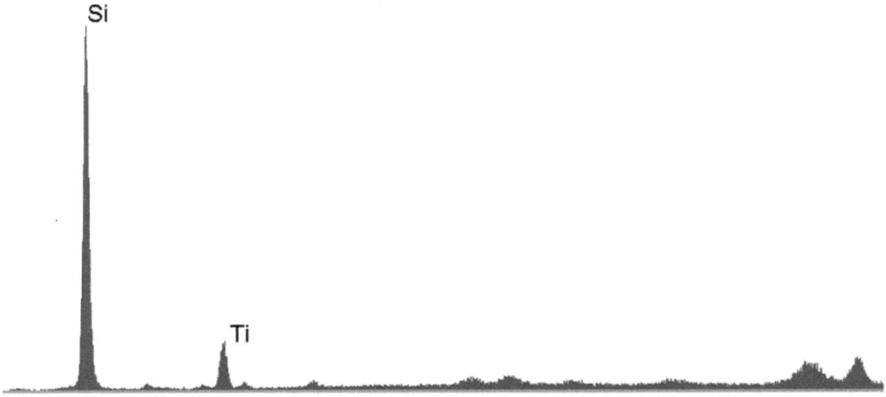

Fig. 33.1 Example of EDXRF spectrum from 500 A TiN film on the Si wafer, acquired in 4 s

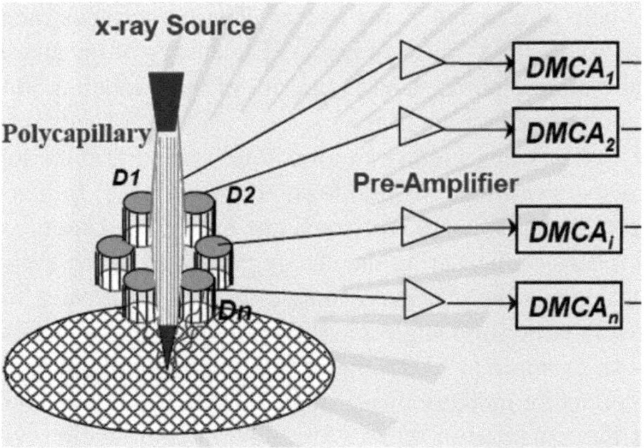

Fig. 33.2 High-luminosity EDXRF scheme with polycapillary excitation and few Si detectors

of SiGe films. The thickness range is 10 A to 10 μm. Pattern recognition and small spot capability enable analysis on product wafers with a 20–60 WPH throughput.

The main advantages of the XRF method are non-destructiveness, non-ambiguity in elements identification, non-sensitivity to film roughness, straightforward calibrations. Using low-power X-ray sources ensures no radiation damage on product wafers.

33.2 X-Ray Reflectometry (XRR)

X-rays striking a polished surface at a very low angle (less than the so-called critical angle θ_c) will be totally reflected. Critical angles depend on density and usually are in the range 0.15–0.4° for 8 keV radiation. At the angles $> \theta_c$ reflectivity drops as $(\theta/\theta_c)^{-4}$, that is, about 6 decimal orders drop at $\theta \sim 3°$. If a thin film is present on the surface, interference from interfaces will cause fringes on the reflectivity curve. It is essential to use a monochromatic radiation in order to obtain fringes in the angle domain. Most often CuKα (8.05 keV) or WLα (8.4 keV) lines are used – relatively low absorption in air allows operation in the regular atmosphere. The period of the fringes carries information about the film thickness. In the case of more than one film, reflectivity curve becomes more complex (Fig. 33.3); however, using special fitting techniques it is still possible to reconstruct thickness, density, and roughness for each film of the stack. Practically, XRR capabilities are determined by the electron density contrast between neighboring films, and reliable decoupling of signal becomes possible starting about 10% contrast.

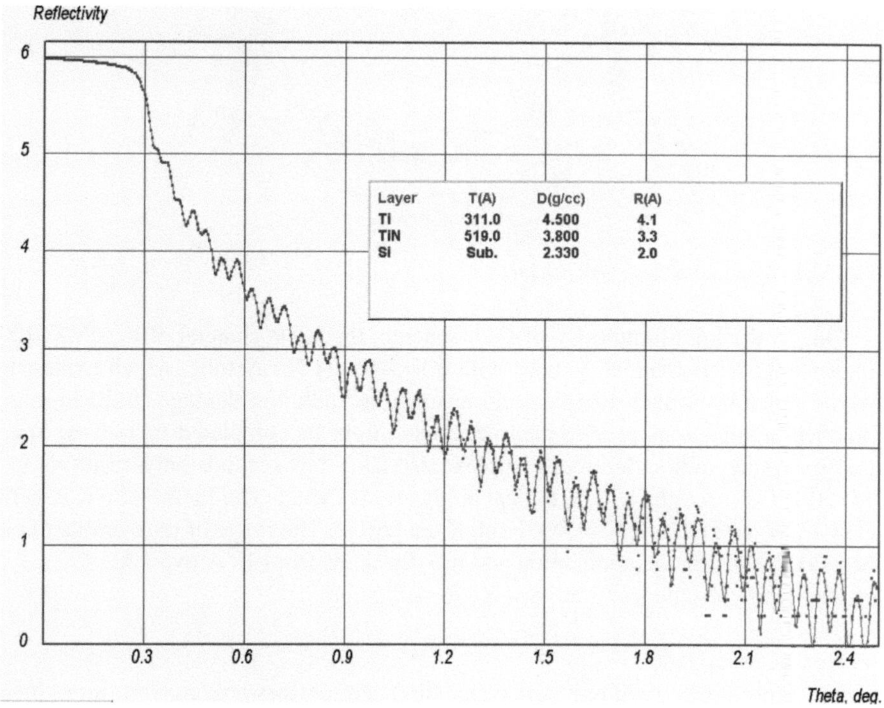

Fig. 33.3 XRR spectrum example. Dots represent experiment, solid line curve presents simulation. The vertical scale is logarithmic

Since low roughness (few Å) surfaces are very common in the semiconductor industry, the XRR method became widely used there since mid-1990s. The classical XRR devices are based on diffractometers which scan mechanically the incidence angle, simultaneously moving the detector to maintain the condition θ–2θ. As usually few hundred points should be covered, such measurements are slow. Another drawback of the classical XRR is relatively large beam cross section of the beam (few square millimeters). Due to the small incidence angle, the measurement spot elongates up to many millimeters, and then analysis becomes limited to blanket wafers only.

The alternative approach is the so-called fast XRR. Such devices, for example the JVX6200 tool manufactured by Jordan Valley Semiconductor, incorporate special X-ray mirrors providing X-ray beam converging (in two dimensions) onto the measurement point. The convergence in vertical plane covers the whole working XRR angle range. The convergence in horizontal plane makes the measurement spot narrow enough (~60 μm) to fit into the scribe line of a product wafer. On the side of detection (Fig. 33.4), the reflectivity values at all angles are recorded simultaneously with the help of a special array having sufficient dynamic range in order to cover both high- and low-intensity segments of the reflectivity curve.

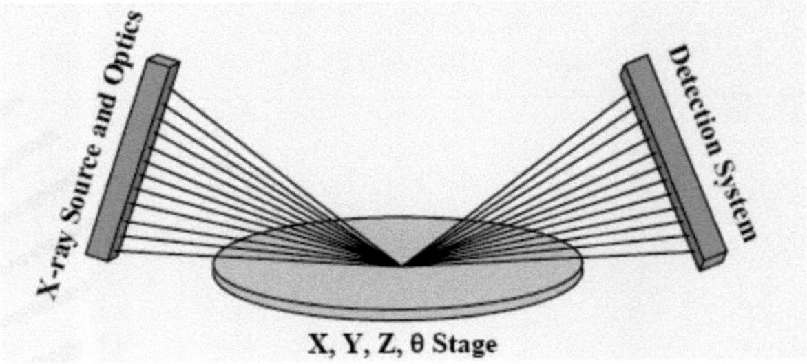

Fig. 33.4 A simplified fast XRR scheme

Due to the high luminosity of such scheme, the XRR channel of the JVX6200, likewise its XRF channel, is built with a low-power X-ray tube, which makes the whole assembly rather compact and ensures no radiation damage. Still, in many practical applications precise measurements might be completed in 1–10 s. Spectra processing, although rather complex, also takes few seconds only, resulting in a 15–50 WPH throughput. The typical applications include Cu/Ta/TaN, SiOC/Cu/Ta, Low *K*, W, Ti/TiN, SiON, ONO, High *K*, and SiGe. The range of measurable thicknesses depends on the application and usually spans from 10 A to 5 kA.

The main advantages of the fast XRR method are

- Non-destructiveness;
- No calibrations required – spectra R(θ) straightforward convert into thicknesses/densities absolute output values;
- Product wafers capability.

33.3 Small-Angle X-Ray Scattering (SAXS)

Porous fims are intensively pursued in the semiconductor industry as a strategy for reducing the dielectric constant of interlayer insulators (low-*K* polymers). Therefore there is a need in characterization methods capable to measure pores size distribution (PSD) for nanometer-sized pores (typically 10–40 A in diameter).

Small-angle X-ray scattering is one of the very few methods allowing measuring PSD non-destructively. Likewise in XRR, the intensity of scattered radiation depends on the electron density contrast $(\Delta\rho)^2$, in this case between the substance within pores (usually vacuum) and low-K material skeleton density:

$$\frac{d\sigma}{d\Omega}(q) = (\Delta\rho)^2 \int_0^\infty N(R)(\Phi(q,R))^2 dR$$

where $N(R)$ – PSD,

R – radius (for spherical pores),
q – transfered momentum: $q = \frac{4\pi}{\lambda}\sin\theta$,
2θ – angle of scattering relative to the primary beam, and
$\Phi(q,R) = 3V_0(\sin(qR) - qR\cos(qR))/(qR)^3$ – form factor for spherical pores.

To increase the length of X-ray beam interaction with the low-K film, the grazing angle mode of measuring might be applied. In this case the incidence beam is directed onto the measurement point at a small angle close to the critical angle for Si, and then the beam totally reflected from the substrate will again interact with the film. The intensity of SAXS should be axially symmetric (Fig. 33.5), however, to avoid contribution from scattering from the surface roughness, the measured 2θ angle of scattering is chosen usually in the horizontal plane.

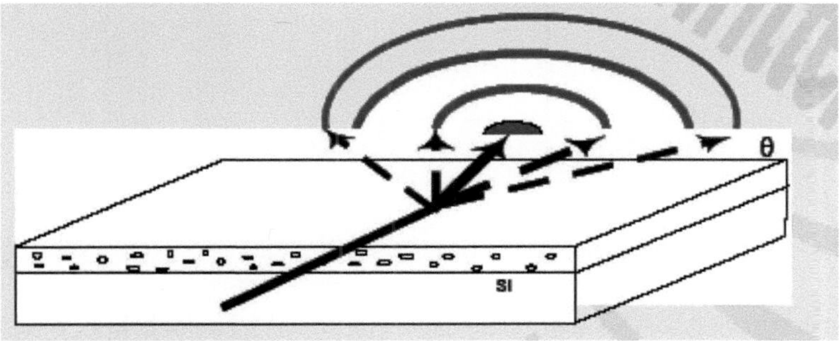

Fig. 33.5 A simplified grazing incidence SAXS scheme

To record the SAXS spectrum one can scan mechanically over the range of 2θ values, or, alternatively, to capture this range simultaneously using a special array. The latest approach is adopted in the JVX6200 tool manufactured by Jordan Valley Semiconductor. The SAXS channel is optional in this device. It includes a low-power X-ray tube and a special collimating monochromator allowing a narrow measuring spot (~80 μm). The simulation software extracts the average pore diameter and the distribution width (most often the shape of the distribution is assumed log-normal). In the combination with XRR measurement which outputs densities, the tool also can measure porosity values (relative volume occupied by pores). Figure 33.6 shows an example of the SAXS spectrum; typical measurement time is 300 s. Although it seems rather slow in terms of routine measurements on production wafers, there is no alternative method to measure pores size in few minutes and non-destructively.

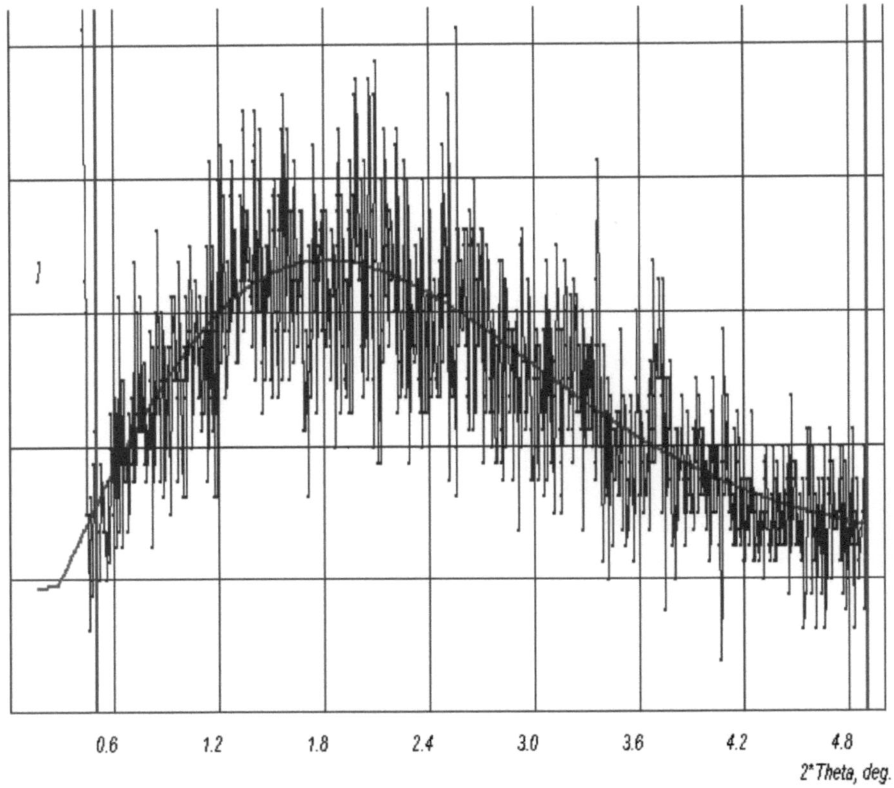

Fig. 33.6 SAXS spectrum example. Average pores size $D = 28$ Å

The accuracy of SAXS was verified and confirmed with the help of positronium annihilation lifetime spectrometry (PALS), in the range of pores diameter 10–80 A. Precision of the SAXS method is about 5% (RSD).

Part IX
Summary and Foresight

This Part includes nanoscale interconnect processing technologies and self-assembly of short aromatic peptides. Chapter 34 "Emerging Nanoscale Interconnect Processing Technologies: Fundamental and Practice" begins with a brief review of current copper-based interconnect technologies, with focus on the role of conventional electroplating techniques as enabler in the reliable and reproducible fabrication of the nanoscale copper wiring architectures necessary to ensure continuation of the historical rate of progress of the semiconductor industry. This review is followed by a discussion of future nanoscale interconnect technologies, including hyper-integration circuitry (HIC), spintronics, moletronics (including carbon nanotubes), and nano-bio interconnects. Particular emphasis is placed on the implications of these technologies on interconnect fabrication processes, particularly in terms of the emergence of "molecular plating" processes that make use of protocols such as self-assembly, atomic manipulation, and precise management of localized material properties to fabricate the new wiring paradigms needed. In the Chapter 35 "Self-Assembly of Short Aromatic Peptides: From Amyloid Fibril Formation to Nanotechnology," the use of small aromatic peptides as building blocks for the assembly of highly ordered nano-scale structures is being described. These building blocks are derived from recognition modules that mediate the self-assembly of biological amyloid nano-fibrils. We explain the theoretical background that led to the identification of these building blocks and illustrate their technological applications. The author's studies on the mechanism of protein self-assembly into ordered amyloid assemblies had provided important information on the mechanism by which simple biological elements interact and identified new building blocks for nanotechnology. The gained information will lead to a better understanding of the potential use of peptide nanotubes in future applications, such as micro and nano-electromechanics (MEMS and NEMS, respectively) and for medical devices.

Chapter 34
Emerging Nanoscale Interconnect Processing Technologies: Fundamental and Practice

Alain E. Kaloyeros, James Castracane, Kathleen Dunn, Eric Eisenbraun, Anand Gadre, Vincent LaBella, Timothy Stoner, Bai Xu, James G. Ryan, and Anna Topol

34.1 Introduction

The prospects for Gigascale integration and beyond are hindered, in the near term, by increasingly higher *RC* delays in global and semi-global electrical interconnect systems. Long-term, signal transmission delays are projected to become significantly more challenging due to fundamental limits imposed by the basic laws of physics. As feature sizes shrink below the mean free path for electron scattering in conventional metal wires, surface scattering, which is defined as the scattering of electron waves from the boundaries of ultra narrow conductors, severely hinders electronic conductivity and stands as a major roadblock to Moore's Law at the most fundamental level.

In the near term, the choice of electrical interconnect wiring solutions will be most likely mandated by their compatibility with emerging complementary metal-oxide-semiconductor (CMOS) device technologies. This constraint has led to the emergence of three-dimensional hyper-integration circuitry (HIC) schemes, which typically employ multiple wafer stacking approaches to minimize the length of global and semi-global interconnects. Three-dimensional hyper-integration architectures could provide a considerable improvement in intra-chip interconnect performance, while avoiding many of the integration issues of optical structures into electrical components. In particular, the use of conventional metal and insulator material sets as the building blocks of each individual device/interconnect level makes hyper-integration technologies quite desirable from the perspective of adaptability to current semiconductor fabrication flows. In this context, the continued implementation of copper-based wiring in HIC-based architectures is projected to drive further advances in electroplating technology and renewed interest in compatible deposition process technologies including electroless plating.

A.E. Kaloyeros (✉)
College of Nanoscale Science and Engineering, The University at Albany-SUNY, Albany NY 12203, USA
e-mail: akaloyeros@uamail.albany.edu

Y. Shacham-Diamand et al. (eds.), *Advanced Nanoscale ULSI Interconnects: Fundamentals and Applications*, DOI 10.1007/978-0-387-95868-2_34,

The challenges imposed by HIC-based architectures are further compounded by the emergence of high performance "XYZ on a chip" designs for use in a wide range of new, rapidly expanding, "XYZ" applications. Such applications include, for instance, telecommunications, environmental and energy nanosensors, and "biochips" for blood testing and DNA sequencing. The resulting new chip architectures involve the use of a variety of innovative interconnect technologies that combine micro- and nano-systems with wireless communication schemes and non-traditional chemical and biological interconnection materials. Successful integration of these radically different sets of materials requires the development of flexible and manufacturable processing techniques for the hierarchical buildup of three-dimensional, defect-free, wiring systems. In this respect, plating technologies represent a viable candidate for such applications in light of their versatility, scalability, and adaptability.

Longer term, the most promising prospects for generating new enabling electrical interconnect technologies that ensure the continued validity of Moore's Law are expected to be achieved through the vast scientific wealth provided by nanotechnology. The essence of nanotechnology is the ability to "nanoengineer" individual interconnect building blocks at the molecular level, atom by atom, to form high-density interconnects with precisely controlled electronic functionality and customized signal carrying properties and functions. Potential breakthroughs include wireless (nano-optoelectronics), self-assembled molecular (moletronics), and electron spin-wave interconnects networks that could provide terahertz speeds at significantly lower dynamic power dissipation. The ability to fully capitalize on the tremendous potential of nanometer-scale phenomena in wiring technologies necessitates, however, the development of fundamentally new interconnect fabrication strategies that gradually replace the currently prevailing top-down approach to multilevel metallization using lithography with bottom-up protocols based on nanoscale self-assembly and self-organization of smart interconnect materials. Such strategies must also be fully compatible with and integrable into current semiconductor process flows. Again, modifications to current electroplating methodologies, such as molecular plating, are particularly adept at achieving controlled self-assembly on 300 mm wafer platforms.

This chapter begins with a brief review of current copper-based interconnect technologies, with focus on the role of conventional electroplating techniques as enabler in the reliable and reproducible fabrication of the nanoscale copper wiring architectures necessary to ensure continuation of the historical rate of progress of the semiconductor industry. This review is followed by a discussion of future nanoscale interconnect technologies, including hyper-integration circuitry (HIC), spintronics, moletronics (including carbon nanotubes), and nano-bio interconnects. Particular emphasis is placed on the implications of these technologies on interconnect fabrication processes, particularly in terms of the emergence of "molecular plating" processes that make use of protocols such as self-assembly, atomic manipulation, and precise management of localized material properties to fabricate the new wiring paradigms needed.

34.2 Nanoscale Copper Interconnect Architectures

Integrated circuit scaling has enabled improved performance and reduced cost for electronic devices, with revolutionary process and materials changes being required to enable these technological advances. In this context, interconnect technology has seen the most drastic materials changes of any portion of the integrated circuit process flow, including the introduction of copper-based wiring [1] in the 250 nm technology node. The adoption of copper interconnects was followed by the introduction of inter-metal dielectric films with lower dielectric constants than silicon dioxide beginning with fluorinated glasses in the 180 nm node and moving to carbon doped silicon oxides and hybrid schemes including polymers in later nodes [2]. The implementation of copper-based wiring provided a number of significant benefits. Thin film copper wiring, including Ta-based underlayers for technology nodes with minimum dimensions greater than 100 nm, exhibits a resistivity of approximately 2 $\mu\Omega$-cm (bulk resistivity of Cu is 1.65 $\mu\Omega$-cm). Aluminum-alloy-based wiring used in previous generations has a resistivity of approximately 4 $\mu\Omega$-cm (bulk resistivity of Al is 2.7 $\mu\Omega$-cm). The factor of 2 reduction in wiring resistivity can be used either to improve chip speed by keeping wire thickness the same or can be used to lower capacitance, and thereby power consumption, if copper wiring thickness is reduced so that wire resistance is kept constant [3].

The successful manufacturing of the new metal and dielectric material sets has also driven improvements to existing process technologies, such as physical vapor deposition (PVD). More importantly, it led to the adaptation of new film fabrication technologies not previously used in semiconductor chip fabrication, with electroplating being arguably the primary enabler for the successful realization of copper metallization schemes. In this respect, various copper deposition methods were evaluated [4] with electrochemical deposition or electroplating being selected as the preferred method. Electroplated copper films exhibit low resistivity and excellent resistance to electromigration and stress migration effects [5]. The fabrication process for an advanced Cu-based interconnect is a Damascene method [6] involving the formation of trench and/or via structure in a dielectric. A layered structure of tantalum-based films is then deposited. The tantalum-based films serve several functions, including making electrical contact to underlying wiring, serving as an adhesion layer, and acting as a barrier layer to prevent copper diffusion into the dielectric films and silicon substrate. A thin layer of copper is then deposited (usually by PVD) on the tantalum-based film to serve as a seed layer. Copper is then electroplated on the seed layer. Excess copper is removed using chemical mechanical polishing (CMP) in order to make the surface of the wire coplanar with the dielectric. This process is repeated as many times as needed to produce the desired number of wiring levels.

Fabrication techniques for copper metallization schemes have undergone continuous improvement since they were widely adopted in the 1990s. The PVD techniques for tantalum-based films that are in widespread use at the time of this writing will be replaced by methods such as chemical vapor deposition (CVD) or atomic layer deposition (ALD) that are capable of producing thinner, more conformal barrier films. The seed layer may also change from copper to elements such as ruthenium that may play a critical role in improving the copper interconnect′s resistance to phenomena such as electromigration and stress migration [7].

In any case, conventional electroplating will continue playing a major role as the key process technology for copper interconnects. As such, there is a great deal of research and development underway to further enhance the process and ensure its applicability in sub-65 nm generations of electrical wirings. Electroplating equipment with improved plating bath control systems has been developed. Electroless plating of materials like CoWP is also being evaluated for incorporation into the copper interconnect process flow. CoWP and other similar alloys may have utility as a protective cap for the copper interconnect that also enhances the reliability of the copper interconnect [8]. Extensive work is also underway to further improve plating baths that allow deposition into high aspect ratio features and produce smoother, more planarized copper films.

34.2.1 Hyper-integration Interconnects

Even with the gradual reduction in device feature size that has been accomplished over the last few decades, further scaling to higher complexity has demanded that innovative approaches be developed to ensure the survivability of Moore's Law. These approaches are focused along two main thrusts. Under a first thrust, recent advances in materials processing have allowed atomic level control of deposited materials and precise ability to tailor the performance of these materials for specific functions [9]. This nanoscale control forms the foundation of future, longer term, advancements in interconnect science and technology, and will be discussed in more detail in subsequent sections.

Shorter term, the perpetual demand for increased functionality and performance has necessitated a move beyond traditional architectural solutions that have maximized the utility of the available chip surface area in a given architecture. These solutions have run out of steam due to the fact that they are typically confined to a two-dimensional (2D) spatial distribution. Instead, a second thrust has aimed to explore various "3D" methods for component/wafer integration including die-to-die, die-to-wafer, and wafer-to-wafer techniques [10–15]. The utilization of the third dimension promises to have a large impact on the nanoelectronics industry by providing viable process flow options to combine disparate functions such as sensing, logic, communication, etc. throughout a stacked architecture. It is expected that an optimal combination of such nanoscale materials control and the stacked architecture approaches will provide an avenue for progress in advanced interconnect development.

One of the fundamental advantages of the 3D technique is to minimize the speed limiting interconnect lengths between components by utilizing pass-through connections and establishing signal/clock pathways throughout the volume of the stack. Achieving this goal means developing process technologies that enable a mixture of otherwise incompatible circuit technologies to be combined into systems. More specifically, successful processes for the construction of 3D systems need to fit into the present production models established for the semiconductor industry in order to be accepted, implemented, and ultimately, commercially viable.

34.2.2 Die-to-Die Integration

Three-dimensional integration has been successfully achieved at the die level for many years [16, 17]. The development of bump and edge-wire bonding methods has moved the technology ahead to reach the point of multiple-level stacks. Individual die can be attached to each other using polymeric "glue" that compensates for die surface topology and provides adhesion between individual die. Interconnects between die have generally been constructed at the die periphery between standard contact pads. More recent strategies have enabled interconnects to be constructed within the die area, but circuit fabrication is still handled at the die level. Three-dimensional integration at the die level requires the expensive handling of individual die for both assembly and test. Apart from the larger number of operations that must be accomplished to assemble a die level system, additional production issues of die size variation lead to handling issues that will not likely be addressed by process equipment manufacturers. Both of these concerns will keep the cost of die level system assembly too high for mainstream commercial semiconductor manufacturers. Figure 34.1 shows an example of a stacked, die-to-die integration approach [16].

Die level 3D interconnect strategies do not enable the real performance potential of 3D integration to be realized. Ideally, true 3D system integration will not only enable interconnects to be constructed within the die at locations to enhance system performance, such as to directly link individual image sensor elements to preamplifiers, but also enable the economies of wafer scale fabrication to be realized for the production of large numbers of devices. Figure 34.1 shows an overview of the two distinct approaches to wafer level integration [16].

34.2.3 Die-to-Wafer Integration

An alternate approach for 3D integration is the die-to-wafer (or chip-to-wafer) interface method. In this approach, known good dies (KGD) are chosen and directly interfaced with a substrate wafer containing selected functionalities [10]. To achieve this level of integration complexity, multiple critical steps need to be perfected and combined including bonding, adhesive selection/application, and chip thinning. The main advantage of this integration approach is that complete IC processing is done on the candidate wafer including testing and separation. Subsequent processing works with only KGDs so that yield can be maximized. This method is useful when the handling of the multiple dies does not impact yield. Following this "pick and place" step, the remaining process steps are done at the wafer level including via

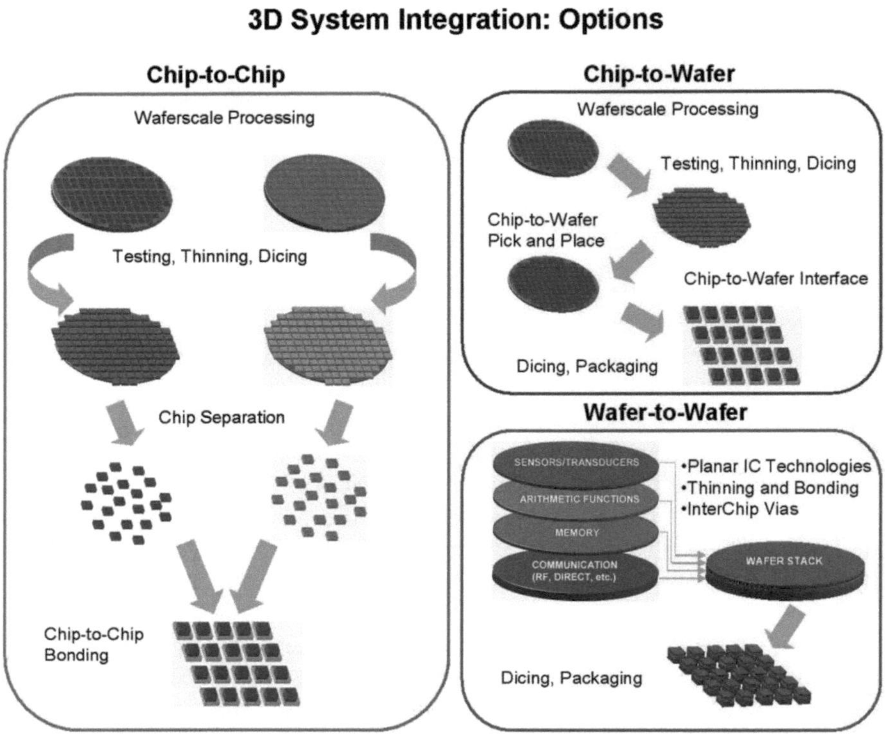

Fig. 34.1 (**a**) Example of a stacked, die-to-die, integration approach. (**b**) Comparison of the wafer-to-wafer and die-to-wafer (also known as chip-to-wafer) integration approaches. Adapted from [16]

production and interchip metallization. Overall fabrication efficiency based on this approach is dominated by the fact that multiple chips which must be independently aligned to the substrate wafer. These repeated tolerance limits on such alignment and integration of the chips to the wafer makes this method suitable for only selected applications. For high-volume throughput, other methods must be perfected.

34.2.4 Wafer-to-Wafer Integration

The advantages of wafer level assembly include fewer operations in system assembly, an extension of the economy of scale enjoyed by traditional semiconductor manufacturers, and the use of a standard substrate size for which fabrication and test equipment already exist. Successful development of a wafer level integration process requires the realization of several new procedures for wafer handling and processing. Anodic and metallic [18, 19] bonding have both been used successfully for die level system assembly. These bonding techniques are susceptible to contamination at the wafer level, however, and can leave large areas of two wafers un-bonded. In addition, the anodic and metallic bonding techniques require planarization steps to bond the individual die and cannot readily compensate for circuit topology that can result in more than 1 μm variations in surface height.

A "glue" layer is the best solution for overcoming these difficulties in wafer level system assembly. Many polymers can be applied by spinning and provide rapid, low-cost planarization of the wafer surfaces being bonded. One choice is Benzocyclobutene (BCB) and has been selected as the inter-wafer "glue" material because it possesses outstanding mechanical and thermal properties that enable existing 200 mm wafer processes to be used in the 3D integration process. Figure 34.2a,b display, respectively, an infrared image of bonded 200 mm silicon wafers using an optimized BCB integration protocol and a scanning electron microscope (SEM) cross section of the stack [15]. The (∼5 μm) bond is void-free and mechanical/cryogenic test data show that the strength of the bond is uniform over the entire wafer area. This optimized bonding method sets the stage for the remainder of the stack processing. The full process flow encompasses multiple steps including via etching, isolation, filling, CMP, and complete integration [15]. Each of these individual steps requires nanoscale control over the deposition/etching and performance of the required material set. Figure 34.3a,b show, respectively, a completed stack with an integrated 256 × 256 pixel focal plane array (FPA) and the resulting imager performance [15].

34.3 Nanoscale Interconnects Technology: Moletronics

While conventional interconnects are undergoing continuous advances, these advances are evolutionary in nature and will eventually run out of steam due to the limits imposed by the basic laws of physics. As discussed earlier, new revolutionary approaches to signal transport will ultimately be required if the historical rate of progress of the nanoelectronics industry is to continue. One such approach under consideration, and the subject of considerable active research, is based on molecular electronics, whereby specifically designed molecular systems are used to transport charge (molecular wires) or to switch electrical signals (molecular devices) [20].

Probably the best known example of molecular wires is metallic carbon nanotubes (CNTs), with their potential for ballistic (scatter-free) electron transport over practical length scales. CNTs have received much interest due to a series of impressive performance milestones, including a recent report of a metallic multiwall CNT carrying a current density on the order of 10^8 A/cm^2 and a dissipated power of 1.82 mW [21]. However, considerable practical challenges for CNTs remain, in particular with regard to CNT fabrication and integration protocols [22]. As a result, alternative molecular systems have also been researched for nanoelectronics applications. These systems offer tailored electrical behavior, controllable attachment moieties, and the potential of low-cost fabrication. What follows is a brief overview of the current trends in molecular interconnects, including carbon nanotubes and alternative molecular systems.

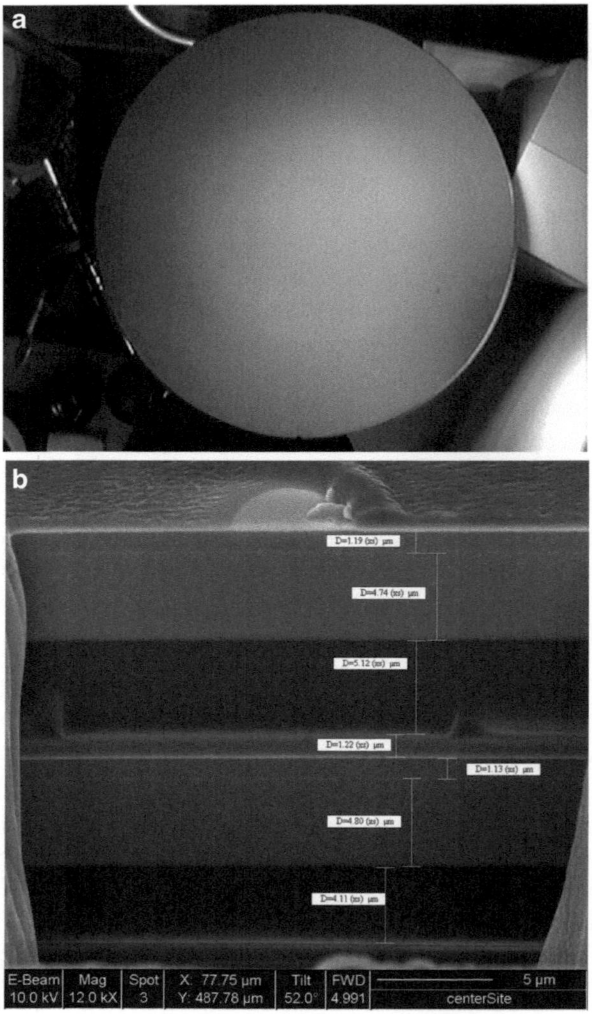

Fig. 34.2 (**a**) Infrared Image of BCB bonded wafers. (**b**) Cross-section scanning electron microscope (SEM) of a three layer stack of silicon wafers with BCB bonds. Adapted from [15]

34.3.1 Carbon Nanotubes

From time of their discovery by *Iijima* in 1991 [23], carbon nanotubes (CNTs) have drawn intense interest among researchers who were fascinated by unique structural, mechanical, chemical, and electronic properties of this new class of materials. Improvements in nanotube fabrication approaches have enabled further study of electrical transport and other new phenomena [24, 25] and increased interest in CNTs as materials for future nanoelectronics applications [26, 27].

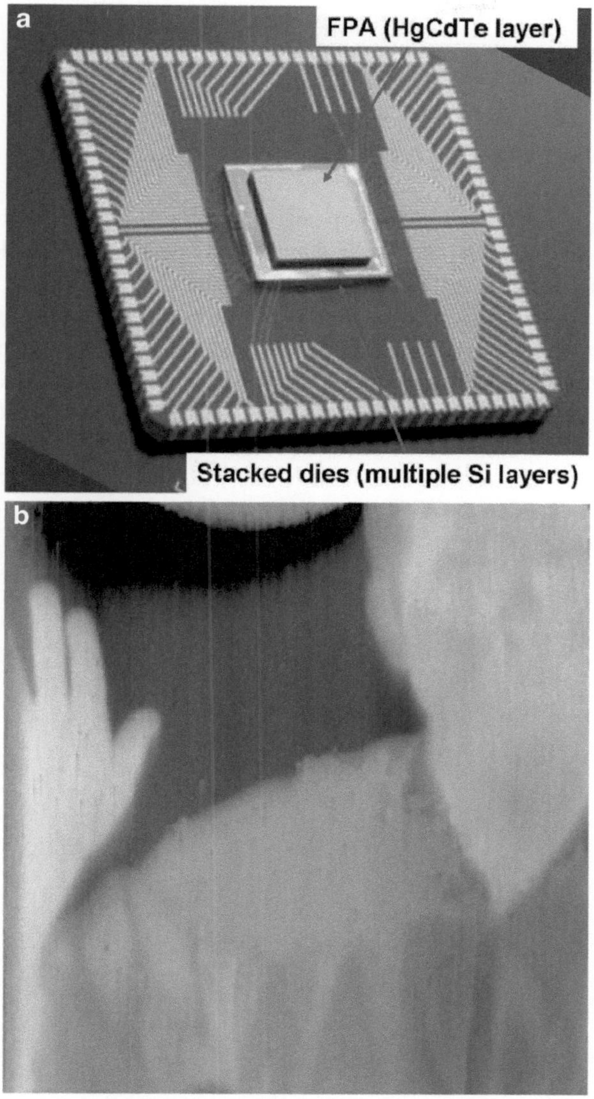

Fig. 34.3 (**a**) Completed stack with an integrated 256×256 pixel focal plane array (FPA). (**b**) Resulting imager performance. Adapted from [15]

Ideal carbon nanotubes exhibit ballistic (scatter-free) electron transport with little energy dissipation along the tube. In spite of significant progress in developing CNT networks [28, 29] and successfully growing aligned carbon tube films [30, 31] the inability to control the selection of specific types of CNTs in large quantity, and the tendency of nanotubes to bundle together have continued to act as the key limiting factors in preventing widespread use of CNT-based device and wiring technologies

[32, 33]. If these difficulties are overcome, CNT-based device performance enhancements and innovative interconnect architectures are possible.

34.3.2 Geometric and Electronic Structures of CNTs

The behavior of carbon nanotubes depends concurrently on their diameter and chirality. Individual cylinders, 1–2 nm in diameter, defined as single-walled nanotubes (SWNTs) [34–37] represent simultaneously a macro-molecule and a crystal, and are treated as quasi-one-dimensional (1D) conductors. Bundles of aligned SWNTs (tens to hundreds of tubes), typically ~0.1 mm long (Fig. 34.4), are defined as ropes [38]. Tubes in a rope are spaced ~3.2 Å apart in a close-packed triangular lattice, with nearly identical diameters [39, 40]. Multi-walled nanotubes (MWNTs), typically 10 nm in diameter and ~1 μm long (Fig. 34.5), represent two-dimensional (2D) conductors, and contain coaxial SWNTs with interlayer spacing of ~3.4 Å. The various shells of the MWNT have different diameters and hence may exhibit a metallic or semiconducting nature [41].

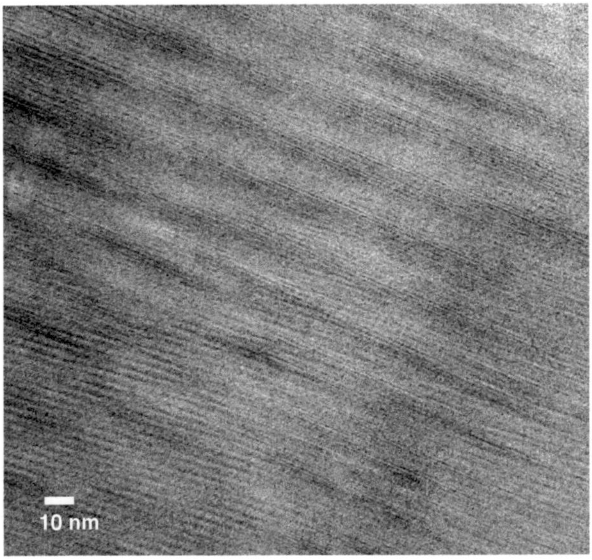

Fig. 34.4 TEM image of typical ropes of aligned CNTs. Adapted from [33]

The advantage of CNTs over other 1D systems (such as semiconducting quantum wires) is atomic uniformity, as characterized by strong confinement around the circumference. This strong confinement results in a large spacing between 1D subbands, namely, ~1 eV for 1 nm tube in contrast to ~10 eV for a quantum wire [42]. However, a frequent hindrance in nanotube resistance and transport studies is high contact resistance, resulting in charging and Coulomb blockade at low temperature [43, 44]. Successful measurements provided resistivity values in agreement

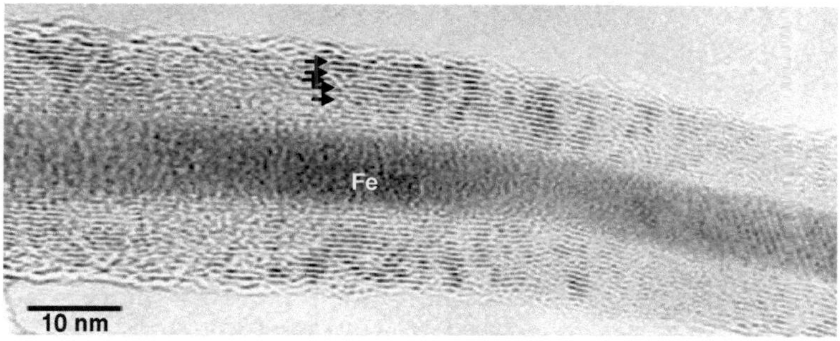

Fig. 34.5 HRTEM image of a MWNT with iron-filled core. Adapted from [33]

with theoretical prediction, thereby proving the existence of both metallic and semiconducting tubes [41]. At low temperatures, metallic nanotubes (SNTs or small ropes) act like quantum wires with conduction via discrete, quantum-mechanically coherent (over long-hundreds of nanometers-distances) electron states. Upon further decrease of temperature, their behavior resembles that of the elongated quantum dots. These geometrically enabled electronic properties and charge transport capability of CNTs is matched with their unsurpassed mechanical strength.

34.3.3 Mechanical and Thermal Properties of CNTs

The nanoscale diameter and macro-scale length of CNT structures, combined with their unique topology and low density, result in distinct mechanical properties, such as strength, stability, and minimal elastic deformation [45]. The exceptional strength of these tubes along their axes is due to the strength of the carbon–carbon bonds. The experimental Young's modulus value for CNTs is ~1.8 TPa which is an order of magnitude higher than that of steel and several times higher than that of common commercial fibers [46]. Theoretical estimation of tensile strength for an individual SWNT is ~300 GPa and the best current MWNT experimental values are 50 GPa [47]. However, this result is an order of magnitude higher than that of carbon fibers [48]. Nanotube-based fibers exhibit plastic behavior under strong loading at room temperature, possibly due to the displacement of nanutobes within the fiber (sliding of individual layers in MWNT and shearing of individual tubes in ropes) [49]. CNTs withstand large strains (40%) in tension without showing signs of fracture [50]. Theoretical and experimental evaluation of CNTs suggests the ability of nanotubes to reversibly alter shape while accommodating external forces.

Carbon nanotubes exhibit good thermal conductivity and very high thermal stability [24]. Thermal Gravimetric Analysis (TGA) studies [51] indicate that nanotubes are more resistant to oxidation than graphite. The oxidation occurs primarily at the tips of the tubes where there is a high concentration of pentagonal defects.

34.3.4 Interconnect Applications of CNTs

As future interconnects shrink in dimension to allow gigascale integration, signal delay, and signal fidelity problems associated with interconnects become significant limiters of overall system performance (e.g., maximum supportable chip clock frequencies) and hence new solutions such as nanotube-based interconnect systems are of great relevance. In particular, one-dimensional CNT conductors with ballistic electron transport would be ideal candidates for advanced interconnect networks. Experimentally, metallic nanotubes were shown to withstand extremely large current densities, exceeding 10^8 A/cm^2 [52]. Additionally, experimentally observed long mean free path and measured long coherence length of the metallic nanotubes indicate robustness against defects and long-range perturbations near the Fermi Energy (E_F). Due to this insensitivity toward long-range disorder, metallic tubes indeed represent near perfect 1D conductors, and their synthesis, separation, and alignment remain a primary challenge in the implementation of CNTs in interconnect technologies. An additional challenge is related to the demonstration of reliable tube–tube connections and junctions (Figs. 34.6 and 34.7) and tube–metal contacts (Fig. 34.8). In particular, since calculations show that perfectly straight tubes may be required to achieve ballistic transport in interconnects [53], formation of minimal loss junctions and contacts is critical to the ultimate fabrication of interconnect structures. The following two subsections present a brief discussion of the opportunities and challenges associated with tube–tube and tube–metal connections and junctions, and resulting implications for signal transmission in resulting wiring systems.

34.3.4.1 Tube-Tube Junctions

In the crossed-tube junction, physical contact between two tubes forms the basis of a connection and exists in multi-tube structures, such as ropes, MWNTs, and crossed-over tubes. Tubes in such structures are in a close proximity, but are not chemically bonded to one another [54]. When pressure is applied on the top of the cross-over region (via an AFM tip, for example) [55], an overlap of the electronic wave functions is created across the tube junctions, thus enabling intertube tunneling. In this system, a small structural deformation of the tubes (smooth bending) will induce minimal scattering [56, 57], while significant deformations will enhance scattering and result in resistive losses. Under low temperature conditions, bundles of parallel tubes were also investigated, with a measured penetration depth of $\sim$1.25 nm indicating direct tunneling transport between tubes [58]. However, calculations suggest that intertube coupling in the ropes alters fundamental electronic transport properties of individual tubes [41].

Finally, in the case of MWNT, only weak coupling exists between various SWNTs as interplanar stacking disorder of adjacent shells decreases intertube electronic coupling [59]. Is it then possible to create the world's smallest coaxial wire within which each insulated wire – a metallic nanotube within a semiconducting one – could operate independently of the others? Experiments show that at low

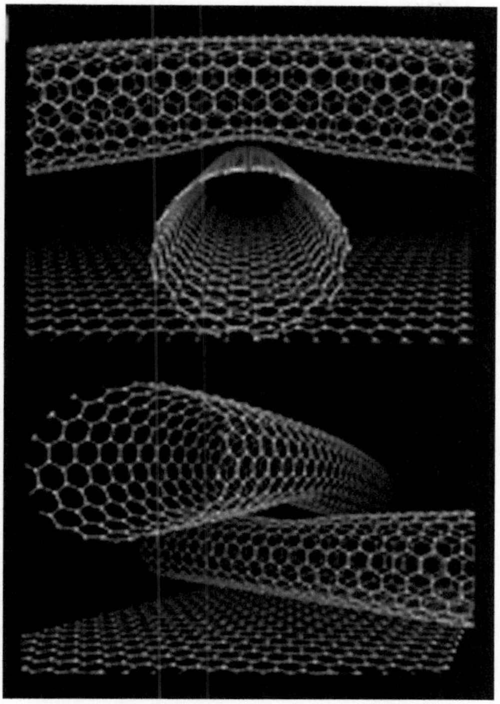

Fig. 34.6 Schematic diagram of CNTs crossing over each other. Reprinted figure with permission from [54]. ©1998 by the American Physical Society

temperature or low bias conditions, current is transported through the outermost layer of the MWNTs – the only shell in contact with electrodes [53]. At room temperature and low bias leakage, current through the inner metallic shells contributes to the electron transport. Alternatively, under high-bias conditions, all shells of the MWNT contribute to electronic transport [29], indicating that transport in the innermost shells is tunneling barrier-dominated. Hence, the problem of contact creation between metallic inner tubes and electrical contacts remains a central obstacle to the realization of MWNT-based interconnect technologies.

In the tube–tube junction called on-tube junction, chemical bonding provides the basis of the connection as interfacial chemical bonds within individual tubes are rearranged to join to constituent tube. Theoretical predictions indicate that when metallic nanotubes with appropriate diameters are joined in the form of "X", Y," and "T" structures (Fig. 34.8), their connection is stable and can serve as multi-terminal electronic devices or wiring interconnection [60–63]. Experimentally, several such junctions have been observed [29, 64], but the on-tube fabrication process continues to be a challenge. Successful synthesis of "Y"-shaped nanotubes using template structures [65] and nanofabrication manipulation techniques [66, 67] has been achieved. The majority of current junction research focuses on post-growth formation methods, such as heat-assisted irradiation. Using these techniques, tubes that

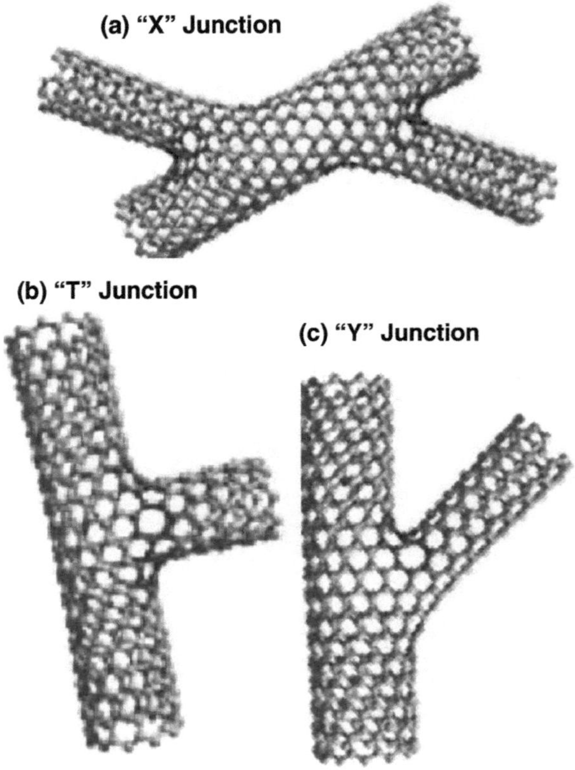

Fig. 34.7 Atomic model of on-tube junctions. Reprinted figure with permission from [54]. © 2002 by the American Physical Society

were not connected during synthesis can be joined in various junction geometries by tuning the irradiation conditions.

34.3.4.2 Tube–Metal Contacts

Early attempts to create contacts between CNTs and metal pads focused on dispersing nanotubes on lithographically defined contact pads on an insulating surface. Transfer studies using this integration method were hindered because a tunneling barrier was formed at the nanotube–metal interface [57, 68]. Since then, improved physical and electrical contacts have been achieved by partially collapsing the ends of the tubes [69]. In these cases, deposition of electrodes was performed using ion-assisted chemical vapor deposition (I-CVD) in a FIB, thus enabling the formation of a gradual transition region between the tube and the metal by nanomodulating the interfacial structure and electrical characteristics through doping of narrow transition region between these two dissimilar systems [70]. Recently, a method that combines e-beam lithography and subsequent liftoff of metal electrodes has also been utilized to successfully create nanocontacts [71]. To achieve an Ohmic CNT–metal

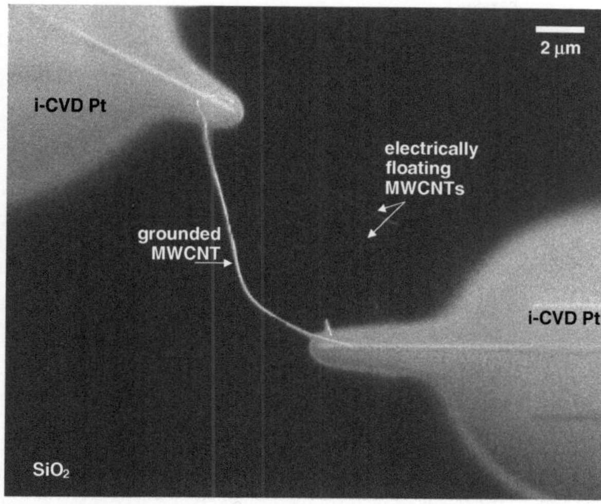

Fig. 34.8 FIB image of CNT-Pt contacts fabricated using ion-assisted CVD (i-CVD) in a FIB system. Adapted from [33]

junction, samples processed by these various deposition methods are often post-treated using rapid thermal annealing (RTA) [65]. Also, to reduce contact resistance to nanotubes, a gold (Au) coating could be used on the electrodes [72].

34.3.5 Alternatives Molecular Systems

34.3.5.1 Charge Transport Mechanisms

In the case of molecular wires, charge transport can be achieved in several ways. It is important to recognize that molecular systems conduct charge in a fundamentally different way than do conventional electrical wires (such as metal lines), and that the specific transport mechanism varies depending on the nature of the molecule. In particular, molecules do not obey Ohm's Law, owing to the fact that molecular orbitals are spatially localized and do not form energy bands. This feature plays a key role in the scalability and performance considerations for molecular interconnects, since such systems will not be susceptible to the same scattering events as is the case with conventional interconnects.

Simple molecules, such as alkanes, generally act electrically as barriers, and charge transport occurs via tunneling, with an exponential dependence on the length of the molecule [73]. On the other hand, most commonly researched approaches to overcoming this challenge employ the intrinsically conductive nature of aligned, overlapping molecular orbitals. The typical embodiment combines alkanes with more complex molecules containing charge donors and acceptor groups that have correspondingly more complex charge transport mechanisms. Thus, conduction in these molecules may include coherent transport via a process called electron

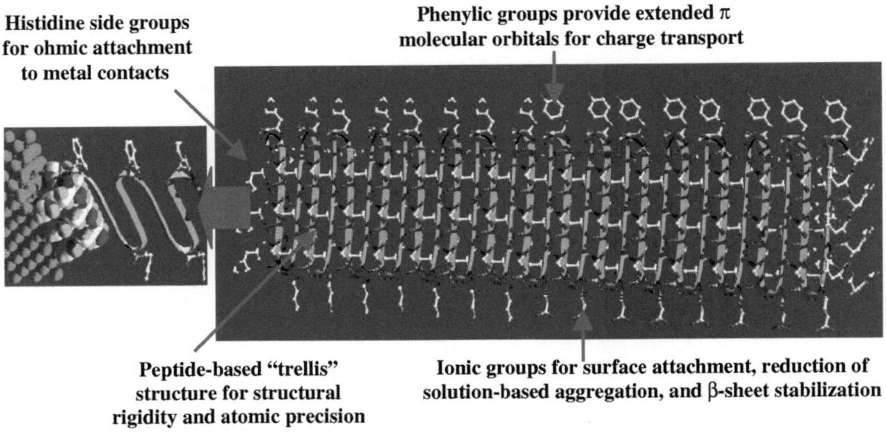

Fig. 34.9 Depiction of a mechanically robust β-sheet polypeptide "trellis" structure which can be used as a fabrication platform for molecular interconnects. Charge transport and self-assembly features can be genetically engineered on neighboring molecular turns

type superexchange, in addition to tunneling events to and from the donor and acceptor sites [74]. If the molecule is designed such that the donor and acceptor sites are separated by a small bridging moiety (i.e., <10 Å), tunneling effects can be minimized and the molecular behavior can more closely resemble perfect coherent transport through a single overlapping orbital, while longer bridging molecules tend to make incoherent hopping conduction more prevalent [75, 76]. In addition, molecular systems may be employed as "scaffolds" upon which extrinsic charge transport functionality can be fabricated, such as those based on metal nanoparticles [77].

Since any shift in the relative position and overlap of the donor and acceptor moieties will degrade the charge transport behavior of these molecules, it is imperative that a mechanically robust platform which prevents shifting of these groups be employed. One approach to achieve this goal involves the use of folded organic molecules, such as a β-sheet polypeptide "trellis" structure as shown in Fig. 34.9, where charge transport features can be genetically engineered on neighboring molecular turns. These types of molecules employ H-bonding between neighboring antiparallel strands to strengthen the β-sheet structure, and thus help mechanically constrain the benzene-like charge transport features to enhance molecular orbital overlap.

34.3.5.2 Self-assembly Techniques

One of the fundamental challenges to scaling of conventional interconnects involves the accurate and reproducible patterning of nanoscale structures. Self-assembly, whereby the molecular system aligns directly to the surface, offers promise not only for the "programmed" formation of aligned supermolecular networks, but also serve to allow the molecule to assemble so as to provide the optimal charge transport

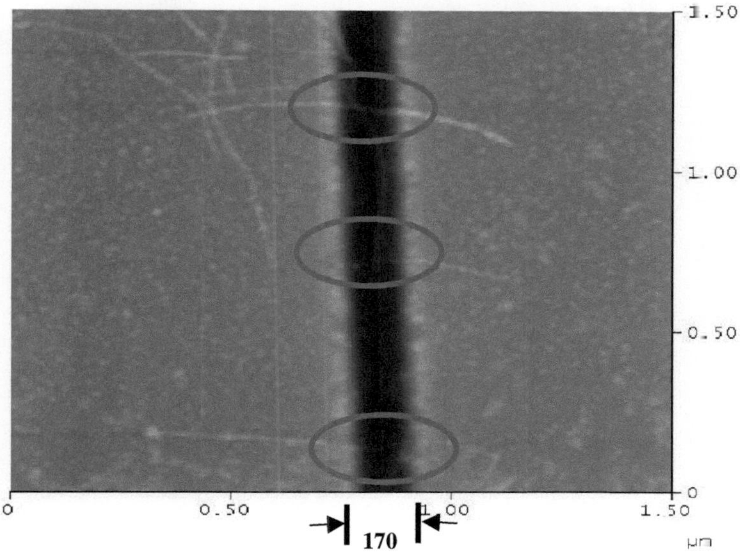

Fig. 34.10 AFM image demonstrating controlled assembly of polypeptide-based molecular interconnect fibrils across a fabricated Ni contact structure

conformation. In contrast to other atomic and molecular assembly techniques which are, essentially, serial in nature, self-assembly offers the potential for massively parallel fabrication with the promise of substantially lower costs – essentially exploiting Nature's own fabrication machinery [78, 79]. Examples of self-assembling organic molecules are extensive; what remains, then, is the development of a site-selective self-assembly such that molecules arrange themselves only in specific locations [80]. In particular, controlled assembly across electrode or contact structures, as shown in the atomic force micrograph (AFM) in Fig. 34.10, is expected to be critical for the effective use of molecular interconnects.

One approach involves designing the molecular system with specific surface tethering functionality, in much the same way as charge transport moieties can be designed onto molecules. These tethers could be engineered to attach only to specific surface chemical sites in order to achieve the desired site selectivity. Alternatively, templated self-assembly approaches exploit the potential of molecular systems for assembling into supramolecular structures either in solution phase or on an appropriate surface. Such methods allow spatial placement accuracy of better than 5 nm, as shown in the scanning tunneling micrograph (STM) image in Fig. 34.11, which demonstrates the directed assembly of aligned polypeptide fibrils on a graphite surface. Taken together, these approaches offers the possibility of a complete "bottoms-up" design of entire nanoscale interconnect networks.

Fig. 34.11 STM micrograph exhibiting the directed assembly by molecular plating of aligned polypeptide fibrils on a graphite surface

34.3.6 Self-assembly Techniques

While it is expected that any molecular interconnect technology will first be implemented on a silicon CMOS-type platform, a number of intrinsic challenges arise from this reality, including reliability and reproducibility of large area assemblies. For example, since many of the popular moletronic candidate materials are organic based, exposure to high-processing temperatures can severely degrade the performance of these molecular systems. Moreover, CMOS structures generally do not fully exploit the inherent potential of molecular systems. A more relevant consideration in the long-term potential performance and extendibility of molecular interconnects is the realization of complete molecular electronic interconnected device networks. While such devices are predicted to operate more slowly at the switch level, such neural-network type systems are expected to be inherently parallelizable, much more so than equivalently scaled CMOS-based designs [81].

34.4 Nanoscale Interconnects Technology: Spintronics

The ability to create and manipulate coherent populations of spin-polarized carriers in semiconductors has promising potential for nanoscale interconnects due to the fact that spin polarization is generally less sensitive to scattering than charge. In addition, spin coherence times in semiconductors are much greater than in metals. Recent experiments using optically generated spin-packets in semiconductors have demonstrated the ability to drag coherent populations of spins up to 100 µm in materials such as GaAs [82]. This result demonstrates that these coherent spin packets (or spin excitons) exhibit qualitatively different transport properties than conventional charge-based excitons in doped semiconductors [83]. These differences arise from the differences in mobility and diffusion of packets of electron spins.

A spin exciton or spin packet is a redistribution of electron spin into non-equal amounts of spin up and down. In an undoped semiconductor this type of exciton involves both electrons and holes. Therefore, these types of spin packets exhibit similar transport properties as charge excitons. However in an n-type doped semiconductor the spin packet can consist of a redistribution of *only* the conduction band electrons. It is this novel ability of electron spin that gives these packets much different transport properties in an electric field since they do not need to carry holes along with them. These spin packets (either spin up or spin down) can be dragged by moving this packet in the conduction band by an applied electric field. In contrast, in an electric field a charge packet must move both electrons and holes in different directions which creates a space charge field forcing the electrons and holes back together opposing their motion. However, this space charge field does not occur when dragging a conduction band spin packet in a doped semiconductor. It is this fundamental difference that gives spin packets improved transport properties.

The motion of the semiconductor spin packet in a sea of conduction band spins is somewhat analogous to motion of electrons in a metal. In a metal wire, the charge signal travels near the speed of light, not due to the motion of a single electron at that speed, but due to the collective electromagnetic disturbance traveling at that speed. One wonders if this phenomenon can be carried over to conduction band spin packets in a semiconductor. That is, can conventional charge-based Cu-metal interconnects be replaced with spin-packet-based interconnects using a doped semiconductor? This would result in monolithic circuit design where the devices and interconnects are made from the same semiconducting material.

Numerous hurdles must be overcome to achieve such a vision. First, creating a spinpacket in a semiconductor using non-optical techniques necessitates the use and integration of ferromagnetic elements into the circuit design to couple to electron spin. Diluted magnetic semiconductors (DMS) have some potential here as they are conventional semiconductors made ferromagnetic via doping with a transition metal impurity such as Mn [84]. These DMS materials have the unique property that an electric field can be used to control their magnetization, like a solid state electromagnet. This property might be able to be harnessed for creation of spin packets. However, to date the best DMS materials have been made out of Mn-doped GaAs, while work on doping Si with Mn has achieved ferromagnetic behavior above room temperature, more work is needed to determine if this material can be made into a true DMS [85].

A theoretical understanding of the transport of these spin-packets is needed to establish appropriate performance metrics of these interconnects. For example, what are the speed limiting factors? What speeds can be reached in an ideal material? Can speeds close to the speed of light be reached? Two characteristic times T_1 and T_2 enter the spin-transport equations as fundamental parameters. These parameters govern the relaxation of the longitudinal and transverse spin polarization, respectively. The other important parameters include the spin-dependent diffusion constant and mobility of the polarized carriers. In addition, experimental instruments that can measure spin injection and transport on the nanoscale are needed to probe and characterize these interconnects. Scanning probe instruments such as spin-polarized scanning tunneling microscopy and spin-polarized ballistic electron emission microscopy have potential to provide needed insight.

The use of spin packets in a doped semiconductor can provide a monolithic interconnect technology that has potential to replace or augment conventional Cu-metal interconnects. While complex, this monolithic spin-packet technology does have advantages for nanoscale device nodes where the resistance of Cu interconnects increases. In addition, this technology utilizes the conventional integrated circuit material set and tool chain.

34.5 Prolog: Electroplating as Enabling Technology

For the foreseeable future, electroplating (EP) or electrochemical deposition (ECD) will continue to play an enabling role in the fabrication of high-speed, copper-based interconnects beyond the first level of contacts where chemical vapor deposition (CVD) or atomic layer deposition (ALD) tungsten (W) will remain dominant. As copper-based interconnect processing continues to face the well-documented challenges associated with scaling down of device nodes, additive packages used in EP to support "super filling" of ever decreasing size and ever increasing aspect ratio features will have to continue to improve. The two component additive packages – accelerator and suppressor – used in pre-65 nm half-pitch technology nodes will be replaced with three component additive packages – accelerator, suppressor, and leveler – in sub-65 nm half-pitch generations in order to minimize "overburden" above the dense feature topographies. Also, the chemical stability of all the components of the additive package needs to be enhanced to ensure process stability in manufacturing, prevent contaminant incorporation and subsequent resistivity increases in the resulting interconnects, and promote the formation of larger grains within the interconnect following annealing for enhanced electromigration performance.

It is suggested that another component may also have to be incorporated into the additive package – a "nucleator" – to support adequate nucleation density directly on "zero-thickness" liner/seed layer systems, such as TaN/Ru via ALD, or to protect a zero-thickness liner/seed layer system, such as ALD Ta/TaN/Cu, from oxidation or in situ etching. Additive packages might also have to completely change in order to work with another ultra-thin liner/seed layer system that could exhibit surface characteristics that are substantially different from the present PVD Ta/TaN/Cu liner seed layer systems.

Processing tools used to support EP copper interconnects in ever decreasing size and ever increasing aspect ratio features will also have to continue to evolve. EP equipment will have to be capable of uniform processing across large (300 mm or even 450 mm) wafers with ultra-thin liner/seed layer systems that will result in substantial potential drops at points on the wafer surface that are distant from the electrical contact points. EP equipment will have to include increasingly complex diagnostic and dosing devices capable of accurately and precisely maintaining the proper concentration of each necessary additive component in the electrolyte. It may also be necessary to remove residual spent additives from the electrolyte to minimize the amount of hazardous wastes generated. Furthermore, the EP equipment might also have to cope with porous, ultra-low k dielectrics. Thus, it would have to prevent the retention of electrolyte and additives within the porous material, while promoting wetting of the porous surface to prevent EP copper defects produced by trapped air bubbles, and it would have to produce interconnects with minimal surface roughness to limit surface scattering.

Finally, EP copper fabrication equipment might need to be clustered with pre-EP surface conditioning apparatus for surface cleaning of sensitive low-*k* dielectric films. It might also need to be clustered with post-EP copper electroless plating equipment to selectively cover copper interconnects with an ultra-thin, highly adherent capping layer system such as CoWP or CoWB. This step will be required in order to provide the prerequisite level of reliability via minimization of electromigration and line-to-line shorts, and/or with post-EP copper electrochemical mechanical polishing (ECMP) equipment to remove the bulk of the excess EP copper "overplate" while protecting the underlying fragile ultra-low-*k* dielectric.

As computer chip technologies transition toward new nanotechnology enabled device paradigms that replace CMOS-based architectures, the development of the fundamentally new interconnect schemes will require smart modifications to current fabrication methodologies. These modifications must enable the realization of bottom-up fabrication protocols based on nanoscale self-assembly

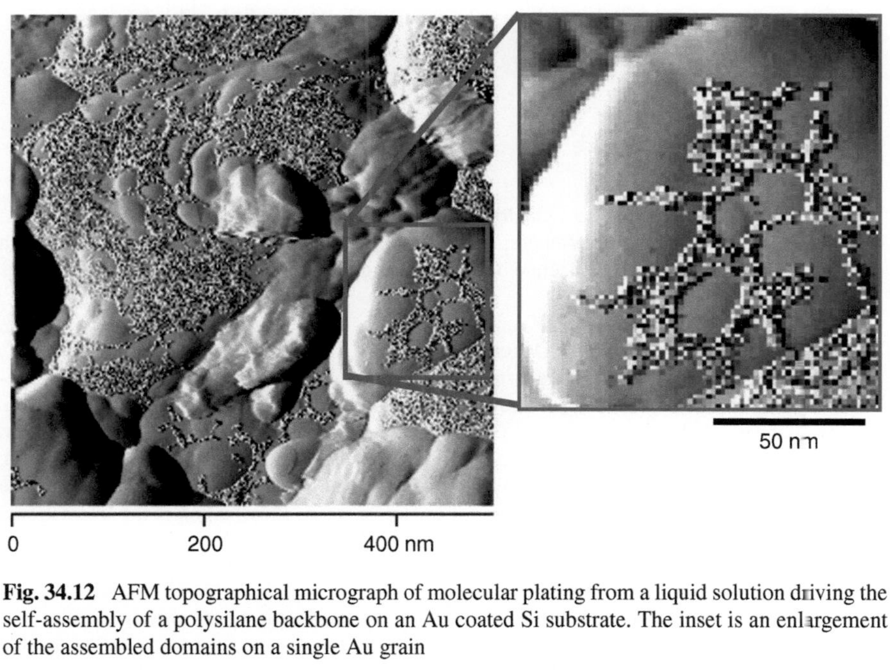

Fig. 34.12 AFM topographical micrograph of molecular plating from a liquid solution driving the self-assembly of a polysilane backbone on an Au coated Si substrate. The inset is an enlargement of the assembled domains on a single Au grain

and self-organization of smart interconnect materials, while demonstrating full compatibility with and integrability into prevailing semiconductor process flows. In this respect, modifications to current EP methodologies, such as molecular plating, are particularly adept at achieving controlled self-assembly on 300 mm wafer platforms and, accordingly, are projected to play a major role in the development and incorporation of nanoscale metallization schemes. An example of emerging molecular plating technologies is displayed in Fig. 34.12. In this case, aqueous chemisorption was demonstrated through the application of molecular plating from a liquid solution to drive the self-assembly of a polysilane backbone on a Au coated Si wafer platform.

34.6 Conclusions

Over the last 50 years or so, computer chip interconnect technology has experienced revolutionary changes in materials, processing and integration techniques. EP copper has become one of the key technology elements in the fabrication of devices. Continued advances in electroplating as well as other process technologies such as physical vapor deposition, chemical vapor deposition, atomic layer deposition, and electroless plating are expected to support device scaling at least through the 22 nm half-pitch technology node. Beyond the 22 nm device node, continued performance requirements will drive the incorporation of nanotechnologies such as 3D hyper-integration, moletronics, or spintronics into interconnect architectures. Even then, emerging modifications to conventional EP technologies will ensure that plating will continue to play an enabling role in the realization of ultra-high performance wiring systems.

References

1. Edelstein, D.; Heidenreich, J.; Goldblatt, R. D.; Cote, W.; Uzoh, C.; Lustig, N.; Roper, P.; McDevitt, T.; Wachnik, R.; Rathore, H.; Luce, S.; and Slattery, J.: Full Copper Wiring in a Sub-0.25 μm CMOS ULSI Technology. Tech. Digest IEEE, International Electron Devices Meeting, 773–776 (1997)
2. Goldblatt, R. D.; Agarawala, B.; Anand, M. B.; Barth, E. P.; Biery, G. A.; Chen, Z. G.; Cohen, S.; Connolly, J. B.; Cowley, A.; Dalton, T.; Das, S. K.; Davis, C. R.; Deutsch, A.; DeWan, C.; Edelstein, D. C.; Emmi, P. A.; Faltermeier, C. G.; Fitzsimmons, J. A.; Hedrick, J.; Heidenreich, J. E.; Hu, C. K.; Hummel, J. P.; Jones, P.; Kaltalioglu, E.; Kastenmeier, B. E.; Krishnan, M.; Landers, W. F.; Liniger, E.; Liu, J.; Lustig, N. E.; Malhotra, S.; Manger, D. K.; McGahay, V.; Mih, R.; Nye, H. A.; Purushothaman, S.; Rathore, H. A.; Seo, S. C.; Shaw, T. M.; Simon, A. H.; Spooner, T. A.; Stetter, M.; Wachnik, R. A.; and Ryan, J. G.: A High Performance 0.13 pm Copper BEOL Technology with Low-k Dielectric. Presentation at the International Interconnect Technology Conference, Burlingame, CA (2000)
3. El-Kareh, B.: Fundamentals of Semiconductor Processing Technologies, Kluwer Academic Publishers, Boston, 552 (1995)
4. Singer, P.: Changing the Promise of Faster Chips. Semicond. Int. **11**, 52 (1994)
5. Hu, C.-K.; Luther, B.; Kaufman, F. B.; Hummel, J.; Uzoh, C.; and Pearson, D. J.: Copper interconnection integration and reliability. Thin Solid Films **262**, 84 (1995)

6. Kaanta, C. W.; Bombardier, S. G.; Cote, W. J.; Hill, W. R.; Kerszykowski, G.; Landis, H. S.; Poindexter, D. J.; Pollard, C. W.; Ross, G. H.; Ryan, J. G.; Wolff, S.; and Cronin, J. E.: Dual-Damascene: a ULSI wiring technology. Proceedings of the 8th International VLSI Multilevel Interconnection Conference, 144 (1991)
7. Zhu, Y.: Integration of Atomic Layer Deposition Tantalum Nitride and Platinum with Electrochemical Deposition of Copper for Interconnect Technology, Ph.D. Thesis College of Nanoscale Science and Engineering of the University at Albany-SUNY, (2006)
8. Lee, B.: Electroless CoWP boosts copper reliability, device performance. Semicond. Int. **7**, 95 (2004)
9. Ritala, M. and Leskela, M.: In Handbook of Thin film Materials. Nalwa, H., Ed. Deposition and Processing of Thin Films, Academic Press **1**, 103 (2002)
10. Ramm, P.; Klumpp, A.; Merkel, R.; Weber, J.; Wieland, R.; Ostmann A.; and Wolfe, J.: 3D System Integration Technologies. Mat. Res. Soc.**766**, E5.6.1 (2003)
11. Fukushima, T.; Yamada, Y.; Kikuchi, H.; and Koyanagi, M.: New 3D Integration technology using chip to wafer bonding to achieve ultimate super-chip integration. Jap. J. Appl. Phys. **45(4B)**, 3030–3035, (2006)
12. Niklaus, F.; Stemme, G.; Lu, J.-Q.; and Gutmann, R. J.: Adhesive wafer bonding. J. Appl. Phys. **99**, 031101-01-031101-28 (2006)
13. Islam, R.; Brubaker, C.; Lindner P.; and Schaefer, C.: Wafer Level Packaging and 3D Interconnect for IC Technology. IEEE/SEMI Advanced Semiconductor Manufacturing Conference, 212–217 (2002)
14. Xu, B.; Gracias, A.; Tokranova, N.; and Castracane, J.: Wafer Bonding for 3D Integration of MEMS/CMOS. to be published, MOEMS and Miniaturized Systems, (2006)
15. Fletcher, C.; Skele, M.; and Castracane, J.: Recent Developments in Vertically Integrated Sensor Arrays. Proceedings-GOMAC, (2005)
16. Reichl, H. and Ramm, P.: 3D System Integration. Fraunhofer IZM Bulletin, (2006); Wieland, R.; Ramm, P.; and Schulz, S.: Fraunhofer IZM Annual Report, 115 (2002); Ramm, P.; Klumpp, A.; Merkel, R.; Weber, J.; Weiland, R; Ostmann, A.; and Wolf, J.: 3D system integration technologies. Proc. Mat. Res. Soc. Symp. **766**, 3 (2003)
17. Pascual, D.: Fabrication and Assembly of 3D MEMS Devices. Solid State Technol. **48**, 22 (2005)
18. Alexe, M. and Gosele, U.: Wafer Bonding Applications and Technology, Springer-Verlag, Berlin, (2004)
19. Iyer, S. S. and Auberton-Herve, A. J.: Silicon Wafer Bonding Technology for VLSI and MEMS, INSPEC, London, (2002)
20. Heath, J.: The Bridge (National Academy of Engineering) **33(4)**, 197 (2003)
21. Li, H. J.; Ly, W. G.; Li, J. J.; Bai, X. D.; and Gu, C. Z.: Multichannel Ballistic Transport in Multiwall Carbon Nanotubes. Phys. Rev. Lett. **95**, 086601 (2005)
22. Kaloyeros, A. E.; Dunn, K. A; Carlsen A. T.; and Topol, A. W.: Carbon Nanotube Interconnects invited article for the Marcel Dekker Encyclopedia of NanoScience and NanoTechnology. Schwarz, J. A.; Contescu, C. I.; and Putyera, K.Eds. **1**, 435–446 (2004)
23. Iijima, S.: Helical microtubules of graphitic carbon. Nature (London) **354(6348)**, 56 (1991)
24. Dresselhaus, M. S.; Dresselhaus, G.; and Eklund, P. C.: Science of Fullerenes and Carbon Nanotubes; Academic Press.; San Diego, US (1996)
25. Ajayan, P. M. and Ebbesen, T. W.: Nanometre-size tubes of carbon. Rep. Prog. Phys. **60**, 1025 (1997)
26. Martel, R.; Schmidt, T.; Shea, H. R.; Hertel, T.; and Avouris, Ph.: Single and multi-wall carbon nanotube field-effect transistors. Appl. Phys. Lett. **73(17)**, 2447 (1998)
27. Tans, S. J.; Verschueren, R. M.; and Dekker, C.: Room-temperature transistor based on a single carbon nanotube. Nature **393**, 49 (1998)
28. Charlier, J.-C.; and Iijima, S.: Electronic properties, junctions, and defects of carbon nanotubes. In Growth mechanisms of Carbon Nanotubes, Dresselhaus, M. S.; Dresselhaus, G.; Avouris, Ph., Eds. Topics A Physics; Springer-Verlag Heidelberg, Berlin, **80**, 55 (2001)

29. Ebbesen, T. W. and Ajayan, P. M.: Large-scale synthesis of carbon nanotubes. Nature **358**, 220 (1992)
30. Li, W. Z.; Xie, S. S.; Qian, L. X.; Chang, B. H.; Zou, B. S.; Zhou, W. Y.; Zhao A.; and Wang, G.: Large-scale synthesis of aligned carbon nanotubes. Science **274**, 1701 (1996)
31. Ren, Z. F.; Huang, Z. P.; Xu, J. W.; Wang, J. H.; Bush, P.; Siegal, M. P.; and Prevencio, P. N.: Synthesis of large arrays of well-aligned carbon nanotubes on glass. Science **282**, 1105 (1998)
32. Collins, P. G.; Arnold, Ml S.; and Avouirs, P.: Engineering carbon nanotubes and nanotube circuits using electrical breakdown. Science **292(5517)**, 706 (2001)
33. Kaloyeros, A. E; Dunn, K. A; Carlsen, A. T.; and Topol, A. W.: Carbon Nanotube Interconnects. In Dekker Encyclopedia of Nanoscience and Nanotechnology, Marcel Dekker, Inc., New York, 435 (2003)
34. Iijima, S. and Ichihashi, ST.: Single-shell carbon nanotubes of 1-nm diameter. Nature (London) **363(6430)**, 603 (1993)
35. Bethune, D. S.; Kiang, C. H.; Devries, M. S.; Gorman, G.; Savoy, R.; Vazquea, J.; and Beyers, R.: Cobalt-catalysed growth of carbon nanotubes with single-atomic-layer walls. Nature (London) **363(6430)**, 605 (1993)
36. Ajayan, P. M.; Lambert, J. M.; Bernier, P.; Barbedette, L.; Colliex, C.; and Planeix, J. M.: Growth morphologies during cobalt-catalyzed single-shell carbon nanotube synthesis. Chem. Phys. Lett. **215(5)**, 509 (1993)
37. Kong, J.; Soh, H. T.; Cassell, A. M.; Quate, C. F.; and Dai, H.: Synthesis of individual single-walled carbon nanotubes on patterned silicon wafers. Nature **395(6705)**, 878 (1998)
38. Saito, R.; and Dresselhaus, M. S.; and Dresselhaus, M. S.: Electronic structure of double-layer graphene tubules. J. Appl. Phys. **73(2)**, 494 (1993)
39. Terrones, M.; Grobert, N.; Olivares, J.; Zhang, J. P.; Terrones, H.; Kardatos, K.; Hsu, W. K.; Hare, J. P.; Townshend, P. D.; Prassides, K.; Cheetham, A. K.; Kroto, H. W.; and Walton D. R. M.: Controlled production of aligned-nanotube bundles. Nature **388**, 52 (1997)
40. Guo, T.; Jin, C.-M.; and Smalley, R. E.: Catalytic growth of single-walled nanotubes by laser vaporization. Chem. Phys. Lett. **243(1–2)**, 49 (1995)
41. Louie, S. G.: Electronic properties, junctions, and defects of carbon nanotubes. In Carbon Nanotubes: Synthesis, Structure, Properties and Applications, Dresselhaus, M. S., Dresselhaus, G., Avouris, Ph., Eds. Topics App. Physics; Springer-Verlag Heidelberg, Berlin **80**, 113 (2001)
42. Stahl, H.: Electronic transport in ropes of single wall carbon nanotubes. In Dissertation approved by the Faculty for Mathematics, Informatics and Natural Sciences at the Aachen University of Technology (2000)
43. Tans, S. J.; Devoret, M. H.; Dai, H., Thess, A. Smalley, R. E.; Geerligs, L. J.; and Dekker, C.: Individual single-wall carbon nanotubes as quantum wires. Nature (London) **386(6624)**, 474 (1997)
44. Bockrath M.; Cobden, D. H.; McEuen, P. L.; Chopra, N. G.; Zettl, A.; Thess, A.; and Smalley, R. E.: Single-electron transport in ropes of carbon nanotubes. Science **275(5308)**, 1922 (1997)
45. Yakobson, Boris I. and Avouris, Ph.: Mechanical properties of carbon nanotubes. In Carbon Nanotubes: Synthesis, Structure, Properties and Applications, Dresselhaus, M. S., Dresselhaus, G., Avouris, Ph., Eds; Topics App. Physics; Springer-Verlag Heidelberg, Berlin **80**, 287 (2001)
46. Wong, E. W.; Sheehan, P. E.; and Lieber, C. M.: Nanobeam mechanics: Elasticity, strength, and toughness of nanorods and nanotubes. Science **277(5334)**, 1971 (1997)
47. Yu, M. F.; Lourie, O.; Dyer, M.; Moloni, K.; and Rouff, R. S.: Strength and breaking mechanism of multiwalled carbon nanotubes under tensile load. Science **287(5453)**, 637 (2000)
48. Dresselhaus, M. S. and Endo, M.: Relation of carbon nanotubes to other carbon materials. In Carbon Nanotubes: Synthesis, Structure, Properties and Applications, Dresselhaus, M. S.,

Dresselhaus, G., Avouris, Ph., Eds. Topics App. Physics; Springer-Verlag Heidelberg, Berlin **80**, 11 (2001)

49. Ajayan, P. M.; Schadler, L. S.; Giannaris, C.; and Rubio, A.: Mechanical response of single-walled carbon nanotubes in polymer nanocomposites. Adv. Mater. **12**, 750 (2000)
50. Yakobson, B. I.; Brabec, C. J.; and Bernholc, J.: Nanomechanics of carbon tubes instabilities beyond linear response. Phys. Rev. Lett. **76(14)**, 2511 (1996)
51. Ajayan, P. M.; Ebbesen, T. W.; Ichihashi, T.; Iijima, S.; Tanigaki, K.; and Hiura, H.: Opening carbon nanotubes with oxygen and implications for filling. Nature **362(6420)**, 522 (1999)
52. Fischer, J. E.; Dai, H.; Thess, A.; Lee, R.; N. Hanjani, M.; Dehaas, D. L.; and Smalley R. E.: Metallic resistivity in crystalline ropes of single-wall carbon nanotubes. Phys. Rev. B **55**, R4921 (1997)
53. Frank, S.; Poncharal, P.; Wang, Z. L.; and de Heer, W. A.: Carbon nanotube quantum resistors. Science **280**, 1744 (1998)
54. Hertel, T; Walkup, R. E; and Avpuris, P: Deformation of carbon nanotubes by surface van der Walls forces. Phys. Rev. B **58(20)**, 13870 (1998)
55. Fuhrer, M. S.; Nygård, J.; Shih, L.; Ferero, M.; Yoon, Y.-G.; Mazzoni, M. S. C.; Choi, H. J.; Ihm, J.; Louie, S. G.; Zettl, A.; and McEuen, P. L.: Crossed nanotube junctions. Science **288 (5465)**, 494 (2000)
56. Kane, C. L.; and Mele, E. J.: Size, shape, and low energy electronic structure of carbon nanotubes. Phys. Rev. Lett. **78**, 1932 (1997)
57. Terrones, M.; Banhart, F.; Grobert, N.; Charlier, J.-C.; Terrones H.; and Ajayan, P. M.: Molecular junctions by joining single-walled carbon nanotubes. Phys. Rev. Lett. **89** 075505-1 (2002)
58. Stahl, H.; Appenzeller, J.; Martel, R.; and Avouris, Ph.: Intertube coupling in ropes of single-wall carbon nanotubes. Phys. Rev. Lett. **85(24)**, 5186 (2000)
59. Farró, L. and Schönenberger, C.: Physical properties of multi-wall nanotubes. In Carbon Nanotubes: Synthesis, Structure, Properties and Applications, Dresselhaus, M. S., Dresselhaus, G., Avouris, Ph., Eds. Topics App. Physics; Springer-Verlag Heidelberg, Berlin, **80**, 329 (2001)
60. Vajtai, R.; Wei, B. Q.; Zhang, Z. J.; Jung, Y.; Ramanath G.; and Ajayan, P. M.: Building carbon nanotubes and their smart architecture. Smart Mater. Struct.**11**, 691 (2002)
61. Scuseria, G. E.: The equilibrium structures of giant fullerenes: Faceted or spherical shape? An ab initio Hartree-Fock study. Chem. Phys. Lett. **195**, 534 (1992)
62. Chico, L.; Crespi, V. H.; Benedict, L. X.; Louie, S. G.; and Cohen, M. L.: Pure carbon nanoscale devices: Nanotube heterojunctions. Phys. Rev. Lett. **76(6-7)**, 971 (1996)
63. Menon, M.; and Srivastava, D.: Carbon nanotube t junctions: Nanoscale metal semiconductor metal contact devices. Phys. Rev. Lett. **79(22)**, 4453 (1997)
64. Yao Z.; Postma, H. W. Ch.; Balents, L.; and Dekker, C.: Carbon nanotube intermolecular junctions. Nature (London) **402**, 273 (1999)
65. Choi W. B. and Lee, Y. H.: Carbon nanotube and its application to nanoelectronics. Industrial Applications of Electron Microscopy; Li, Z. Ed., Marcel Dekker, New York, Chapter **14**, 614 (2002)
66. Kong, J.; Soh, H. T.; Cassell, A. M.; Quate, C. F.; and Dai, H. J.: Synthesis of individual single-walled carbon nanotubes on patterned silicon wafers. Nature **395**, 878 (1998)
67. Tans, T. J.; Verschueren, R. M.; and Dekker, C.: Room-temperature transistor based on a single carbon nanotube. Nature **393**, 49 (1998)
68. Martel, R.; Schmidt, T.; Shea, H. R.; Hertel, T.; Avouris, Ph.: Single and multi wall nanotube field effect transistors. Appl. Phys. Lett. **73(17)**, 2447 (1998)
69. Wei, B.-Q.; Kohler-Redlich, P.; Bader, U.; Heiland, B.; Spolenak, R.; Arzt E.; Rühle, M.: Selective specimen preparation for TEM observation of the cross section of individual carbon nanotube/metal junctions. Ultramicroscopy **85(2)** 93 (2000)

70. Kaloyeros, A. E, Welch, J.; Castracane J.; Oktyabrsky, S.; Geer, R.; and Dovidenko, K.: Interconnect Nanotechnology. Overarching Concepts and Demonstration Vehicles. Annual review of the Interconnect Focus Center 2002, Atlanta, GA, (2002)
71. Sagnes, M.; Broto, J.-M.; Raquet, B.; Ondarçuhu, T.; Laurent, Ch.; Flahaut, E.; Vieu, Ch.; and Carcenac, F.: Alignment and nano-connections of isolated carbon nanotubes. Microelectron. Eng. **67–68**, 683 (2003)
72. Austin, D. W.; Puretzky, A. A; Geohegan, D. B.; Britt, P. F.; Guillorn M. A.; and Simpson, M. L.: The electrodeposition of metal at metal/carbon nanotube junctions. Chem. Phys. Lett. **361**, 525 (2002)
73. Boulas C.; Davidovits, J. V.; Rondelez, F.; and Vuillaume, D.: Suppression of charge carrier tunneling through organic self-assembled monolayers. Phys. Rev. Lett. **76**, 4797 (1996)
74. Mujica, V. and Ratner, M. A.: In Handbook of Nanoscience, Engineering, and Technology. Goddard III W. A. et al., eds. CRC Press, Boca Raton, Fla. (2002)
75. Rochefort, A.; Martel, R.; and Avouris, P.: Electrical Switching in π-Resonant 1D ntermolecular Channels. Nano Lett. **2(8)**, 877 (2002)
76. Heath, J. and Ratner, M.: Molecular electronics. Phys. Today **56(5)**, 43 (2003)
77. Fishelson, N.; Shkrob, I.; Lev, O.; Gun, J.; and Modestov, A. D.: Studies on charge transport in self-assembled gold-dithiol films: Conductivity, photoconductivity, and photoelectrochemical measurements. Langmuir **17(2)**, 403 (2001)
78. Eigler, D. M. and Schweizer, E. K.: Positioning single atoms with a scanning tunneling microscope. Nature **344**, 524 (1990)
79. Piner, R. D.; Zhu, J.; Xu, F.; Hong, S.; and Mirkin, C. A.: Dip pen nanolithography. Science **283**, 661–663 (1999)
80. Hodneland, C. D.; Lee, Y.-S.; Min, A.-H.; and Mrksich, M.: Supramolecular chemistry and self-assembly special feature: Selective immobilization of proteins to self-assembled monolayers presenting active site-directed capture ligands. PNAS **99**, 5048 (2002)
81. Molecular Electronics: Biosensors and Biocomputers Hong, F., Ed. Plenum Press, New York (1989)
82. Kikkawa, J. M. and Awschalom, D. D.: Lateral drag of spin coherence in gallium arsenide. Nature **397**, 139 (1999)
83. Flatté, M. E. and Byers, J. M.: Spin diffusion in semiconductors. Phys. Rev. Lett. **84(18)**, 4220 (2000)
84. Ohno, H.: Making nonmagnetic semiconductors ferromagnetic. Science **281(5379)**, 951 (1998)
85. Bolduc, M.; Awo-Affouda, C.; Stollenwerk, A.; Huang, M. B.; Ramos, F. G.; Agnello, G.; and LaBella, V. P.: Above room temperature ferromagnetism in Mn-ion implanted Si. Phys. Rev. B **71**, 033302 (2005)

Chapter 35
Self-Assembly of Short Aromatic Peptides: From Amyloid Fibril Formation to Nanotechnology

Ehud Gazit

35.1 The Formation of Nanostructures by Short Aromatic Peptides

35.1.1 Bio-inspired Biological Nano-assemblies

Molecular self-assembly offers unique directions for the fabrication of novel supramolecular structures and advance materials. The inspiration for the development of such structures is often derived from self-assembling modules in biology, as natural systems form complex structures from simple building blocks such as amino acids, nucleic acids, and lipids. Self-assembled peptide nanostructures are assumed to serve as key building blocks in future nanotechnological devices and assemblies. These peptide structures offer the advantage of high biological systems specificity while allowing numerous chemical modifications. Short peptides self-assemble into various forms in the nano-scale, including tapes, fibrils, tubes, and spheres. Yet, the inherent thermal and chemical instability of many protein structures raise a question regarding the compatibility of peptide structure with common lithographic techniques and the long-term durability of such devices.

35.1.2 The Role of Aromatic Residues in the Self-Assembly of Amyloid Nano-fibrils

Amyloid fibrils are naturally occurring protein and peptide nano-fibrils that have various pathological and physiological roles in various biological systems. The formation of these 7–10 nm diameter fibrils is associated with major human disorders such as Alzheimer's disease, Parkinson's disease, and Type II diabetes. There are

---

E. Gazit (✉)
Department of Molecular Microbiology and Biotechnology, Life Sciences faculty,
Tel Aviv University, Tel Aviv 69978, Israel
e-mail: ehudg@post.tau.ac.il

Y. Shacham-Diamand et al. (eds.), *Advanced Nanoscale ULSI Interconnects: Fundamentals and Applications*, DOI 10.1007/978-0-387-95868-2_35,

more than 20 human disorders that are associated with the formation of ordered amyloid fibrils. Yet, amyloid assemblies also have physiological role in the formation of bacterial biofilms and aerial hyphae. Thus these ubiquitous nano-assemblies are one of the most important biological nano-fibrils.

In all these cases well-ordered nano-fibrils are being identified by electron microscopy or atomic force microscopy. The fibrils are remarkably ordered quasi-crystalline assemblies with a very clear 0.44–0.46 nm reflection on the meridian studied by X-ray fiber diffraction studies. The fibril deposits also show high macroscopic order as reflected by very strong birefringence upon staining with the Congo Red dye. The fibrils are formed very efficiently by a self-assembly process at critical protein concentrations as low as the nanomolar ranges. A very interesting point is that fibrils of different origins show remarkable biophysical and ultrastructural properties. For example fibrils form the brains for Alzheimer's disease patients or those suffering from Type II diabetes show very similar molecular properties.

There is much interest in the mechanism that leads to the formation of these nanostructures due to their grave importance on the one hand and lack of any therapy on the other. Much effort is being invested in the determination of the molecular events that lead to the formation of these well-ordered structures. Yet, the exact mechanism of amyloid formation is not fully understood.

Our approach for the study of amyloid fibrils formation is a reductionist one. While amyloid fibrils are being formed in most cases by polypeptides of 30–40 amino acids or longer, it was demonstrated that peptides, as short as hexapeptides, form amyloid fibrils of similar physical and ultrastructural properties. We systematically analyzed short peptide fragments to pinpoint residues that play a role in the molecular recognition and self-assembly process. Based on the mechanistic insights, we identified novel fragments, as short as tetrapeptide, that can form amyloid-like nanostructures

Our earlier studies of the mechanism of molecular recognition and self-assembly that lead to the formation of amyloid nano-fibrils have pinpointed the key role of aromatic amino-acids in many cases of amyloid formation. By the use of peptide models and peptide array technology, we demonstrated the role of aromatic amino-acids both in the recognition steps and the self-assembly of the fibrils. We suggested that the geometrically restricted interactions with aromatic moieties may provide not only energetic contribution but also order and directionality in the formation of amyloid nano-fibrils. Our model is supported by experimental high-resolution studies as well as molecular modeling and free-parameter theoretical modeling. The stacking hypothesis suggests a new approach to understand the self-assembly mechanism, enables the identification of novel very short motifs, and indicates new ways to control this process.

Our aromatic interaction model had led us to the exploration of the self-assembly process of very short aromatic fragments of known amyloid fibril-forming proteins and polypeptides. We had demonstrated that peptide fragments as short as pentapeptides can form typical amyloid fibrils with all the biophysical and ultrastructural properties of amyloid nano-fibrils as described above [1].

35.1.3 The Identification of Aromatic Dipeptide Nanotubes (ADNT)

As part of our search for the shortest amyloid forming peptide fragments, we had studied the dipeptide core motif of the β-amyloid (Aβ) polypeptide. This polypeptide is involved in the formation of amyloid plaques in Alzheimer's disease patients. We identified the aromatic diphenylalanine core as the common denominator between former and inhibitor fragments of the pathological polypeptide. When we studied the ultrastructural properties of assemblies being formed by the dipeptide we had identified its ability to form nanotubular assemblies (Fig. 35.1) [2]. This motif is of a special interest since several studies identified the ability of larger peptides and conjugated organic molecules that contain this motif to inhibit fibrils formation by Aβ. Some of those inhibitors are currently under clinical trials as potential drugs to treat Alzheimer's disease.

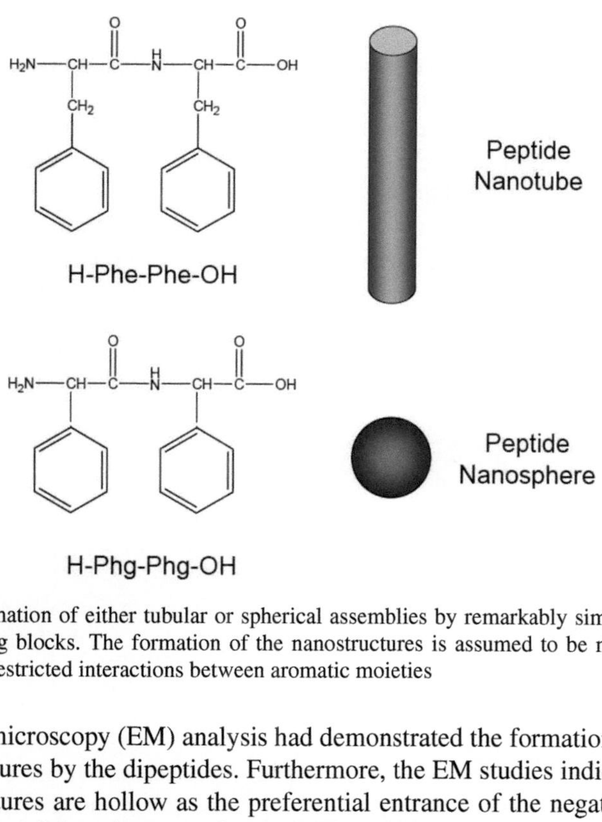

Fig. 35.1 Formation of either tubular or spherical assemblies by remarkably similar and simple peptide building blocks. The formation of the nanostructures is assumed to be mediated by the geometrically restricted interactions between aromatic moieties

Electron microscopy (EM) analysis had demonstrated the formation of seamless tubular structures by the dipeptides. Furthermore, the EM studies indicated that the tubular structures are hollow as the preferential entrance of the negative stain dye into the lumen of the tubes was observed. The tubular peptide structures are being formed readily and efficiently under mild conditions in aqueous solutions. Diluting a concentrated fluorinated alcoholic solution of the dipeptide into aqueous solution resulted in its efficient assembly into long and hollow tubular nanostructures. A follow-up study by another group demonstrated that similar structures could be formed by solubilizing the peptide in water at 65°C, followed by cooling the solution to room temperature [3].

The assemblies have remarkable persistence length at the micron range which was assumed to reflect a very rigid structure at the nano-scale. Indeed when the stiffness of the peptide nanotubes was directly determined by atomic force microscopy indentation experiments, the structures were determined to be among the most rigid self-assembled entities with Young's modulus of about 20 GPa [4].

The peptide nanotubes also show remarkable thermal and chemical stability. The tubes withstand dry and wet heating for over 100°C as indicated by electron microscopy studies [5]. The tubes are also resistant to many common solvents including ethanol, methanol, and acetone. The chemical and thermal stability of the peptide tubes suggest that they may be integrated into various nano-devices.

35.2 Technological Applications of the Peptide Tubes

The hollow nature of the peptide nanotubes was used for the fabrication of metal objects at the nano-scale. Ionic silver that preferentially entered into the lumen of tubes was reduced into element silver to end-up with silver-filled peptide tubes. When the peptide envelope was removed by enzymatic degradation, silver nanowires of 20 nm were obtained [2]. The diameter of the silver wires reflects the inner diameter of the peptide tubes that served as a nano-scale casting mode. More recent study had utilized the peptide template to form trilayer coaxial nanowires. In these experiments, silver-filled peptide nanotubes were further coated with gold to form a trilayer silver-peptide-gold cables at the nano-scale (Fig. 35.2) [6].

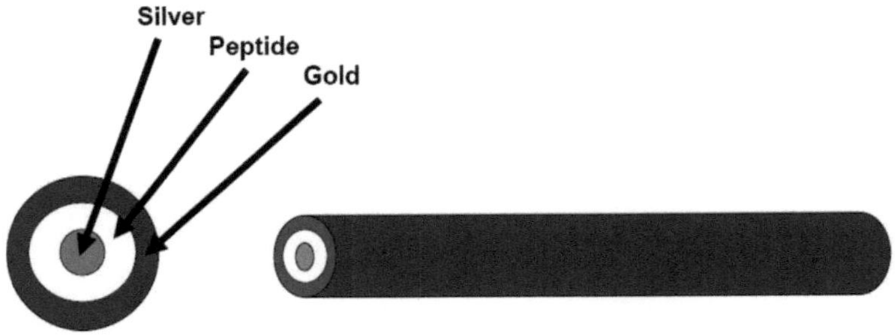

Fig. 35.2 Formation of coaxial nano-cables by the selective filling of peptide nanotubes with silver following their coating with gold

The peptide nanotubes were also used for electrochemical biosensors. Printed carbon electrodes were modified with the peptide tubes and the electrochemical activity was determined by cyclic voltametry [7]. The modified electrode had significantly better performance as compared to the non-modified ones. In a follow-up study, gold electrodes were modified with peptide tubes to which a glucose oxidase enzyme was covalently attached [8]. In this case an even higher improvement of the electrochemical sensing was observed.

Other envisioned applications of the peptide nanotubes include their integration into nanoelectromechanical systems (NEMS) and nanofluidic devices. For these purposes chemically and biologically stable tubes are needed. To achieve such stable assemblies, dipeptides that are composed of d-isomer amino acids were designed. The d-isomer amino acids are the mirror image of the naturally occurring amino acids that are not recognized by the common peptide degrading enzymes. Indeed it was demonstrated that d-amino acid aromatic dipeptides form nanotubes that are ultrastructurally identical to the l-amino acid peptide nanotubes, but are not degraded by proteolytic enzymes [2].

35.3 The Formation of Nano-spheres by Related Aromatic Dipeptides

Diphenylalanine represents the simplest natural aromatic peptide fragments. To explore even simpler peptide building blocks, we had studied the diphenylglycine peptide that is the simplest aromatic dipeptide. We had revealed the diphenylglycine peptide forms spherical rather than tubular structures at the nano-scale (Fig. 35.1) [9]. Interestingly, similar tubular assemblies were observed when the diphenylalanine peptide was modified to add a thiol group to the peptide [9].

The formation of either tubular or spherical assemblies by remarkably similar peptide building blocks is very intriguing. It is of course well known that carbon can form either nanotubes or closed-cage fullerenes at the nano-scale. Such alternative arrangement of tubular and spherical nano-assemblies is also being observed with layered inorganic materials such as WS_2, MoS_2, and NbS_2 [10]. Also biological membranes' phospholipids building blocks that form spherical nano-vesicles in aqueous solutions can be engineered to form tubular structures. The formation of these two types of peptide nanostructures, also the short peptide building blocks, hints that they might also have a layered two-dimensional conformation as these assemblies.

The peptide nano-spheres are also envisioned to have various nanotechnological applications due to their biocompatibility and the facile way for their chemical and biological modifications. Such applications include targeted drug delivery by modified nano-spheres which are set to bind specific molecular epitopes. Other direction is the use of the nano-spheres as contrast agents for magnetic resonance imaging (MRI). For this application the peptide nano-spheres will be loaded with ferromagnetic materials for nuclear magnetic resonance (NMR) signal.

35.4 Other Modified Aromatic Homo-dipeptides

The self-assembly properties of the diphenylalanine and diphenylglycine peptide had led to the study of other aromatic homo-dipeptides. One peptide, the dinaphthylalanine peptide was also observed to form nanotubes [11]. Yet these nanotubes appeared to be less rigid as compared to the diphenylalanine tubes. Other aromatic and hetero-aromatic systems are currently being explored for molecular electronic applications that will be based on the peptide nanostructures.

Other analogues of the diphenylalanine include the termini-modified peptides. In these peptides either the carboxyl or the amine termini were modified with various chemical groups [12]. The various peptides including Boc-diphenylalanine, Fmoc-diphenylalanine, acetyl-diphenylalanine, and diphenylalanine-amide were shown to form various assemblies at the nano-scale. The most interesting peptide is Fmoc-diphenylalanine which forms rigid macroscopic hydrogels with nano-scale order [13]. This hydrogel was already studied for the encapsulation and release of drugs and for the support of the growth of cells for tissue engineering applications.

35.5 The Technological Advantages of Peptide Building Blocks

Peptide assemblies are very attractive building blocks. Their inherent biocompatibility and the easy ability to modify them chemically or biologically make these building blocks very unique components in future nanotechnological applications. These peptides combine the advantages of biologically derived components on the one hand and the chemical properties of advanced functional materials on the other. These biomolecular blocks could be chemically synthesized in large amounts, from milligrams to kilograms or even tons. Peptides are also quite inexpensive materials when their large-scale synthesis is being optimized. One dipeptide (a peptide composed of two amino-acids) that is routinely being synthesized in large amounts is l-alpha-aspartyl-l-phenylalanine methyl ester. This compound that is better known as the sweetener aspartame is being annually synthesized in ton scale. The price of 1 gram of aspartame is only few cents. A quite similar peptide, diphenylalanine, was shown to form peptide nanotubes in high efficiency. Therefore, this peptide could allow the assembly of functional nanotubes for a price of cents per gram as compared to tens of dollars per gram for carbon nanotubes, the current major building blocks for nanotechnology.

35.6 Summary

Recent studies had brought new challenges and directions toward the understanding of peptide self-assembly at the molecular level. It was demonstrated that peptides could be used in various ways to engineer complex nanostructures for different applications such as molecular electronics, tissue engineering, and drug delivery. Peptide holds a great promise in the "bottom-up" approach due to their low cost, their simplicity, and the ability to easily decorate them with chemical and biological elements.

Our studies on the mechanism of protein self-assembly into ordered amyloid assemblies had provided important information on the mechanism by which simple biological elements interact and identified new building blocks for nanotechnology. We believe that the gained information will lead to a better understanding of the potential use of peptide nanotubes in future applications, such as micro and nano-electromechanics (MEMS and NEMS, respectively) and for medical devices. Several applications were already suggested for these biological nanostructures and these will be explored in years to come.

References

1. Reches, M.; Porat, Y.; and Gazit, E.: Amyloid fibrils formation by pentapeptide and tetrapeptide fragments of human calcitonin. J. Biol. Chem. **277**, 35475 (2002)
2. Reches, M. and Gazit, E.: Casting metal nanowires within discrete self-assembled peptide nanotubes. Science **300**, 625 (2003)
3. Song, Y.; Challa, S. R.; Medforth, C. J.; Qiu, Y.; Watt, R. K.; Peña, D.; Miller, J. E.; van Swol, F.; and Shelnutt, J. A.: Synthesis of Peptide-nanotube Platinum-nanoparticle Composites, Chem. Commun. **9**, 1044 (2004)
4. Kol, N., et al.: Self-assembled peptide nanotubes are uniquely rigid bioinspired supramolecular structures. Nano Lett. **5**, 1343 (2005)
5. Adler-Abramovich, L.; Reches, M.; Sedman, V. L.; Allen, S.; Tendler, S. J. B.; and Gazit, E.: Thermal and chemical stability of diphenylalanine peptide nanotubes: implications for nanotechnological applications. Langmuir **22**, 1313 (2006)
6. Carny, O.; Shalev, D.; and Gazit, E.: Fabrication of coaxial metal nanowires using self-assembled peptide nanotube scaffold. Nano Lett. (2006)
7. Yemini, M.; Reches, M.; Rishpon, J.; and Gazit, E.: Novel electrochemical biosensing platform using self-assembled peptide nanotubes. Nano Lett. **5**, 183 (2005)
8. Yemini, M.; Reches, M.; Gazit, E.; and Rishpon, J.: Peptide nanotubes modified electrodes for enzyme-biosensors applications. Anal. Chem. **77**, 5155 (2005)
9. Reches, M. and Gazit, E.: Formation of closed-cage nanostructures by self-assembly of aromatic dipeptides. Nano Lett. **4**, 581 (2004)
10. Margulis, L.; Salitra, G.; Tenne, R.; and Talianker, M.: Nested fullerene-like structures. Nature **365**, 113 (1993)
11. Reches, M. and Gazit, E.: Designed aromatic homo-dipeptides: Formation of ordered nanostructures and potential nanotechnological applications. Phys. Biol. **3**, S10 (2006)
12. Reches, M. and Gazit, E.: Self-assembly of peptide nanotubes and amyloid-like structures by charged-termini capped diphenylalanine peptide analogues. Israel J. Chem. **45**, 363 (2005)
13. Mahler, A.; Reches, M.; Rechter, M.; Cohen, S.; and Gazit, E.: Rigid, self-assembled hydrogel composed of a modified aromatic dipeptide. Adv. Mater. **18**, 1365–1370 (2006)

Index

Note: Page numbers followed by '*f*' and '*t*' refer to figures and tables respectively.

Y. Shacham-Diamand et al. (eds.), *Advanced Nanoscale ULSI Interconnect: Fundamantels and Application*, DOI 10.1007/978-0-387-95868-2,

N

MIX
Papier aus verantwortungsvollen Quellen
Paper from responsible sources
FSC® C105338

If you have any concerns about our products,
you can contact us on
ProductSafety@springernature.com

In case Publisher is established outside the EU,
the EU authorized representative is:
Springer Nature Customer Service Center GmbH
Europaplatz 3, 69115 Heidelberg, Germany

Printed by Libri Plureos GmbH
in Hamburg, Germany